Lecture Notes in Artificial Intelligence 8608

Subseries of Lecture Notes in Computer Science

LNAI Series Editors

Randy Goebel
University of Alberta, Edmonton, Canada
Yuzuru Tanaka
Hokkaido University, Sapporo, Japan
Wolfgang Wahlster
DFKI and Saarland University, Saarbrücken, Germany

LNAI Founding Series Editor

Joerg Siekmann
DFKI and Saarland University, Saarbrücken, Germany

Armin Duff   Nathan F. Lepora   Anna Mura
Tony J. Prescott   Paul F.M.J. Verschure (Eds.)

Biomimetic and Biohybrid Systems

Third International Conference, Living Machines 2014
Milan, Italy, July 30 – August 1, 2014
Proceedings

Springer

Volume Editors

Armin Duff
Universitat Pompeu Fabra, Barcelona, Spain
E-mail: armin.duff@upf.edu

Nathan F. Lepora
University of Bristol, UK
E-mail: n.lepora@bristol.ac.uk

Anna Mura
Universitat Pompeau Fabra, Barcelona, Spain
E-mail: anna.mura@upf.edu

Tony J. Prescott
University of Sheffield, UK
E-mail: t.j.prescott@sheffield.ac.uk

Paul F.M.J. Verschure
Universitat Pompeu Fabra and
Catalan Institution for Research and Advanced Studies
Barcelona, Spain
E-mail: paul.verschure@upf.edu

ISSN 0302-9743   e-ISSN 1611-3349
ISBN 978-3-319-09434-2   e-ISBN 978-3-319-09435-9
DOI 10.1007/978-3-319-09435-9
Springer Cham Heidelberg New York Dordrecht London

Library of Congress Control Number: 2014944068

LNCS Sublibrary: SL 7 – Artificial Intelligence

Typesetting: Camera-ready by author, data conversion by Scientific Publishing Services, Chennai, India

Printed on acid-free paper

Springer is part of Springer Science+Business Media (www.springer.com)

Preface

These proceedings contain the papers presented at Living Machines: The Third International Conference on Biomimetic and Biohybrid Systems, held in Milan, Italy, July 30 to August 1, 2014. This followed the first and second Living Machines conferences that were held in Barcelona, Spain, in July 2012 and in London, UK, in July 2013. These international conferences are targeted at the intersection of research on novel life-like technologies based on the scientific investigation of biological systems, or *biomimetics*, and research that seeks to interface biological and artificial systems to create *biohybrid* systems. The conference aim is to highlight the most exciting international research in both of these fields united by theme of "living machines."

Living Machines promotes the idea that in order to build novel advanced artifacts, such as robots, we need to understand, not only mimic, nature and life and to base technology on the same fundamental principles. This idea is also what dominated the interest and curiosity of one of the greatest geniuses of the Renaissance, Leonardo Da Vinci, establishing a synergy between fundamental knowledge and engineering:

> "Knowing is not enough; we must apply. Being willing is not enough; we must do."

In admiration of this champion of knowledge and creativity, the 2014 edition of Living Machines took place at the Museo Nazionale della Scienza e della Tecnologia Leonardo da Vinci in Milan.

In between science and art, Leonardo considered both these disciplines instruments with which to exercise the same objective, understanding nature and humans. In doing so he mastered engineering, physics, anatomy, drawing, and painting to explore technical solutions shared by living beings to build machines of any kind. Besides his interest in observing nature and life forms (for instance, he studied the anatomy and behavior of birds, in their natural habitat, in order to build his flying machines), Leonardo was also a great intellectual with a passion for writing. Leonardo's sketches and manuscripts were a vehicle to advance his reasoning, to report on his obsession for understanding, as well as to give meaning to his drawings. Through his writings he articulated reflections on nature, reports of experiments, calculations, drawings, plans, discoveries, inventions, and commentaries on literature. As a man of the Renaissance, Leonardo da Vinci used interdisciplinary methods, writing treatises on mechanics, anatomy, cosmology, hydraulic, and earth sciences. In this sense, his work can be considered as a true precursor of the modern form of biomimetics expressed in the Living Machines conference.

The development of future real-world technologies will depend strongly on our understanding and harnessing of the principles underlying living systems and the flow of communication signals between living and artificial systems. In

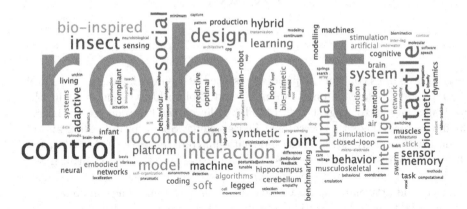

Fig. 1. Word cloud of the most frequent terms appearing in the contributions to Living Machines 2014

this context, research areas presented at the third edition of the Living Machines conference included brain-based systems, active sensing, soft robotics, learning, memory, control architectures, self-regulation, movement and locomotion, sensory systems, and perception (Fig. 1).

Biomimetics: Past, Present, and Future

As was already clear to da Vinci 500 years ago, nature is a good source for finding solutions to technological problems as it continually reinvents itself to solve challenges in the natural world. Nature's improvements have evolved over hundreds of millions of years in plants and animals resulting in the myriad of natural design solutions around us. Nevertheless, although engineers are able to build incredible robots with some of the attributes and abilities of animals (including humans), we still fall short of reproducing the dexterity and adaptability of animal perception, cognition, consciousness, and action. Progress here may depend on advances in the reverse-engineering of natural control systems implemented by biological brains. In this sense Living Machines moves beyond mere "inspiration" to the understanding of principles. The Living Machines conference anticipates a future in which there will be many artificial devices with abilities based on biomimetic solutions, and biohybrid systems that combine synthetic and natural components. The 2014 conference reports on a range of such systems from flying micro devices, based on insects, to robotic manipulators modeled on the human hand, to fish-like swimming robots. Such systems are expected to have a role in our society, economy, and way of life in years to come. and the Living Machines conference also seeks to anticipate and understand the impacts of these technologies before they happen.

The Living Machines Conference in Milan

The main conference, July 30 to August 1, 2014, took the form of a three-day single-track program including 18 oral, 10 poster spotlights, and 40 poster presentations and six plenary lectures from leading international researchers in biomimetic and biohybrid systems: Mandyam Srinivasan, Queensland Brain Institute, Australia: "Insect-inspired cognition and vision"; Andrew Schwartz, University of Minnesota, Pittsburgh, USA: "Neural Control of Prosthetics"; Sarah Bergbreiter, University of Maryland, USA: "Microrobotics"; Darwin Caldwell, Italian Institute of Technology, Genova, Italy: "Legged Locomotion and Revise"; Ricard Sole, Universitat Pompeu Fabra, Barcelona, Spain: "Evolution of Complex Networks"; Minoru Asada, Osaka University, Japan: "Cognitive and Affective Developmental Robotics." The program also included an invited talk on "Biomimetics in Design" by Franco Lodato the Creative Director of VSN Mobil. Session themes included: soft robotics, neuromechanic systems, locomotion, biohybrid systems, biomimetic systems, active sensing, and social robotics.

The conference was complemented with a day of workshops on July 29 around the themes: Embodied Interaction and Internal Models (Volker Dürr and Paolo Arena); The Robot Self (Tony Prescott and Paul Verschure); Emergent Social Behaviours in Bio-hybrid Systems (José Halloy, Thomas Schmickl, and Stuart Wilson); and Biomimetics in Design (Pino Trogu and Franco Lodato).

The main meeting, together with a poster and demo session and reception, was hosted at the Museo Nazionale della Scienza e della Tecnologia Leonardo da Vinci, Milan, Italy. This wonderful museum holds the greatest collection in the world of machine models based on the designs and drawings of Leonardo da Vinci and it is the largest science and technology museum in Italy. This venue, with its outstanding collections of natural and human-made technologies provided an ideal setting to host the third Living Machines conference.

We wish to thank the many people that were involved in making LM 2014 possible. Tony Prescott and Paul Verschure co-chaired and co-planned the meeting, with Armin Duff chairing the Program Committee and editing the proceedings volume. Anna Mura was responsible for the overall organization of the conference and its communication. The workshop program was chaired by Barbara Mazzolai and Anna Mura. Nathan Lepora contributed to the conference program and communication, Carme Buisan and Pedro Omedas, from UPF, Barcelona, Silvia Matti from IIT Milan, and Samantha Johnson from the University of Sheffield provided administrative support including registration, booking, and financing. Additional guidance and support were provided by Roberto Cingolani, Giorgio Metta, and the Living Machines International Advisory Board. We would like to thank artist Behadad Rezazadeh for the design of the LM 2014 logo and Sytse Wierenga with Anna Mura for the web graphics. We would also like to thank the authors and speakers who contributed their work, and the members of the international Program Committee for their detailed and considerate reviews. We are grateful to the six keynote speakers who shared with us their vision of the future of Living Machines.

Finally, we wish to thank the sponsors of LM 2014: The Convergence Science Network for Biomimetics and Neurotechnology (CSN II) (ICT-601167), which is funded by the European Union's Framework 7 (FP7) program in the area of Future Emerging Technologies (FET), the University of Sheffield, the University Pompeu Fabra in Barcelona, and the Institució Catalana de Recerca i Estudis Avançats (ICREA). We are grateful for the additional support provided by the Italian Institute of Technology (IIT). The Living Machines reception at the conference venue's Sala delle Colonne featuring a unique exhibition of Leonardo da Vinci's machines was supported by the Taylor & Francis group, publishers of the journal *Connection Science*.

July 2014

Armin Duff
Nathan F. Lepora
Anna Mura
Tony J. Prescott
Paul F.M.J. Verschure

Organization

Committees

Conference Chairs

Tony Prescott	University of Sheffield, Sheffield, UK
Paul Verschure	Universitat Pompeu Fabra and Catalan Institution for Research and Advanced Studies, Barcelona, Spain

Program Chair

Armin Duff	Universitat Pompeu Fabra, Barcelona, Spain

Communication, Organization, and Media

Anna Mura	Universitat Pompeu Fabra, Barcelona, Spain

Communication

Nathan Lepora	University of Bristol, Bristol, UK

Local Organizers

Roberto Cingolani
Giorgio Metta
Barbara Mazzolai

Program Committee

Andrew Adamatzky	Anders Christensen
Robert Allen	Frederik Claeyssens
Sean Anderson	Holk Cruse
Joseph Ayers	Mark Cutkosky
Yoseph Bar-Cohen	Danilo De Rossi
Lucia Beccai	Angel Del Pobil
Frédéric Boyer	Peter Dominey
Dieter Braun	Stéphane Doncieux
Federico Carpi	Marco Dorigo
Hillel Chiel	Volker Dürr
Eris Chinellato	Mat Evans

Benoît Girard
Michele Giugliano
Paul Graham
Roderich Gross
Koh Hosoda
Ioannis Ieropoulos
Auke Ijspeert
Holger Krapp
Cecilia Laschi
Nathan Lepora
Arianna Menciassi
Ben Mitchinson
Martin Nawrot
Stefano Nolfi
Thomas Nowotny
Jiro Okada
Enrico Pagello
Tim Pearce

Martin Pearson
Giovanni Pezzulo
Andrew Philippides
Andrew Pickering
Tony Pipe
Tony Prescott
Roger Quinn
Ferdinando Rodriguez y Baena
Sylvain Saighi
Thomas Schmickl
Reiko Tanaka
Pablo Varona
Eleni Vasilaki
Stefano Vassanelli
Paul Verschure
Stuart Wilson
Hartmut Witte

International Steering Committee

Iain Anderson
Joseph Ayers
Ralph-Etienne Cummings
Mark Cutkosky
Kenji Doya
Armin Duff
Jose Halloy
Koh Hosoda
Auke Ijspeert
Giacomo Indiveri

Maarja Kruusmaa
David Lane
Nathan Lepora
Barbara Mazzolai
Anna Mura
Tony Prescott
Barry Trimmer
Stefano Vassanelli
Paul Verschure

Table of Contents

Full Papers

Extended Abstracts

Monolithic Design and Fabrication of a 2-DOF Bio-Inspired Leg Transmission*

Daniel M. Aukes, Önur Ozcan, and Robert J. Wood

School of Engineering and Applied Sciences
Wyss Institute for Biologically Inspired Engineering
Harvard University, Cambridge MA

Abstract. We present the design of a new two degree-of-freedom transmission intended for micro / meso-scale crawling robots which is compatible both with laminate manufacturing techniques and monolithic, "pop-up" assembly methods. This is enabled through a new design suite called "popupCAD", a computer-aided design tool which anticipates laminate manufacturing methods with a suite of operations which simplify the existing design workflow. The design has been prototyped at three times the anticpated scale to better understand the assembly and motion kinematics, and simulated to establish the basic relationships between the actuator and end-effector transmission ratios.

1 Introduction

The advent of new laminate-manufacturing techniques such as Printed-Circuit MEMS (PC-MEMS) [9, 10], Smart Composite Micro-structures (SCM) [6, 11], and Lamina Emergent Mechanisms (LEM) [4, 8] has enabled the development of a new class of millimeter-scale devices. These manufacturing techniques use a relatively small set of operations such as cutting, lamination, and folding to create a variety of mechanical components, such as hinges, structural elements, and springs. These devices are typically designed through the selective removal of material across neighboring material layers to create planar mechanisms. By using multiple materials in the laminate, the mechanical properties of these devices can be tuned for each component. Highly specialized devices can be developed through the iterative use of these operations, and discrete components can be added throughout the process. Using laser cutters to create precise alignment geometry, highly complex kinematics can also be created between hinged rigid bodies and utilized both for structures and mechanisms.

Through these new manufacturing techniques, a variety of new devices have been realized, from flying micro-robots inspired by bees [10] to crawling robots

* This material is based upon work supported by the National Science Foundation (grant numbers EFRI-1240383 and CCF-1138967) and the Wyss Institute for Biologically Inspired Engineering. Any opinions, findings, and conclusions or recommendations expressed in this material are those of the authors and do not necessarily reflect the views of the National Science Foundation.

A. Duff et al. (eds.): Living Machines 2014, LNAI 8608, pp. 1–10, 2014.

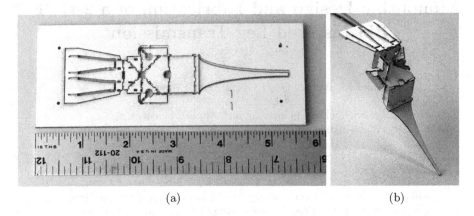

(a)   (b)

Fig. 1. Prototype transmission in (a) flattened and (b) assembled configurations

[2,3,7]. In addition, [9] demonstrates monolithic assembly techniques for faster, more precise manufacturing with a device called "Mobee". For all the potential of this concept, however, only this device has used monolithic "pop-up" fabrication techniques. This is due in part to the lack of design software which encapsulates laminate design and manufacturing rules; consequently the addition of assembly scaffolds adds too much complexity to the average design. To work around such complexities, designers split devices into simpler parts, ultimately relying on manual assembly and locking operations which require dexterity and expertise. Thus, manufacturing remains slow and error-prone.

In order to facilitate the transition from manual manufacturing of discrete components to more automated, monolithic fabrication of entire robots, we present preliminary work on a bio-inspired, two degree-of-freedom transmission to be used in the leg of a crawling robot, which is compatible with the concepts of PC-MEMS manufacturing and monolithic, "pop-up" assembly. This design is composed entirely of elements which begin flat and pre-assembled in a laminate, and by actuating and locking a single degree of freedom, are positioned into their assembled state.

2   Device Overview

The motion of cockroach and centipede legs in robotic systems is often approximated as a two-degree-of-freedom system (lifting the foot up and down, and swinging the foot forward and backward). We have previously built robots inspired from these organisms using PC-MEMS fabrication methods with manual [5] or popup assembly techniques [3] that can achieve two-degree-of-freedom leg motion.

The device shown in Figure 2a consists of many individual linkages connected through hinges on two sub-laminate layers. These linkages are arranged around four spherical linkages: two six-bar rotational linkages (6R) and two four-bar

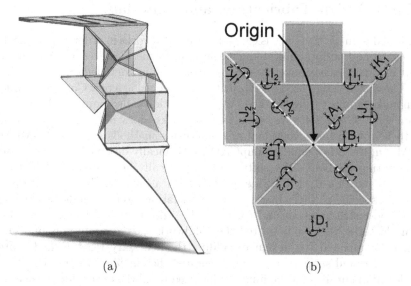

(a)   (b)

Fig. 2. The full leg design is shown in in (a). Links of the upper 6R spherical mechanism are translucent . Reference frames for the three linkages on the bottom sublaminate are shown in (b).

rotational linkages(4R). Both linkages start in their flattened state, but rotate in three-dimensional space about one point when actuated.

In combining the input and output stages of such linkages in various ways, interesting kinematic properties can be exploited. For example, the output stage of the 6R spherical linkage used in Figure 1 has the properties of a spherical joint, with three degrees of freedom in rotation. Two grounded spherical linkages whose output stages are rigidly connected, however, are more constrained in their motion, acting as a single degree-of-freedom rotational hinge. Similarly, four-bar linkages exhibit only a single degree of freedom between input and output stages, yet can impart highly nonlinear motion relationships between the two. Yet all these devices start flat and are thus compatible with PC-MEMS manufacturing processes, making them ideal building blocks for pop-up compatible designs.

In this design, the connection of these elementary building blocks mimics structures seen in biology. Multiple degrees of freedom can be seen in the human hip joint, for example, where ball-and-socket mechanisms in the bones and cartilege provide smooth rotation across a wide range of angles. Yet the motion of joints such as this is rarely constrained by a single mechanism but are complemented by a redundant set of muscles and tendons which route across and between joints, guiding, supporting and restricting motion. In this way, the leg mechanism, with its redundant-yet-constrained six bar linkgages, provides a more-limited degree of motion than each subcomponent independently would capable of delivering – and in a way which is inherently more manufactuable. Parallels to this concept are prevalent in biological systems as well, where the capabilities of the integrated system are greater than the sum of its parts.

3   CAD Design, Fabrication, and Materials

The initial design of the device is carried out in Solidworks in order to test the design concept and perform initial kinematic simulations. However, the Solidworks design only includes basic sketches of the layers, and not detailed drawings of features. The actual design used for fabrication is done using a new Computer Aided Design and Manufacturing (CAD/CAM) application called popupCAD [1]. Traditional CAD/CAM software, such as Solidworks, does not have any built-in tools to help with unconventional manufacturing methods such as PC-MEMS or SCM. On the other hand, popupCAD is specifically designed for layer-based manufacturing methods. The software automatically generates cut files using PC-MEMS and Pop-Up MEMS design and fabrication rules. It also has several operations specific to layer-based design and fabrication methods (e.g. generating support structures around features). In order to take advantage of these software features, we have imported the Solidworks sketches into popupCAD as guidelines, completed the design by adding the necessary PC-MEMS features such as hinges and support web, and generated the cut files in popupCAD. An example initial cut file and the final cut file used to fabricate the device is shown in Figure 3.

(a) First Pass Cuts   (b) Second Pass Cuts

Fig. 3. First and Second Pass Cuts

The device is fabricated using SCM [6,11] and Pop-Up Manufacturing / Assembly methods [9,10], outlined in Figure 4. Due to the kinematic loops in the mechanism, the device cannot fit on a single linkage sublaminate. Therefore the complete device requires two linkage sublaminates. Each linkage sublaminate consists of five layers: two outer cardboard layers for rigid links, a polyimide (Kapton) layer that forms the joints, and two adhesive layers made of double sided acrylic tape that bond these functional layers. In order to generate the two linkage sublaminates needed for the device, we need 11 layers: five layers per linkage sublaminate and a layer of adhesive that connects the two linkage sublaminates. The linkage sublaminate layers are first bulk micro-machined individually using a CO_2 laser (Universal Laser Systems, PL3.50) and a layup is formed using pin alignment. Rigid links can be formed by retaining stiff material layers; likewise, hinges are defined by regions where the stiff cardboard is removed and polyimide is retained, allowing neighboring stiff regions to move relative to each other. To bond the two sublaminates together, an adhesive layer is defined by the intersecting geometries between neighboring sublaminates, forming islands of material which must be selectively added during the stacking procedure; a negative of this pattern is thus cut from polyimide film and used as a mask to selectively spray adhesive (3M Hi-Strength 90 Spray Adhesive) in the required pattern. The cut files for these all these geometries are automatically generated by popupCAD, which is able to account for any material which becomes obscured after lamination.

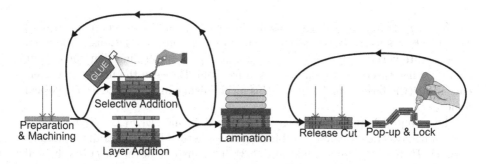

Fig. 4. The Device Fabrication Process.

4 Kinematics

Four kinematic loops determine the transmission characteristics between the actuators and the leg. The two 6R spherical linkages establish one kinematic loop each, and the two 4R spherical linkages define two additional kinematic loops. One of the six-bar linkages is shown in Figure 2b. Frames $A_1,B_1,C_1,D_1,A_2,B_2,C_2$, and D_2 belonging to this linkage can be used to generate the kinematic loop equations, which are generated by aligning the basis vectors of frames D_1 and D_2 using the equations

Table 1. Denavit-Hartenberg Parameters

Frame	A_1	B_1	C_1	D_1	A_2	B_2	C_2	D_2	E_1	F_1	G_1	H_1
Parent	Base	A_1	B_1	C_1	Base	A_2	B_2	C_2	Base	E_1	F_1	G_1
a	0	0	0	0	0	0	0	0	1	0	0	0
a_i	0	0	0	0	0	0	0	0	0	0	0	0
α_i	$-\pi/4$	$\pi/4$	$\pi/4$	$-\pi/4$	$-3\pi/4$	$-\pi/4$	$-\pi/4$	$-3\pi/4$	$-\pi/4$	$\pi/4$	$\pi/4$	$-\pi/4$
θ_i	$-q_{a1}^*$	q_{b1}^*	$-q_{c1}^*$	0	q_{a2}^*	$-q_{b2}^*$	q_{c2}^*	0	$-q_{e1}^*$	q_{f1}^*	$-q_{g1}^*$	0
Frame	E_2	F_2	G_2	H_2	I_1	J_1	K_1	L_1	I_2	J_2	K_2	L_2
Parent	Base	E_2	F_2	G_2	Base	A_1	I_1	K_1	Base	A_2	I_2	K_2
a	1	0	0	0	1	0	0	0	1	0	0	0
a_i	0	0	0	0	0	0	0	0	0	0	0	0
α_i	$-3\pi/4$	$-\pi/4$	$-\pi/4$	$-3\pi/4$	0	$-\pi/4$	$-\pi/4$	$-\pi/4$	0	$-\pi/4$	$-3\pi/4$	$\pi/4$
θ_i	q_{e2}^*	$-q_{f2}^*$	q_{g2}^*	0	$-q_{i1}^*$	q_{j1}^*	q_{k1}^*	0	$-q_{i2}^*$	$-q_{j2}^*$	$-q_{k2}^*$	0

* indicates state variable.

$$0 = \widehat{x}_1 \cdot \widehat{x}_2 - 1 \tag{1}$$

$$0 = \widehat{y}_1 \cdot \widehat{y}_2 - 1 \tag{2}$$

$$0 = \widehat{z}_1 \cdot \widehat{z}_2 - 1, \tag{3}$$

where x_1, y_1, z_1 and x_2, y_2, z_2 represent the orthonormal basis vectors for frames D_1 and D_2, respectively. These three equations establish a relationship between the six state variables $q_{a1}, q_{b1}, q_{c1}, q_{a2}, q_{b2}$, and q_{c2}, establishing that the output frame D_1 has three rotational degrees of freedom. Between the newtonian reference frame(the base frame) and D_1, this mechanism can be treated as a spherical joint.

The second 6R spherical linkage consisting of frames $E_1, F_1, G_1, H_1, E_2, F_2, G_2$, and H_2 behaves like the first, establishing the spherical loop constraint equations between H_1 and H_2 using Equations (1-3). The outputs of the two 6R spherical linkages are connected by a 2R linkage which enforces orientation between the frames D_1 and H_1 using Equations (1-3) and

$$0 = (\widehat{n}_y \times \widehat{d}_{1z}) \cdot (\widehat{n}_y \times \widehat{d}_{1z}) - 1, \tag{4}$$

where n_z represents the z-oriented basis vector of the Newtonian reference frame. Figure 5 highlights the output degrees of freedom produced by the above constraint equations. The six-bar linkages have been replaced with equivalent spherical joints and input links have been omitted for clarity.

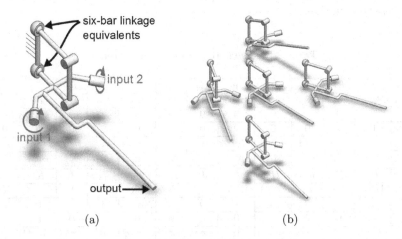

(a)   (b)

Fig. 5. An equivalent leg mechanism. In (a), the output stage of the leg is represented by spherical and revolute joints. Input linkages are omitted for clarity. In (b), the two degrees of freedom are highlighted by a variety of leg positions.

The two 4R spherical linkages, defined by the frames $\{A_1, I_1, J_1, K_1\}$ and $\{A_2, I_2, J_2, K_2\}$ respectively, are used to transmit the linear forces from each piezo-electric actuator into frames A_1 and A_2 of the bottom 6R spherical linkage. Frames J_1 and J_2 are aligned with K_1 and K_2 respectively using Equations (1-3). Thus for the four kinematic loops, 10 constraint equations can be established for the 12 state variables, establishing a two degree-of-freedom system.

Using the principle of virtual work, two Jacobians of the constraint equations may be obtained for the independent and dependent state variables by taking the partial derivative of the vector of constraint equations $\mathbf{f}(q)$ with respect to to each variable in $\mathbf{q}_{ind}$ and $\mathbf{q}_{dep}$ to obtain $\mathbf{J}_{ind}$ and $\mathbf{J}_{dep}$, respectively. Two state variables must be selected for the independent state vector, so in this case we pick $\mathbf{q}_{ind} = [q_{i1}, q_{i2}]^T$. The rest are put in $\mathbf{q}_{dep}$. The Jacobians can then be derived according to

$$0 = J_{ind}\dot{\mathbf{q}}_{ind} + J_{dep}\dot{\mathbf{q}}_{dep}, \tag{5}$$

$$-J_{dep}\dot{\mathbf{q}}_{dep} = J_{ind}\dot{\mathbf{q}}_{ind}, \text{ and} \tag{6}$$

$$\dot{\mathbf{q}}_{dep} = \underbrace{-J_{dep}^{-1}J_{ind}}_{T}\dot{\mathbf{q}}_{ind}. \tag{7}$$

The position of the actuator input and leg output can be represented in terms of the independent and dependent state variables $\mathbf{q}_{ind}$ and $\mathbf{q}_{dep}$, respectively. This permits the calculation of input and output Jacobians as

$$\dot{\mathbf{q}}_{in} = A\dot{\mathbf{q}}_{ind} + B\dot{\mathbf{q}}_{dep} \tag{8}$$

$$\dot{\mathbf{q}}_{out} = C\dot{\mathbf{q}}_{ind} + D\dot{\mathbf{q}}_{dep}. \tag{9}$$

By combining Equations (7-9), a direct relationship between input and output velocities can be determined, as

$$\dot{\mathbf{q}}_{in} = \underbrace{(A + BT)}_{E}\dot{\mathbf{q}}_{ind} \tag{10}$$

$$\dot{\mathbf{q}}_{out} = \underbrace{(C + DT)}_{F}\dot{\mathbf{q}}_{ind} \tag{11}$$

$$\dot{\mathbf{q}}_{out} = FE^{-1}\dot{\mathbf{q}}_{in}. \tag{12}$$

These equations permit the calculation of the input/output transmission ratios and the resulting output path as a function of a valid initial state, as shown in Figure 6.

5   Discussion

A motion study has been performed for the constrained mechanism, determining output trajectories and velocities as a function of actuator input signal. A variety of input signals were tested, with the results from three simulations shown

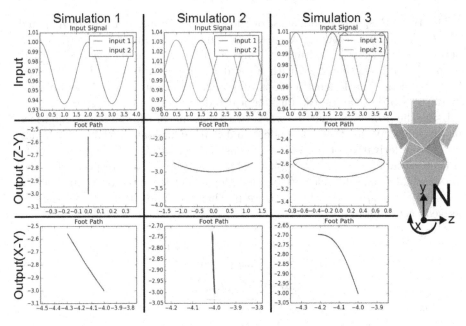

Fig. 6. Transmission Plots for three different input signals. In the first simulation, inputs are driven in phase, highlighting the lift degree of freedom. In the second simulation, a phase shift of π is used to highlight the swing degree of freedom. In the third simulation, a phase shift of $\pi/2$ shows a more-circular foot path typical of that likely to be used during walking. In these simulations, the x-axis is into the page.

in Figure 6. As can be seen in the plots of the three simulations, driving the signals in phase with each other produces lift motion in the leg (along $\widehat{n}_y$), and differential signals produce swing motion (motion along $\widehat{n}_z$). By supplying combinations of input and phase offset, a variety of trajectories can be produced, as seen in the third simulation of Figure 6.

6   Future Work

Several initial prototypes were constructed in order to understand the kinematics of this rather non-intuitive device. While these prototypes were useful for confirming the basic kinematics of the device, they were not ideal in terms of their hinges' functionality. This is due to the large gap between rigid elements, as well as to the non-zero thickness of the stacked cardboard, kapton, and adhesive. Future prototypes will be constructed with the materials used in the PC-MEMS paradigm, which will allow these hinges to perform more ideally. At that point, a full characterization of leg kinematics will be performed.

In addition, while the leg design presented in this paper utilizes concepts often seen in biology, we have not yet optimized it to match any biological system seen in nature. This will occur later in the design process, at which point it will accompany a much more in-depth performance analysis.

7   Conclusions

In this paper we have presented the initial design and kinematic evaluation for a bio-inspired leg transmission, suitable for integration into new crawling robot designs. Work is continuing in this direction, both on leg design and integration issues. While this transmission offers similar capabilities to the legs employed in current-generation legs, it offers the additional compatibility with monolithic assembly methods which can help reduce errors during assembly and speed up the process. As design criteria for these new robots become more developed, the models developed here can become the basis for further design and optimization of the transmission. In particular, the study of the dynamics of this new device will be crucial in evaluating and comparing current and future designs. In addition, by deriving the inverse kinematics of the leg, complex custom foot trajectories may be generated as a function of the two actuators.

The mechanism presented in this work is also ideal for insect-inspired robots due to its size and two-degree-of-freedom output. Since the mechanism includes the actuators, transmission, and the leg in a single system, it can be considered modular; i.e. several of them can be connected together to form a quadruped, hexapod, or a centipede-inspired multi-legged modular robot. We plan to use the mechanism presented in a centipede-inspired modular robot in the near future.

References

1. Aukes, D.M., Goldberg, B., Cutkosky, M.R., Wood, R.J.: An Analytic Framework for Developing Inherently-Manufacturable Pop-up Laminate Devices. Smart Materials and Structures (to appear, 2014)
2. Baisch, A.T., Heimlich, C., Karpelson, M., Wood, R.J.: HAMR3: An autonomous 1.7g ambulatory robot. In: 2011 IEEE/RSJ International Conference on Intelligent Robots and Systems, pp. 5073–5079. IEEE (September 2011)
3. Baisch, A., Ozcan, O., Goldberg, B., Ithier, D., Wood, R.: High Speed Locomotion for a Quadrupedal Microrobot. International Journal of Robotics Research (2014)
4. Gollnick, P.S., Magleby, S.P., Howell, L.L.: An Introduction to Multilayer Lamina Emergent Mechanisms. Journal of Mechanical Design 133(8), 081006 (2011)
5. Hoffman, K.L., Wood, R.J.: Passive undulatory gaits enhance walking in a myriapod millirobot. In: 2011 IEEE/RSJ International Conference on Intelligent Robots and Systems, vol. 2, pp. 1479–1486. IEEE (September 2011)
6. Hoover, A.M., Fearing, R.S.: Fast scale prototyping for folded millirobots. In: 2008 IEEE International Conference on Robotics and Automation, pp. 1777–1778 (May 2008)
7. Hoover, A., Steltz, E., Fearing, R.: RoACH: An autonomous 2.4g crawling hexapod robot. In: 2008 IEEE/RSJ International Conference on Intelligent Robots and Systems, pp. 26–33. IEEE (September 2008)
8. Jacobsen, J.O., Winder, B.G., Howell, L.L., Magleby, S.P.: Lamina Emergent Mechanisms and Their Basic Elements. Journal of Mechanisms and Robotics 2(1), 011003 (2010)

9. Sreetharan, P.S., Whitney, J.P., Strauss, M.D., Wood, R.J.: Monolithic fabrication of millimeter-scale machines. Journal of Micromechanics and Microengineering 22(5), 055027 (2012)
10. Whitney, J.P., Sreetharan, P.S., Ma, K.Y., Wood, R.J.: Pop-up book MEMS. Journal of Micromechanics and Microengineering 21(11), 115021 (2011)
11. Wood, R.J., Avadhanula, S., Sahai, R., Steltz, E., Fearing, R.S.: Microrobot Design Using Fiber Reinforced Composites. Journal of Mechanical Design 130(5), 052304 (2008)

Optimization of the Anticipatory Reflexes of a Computational Model of the Cerebellum

Santiago Brandi[1], Ivan Herreros[1], and Paul F.M.J. Verschure[1,2]

[1] SPECS, Technology Department, Universitat Pompeu Fabra,
Carrer de Roc Boronat 138, 08018 Barcelona, Spain
[2] ICREA, Institucio Catalana de Recerca i Estudis Avan cats,
Passeig Llu s Companys 23, 08010 Barcelona
{santiago.brandi,ivan.herreros,paul.verschure}@upf.edu

Abstract. The cerebellum is involved in avoidance learning tasks, where anticipatory actions are developed to protect against aversive stimuli. In the execution and acquisition of discrete actions we can distinguish errors of omission and commission due to a failure to execute a required defensive Conditioned Response (CR) to avoid an aversive Unconditioned Stimulus (US), and the energy expenditure of triggering an unnecessary CR in the absence of a US respectively. Hence, a motor learning cost function must consider both these components of performance and energy expenditure. Unlike remaining noxious stimuli, unnecessary actions are not directly sensed by the cerebellum. It has been suggested that the Nucleo-Olivary Inhibition (NOI) serves to internally rely information about these needless protective actions. Here we argue that the function of the NOI can be interpreted in broader terms as a signal that is used to learn optimal actions in terms of cost. We work with a computational model of the cerebellum to address: (i) how can the optimum balance between remaining aversive stimuli and preventing effort be found, and (ii) how can the cerebellum use the overall cost information to establish this optimum balance through the adjusting of the gain of the NOI. In this paper we derive the value of the NOI that minimizes the overall cost and propose a learning rule for the cerebellum through which this value is reached. We test this rule in a collision avoidance task performed by a simulated robot.

Keywords: cerebellum, nucleo-olivary inhibition, adaptive reflexes, cost minimization.

1 Introduction

When we avoid a specific disturbance, e.g., avoiding getting hit when boxing, we want our actions to be not only effective but also efficient. For example, excessive ducking would serve to elude the hit but would also mean a waste of energy. A general principle underlying this type of problems is that the way we behave carries a cost composed of (1) the disturbance we fail to prevent -in this case getting hit- and (2) the preventing action itself -the evading movement-. The optimum actions will be those that minimize the overall cost.

A. Duff et al. (eds.): Living Machines 2014, LNAI 8608, pp. 11–22, 2014.

One of the characteristics of avoidance tasks is that when we fail to prevent a disturbance (US) we receive direct feedback, but there is no sensory input when we perform an excessive preventive action, i.e., there is no negative US when aversive stimuli don't arrive. In other words, to a great extent our sensory system might not sense the lack of an *expected* disturbance as something different from the lack of an *unexpected* disturbance.

The cerebellum takes part in Avoidance Learning (AL) tasks where adaptive reflexes are developed to prevent disturbances caused by a noxious stimulus [Thompson, 1976]. One example of avoidance learning is the eye-blink conditioning paradigm, where subjects learn to develop predictive blinking to protect the eye from an air puff announced by a cue such as a tone [Hesslow and Yeo, 2002]. One of the characteristics of eye-blink conditioned responses is that after acquisition they are extinguished when no longer necessary. Therefore, the cerebellum must have access to an internally generated signal coding for unnecessary actions. It has been suggested that the Nucleo-Olivary Inhibition (NOI) serves as such a signal in the absence of aversive stimuli [Medina et al., 2002].

The Inferior Olive (IO), which activity serves as a teaching signal for the cerebellum, is excited by the US and inhibited by the deep cerebellar nuclei through the NOI. The bigger the gain of the NOI (k_{noi}), the smaller the output necessary to cancel the signal produced by the US. The learning stops when these two signals cancel out and the teaching signal is zero. The aversive stimulus that remains triggers a reactive reflex. Therefore, the balance between the amplitude of cerebellar -adaptive- responses and the reactive reflexes is modulated by k_{noi}. A detailed explanation on how the value of k_{noi} can determine the balance between adaptive and reactive reflexes can be found in [Herreros and Verschure, 2013]. As a short note, adaptive reflexes are learned preventive actions that anticipate aversive stimuli and are triggered by cues (e.g. blinking after hearing the tone). Reactive reflexes respond to the noxious stimuli (e.g., blinking after perceiving the air-puff) and their amplitude depends on the intensity of the latter.

This implies that k_{noi} can establish the final balance between the disturbance perceived and the effort of the preventing actions. Since both the disturbances and the preventing actions carry a cost we propose that the modulation of the k_{noi} serves to minimize the overall cost of our behaviour in an AL task.

In this paper we will investigate (i) *how can the optimum balance between anticipatory and reactive actions be found* and, assuming that the balance between both types of actions is determined by the gain of the NOI; (ii) *what plasticity mechanism might allow this optimal gain to be reached on-line using global cost information.* We will first arrive to an analytic relation between k_{opt} (the value of k_{noi} that minimizes the cost) and the parameters of our cost function when the task is the control of a zero-order plant. Next we will heuristically define an on-line learning rule and verify that it makes k_{noi} converge to k_{opt} in the same control task. Finally we will replicate this results in a simulated collision avoidance task to prove that the learning rule results in an approximated convergence of the gain of NOI to its optimal value.

2 Methods

2.1 Basic Cerebellar Model

We use a rate-based computational model of the cerebellar microcircuit as implemented in [Herreros, 2013]. The signal conveyed by the mossy fibres pathway (cue) is decomposed into several signals to resemble the expansion of information into cortical basis that occurs in the cerebellar granular layer.

In our model the cortical basis are generated by convolving the signal coming from the mossy fibres with two exponentials, an economic way to model the dynamics of granule cells responses. Therefore the profile of the basis response to a pulse input is that of an alpha function, where the time constants of the exponentials determine the temporal span and profile of the basis. The time constants of the exponentials are randomly drawn from two different ranges as to generate basis with different profiles, this variety allows the model to produce a wide range of output shapes. For more details on construction of the cortical basis see [Herreros et al., 2013].

The output of the microcircuit (adaptive response) is the sum of the half-wave rectified weighted components [Lepora et al., 2010]:

$$y(t) = [\mathbf{w}^T \mathbf{x}(t)]^+ \qquad (1)$$

where $\mathbf{x}(t)$ and $\mathbf{w}$ are the (column) vectors of basis and weights respectively.

The bank of dynamic weights $\mathbf{w}$ is updated according to the decorrelation rule [Fujita, 1982]:

$$\Delta \mathbf{w} = \beta e(t) \mathbf{x}(t - \delta) \qquad (2)$$

where β is the learning factor, $e(t)$ is an error signal coming from the inferior olive (see fig. 1) and the delay δ corresponds to the latency of the error feedback [Miall et al., 2007]. The decorrelation rule assures that $\mathbf{w}$ will converge to the values that minimize the error signal squared.

2.2 Cost Function

The cost function is a weighted sum of both the norm squared of the performance error $(\hat{y}(t))$ and the effort of the preventing actions, c_{error} and $c_{actions}$ are the respective cost weights:

$$C = \int_T c_{error} (\hat{y}(t))^2 + c_{actions} (y(t))^2 dt \qquad (3)$$

Note that this is the same cost function used in the framework of controller design with Linear Quadratic Regulators [Aström and Murray, 2010].

2.3 Zero-Order Plant Control Task

To analytically study the properties of the cerebellar controller with NOI, we use a set-up where the controller is connected to a stateless plant (zero-order).

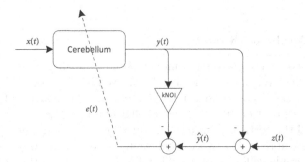

Fig. 1. *Zero-order plant* Process diagram of the cerebellar controller with NOI for a task that consists in tracking the target signal $z(t)$. In this case $x(t)$ is the cue signal and it is directly weighted by a gain $\mathbf{w}$ (see text) to generate the output of the cerebellum $(y(t))$. $y(t)$, scaled by k_{noi}, is compared with the performance error $\hat{y}(t)$ and the mismatch generates an error signal $(e(t))$.

In this case the controller works in isolation, and the task simply becomes to track an incoming signal (Fig. 1). For analytical purposes we remove the rectification from eq. 1, $z(t)$ is the target signal and the cost function becomes:

$$C = \int_T c_{error}(z(t) - \mathbf{w}^T \mathbf{x}(t))^2 + c_{actions}(\mathbf{w}^T \mathbf{x}(t))^2 dt \qquad (4)$$

2.4   Simulated Robot Setup and Control Architecture

We use our cerebellar controller in a collision avoidance task where a simulated robot learns to traverse a track with a single turn. The environment simulates a Mixed Reality Robot Arena (MRRA) [Fibla et al., 2010], see Fig. 2 (*Right*). For this task we implemented a computational architecture with two layers of control (Fig. 2, *left*). In the reactive layer the robot's proximity to the walls of the track -US- is mapped into turns -Unconditioned Response (UR)- to avoid collisions. The second layer consists of a computational model of the cerebellum as an adaptive filter (see [Dean et al., 2009]). This is the adaptive controller and it learns to link the green stripe on the floor -Conditioning Stimulus (CS)- with the later turn of the track, developing a predictive turn -CR- as trials progress.

3   Results

3.1   Analytical Solution of the Zero-Order Task

We begin by considering a scenario that allows for a closed-form solution for the problem of finding the optimal k_{noi} value. For simplicity, we address only the case where the cerebellum contains a single basis, thus there is only one weight

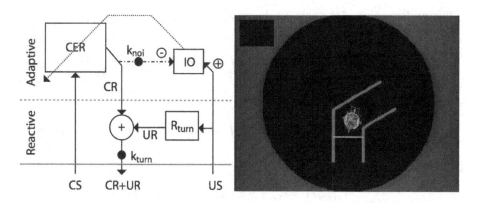

Fig. 2. : *(Left)* The control architecture has a reactive and an adaptive layer. CS, conditioning stimulus (green stripe on the floor); US, unconditioning stimulus (proximity to the walls); UR, unconditioned response (reactive turn); CR, conditioned response (predictive or adaptive turn). *(Right)* Experimental setup: Virtual environment with the robot avatar.

w to adjust. However the result in terms of the optimal value of k_{noi} also holds in the multidimensional input case, using gradients instead of a derivative.

We know that the decorrelation rule implemented in the computational model of the cerebellum minimizes the error squared. It is easy to see from the process diagram given in Fig. 1 that for a complete trial T the error signal minimized by the cerebellum is

$$e^2 = \int_T (z(t) - (1 + k_{noi})wx(t))^2 dt \qquad (5)$$

On the other hand we look for the weight **w** that also minimizes the cost function. For this we equate the **w** derivatives of equations 5 and 4 to zero. Note that the error terms in both equations do not match, as they refer to different concepts. The error term in Eq. 4 refers to the error in the prediction task whereas the error in Eq. 5 is the teaching signal that reaches the cerebellum and integrates the effect of the internal negative feedback.

The **w** derivative of the error yields

$$\frac{de^2}{dw} = \int_T -2(z(t) - (1 + k_{noi})wx(t))(1 + k_{noi})x(t)dt = 0$$

which can be further simplified to

$$-\int_T z(t)x(t)dt + w(1 + k_{noi})\int_T x(t)^2 dt = 0 \qquad (6)$$

The **w** derivative of the cost produces

$$\frac{dC}{dw} = -2c_{error}\int_T z(t)x(t)dt + 2c_{error}w\int_T x(t)^2 dt + 2c_{actions}w\int_T x(t)^2 dt = 0$$
(7)

subtracting $2c_{error}$ * Eq. 6 from Eq. 7 we obtain

$$-2c_1w(1+k_{noi})\int_T x(t)^2 dt + 2c_1w\int_T x(t)^2 dt + 2c_2w\int_T x(t)^2 dt = 0$$

that simplifies to

$$-c_{error}(1+k_{noi}) + c_{error} + c_{actions} = 0$$

This can be solved for k_{noi} as a function of the cost weight factors.

$$k_{noi} = \frac{c_{actions}}{c_{error}}$$
(8)

This result demonstrates that for the scenario in Fig. 1 it is possible to determine the value of k_{noi} that will minimize the cost function in Eq. 4. The utility of this result is two-fold. First, it shows that in general, the k_{noi} gain allows to control the trade-off between behavioural error and amplitude of the output signal, this is the performace/effort trade-off. Secondly, by giving a closed form solution for this particular set-up it will allow us to check for the correct convergence of an on-line learning rule for adjusting k_{noi}.

3.2   Finding the Optimum k_{noi} 'on-line'

The previous analysis show that the value of k_{noi} that minimizes the cost is a function of the weight parameters of the given cost function. Here our goal is to check whether it is possible to design an update rule that ensures that k_{noi} converges to the k_{opt} when the cerebellar controller has access only to the cost estimate, with no information about the parameters of the cost function.

We design the update rule based using the following assumptions.

1. The overall cost function is a sum of two terms. One that depends on the amplitude of the output and one that depends on the amplitude of the error in the performance.
2. The gain of the NOI controls the amplitude of the response of the cerebellum ($y(t)$) by scaling it before the comparison with the performance error [Herreros and Verschure, 2013]. In general, increasing k_{noi} will decrease the amplitude of $y(t)$, as seen in Fig. 4.
3. After convergence of the cerebellar output, increasing the amplitude of the response y will decrease the error in performance and vice versa.

Considering these assumptions, we will use the information about the evolution of the cost and the adaptive response amplitude to dynamically vary k_{noi}. We propose the following rule:

$$k_{noi_n} = (1 + \alpha(\text{sign}(C - \hat{C})\text{sign}(y - \hat{y})))k_{noi_{n+1}} \qquad (9)$$

Where C and y are the values of the overall cost and output amplitude for the last trial, respectively. $\hat{C}$ and $\hat{y}$ are the corresponding mean values for a number of past trials. Therefore $\hat{C} - C$ and $\hat{y} - y$ indicate whether the cost and amplitude of the response respectively have increased or decreased in the the past trial. α is the learning rate, namely, it is the size of the variation of k_{noi} from trial to trial. It can be appreciated that (1) the bigger α is the faster k_{noi} will converge to its optimum value, but also (2) the less precise will be the oscillation around this convergence point.

Informally, the logic of the update rule is the following. If the cost is decreasing, it modifies k_{noi} so that the amplitude of the response prolongs its current evolution (it keeps increasing or decreasing). On the other hand, if the cost is increasing, it modifies the value of the k_{noi} such that the tendency on the evolution of the response amplitude is reversed.

We tested the learning rule in a set-up where the target signal $(z(t))$ is a Gaussian that has to be approximated by weighting multiple bases functions. Since the result in Eq. 8 does not depend on $z(t)$ or $\mathbf{x}(t)$, we know that the optimal value for k_{noi} (k_{opt}) is the ratio of the cost weights $(\frac{c_{action}}{c_{error}})$.

We performed simulations with two pairs of values for c_{error} and $c_{actions}$, and for each pair we tested convergence for the optimal value when the initial value was either above or below the optimal. We can see that in all four cases k_{noi} converges to k_{opt} and oscillates around its value (Fig. 3 left), meaning that the learning rule controls the system successfully. Additionally, we can further verify that k_{opt} does minimize the overall cost, because in each case the cost is minimum at the end of the simulation (Fig. 3 right). Interestingly, neither the evolution of the k_{noi} nor the evolution of the cost are monotonous during the simulation. This occurs because two adaptive processes are concurrently acting and they interact in a non-trivial way. Namely, while the on-line update rule is modifying the value of the k_{noi}, the decorrelation learning rule is converging to the set of weights that approximate the target function. In other words, not only the correct weighting of the error and the output signals but also the output signal itself is being learned. In all but one case, at some point in the simulation, the overall cost sustainedly increases for some period before finally decreasing to the minimum value. Note that the final oscillations around the optimal value are due to the sign function in the update rule.

In summary, we showed that the update rule for the k_{noi} results in convergence to the optimal value for the zero-order plant, relying only on global information regarding the evolution of the cost the amplitude of cerebellar actions.

3.3   Collision Avoidance Task

We now move to a simulated environment where a robot provided with the cerebellum model learns to traverse a track without colliding with the walls. As the trials progress, the cerebellum learns to perform a predictive turn after a

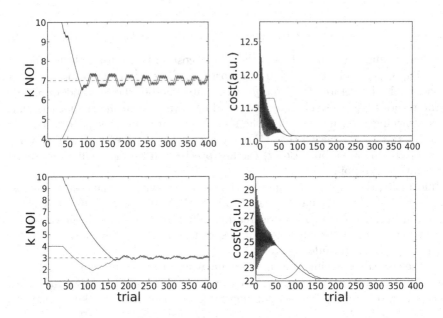

Fig. 3. *Learning rule evolution 1:* *(Above)* $k_{opt} = 7$ *(Red dashed line).* *(Left)* Two scenarios: The system starts with k_{noi} values 10 and 4 *(Dark blue and light green respectively)* and in both cases it converges to k_{opt}. For both cases the cost finds its minimum around the trial in which k_{noi} gets to k_{opt}. *(Below)* Same two scenarios with $k_{opt} = 3$.

cue signal (a green stripe in the ground), anticipating and avoiding the collision with the walls. Note that in this setup there is a plant (the simulated robot) that has a state (position and orientation), and that the coupling between the output of the cerebellum (turning action) and the target signal (proximity signal) is non-linear. Also, note that contrary to the previous example, in this case we do have the asymmetry in the error signal. Namely, the performance error signal is only triggered when the robot is too close to the wall, but not when it is too far away. For this reason, unnecessary turning has to be internally detected.

We first confirm that in this setup k_{noi} controls the balance between the adaptive and the reactive responses. Indeed, performing simulations with different values for k_{noi} showed that the amplitude of the adaptive response, directly controlled by the cerebellum, is inversely related to the gain of the NOI (Fig. 4). Smaller cerebellar responses result in a lesser avoidance of the noxious stimulus (i.e., more proximity to the wall), that, in turn, produces a bigger reactive response. Therefore, it is clear that in this scenario, there must be a particular k_{opt} that minimizes the cost function in Eq. 3.

However, since we don't have a closed-form solution for the k_{opt} we begin by searching an approximate solution numerically. We perform a number of simulations with a different k_{noi} values and compute the cost at the last trials. Again, the cost function is computed using Eq. 3. The relation between final

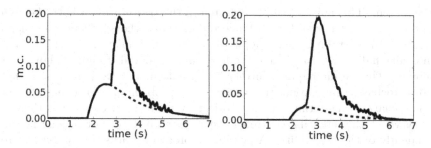

Fig. 4. k_{noi} *and adaptive response:* Total motor command (solid line) for k_{noi} values of 1 (*Left*) and 5 (*Right*). The dashed line divides the adaptive *below* from the reactive *above* components of the response.

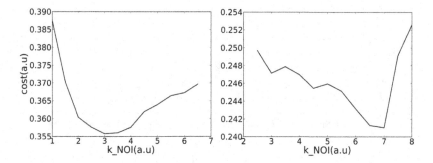

Fig. 5. Cost as function of k_{noi} for different cost parameters for the collision avoidance task. $c_{error} = 0.2$, $c_{actions} = 1$ (*Left*), $c_{error} = 0.125$, $c_{actions} = 1$ (*Right*).

cost and k_{noi} for two different pairs of c_{error} and c_{action} is shown in Fig. 5. From this figure we can confirm that a) there is indeed an optimal value for k_{noi} that minimizes the overall cost even when we are dealing with complex plants, and b) this k_{noi} is a function of the weights of the error and action costs, such that it increases with the ratio $\frac{c_{action}}{c_{error}}$. In any case, k_{opt} does not satisfy Eq. 8 because the solution was only valid for the zero-order plant.

Having portrayed the relation between k_{noi} and the minimization of the overall cost for the collision avoidance scenario, we validate now the convergence of the on-line update rule in Eq. 9. We expect a similar convergence as the one shown in Fig. 3 because the assumptions used to derive the update rule for k_{noi} hold in the current scenario. The results can be seen in Fig. 6. As in the previous section we run the simulations for two different pairs of cost parameters (the same ones used in Fig. 5). In all cases the value of k_{noi} converges. The horizontal dashed lines represent the value of k_{opt} previously estimated (Fig. 5). We observe a general trend of convergence, even though the results now are noisier than the ones for the zero-order plant. We note that the simulation and the computational cerebellar model were asynchronously interfaced, which introduced noise

in the system. This noise is clearly reflected in the trajectories of both the gain
parameter and of the overall cost. We also note that although there is a general
decrease in the cost during learning, it seems that all simulations reach the minimum value half-way through, and the final value achieved is slightly above that
minimum. This result requires further investigation, but we hypothesize that
besides stochastic fluctuation, there might be a problem of overfitting inherent
to the scenario. The optimal robot's trajectory in terms of cost is not the one
towards which the cerebellar controller converges with the decorrelation rule (independently of the k_{noi} value). A possible approach towards this problem is to
constrain the cerebellar outputs so that the final profile of the responses allows
for the cost to be kept at its minimum. Again, this problem did not exist in the
zero-order plant case.

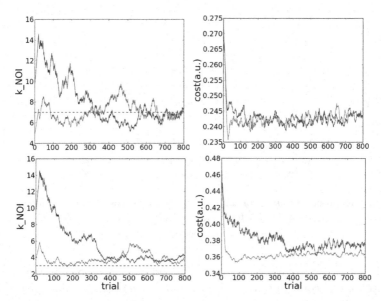

Fig. 6. *Learning rule evolution 2: Up $k_{opt} = 7$, Down $k_{opt} = 3$.* Simulations start
above (*Dark blue line*) and below (*Light green line*) the value of k_{opt} found empirically
in section 3.3 (*Red dashed line*). Although the results resemble the desired overall
behaviour, the learning rule is still vulnerable to the noise in the system, the cost
approaches its minimum value.

4   Discussion and Conclusions

We have shown, using a model of the cerebello-olivary system, that it is possible
to describe the function of the NOI in terms of cost minimization, defining
the cost as a combination of the error in performance and the control effort.
Specifically, we have proven that through the manipulation of the gain of the
NOI (k_{noi}) it is possible to adjust the operation of a cerebellar controller to
achieve different error/performance trade-offs.

In avoidance learning tasks, this feature of the k_{noi} serves to balance the remaining noxious stimuli and the learned preventive actions. We propose that there is an optimal point for this balance and it is found when the cost of the unprevented (portion of) noxious stimuli plus the cost of the preventive actions themselves is minimum. In nature, this trade-off should be reflected in the strength of NOI.

We demonstrated that we can analytically determine the optimal value for k_{noi} when the cerebellar output controls a plant with zero-order dynamics (or in other words, assuming that the function of the cerebellum is simply to track or predict a given input signal). Moreover, we proposed a learning rule that updates k_{noi} and converges towards its optimal value making use of information about the evolution of the cost (or evolution of the reward, if we consider cost as negative reward) and the evolution of the cerebellar output.

Regarding the plausibility of such a learning mechanism in nature, that we hypothesize should reside in the potentiation (or depression) of the nucleo-olivary synapses, we find a clear pre-requisite: the nucleo-olivary system should receive projections of a neuromodulator coding reward. Many studies propose that dopamine neurons display phasic signals that encode differences between expected and actual rewards [Spanagel and Weiss, 1999, Schultz, 2002]. Recently, it has been shown that dopaminergic neurons from the ventral tegmental area provide a robust projection to the medial column of the inferior olive in mammals [Winship et al., 2006]. Therefore, it can be argued that inferior olive neurons may have enough information to implement a learning rule coherent with the one proposed here. If that is the case, one would expect that dopamine should control the direction of nucleo-olivary plasticity induced by presynaptic activity.

On a different note, we believe that our work may have provided some insight into the functioning of the cerebellum. Our results advance the view that the cerebello-olivary system function can go beyond the usual supervised learning [Doya, 1999]. As shown in our collision avoidance simulation, the cerebellar controller has now two levels of adaptability: the first level, residing in the cerebellar cortex and embodied in the decorrelation learning rule, is able to match the temporal profile of an incoming signal using a locally generated error signal. The second, at the level of the nucleo-olivary connection, adjusts the general amplitude of the cerebellar output using global reward information. This two-level learning can be of particular interest in the field of robotics since it would extend the range of application and the versatility of cerebellum-based controllers. Future work should include a performance comparison between the present model and non-biologically-inspired approaches.

References

[Aström and Murray, 2010] Aström, K.J., Murray, R.M.: Feedback systems: an introduction for scientists and engineers. Princeton University Press (2010)

[Dean et al., 2009] Dean, P., Porrill, J., Ekerot, C.-F., Jörntell, H.: The cerebellar microcircuit as an adaptive filter: experimental and computational evidence. Nature Reviews Neuroscience 11(1), 30–43 (2009)

[Doya, 1999] Doya, K.: What are the computations of the cerebellum, the basal ganglia and the cerebral cortex? Neural Networks 12(7), 961–974 (1999)

[Fibla et al., 2010] Fibla, M.S., Bernardet, U., Verschure, P.F.: Allostatic control for robot behaviour regulation: An extension to path planning. In: 2010 IEEE/RSJ International Conference on Intelligent Robots and Systems (IROS), pp. 1935–1942. IEEE (2010)

[Fujita, 1982] Fujita, M.: Simulation of adaptive modification of the vestibulo-ocular reflex with an adaptive filter model of the cerebellum. Biological Cybernetics 45(3), 207–214 (1982)

[Herreros, 2013] Herreros, I.: Data-Driven Batch Scheduling. PhD thesis, Universitat Pompeu Fabra. Departament de Tecnologies de la Informaci i les Comunicacions (2013)

[Herreros et al., 2013] Herreros, I., Maffei, G., Brandi, S., Sanchez-Fibla, M., Verschure, P.F.: Speed generalization capabilities of a cerebellar model on a rapid navigation task. In: 2013 IEEE/RSJ International Conference on Intelligent Robots and Systems (IROS), pp. 363–368. IEEE (2013)

[Herreros and Verschure, 2013] Herreros, I., Verschure, P.F.: Nucleo-olivary inhibition balances the interaction between the reactive and adaptive layers in motor control. Neural Networks (2013)

[Hesslow and Yeo, 2002] Hesslow, G., Yeo, C.: The functional anatomy of skeletal conditioning. In: A Neuroscientist's Guide to Classical Conditioning, pp. 86–146. Springer, Heidelberg (2002)

[Hollerman and Schultz, 1998] Hollerman, J.R., Schultz, W.: Dopamine neurons report an error in the temporal prediction of reward during learning. Nature Neuroscience 1(4), 304–309 (1998)

[Lepora et al., 2010] Lepora, N.F., Porrill, J., Yeo, C.H., Dean, P.: Sensory prediction or motor control application of marr–albus type models of cerebellar function to classical conditioning. Frontiers in Computational Neuroscience 4 (2010)

[Medina et al., 2002] Medina, J., Nores, W., Mauk, M.: Inhibition of climbing fibres is a signal for the extinction of conditioned eyelid responses. Nature 416(6878), 330–333 (2002)

[Miall et al., 2007] Miall, R.C., Christensen, L.O., Cain, O., Stanley, J.: Disruption of state estimation in the human lateral cerebellum. PLoS biology 5(11), e316 (2007)

[Schultz, 2002] Schultz, W.: Getting formal with dopamine and reward. Neuron 36(2), 241–263 (2002)

[Spanagel and Weiss, 1999] Spanagel, R., Weiss, F.: The dopamine hypothesis of reward: past and current status. Trends in Neurosciences 22(11), 521–527 (1999)

[Thompson, 1976] Thompson, R.F.: The search for the engram. American Psychologist 31(3), 209 (1976)

[Winship et al., 2006] Winship, I., Pakan, J., Todd, K., Wong-Wylie, D.: A comparison of ventral tegmental neurons projecting to optic flow regions of the inferior olive vs. The Hippocampal Formation 141(1), 463–473 (2006)

Evolving Optimal Swimming in Different Fluids: A Study Inspired by *batoid* Fishes

Vito Cacucciolo*, Francesco Corucci*, Matteo Cianchetti, and Cecilia Laschi

The BioRobotics Institute, Scuola Superiore Sant'Anna, Pisa, Italy
{v.cacucciolo,f.corucci,m.cianchetti,c.laschi}@sssup.it

Abstract. For their efficient and elegant locomotion, *batoid* fishes (e.g. the manta ray) have been widely studied in biology, and also taken as a source of inspiration by engineers and roboticists willing to replicate their propulsion mechanism in order to build efficient swimming machines. In this work, a new model of an under-actuated compliant wing is proposed, exhibiting both the oscillatory and undulatory behaviors underlying *batoid* propulsion mechanism. The proposed model allowed an investigation of the co-evolution of morphology and control, exploiting dynamics emergent from the interaction between the environment and the mechanical properties of the soft materials. Having condensed such aspects in a mathematical model, we studied the adaptability of a *batoid*-like morphology to different environments. As for biology, our main contribution is an exploration of the parameters linking swimming mechanics, morphology and environment. This can contribute to a deeper understanding of the factors that led various species of the *batoid* group to phylogenetically adapt to different environments. From a robotics standpoint, this work offers an additional example remarking the importance of morphological computation and embodied intelligence. A direct application can be an under-water soft robot capable of adapting morphology and control to reach the maximum swimming efficiency.

Keywords: Bio-mimetics, embodied intelligence, evolutionary robotics, genetic algorithms.

1 Introduction

Batoid fishes (such as the manta ray) have been object of several studies in the past years, not only in biology [1] [2] [3] but also in the engineering field [4] [5]. The reasons for the interest in these fishes lie in some peculiar aspects they exhibit, such as their unique morphology and locomotion strategy (sometimes identified as a form of underwater flight). Those aspects confer them interesting capabilities, that are desirable also for artificial machines. The smallest species are extremely agile and quick. The biggest ones still preserve agility and high maneuverability (they can, for example, perform turn-on-a-dime maneuvers), in addition to being very efficient swimmers capable of cruising for very long

---

* These authors contributed equally to this work.

A. Duff et al. (eds.): Living Machines 2014, LNAI 8608, pp. 23–34, 2014.

distances. In related literature biologists focused on the odd anatomy of these fishes [1] [2], engineers investigated the fluid-dynamics of their propulsion mechanism [4] [3], while robotics studies tried to replicate their compliant morphology and actuation [5]. Despite the large amount of works on this topic, most of contributions embrace an experimental approach, while less efforts were put in modeling the interaction between morphology and environment. In this work we study the adaptability of a *batoid*-inspired wing to different environments, using a model-based synthetic methodology. The objective is twofold: investigate how different environments can lead to the evolution of different morphologies and behaviors, while offering an additional example of morphological computation and embodied intelligence. Possible applications are novel adaptable under-water soft robots.

The paper is organized as follows. In Sect. 2 we summarize generalities on *Batoid* fishes, focusing on peculiarities related to their locomotion mechanism. Section 3 illustrates how the fundamental aspects related to our study are synthesized in a mathematical model, with complex features condensed in few, elegant, parameters. Then we describe the genetic optimization set-up used to investigate the co-evolution of morphology and control in order to achieve optimal swimming in different environments (Sect. 4). The paper ends presenting simulation results (Sect. 5) and discussing future perspectives (Sect. 6).

2   Biological Inspiration

Batoid fishes (e.g. electric rays, saw-fishes, guitar-fishes, skates and stingrays) are a monophyletic group of over 500 elasmobranch species nested within sharks [4]. They are distinguished by their flattened bodies, enlarged pectoral fins that are fused to the head, and gill slits that are placed on their ventral surfaces [6] [2] [7]. *Batoid* fishes can cruise at high speed and can perform turn-on-a-dime maneuvering, making them rich source of inspiration for bio-mimetic artificial underwater locomotion (e.g., AUV, ROV).

Most of the investigators found that *Batoids* propel themselves by moving their enlarged pectoral fins in a flapping motion, combined with an undulatory motion (a traveling wave moving in the downstream direction) [4]. Rosenberger [3] analyzed the kinematics of eight species of ray, each with a different swimming motion. He identified a continuous spectrum of motion ranging from rajiform undulation (where multiple traveling waves pass down the fins and body) to mobuliform oscillation (characterized by a broad flapping of the pectoral fins). The combination of these two mechanisms appear to be a key element in the swimming dynamics and it is directly connected with the animal morphology.

In the next section we illustrate how these peculiarities were synthesized in a mathematical model.

3 Model

This work follows a so called *synthetic methodology*. In this section we describe how the synthesis was performed, both for what concerns the definition of a *batoid*-like morphology and the physical modeling of the swimming dynamics.

3.1 Morphology

As for morphology, two features being exhibited by *batoid* fishes are of particular interest for roboticists. *a*) Compliance: *batoids* have elastic, flexible, compliant bodies; *b*) Under-actuation: most of *batoid* fishes swim using pectoral fin loco-motion [3], i.e. they actuate only the frontal edge of their wings. From a robotics stand-point, this means that their body, featuring – theoretically – an infinite number of degrees of freedom (DOFs), is under-actuated. Moreover, the control of pectoral fins appears to be quite simple, consisting just in oscillatory move-ments. Compliance and under-actuation make *batoids* interesting case studies also for *embodied intelligence* and *morphological computation*: a rich behavior emerges from the strong interaction with the environment, without the need of a complex control. The above mentioned features are highly relevant to our study, being involved in the formation of the oscillatory and undulatory behav-iors responsible of thrust production. Those traits were thus synthesized in our model. First of all, modeling a single wing was sufficient for our intended study. Furthermore, in order to simplify the continuous structure without losing the essential features, we discretized the morphology by means of small rigid seg-ments (Fig. 1). We modeled the wing by means of 10 adjacent radial chains, each being composed by a variable number (10 in the longest chain) of rigid prismatic segments (22 x 24 x 5 *mm*). This discretization was selected as a trade-off be-tween accuracy and computational cost. The resulting fin shape is approximately semi-elliptic, with root chord of 240 *mm* and span of 220 *mm*, following the ex-perimental set-up presented in [4]. The segments are interconnected by means of hinge joints with torsion springs (modeling bending elasticity). Adjacent chains are free to move one respect to the others, with interactions among them being modeled by means of linear springs (Fig. 2). This structure is able to mimic the passive compliance using the minimum number of DOFs. No structural damp-ing was included as it was assumed to be negligible with respect to the one provided by fluid drag forces. Only the joints belonging to the frontal chain are actuated, by means of simple sinusoidal oscillators (the same oscillation is ap-plied at each actuated joint). Although it may seem oversimplified, the resulting global actuation mimics the one observed in *batoids*, that do not appear to adopt more complex control strategies (traveling waves are not directly actuated, in-stead they emerge from the interaction with the environment). This modular, under-actuated, structure is able to reproduce both the undulatory (longitudinal waves) and oscillatory (radial waves) modes involved in thrust production, i.e. waving and flapping.

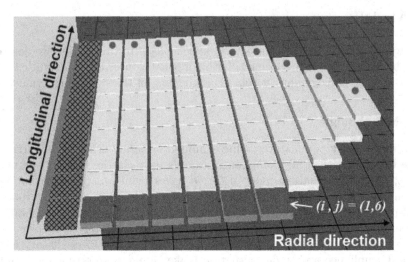

Fig. 1. Discretization of the wing by means of small rigid segments. Segments in the radial direction (index j) are interconnected by hinge joints: the ones of the first chain are actuated (blue segments), while the others are modeled by passive torsion springs. In the longitudinal direction (index i) there are no rigid joints: spring forces, better described in Fig. 2, tend to keep adjacent chains on the same plane. Red dots denote the characteristic points of the trailing edge where the amplitude A is sampled. Crossed segments are, instead, fixed.

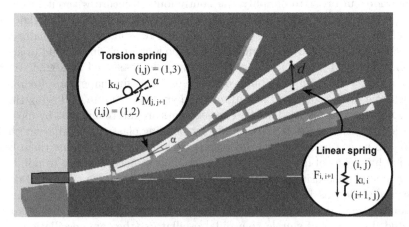

Fig. 2. Displacements and spring forces. The figure shows an angular displacement (α) between two adjacent segments (segment j and $j+1$ respectively) in the radial direction. A torsion spring couple $M_{j,j+1} = k_{t,j} \cdot \alpha$ is applied between the two. $k_{t,j}$ is fixed for all the hinge joints having the same distance from the body. The figure also shows a displacement (d, vector) between the center of mass of two adjacent segments in the longitudinal direction (segments i and $i+1$). A force $F_{i,i+1} = k_{l,i} \cdot d$ is applied between the two centers of mass, with $k_{l,i}$ being fixed for all segments at the same distance from the body.

3.2   Batoids Swimming Dynamics

As for most fishes, swimming locomotion of *batoids* is dominated by vortices dynamics. Modeling this kind of phenomena can be very hard, however various experimental investigations have been done, showing that these complex fluid phenomena can all be described by means of some elegant non-dimensional parameters. The most important one is arguably the Strouhal number. Firstly introduced in 1915 by Lord Rayleight in order to describe vortex phenomena, it was then widely used in the description of fishes locomotion, and it is defined as [8] [9]

$$St = \frac{fA}{U},$$
(1)

where U is the average swimming speed and f and A are respectively the frequency and peak-to-peak amplitude of the characteristic locomotion swimming movement (e.g. tail beating for fishes, fins flapping for *batoids*). In [8] different fishes over a large range of Reynolds numbers have been studied, finding that they all tend to cruise (i.e. swim at constant speed) in the same Strouhal number range ($0.25 < St < 0.35$). Authors suggested that this may correspond to the most efficient cruising condition (confirmed by other studies [9]). The other relevant parameters governing the phenomenon are the Reynolds number (Re) and the phase (ϕ) of the undulatory waves. In order to describe the phenomenon we built a mathematical model able to describe the dynamics of the *batoid* fin and its interaction with the fluid, modeled through local drag forces (more details in Sect. 5). The main goal of this model is to show how the interaction between the fluid (mainly described by viscosity μ and density ρ) and the compliant, under-actuated, body structure leads to the emergence of different combinations of flapping and waving, as observed among *batoid* species.

In order to compute the Strouhal number, the amplitude A is estimated as the mean transverse oscillation of the trailing edge, sampled in a number of discrete points (Fig. 1). On the other hand, the frequency f is a parameter, while the mean cruise speed U is assumed to be constant ($U = 0.11\ m/s$), referring to a similar set-up present in literature [4].

4   Optimization Setup

The model was implemented in the Webots simulation environment [10]. As for the optimization study, we used genetic algorithms offered by the *MATLAB 2011b Global Optimization Toolbox* (The MathWorks Inc., Natick, MA, 2000). We set-up the problem as a bounded optimization, adopting a real-coded genetic algorithm based on the *Augmented Lagrangian Genetic Algorithm (ALGA)* [11]. An *adaptive feasible* mutation operator and a *scattered* crossover operator are adopted. The chosen population size is of 50 individuals, the maximum number of generations was set to 1000, and the termination criterion was the best individual reaching a fitness of 0.

Parameters. We co-evolved both control and morphological parameters. The first consist of amplitude (a) and frequency (f) of the oscillators driving the active joints. For the latter we assumed both torsion and linear springs to be constant at a given distance from the body. The adopted discretization consists in 10 torsion springs coefficients (kt_j) and 10 linear springs coefficients (kl_j). The 22 evolved parameters are thus:

$$[A, f, kt_1, ..., kt_{10}, kl_1, ..., kl_{10}].$$

Target. As explained in Sect. 2, we optimized the *Strouhal number* resulting from a given set of parameters. As target value, we chose $St^* = 0.3$. The particular value is motivated by fluid-dynamics studies on fishes [4] [8] [9], showing that most fishes exhibit a Strouhal number close to the above one.

Fitness. Given a set of parameters, the fitness is computed as the normalized, squared difference between the Strouhal number achieved for the current morphology and control (St) and the target Strouhal number (St), thus:

$$fitness = \left(\frac{St - St^*}{St^*}\right)^2. \tag{2}$$

5   Results and Discussion

We investigated the role of density and viscosity as they directly affect drag forces that, interacting with the under-actuated body structure, lead to the emergence of different combinations of flapping and waving. Particularly, we evolved our wing in two different fluids, water ($\rho = 1000\ Kg/m^3$, $\mu = 1.15 \cdot 10^{-3}\ Pa \cdot s$) and Tetrachloroethylene ($\rho = 1622\ Kg/m^3$, $\mu = 0.89 \cdot 10^{-3}\ Pa \cdot s$). It has to be noted that our simulations fall in an interval of Reynolds numbers ($[2 \cdot 10^4, 3 \cdot 10^5]$, turbulent flow) in which the effect of viscosity variation on the drag coefficient is negligible. Thus the adopted local drag forces are: $F_{drag} = \frac{1}{2}C_d\rho S_{ij}v_{ij}^2$, where the terms are respectively the drag coefficient, the fluid density, the exposed surface of the segment (i, j) and its velocity perpendicular to S_{ij}. The optimization was performed with two different choices for the parameters bounds. We kindly invite the reader to watch the accompanying videos at: http://sssa.biineroboticsinstitute.it/papers/FinEvolution

5.1   Wide Bounds

In a first step, parameters bounds were kept wide in order to let the genetic algorithm explore a larger space of wing configurations. Adopted bounds are reported in Tab. 1. In both the fluid environments the genetic algorithm managed to reach the desired optimal Strouhal number St^*, with no residual error (Tab. 2). Analyzing f and a, it resulted that the wing evolved in water (lower density) is actuated at a slightly higher frequency, while the amplitude of the oscillation is slightly smaller. The evolved spring parameters are summarized in Fig. 3. Interestingly, the actuation frequency, the amplitude of emergent waves and their wavelength are consistent with experimental data related to some species of *batoids* (e.g. *Rhinoptera bonasus, Dasyatis violacea* [3]), thus supporting the validity of the proposed model. From a qualitative point of view, the waves emerging in the model resemble the ones observed in some reference animals (Fig. 4).

From the achieved results we observed that the problem is under-constrained, meaning that there are several possible solutions that minimize the fitness. In order to reduce the redundancy we decided to shrink the bounds towards more biologically plausible ranges of values. Moreover, the necessity of imposing the continuity of spring parameters emerged to avoid discontinuous behaviors among parameters.

Table 1. Wide parameters bounds

	$f\ [Hz]$	$a\ [rad]$	$kl_j\ [N/m]$	$kt_j\ [N/rad]$
Lower	0.2	0.0349	0.05	0.005
Upper	6.0	0.349	100.0	5.0

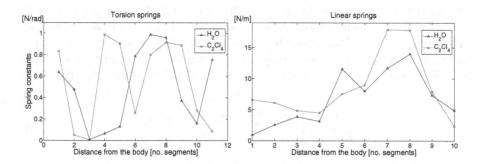

Fig. 3. Evolved spring parameters with wide bounds. Left: torsion springs. The plot highlights that without imposing the continuity of spring parameters results can be irregular, not biologically plausible, and difficult to interpret. Right: linear springs. The wing evolved in the tetrachloroethylene is more rigid (almost everywhere) than the one evolved in water. In this case resulting spring parameters are reasonably smooth.

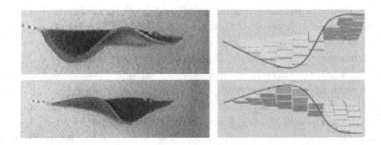

Fig. 4. Comparison of wing profiles in two different instants. Left: *Gymnura micrura* (adapted from [3]). Right: The proposed model, evolved for water environment.

Table 2. Optimization results. Actuation frequency and amplitude, mean emergent amplitude on the trailing edge, Strouhal number are reported.

Fluid		f $[Hz]$	a $[rad]$	A $[m]$	St
Wide bounds	Water	0.944	0.232	0.035	0.300
	Tetra chloro ethylene	0.887	0.245	0.037	0.300
Narrow bounds	Water	0.737	0.261	0.0448	0.300
	Tetra chloro ethylene	1.06	0.262	0.0311	0.300

5.2   Narrow Bounds

A first refinement consists in imposing the continuity of spring parameters. In order to impose such constraint, we assume the thickness of the fin to vary linearly on the radial direction. It is thus reasonable to model the variation of spring parameters with a cubic polynomial, given that the bending elasticity is a cubic function of the thickness. The resulting non-dimensional equations are:

$$\widetilde{kl}_j = 1 + al_3 \cdot \widetilde{x}_j^{\,3} + al_2 \cdot \widetilde{x}_j^{\,2} + al_1 \cdot \widetilde{x}_j$$
$$\widetilde{kt}_j = 1 + at_3 \cdot \widetilde{x}_j^{\,3} + at_2 \cdot \widetilde{x}_j^{\,2} + at_1 \cdot \widetilde{x}_j,$$

(3)

where:

$$\widetilde{kl}_j = kl_j/kl_{max}, \qquad x_j = j \cdot dx,$$
$$\widetilde{kt}_j = kt_j/kt_{max}, \qquad dx = 22 \; mm,$$
$$\widetilde{x}_j = x_j/x_{max}, \qquad j = 0,\ldots,9.$$

(4)

Consequently, the optimization parameters are reduced to 10, being:

$$[f, a, al_3, al_2, al_1, kl_{max}, at_3, at_2, at_1, kt_{max}].$$

The non-dimensional cubic polynomials were built in order to span from a constant behavior ($k_j = k_{max}$) to a full cubic ($a_3 = a_2 = a_1 = -1$) going to zero at

Table 3. Narrow bounds parameters

	f [Hz]	a [rad]	al_i	kl_{max} [N/m]	at_i	kt_{max} [N/rad]
Lower	0.5	0.0873	−1.0	1.0	−1.0	0.01
Upper	5.0	0.2618	0.0	10.0	0.0	0.1

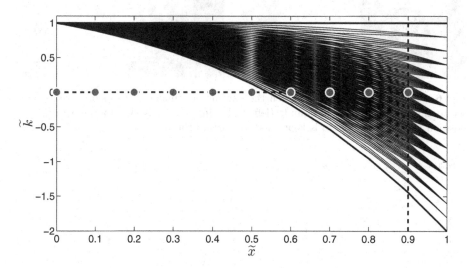

Fig. 5. Non-dimensional cubic polynomials used to model spring constants, plotted with a parameters step of 0.2. Limit curves are thicker. Red dots denote the sampling points x_j. Only the curves that are positive for all the sampling points are admissible.

the tip of the fin, allowing all the possible combinations between these boundaries (Fig. 5). The resulting bounds for the polynomial parameters range from −1 to 0. It has to be noted that among the resulting curves, only those that are positive for all the sampling points are admissible. This was imposed applying a constraint in the genetic algorithm. Limiting the bounds for the a_i coefficients to avoid negative curves could have discarded meaningful solutions.

Narrowing the bounds is motivated by the need to reduce the redundancy of the problem as well as to define a more physically plausible parameters range. Bounds related to actuation frequency (f) and amplitude (a) are consistent with the range of behaviors observed in *batoid* fishes [3]. As for kl_{max} and kt_{max}, the relative bounds were defined by testing several limit conditions for the parameters, and observing the resulting behavior of the wing (Fig. 6). Parameters bounds are summarized in Tab. 3.

Results of the genetic optimization are summarized in Tab. 2. Resulting actuation amplitude (a) does not vary between the two fluids, while there is a significant difference in the frequency (f). Particularly, the actuation frequency

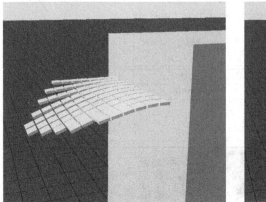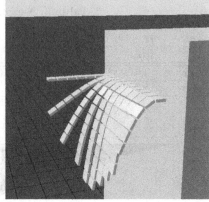

Fig. 6. Free fall of the fin in air when the parameters (linear and torsion springs) are set to the upper and lower bounds (left and right, respectively). In these tests the position of the frontal chain is kept fixed, to sustain the fin.

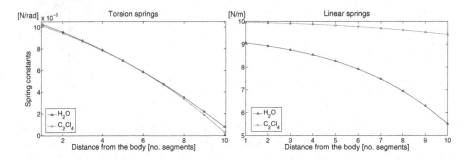

Fig. 7. Evolved spring parameters with narrow bounds. Left: torsion springs. The behavior is almost the same in the two fluid environments, with a steep variation along the span of the fin. Right: linear springs. In this case a pronounced difference can be seen between the two environments, with the wing evolved in water being softer and exhibiting more variation in the radial direction.

results to be higher for the wing evolved in Tetrachloroethylene, showing an inverse trend respect to the simulation featuring wide bounds. This discrepancy arises from the under-constrained nature of the problem: it is thus possible to achieve more than one result exhibiting the optimal fitness. Given the more biologically plausible set-up, results achieved in presence of continuous spring constants and narrow bounds are arguably more reliable.

As for torsion springs (Fig. 7), the behavior is almost identical in the two fluid environments, with a steep variation along the span of the fin. Pronounced differences can be instead observed in linear springs coefficients (Fig. 7). The wing evolved in water is softer, and exhibits more variation in the radial direction. As a result, a greater presence of undulation respect to oscillation is observed for the wing evolved in the Tetrachloroethylene.

6   Conclusions

A model for an under-actuated, compliant wing was presented and optimized by means of genetic algorithms for different fluid properties. The optimization target was motivated by fluid-dynamics studies conducted on fishes, showing a correlation between Strouhal number and swimming efficiency. Simulation results show a good convergence of the genetic algorithm, able to find sets of morphological and actuation parameters which perfectly match the desired Strouhal number in our model. Moreover, the main characteristic quantities are consistent with biological data, thus supporting the validity of the proposed model.

Present results encourage the investigation on speciation of *batoid* fishes in different environments by means of modeling tools. This will require to enrich the fluid dynamics set-up (e.g. modeling the relationship between emergent oscillations and thrust) as well as the capability of defining morphologies able to span the variety of *batoid* shapes. From a robotics perspective, the long-term vision is that of a bio-mimetic robot able to adapt to different fluids and, possibly, to different flow conditions, adjusting both morphology and actuation in order to exploit the interaction with the environment letting an efficient, effective swimming behavior emerge. Besides contributing to the fields of embodied intelligence and morphological computation, this could foster practical soft-robotics applications such as underwater robots for environmental monitoring and inspection of industrial plants.

Acknowledgments. This work is supported by RoboSoft - A Coordination Action for Soft Robotics (FP7-ICT-2013-C # 619319). We would also like to acknowledge the ShanghAI Lectures project and the teaching assistants for the inspiration and assistance.

References

1. Schaefer, J.T., Summers, A.P.: Batoid wing skeletal structure: novel morphologies, mechanical implications, and phylogenetic patterns. Journal of Morphology 264(3), 298–313 (2005)
2. Aschliman, N.C., Claeson, K.M., McEachran, J.D.: Phylogeny of batoidea. In: Carrier, J.C., Musick, J.A., Heithaus, M.R. (eds.) Biology of Sharks and their Relatives, pp. 57–94. Indiana University Press, Bloomington (2012)
3. Rosenberger, L.J.: Pectoral fin locomotion in batoid fishes: undulation versus oscillation. Journal of Experimental Biology 204(2), 379–394 (2001)
4. Clark, R., Smits, A.: Thrust production and wake structure of a batoid-inspired oscillating fin. Journal of Fluid Mechanics 562(1), 415–429 (2006)
5. Suzumori, K., Endo, S., Kanda, T., Kato, N., Suzuki, H.: A bending pneumatic rubber actuator realizing soft-bodied manta swimming robot. In: 2007 IEEE International Conference on Robotics and Automation, pp. 4975–4980. IEEE (2007)
6. Nelson, J.S.: Fishes of the World. John Wiley & Sons (2006)
7. Dunn, K.A., Miyake, T.: Interrelationships of the batoid fishes (chondrichthyes: Batoidea). Interrelationships of Fishes 63 (1996)

8. Triantafyllou, G., Triantafyllou, M., Grosenbaugh, M.: Optimal thrust development in oscillating foils with application to fish propulsion. Journal of Fluids and Structures 7(2), 205–224 (1993)
9. Eloy, C.: Optimal strouhal number for swimming animals. Journal of Fluids and Structures 30, 205–218 (2012)
10. Michel, O.: Webots: Professional mobile robot simulation. Journal of Advanced Robotics Systems 1(1), 39–42 (2004)
11. Conn, A.R., Gould, N.I., Toint, P.: A globally convergent augmented lagrangian algorithm for optimization with general constraints and simple bounds. SIAM Journal on Numerical Analysis 28(2), 545–572 (1991)

Bipedal Walking of an Octopus-Inspired Robot

Marcello Calisti, Francesco Corucci, Andrea Arienti, and Cecilia Laschi

The BioRobotics Institute, Scuola Superiore Sant'Anna, Pisa, Italy
m.calisti@sssup.it

Abstract. In this paper a model is presented which describes an octopus-inspired robot capable of two kinds of locomotion: crawling and bipedal walking. Focus will be placed on the latter type of locomotion to demonstrate, through model simulations and experimental trials, that the robot's speed increases by about 3 times compared to crawling. This finding is coherent with the performances of the biological counterpart when adopting this gait. Specific features of underwater legged locomotion are then derived from the model, which prompt the possibility of controlling locomotion by using simple control and by exploiting slight morphological adaptations.

Keywords: bio-inspired robotics, underwater locomotion, embodied intelligence.

1 Introduction

Legged robotics is the branch of robotics studying the static, quasi-static and dynamic locomotion of robots that move using limbs [1]. Legged locomotion has significant advantages compared to other types of locomotion, e.g. it reduces damages to the environment and is particularly suited for uneven terrains [2].

The investigation of the neural, bio-mechanical and mathematical aspects of legged locomotion, has led computer scientists and engineers to infer a close relationship between locomotion and intelligence [3]. At the same time, biologists and mathematicians have focused their work on basic walking and running models, called templates [4], which describe the locomotion of animals and robots with an arbitrary number of legs [5]. These synergistic efforts among different specializations have brought significant scientific [6] and technological [7] results, with potential for the development of ever more effective and efficient artificial machines.

Despite the vast amount of studies on the subject, there is still a niche that requires further investigation: underwater legged locomotion (ULL). While some biologists analyze aquatic animal walking and running [8], few robotic researchers work in this niche. Marine robotics involves mainly the study of swimming systems, such as remotely operated vehicles (ROVs) or autonomous underwater vehicles (AUVs), or more recently bioinspired fish [9] or cephalopods [10]. These robots usually work near submersed structures; they need to be accurately controlled to avoid damages to fragile surfaces or to the robot itself. Conversely,

A. Duff et al. (eds.): Living Machines 2014, LNAI 8608, pp. 35–46, 2014.

legged robots require a substrate to move on and, as mentioned, they are able to prevent damages and move in unstructured environments. In this context, marine robotics could benefit from the progress made on legged robots, and a new generation of underwater legged robots (ULRs) could arise.

To the best of our knowledge, there are very few ULRs and they are still far from the successes of their terrestrial counterparts. Among them, one of the most advanced is called Crabster200 (CR200) [11], an hexapedal robot equipped with three degrees of freedom (DoF) limbs. Currently there are no reports either on the performances of CR200, or on its control strategies. Another related platform is a bio-inspired robot featuring elastic limbs, developed by the author [12]. By synthesizing mechanical [13] and control [14,15] aspects, the robot mimics the crawling locomotion of the *Octopus vulgaris*. With a pushing-based locomotion strategy the robot moves omnidirectionally, translating the center of mass (CoM) from one position to another in a quasi-static locomotion that alternates pushing phases and recovery phases. This kind of behavior represents one of the basic locomotion strategies of the octopus, which is, however, among the slowest performed by the animal [16].

The work presented here addresses another movement employed by the animal: bipedal walking. Bipedal walking differs from crawling mainly in two aspects. Firstly, when performing this kind of motion the octopus is not sprawled over the substrate, but floats a few centimeters from the ground. Secondly, bipedal walking is considerably faster than crawling. This kind of motion can be performed with a pair of arms pushing alternately or together [16]. With the aim to increase the performances of ULRs, this work investigates underwater bipedal walking. A model based on an extension of the bioinspired octopus-robot previously presented in [12] was developed here and simulations were compared with underwater trials performed by the actual robot. The validated model was then used to explore different morphological configurations of the robot and the resulting locomotion.

2   Robot and Model Description

A variety of models with different levels of complexity were developed to capture the dynamics of robot and animal locomotion. A seminal work is the spring loaded inverted pendulum (SLIP), a simple conservative spring-mass model for sagittal plane locomotion [17]. Despite its simplicity, the SLIP model describes the basic motions and ground reaction forces for a broad range of animals. More complex models exist which describe the locomotion of insects, animals or robots, with accuracy ranging from accurate reproduction of the muscular system [18] to simplified compliant massless legs [19,20]. The model presented in this work comprises massless compliant legs, and introduces some original key components of underwater legged dynamics. It is based on the crawling robotic platform presented in [12], which is briefly recalled in Sect. 2.1 for the reader's convenience.

2.1 Robotic Platform

The robot is made of a swimming, a crawling and a floating module, as shown in Fig. 1a. The swimming module is not actuated in this experimentation, thus completely passive. The crawling module, based on the three-bar mechanism implemented by the authors in [12], comprises four compliant legs radially distributed with respect to a central body. The floating module is oriented (b, β) toward the rear side of the robot, modifying the resting posture from the sprawled posture used for crawling to one that is more suitable for bipedal locomotion (Fig. 1b) and similar to the one assumed by the octopus. The parameter b is the distance between the center of buoyancy (CoB) and CoM, while β represents the orientation of the CoB with respect to the vertical. The floating module can be inflated and deflated, thus varying the robot mean density ρ_r. Due to the posture obtained passively thanks to the floating module, only the frontal legs were activated in the present experiment.

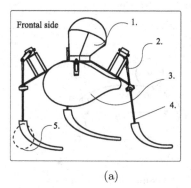
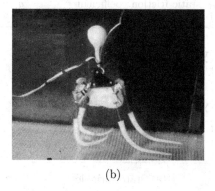

(a)   (b)

Fig. 1. The designed robot (a) and its mechanical components: 1. floating module, 2. leg mechanism, 3. swimming module and 4. compliant limb. The dashed line 5. identifies the end effector trajectory. The actual robot while moving inside the working space (b).

2.2 Model Description

The sagittal plane model, with geometrical parameters selected to match the actual robot, comprises a central body with three DoF and four legs, immersed in water (Fig. 2). Reaction forces are applied to the CoM, while the buoyancy force is applied to the CoB. Legs are approximated as massless spring-damper systems and their kinematics are derived from the mechanism described in [12]. The distal parts of the legs, made of silicone, were neglected. The parameters and variables of the model are summarized in Table 1. The state variable ϑ_l is explicitly considered simply for convenience, however it depends on ϑ_m as follows:

Table 1. Parameters and variables of the proposed model

State variables		Geometric parameters	
ϑ_r	pitch of the robot	m	length of the crank
ϑ_m	angle of crank rotation	l	length of the arm
ϑ_l	angle of leg rotation	d	distance between the crank's two CoRs
x_g	abscess coordinate of the CoM	i	distance from crank's CoR to bearing
y_g	ordinate coordinate of the CoM	b	distance from CoM to CoB
		α	angle between d and i
		β	angle between b and medial plane
Dynamic parameters			
k	stiffness of the leg	M	mass of the robot
c_{da}	damping coefficient of the leg	J	aggregate inertia of the robot
c_{dr}	drag coefficient	V	volume of the robot
X_{uu}	aggregate drag coefficient	ρ_w	density of the water
c_{df}	dynamic friction coefficient	ρ_r	mean density of the robot
c_{sf}	static friction coefficient	g	gravity acceleration
$\overline{M}$	mass of the robot + added mass		

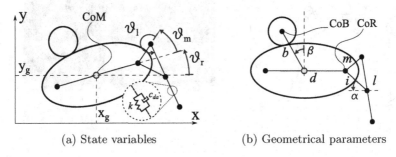

(a) State variables   (b) Geometrical parameters

Fig. 2. Schemes of the robot model: only one frontal leg is shown. Dimensions are not proportional to those of the real robot.

$$\vartheta_l = 2\pi - \arctan\left(\frac{m \cdot \sin(\vartheta_m + \alpha)}{i - m \cdot \cos(\vartheta_m + \alpha)}\right) - \vartheta_m - \alpha \tag{1}$$

By taking Fig. 2 as reference, it is possible to derive the kinematic equations of the legs. The positions of the frontal legs are reported in Eq. 2 and those of the rear legs are derived accordingly. Please note the following abbreviations: $c_r \equiv \cos(\vartheta_r)$, $c_{rm} \equiv \cos(\vartheta_r + \vartheta_m)$, and so on for c_{rml}. The same convention is used for sines, i.e. $s_r \equiv \sin(\vartheta_r)$ etc.

$$\mathbf{L} = \begin{bmatrix} x_g + \frac{d}{2}c_r & m \cdot c_{rm} & l \cdot c_{rml} \\ y_g + \frac{d}{2}s_r & m \cdot s_{rm} & l \cdot s_{rml} \end{bmatrix} \cdot \begin{bmatrix} 1 & 1 & 1 \\ 0 & 1 & 1 \\ 0 & 0 & 1 \end{bmatrix} \tag{2}$$

Given that the notation L_{ij} refers to the element (i, j) of the matrix $\mathbf{L}$, L_{1j} in Eq.2 identifies the x position of joint j, while L_{2j} identifies the y position of joint j. Joint speeds are obtained by analytical derivation of Eq.2, and are identified with the same convention as S_{ij} (this time with respect to the matrix $\mathbf{S}$). Since each leg is considered as a spring-damper system, starting from their positions and speeds, it is possible to derive the forces exerted by the legs to the ground, specifically elastic and damping forces. A touch-down vector $\mathbf{t}$ is used as an auxiliary vector to identify whether the leg is in contact with the ground, i.e. if the condition $L_{23} < 0$ is verified or not. The x position of the touch-down is stored in x_t, and the current length of the leg is derived as $A = \sqrt{L_{22}^2 + (x_t - L_{21})^2}$. Compressions in the x and y directions are, respectively (Eq. 3):

$$dL_x = (x_t - L_{12}) \left(\frac{l}{A} - 1 \right)$$

(3)

$$dL_y = L_{22} \left(\frac{l}{A} - 1 \right)$$

The associated elastic forces are $F_{el_x} = kdL_x$ and $F_{el_y} = kdL_y$. By taking the first derivative of the compressions (Eq. 3), the damping forces of the legs can be evaluated, being respectively $F_{da_x} = c_{da}d\dot{L}_x$ and $F_{da_y} = c_{da}d\dot{L}_y$. In addition, gravity $F_g = Mg$, buoyancy $F_b = \rho_w Vg$ and drag forces $F_{dr_x} = \frac{1}{2}X_{uu}\dot{x}|\dot{x}|$, $F_{dr_y} = \frac{1}{2}X_{uu}\dot{y}|\dot{y}|$ are applied to the body. The parameter X_{uu} is called aggregate drag coefficient as it combines information related to the drag coefficient and the reference area affecting the drag force.

Finally, the following equations describe the dynamics of the body (Eq. 4):

$$\overline{M}\ddot{x} = \sum_{n=0}^{4} t_n (F_{e_x} + F_{da_x})_n + F_{dr_x}$$

$$\overline{M}\ddot{y} = \sum_{n=0}^{4} t_n (F_{e_y} + F_{da_y})_n + F_{dr_y} + F_g + F_b$$

(4)

$$\overline{J}\ddot{\vartheta} = \sum_{n=0}^{4} t_n \left[(x_{t_n} - x_g)(F_{e_y} + F_{da_y})_n - y_g(F_{e_x} + F_{da_x})_n \right] + F_b \cdot b \cdot \sin(\vartheta)$$

The quantity $\overline{J}$ in Eq. 4 is called aggregate inertia coefficient as it embeds the body inertia plus the added inertia of the robot. It is not unusual that a leg slips on the ground, thus a slipping condition is checked at each instant. When the slipping condition $|F_{e_x} + F_{da_x}| > c_{sf}(F_{e_y} + F_{da_y})$ is verified, the force exerted by the leg to the ground is considered $c_{df}(F_{e_y} + F_{da_y})\frac{S_{i3}}{|S_{i3}|}$ as described in [18]. The position of the touch-down x_t is updated by calculating the new positions L_{13} and verifying the touch-down condition $L_{23} < 0$.

3   Experimental Methods

Motion kinematics were derived and used to estimate the unknown parameters of the model by recording the robot while it moved inside a tank. After estimating the parameters, a number of geometrical properties of the model were varied and the resulting locomotion was studied in simulation.

3.1   Robot Bipedal Trials

The robot (equipped with a plate with 3 LEDs) was recorded while moving in a tank with 8 markers that define the working space of the runs. A direct linear transform (DLT) with 11 parameters was used to reconstruct the 3-dimensional positions of the LEDs and accordingly derive the 3-dimensional coordinates of the CoM (the reconstruction procedure is described in detail in [12]).

The floating module was connected to a pneumatic system that was manually actuated; a desired density, i.e. $\rho_r = 1238$ Kg/m^3, was heuristically selected and kept constant during each trial session, comprising of five to ten runs by the robot. Each leg is actuated by a GM12a DC motor, that was properly insulated from the water by an *ad hoc* scaffold. Motors were plugged to a 5V Kert stabilized power supply, and were manually activated by a remote controller. The leg cranks rotate together at a constant speed of about $\dot{\vartheta}_m \simeq 12.57$ rad/s, thus a purely feed-forward control is adopted. Although a phase shift between the legs is achievable and could be interesting to explore, at the moment they are actuated in phase.

Four features were extracted from each run, which characterize the locomotion: amplitude (a), mean value (μ), frequency (f) of the y_g oscillation and mean speed in the x direction (s). The CoM moved approximately onto the x-y plane, so velocity in the z direction was considered to be null. Feature extraction was performed considering the latter part of the test, when the robot achieved stable periodic orbit.

3.2   Parameter Estimation

In order to validate the model, a number of model parameters had to be specified. Geometrical parameters were measured and directly plugged into the model. As for the unknown parameters, a parameter estimation procedure was set up. The parameters relevant to the estimation procedure are listed in Table 2. The decision to identify some parameters as aggregate quantities $(X_{uu}, \overline{J})$ is motivated by the fact that some of the involved quantities are difficult to measure or to estimate individually. For example, the shape of the robot is complex and irregular: it is difficult, therefore, to estimate the reference area (that also changes dynamically) for computing drag forces. Similar considerations apply to the aggregate inertia coefficient $\overline{J}$. The problem was formulated as a bounded minimum optimization problem. The fitness function was defined in terms of a 4-dimensional fitness vector, extracted from the model simulations, enclosing the features mentioned in Sect. 3.1, i.e. amplitude (a), frequency (f), mean value (μ) of the oscillation of the CoM in the y axis, and mean speed (s) in the x axis $(fitnessVector = (a, f, s, \mu))$. A target vector $(targetVector = (a^*, f^*, s^*, \mu^*))$ was extracted from the trial of the actual robot, and the fitness value was computed as the sum of normalized squared errors between target and fitness vectors.

Table 2. Parameters to be estimated and their bounds. The coefficient dr is the damping ratio $dr = c_{da}/2\sqrt{kM}$ and $c_{dfmul} = c_{df}/c_{sf}$.

k	dr	c_{sf}	c_{dfmul}	$\overline{M}$	X_{uu}	$\overline{J}$
[25, 400]	[0, 0.9]	[0.6, 0.9]	[0.6, 0.9]	[0.755, 7.55]	[0.11, 145]	$[2.7 \cdot 10^{-4}, 1.79 \cdot 10^{-2}]$

In our setup, the optimization algorithm must have the ability to cope with discontinuous objective functions, in order to handle situations in which the behavior of a parameter set cannot be quantified. For example, in our simulations, when a set of parameters caused the robot to fall or produce unstable behavior, the fitness was set to NaN. This is usually a problem with gradient-based approaches. For this reason, genetic algorithms were selected as a suitable alternative, as they can simply not consider individuals with NaN fitness for selection and reproduction. Furthermore, they are capable of finding global solutions with no prior assumptions or information about the objective function.

Among the several variants of genetic algorithms, a real-coded version of the Augmented Lagrangian Genetic Algorithm (ALGA) [21] was adopted for its ability to handle bounds and constraints. As for genetic operators, adaptive feasible mutation and scattered crossover were used. Genetic optimization was performed on a population of 500 individuals, with chromosomes composed of 7 genes encoding the parameters to be estimated. Evolution could last for 1000 generations maximum, with additional stop conditions based on the change of average fitness and maximum execution time.

Bounds were defined by considering extreme limit cases of physical feasibility. As an example, the lower bound for the aggregate drag coefficient X_{uu} is the one of a streamlined body (a shape featuring very low drag coefficient) with a circular exposed surface having a radius of just 3 cm, while the upper bound is given by a short cylinder (a shape featuring very high drag coefficient) enclosing the entire robot.

3.3 Model Simulations

The model was validated using the approach described in Sect. 3.2, with the following geometrical parameter values: $m = 0.022$ m, $l = 0.12$ m, $d = 0.15$ m, $i = 0.056$ m, $b = 0.088$ m, $\alpha = 82$ deg, $\beta = 16$ deg. After validation, further simulations were performed varying only parameters b and β, which define the position of the CoB. The model was numerically solved using Matlab®.

4   Results

Despite the high compliance of the legs, the robot is not sprawled on the ground, as happens outside water, due to the buoyancy module and the low density materials composing the legs and the swimming module. Simple feed-forward activation was used to make the robot achieve forward locomotion at a mean speed of $\bar{x} = 0.0411$ m s^{-1} with a standard deviation of s.d. $= 0.0024$. Even when all the parameters were kept constant during the various runs, the robot slightly changed its mean velocity due to small changes in the testing conditions, such as the deposition of material inside the tank that slightly changed the friction to the ground. During locomotion, the frontal legs pushed the body forward and generated a positive momentum that made the frontal part of the robot rise. On the other hand, the buoyancy module, since displaced from the resting position, generated a negative momentum that lowered the frontal part. The designed geometrical configuration of the robot, i.e. the selection of b and β, did not lead to any falls occurring during locomotion.

In order to estimate the parameters, the fastest trial was selected as target, with the extracted vector being $(a^*, f^*, s^*, \mu^*) = (0.0044, 2, 0.045, 0.12)$. The achieved fitness value was $9.1647 \cdot 10^{-6}$ (Table 3), with the model closely matching the behavior of the robot. A comparison between the CoM track of the robot and the one of the model is presented in Fig. 3. The error has been computed as the mean absolute value of the difference between the two CoM tracks.

Table 3. Evolved genome: identified parameters for $\rho_r = 1238$

k	dr	c_{sf}	c_{dfmul}	$\overline{M}$	X_{uu}	$\overline{J}$	Fitness
216.5	0.37	0.65	0.83	4.64	121	0.014	$9.1647 \cdot 10^{-6}$

Once the parameters were estimated, since the actual robot has a fixed geometrical structure, the model was used to investigate the dynamics of the system with respect to variations in the CoB position. Initial conditions and all parameters, apart from b and β, were kept fixed and the resulting locomotion was analyzed. Interestingly, a variety of stable locomotion patterns arose, with different characteristics and speed (Fig. 4). Variations in CoB position led to considerable differences in the resulting locomotion, with some robots proceeding forward and others backward, at different speeds. By taking as reference the configuration of the real robot (for which parameter estimation was performed), simulations highlighted that a variation of $\Delta\beta = +8.8$ deg, $\Delta b = -0.009$ m causes an inversion of motion. Simulations also pointed out that, by changing the position of the CoB, enhanced performances can be achieved with respect to forward locomotion. A variation of $\Delta\beta = +18.9$ deg, $\Delta b = 0.14$ m prompted a significant improvement in speed (by a factor of ~ 1.7) with respect to that exhibited by the robot.

5   Discussion

By keeping a bipedal body posture and using a feed-forward control, the robot's speed was higher than in crawling locomotion. This is coherent with the speed increase observed in the biological counterpart. Quantitatively, average speeds of octopuses are 0.62 BL s^{-1} and 1.34 BL s^{-1} respectively for crawling and

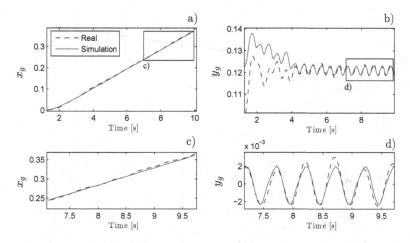

Fig. 3. Trajectories of x_g (a) and y_g (b) for the model executed with the estimated parameters. In c, d) the comparison of real and simulation data is highlighted. The displacement errors in y (meters) are: mean 0.0037, min $1.6 \cdot 10^{-6}$ and max 0.02. The displacement errors in x (meters) are: mean 0.003, min 0.00014, max 0.0068.

bipedal walking [16] (BL stands for body lengths). Bipedal walking appears to be 2.16 times faster than crawling. Analogously, the robot's speed increases from 1.52 cm s^{-1} for crawling [12] to 4.4 cm s^{-1} for bipedal walking. The increase ratio is about 2.9, thus slightly higher for the robot than for the animal. This demonstrates the capability of this type of robot, i.e. a robot with elastic limbs, to perform both crawling and bipedal walking. The features of this kind of locomotion can be further analyzed by considering the results of the model simulations. The parameter estimation methodology proposed was effective despite the significant number of parameters, with small errors in fitness evaluation (Table 3) and a good match between real and simulated signals (Fig. 3).

Moreover, despite the wide bounds, the evolved parameters all look very plausible (Table 3). Notice that the hydrodynamics parameters, $\overline{M}$, X_{uu} and $\overline{J}$, appear to provide a relevant contribution to the dynamics. As an example, in our case, the estimated added mass was $\overline{M} \simeq 6M$ while ROV added masses usually range between 2-3 times the mass of the robot [22,23]. The added mass value increases when there are irregular shapes and sharp transitions between the underwater vehicles' structure and the fluid, thus a higher value of $\overline{M}$ than in traditional ROVs was expected. Moreover the proximity of the substrate entails an additional increase, as known from potential flow analysis of bodies translating close to a fixed boundary [24]. Similar arguments stand for the other parameters, highlighting the difference between terrestrial legged locomotion and ULL.

Another peculiar aspect of ULL is the separation between the CoM and the CoB of the robot. This morphological trait implies, passively, the bipedal posture of the robot, leading to a significant speed increase in the presence of the same feed-forward control. This aspect was further explored through the proposed

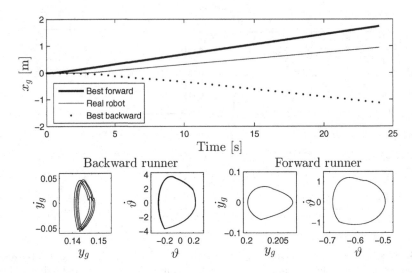

Fig. 4. By varying the position of the CoB (β, b) considerable differences are observed in the resulting locomotion. Top plot: Starting from the configuration of the real robot (middle curve) a variation of $\Delta\beta = +8.8$ deg, $\Delta b = -0.009$ m causes an inversion of motion (lower, dotted curve). A $\Delta\beta = +18.9$ deg, $\Delta b = 0.14$ m (upper, thick curve) entails instead a significant improvement in forward speed (by a factor of ~ 1.7). There are several other trajectories between the plotted curves (not shown for readability). Bottom: stability of the best backward (left) and forward (right) runners. As for the backward runner, what may seem a still unstable orbit is instead a stable one, being composed of four cycles.

model, demonstrating that changes in CoB position can be exploited to achieve different locomotion patterns (backward or forward at different speeds). Given the stressed interaction with the environment, a small change in underwater robot morphology (i.e. CoB position) entails significant changes in the resulting behavior. This has a strong connection with the concept of embodied intelligence and morphological computation. Based upon the stable locomotion highlighted in Fig. 4, we envision the possibility to switch among different stable locomotion patterns by controlling slight morphological adaptations on-line and so allow smooth transition among stable orbits. Above all, the results we have presented offer, for the first time, a starting point for the definition of quantitative design criteria for ULRs. As shown in Fig. 4, geometrical and morphological aspects can be properly designed to improve the robot's performances.

6   Conclusion

In this paper an octopus-inspired robot, capable of multi-gait locomotion, has been presented together with its model. It has been shown that the robot is able to perform bipedal locomotion, with a speed increase (with respect to crawling) that is consistent with biological observations. A model comprising massless spring legs has been proposed to describe the bipedal gait of the robot and has been validated against actual robot trials. The parameter estimation procedure, performed using genetic algorithms, highlighted the prominent role of hydrodynamics effects on the robot's dynamics. Furthermore, the model allowed a preliminary analysis of specific ULL features. The role of the CoB position was investigated, showing that it has a key role in determining the direction and speed of locomotion in the presence of the same feed-forward control, with potential implications on the embodied intelligence framework. This is a peculiar mechanism for ULL that is absent in terrestrial locomotion, where the role of the medium (i.e. air) is usually neglected. Presented methods and results also offer room for the exploration of optimal design for ULRs. By using evolutionary techniques to co-evolve both morphology and control, it is possible to take advantage of the significant body-environment interaction existing in underwater environments and to enhance the performances of ULRs.

Acknowledgments. This work was supported in part by the Fondazione Livorno within the framework of the PoseiDRONE project Grant.

References

1. Raibert, M.H.: Legged robots that balance. The MIT Press (1986)
2. de Santos, P., García, E., Estremera, J.: Quadrupedal locomotion: an introduction to the control of four-legged robots. Springer (2006)
3. Pfeifer, R., Iida, F., Gomez, G.: Morphological computation for adaptive behaviour and cognition. International Congress Series (1291), 22–29 (2006)
4. Full, R., Koditschek, D.: Templates and anchors: neuromechanical hypotheses of legged locomotion on land. The Journal of Experimental Biology 222, 3325–3332 (1999)
5. Blickhan, R., Full, R.J.: Similarity in multilegged locomotion: bouncing like a monopode. Journal of Comparative Physiology A 173(5), 509–517 (1993)
6. Holmes, P., Full, R.J., Koditschek, D., Guckenheimer, J.: The dynamics of legged locomotion: Models, analyses, and challenges. SIAM Review 48(2), 207–304 (2006)
7. Saranli, U., Buehler, M., Koditschek, D.E.: Rhex: A simple and highly mobile hexapod robot. Internation Journal of Robotics Research 20(7), 616–631 (2001)
8. Martinez, M.M.: Running in the surf: hydrodynamics of the shore crab grapsus tenuicrustatus. The Journal of Experimental Biology 204, 3097–3112 (2001)
9. Zheng, C., Shatara, S., Xiaobo, T.: Modeling of biomimetic robotic fish propelled by an ionic polymer–metal composite caudal fin. IEEE Transactions on Mechatronics 15, 448–459 (2010)
10. Serchi, F.G., Arienti, A., Laschi, C.: Biomimetic vortex propulsion: towards the new paradigm of soft unmanned underwater vehicles. IEEE/ASME Transactions on Mechatronics, 1–10 (2013)
11. Shim, H., Jun, B.H., Kang, H., Yoo, S., Lee, G.M., Lee, P.M.: Development of underwater robotic arm and leg for seabed robot, crabster200. In: OCEANS, Bergen. MTS/IEEE (2013)

12. Calisti, M., Arienti, A., Renda, F., Levy, G., Mazzolai, B., Hochner, B., Laschi, C., Dario, P.: Design and development of a soft robot with crawling and grasping capabilities. In: Proc. IEEE Int. Conf. Robotics and Automation, pp. 4950–4955. IEEE (May 2012)
13. Calisti, M., Giorelli, M., Levy, G., Mazzolai, B., Hochner, B., Laschi, C., Dario, P.: An octopus-bioinspired solution to movement and manipulation for soft robots. Bioinspiration & Biomimetics 6(3), 036002 (2011)
14. Li, T., Nakajima, K., Calisti, M., Laschi, C., Pfeifer, R.: Octopus-inspired sensori-motor control of a multi-arm soft robot. In: IEEE Mechatronics and Automation, pp. 948–955. IEEE (August 2012)
15. Calisti, M., Giorelli, M., Laschi, C.: A locomotion strategy for an octopus-bioinspired robot. In: Prescott, T.J., Lepora, N.F., Mura, A., Verschure, P.F.M.J. (eds.) Living Machines 2012. LNCS, vol. 7375, pp. 337–338. Springer, Heidelberg (2012)
16. Huffard, C.: Locomotion by abdopus aculeatus (cephalopoda: octopodidae): walking the line between primary and secondary defenses. The J. of Experimental Biology 209, 3697–3707 (2006)
17. Blickhan, R.: The spring-mass model for running and hopping. Journal of Biomechanics 22(11/12), 1217–1227 (1989)
18. Iida, F., Rummel, J., Seyfarth, A.: Bipedal walking and running with spring–like biarticular muscles. Journal of Biomechanics 21, 656–667 (2008)
19. Ankarali, M.M., Saranli, U.: Control of underactuated planar pronking through an embedded spring–mass hopper template. Auton. Robot. 30, 217–231 (2011)
20. Gorner, M., Albu-Schaffer, A.: A robust sagittal plane hexapedal running model with serial elastic actuation and simple periodic feedforward control. In: IROS (2013)
21. Conn, A.R., Gould, N.I., Toint, P.: A globally convergent augmented lagrangian algorithm for optimization with general constraints and simple bounds. SIAM Journal on Numerical Analysis 28, 545–572 (1991)
22. Caccia, M., Indiveri, G., Veruggio, G.: Modeling and identification of open-frame variable configuration unmanned underwater vehicles. IEEE Journal of Oceanic Engineering 25, 227–238 (2000)
23. Ridao, P., Battle, J., Carreras, M.: Model identification of a low-speed uuv. In: Katebi, R. (ed.) IFAC Proceedings Series, pp. 395–400 (2002)
24. Katz, J., Plotkin, A.: Low-Speed Aerodynamics: From Wing Theory to Panel Methods. Cambridge University Press (2001)

Action Selection within Short Time Windows

Holk Cruse[1,2] and Malte Schilling[1]

[1] Center of Excellence 'Cognitive Interaction Technology' (CITEC),
Bielefeld University, P.O. Box 10 01 31, D-33501 Bielefeld, Germany
[2] Department of Biological Cybernetics and Theoretical Biology, Bielefeld University
holk.cruse@uni-bielefeld.de, mschilli@techfak.uni-bielefeld.de

Abstract. In this paper, we study the expansion of a reactive network that is based on a decentralized, heterarchical architecture able to control a hexapod robot. Within this network, problems may occur if more than one task should be addressed within a short time window. Such situations have been studied in so called dual-task experiments in the field of psychology. We take these results as inspiration to develop a structure that can be integrated into our framework. The model focuses on essential aspects of the basic phenomena observed in human subjects as forward masking, backward masking and PRP. While there are detailed models available concentrating on specific paradigms, those do not cover all three types of paradigms. We are not heading for a detailed simulation of the psychological findings, but show how these effects, on a qualitative level, can be implemented in our framework.

1 Introduction

In this paper, we study an expansion of a reactive network representing a decentralized, heterarchical architecture able to control a hexapod robot, in our case robot Hector [1]. The architecture consists of procedural, or reactive, elements, small neural networks that in general connect sensory input with motor output constituting the procedural memory. Inspired by Maes [2], these procedural elements are coupled by a motivation unit network, a recurrent neural network (RNN), forming the backbone of the complete system. This type of architecture has been termed MUBCA (for Motivation Unit Based Columnar Architecture [3]). To start with a thoroughly tested reactive system, we use an insect-inspired network, Walknet [4], that is able to deal with a specific domain of behavior, namely walking with six legs in an unpredictable environment including climbing over very large gaps — which, when performed in a realistic, natural environment is a non trivial task. This network has been augmented by various components [5,3,6,7] which are not addressed here in further detail.

To explain how the basic network will be expanded, we start with a brief description of a simple but representative part of the network, Fig. 1a shows the controller of a single leg. The two most important procedural elements for controlling the movement of a leg are the Swing-net, responsible for controlling a swing movement, and the Stance-net controlling a stance movement (Fig. 1a).

A. Duff et al. (eds.): Living Machines 2014, LNAI 8608, pp. 47–58, 2014.

These procedural elements might receive direct sensory input and provide output signals that can be used for driving motor elements. Motivation units 'swing' and 'stance' inhibit each other and are coupled via excitatory connections with other motivation units (see [5,3]). As depicted in Fig. 1a, motivation units do not only receive input from other motivation units, but may receive sensory input, too. This is required, in this example, to decide between 'swing' and 'stance' movement of a leg. A sensory input recording ground contact (GC) of the leg actives motivation unit 'stance'. A sensory input monitoring the leg having reached a specific extreme position (PEP) can activate the motivation unit 'swing'.

As long as we remain within the current context of, for example, forward walking and navigation, stimuli able to activate memory elements are normally given with sufficient temporal separation that provides enough time (1) to select the motivation units, (2) to start the procedures and (3) to finish the behavior before the next stimulus is given. Under these conditions a selection of a procedure is possible using a winner-take-all (WTA) network consisting of simple, piecewise linear units with low-pass filter properties [3].

Let us now assume that different procedures have been learnt by the agent that do not belong to a common context. Let us further assume that a stimulation of such a procedure will result in activation of the corresponding motivation unit for some short time ("short term memory") as it is true for many (if not all) living systems. If, in this case two stimuli connected with different procedures are given more or less simultaneously, due to capacity limits, this situation may lead to a bottleneck and thereby to loss of information. How could an agent deal with such a situation? We will show in this article that under these conditions a WTA network being endowed with more complex dynamics is suited to deal with this case.

We will not ask here how a robot could be constructed that would not suffer from such a situation. Instead, we are interested in how the controller of such an agent could be understood by trying to construct a controller endowed with these properties. This may be of particular interest because the paradigms addressed below are tightly connected with the relation between conscious and non-conscious control of behavior. A controller that is able to switch between these two states may well be of interest for constructing advanced robots [8].

To construct such a system, we will focus on how the elements of the short term memory may be structured. To address this question, we take inspiration from psychological experiments. There is a number of experiments performed with human beings, called dual-task experiments, which show specific phenomena, when the subject has to deal with two tasks, T1 and T2, at the same time. If two stimuli, S1 and S2, that trigger two procedures, T1 and T2, respectively, are following each other within a short time, three possible types of reactions can be observed. (1) Either the first task is executed and the second task is discarded (called forward masking), (2) the first task is discarded and the second one is executed (called backward masking), or (3) both tasks are executed but the second one is delayed. The latter case has been described as the psychological refractory period (PRP) paradigm. These findings are interpreted as to result from capacity limits of information processing.

In this article we will try to show how the architecture as applied here can be expanded in order to simulate these three cases at least qualitatively in order to equip an agent with comparable faculties. To this end, we concentrate on the smallest possible structure, a model that contains two procedures only to study how the dual task experiments could be solved within our framework. The fundamental difference between the models depicted in Fig. 1a and the new version discussed in Sect. 3 (Fig. 1b) is that for coping with the dual-task experiments, some kind of short term memory has to be introduced in order to maintain the information representing the stimulus for some time. So we have to deal with the question how such a short term memory may be realized. In this article, we do not attempt to provide a detailed simulation of the three paradigms addressed below which have been studied in a large number of psychological experiments. In the literature there are quite detailed models concerning each of these paradigms on a level that cannot be reached by the granularity used in our approach. Instead our goal is to take these paradigms as inspiration to provide a simple model showing how such dynamics could be addressed within our framework that is formulated by using artificial neurons as basic elements. Sect. 2 explains the three paradigms of which simple versions will be modeled. Sect. 3 details the model structure. Experimental results are given in Sect. 4.

2   Three Dual-Task Paradigms

First, we want to introduce backward masking. We will focus on experiments, where stimuli activating different procedures have been studied which compete for becoming subjectively experienced and can, therefore, be reported. The basic experiment has been performed by Fehrer and Raab [9] which has been followed by detailed later studies (e.g. Neumann and Klotz [10]). Participants first learned to press a button when a square was presented on a screen, but not when two squares were shown at a position on the screen flanking the first square. After learning is finished, in the critical experiment the single square was given for a short period (about 30 ms), which was then followed by a longer presentation of the two square pattern. The participants did not report to have seen the single square, but only reported to have seen the two square pattern. Nonetheless, they pressed the button. This result shows first, that the initial procedure one, ("stimulus single square – motor response") can be executed without being accompanied by subjective experience of stimulus S1, the single square. Second, procedure two ("stimulus double squares – no motor response") appears to influence the first procedure by inhibiting the process leading to subjective experience of stimulus S1. The first stimulus, S1, is "masked" by procedure two. This backward masking effect is observed in a period of 40 - 80 ms after the beginning of the first stimulus.

Another case with a similar effect has been shown by Schmidt et al. [11]. In their experiment, participants were shown two filled circles, a red one and a green one, at opposite sides of the fixation point. Each circle was then replaced by an annular mask. The masks were of the same color or the opposite color compared to the circles. In each trial the participant was asked to move the finger towards one specified stimulus color as fast as possible. Let's assume the participants were asked to move towards the red object. When the circle and the ring showed the same color, the participants moved into the direction of the red shapes, as expected. Interestingly, when the circle and the ring (mask) were of different colors, the finger was first briefly moved in the direction of the red circle and then reversed direction and moved to the then only visible red ring. As in the Fehrer and Raab experiment, the participants reported not to have seen the red circle.

In a second type of dual-task experiments, studying the psychological refractory period (PRP) effect, first a stimulus S1 is given, that triggers a task one. Then, after a variable time lag, another stimulus, S2, triggering another task, task two, is provided. If the time lag between S1 and S2 is short enough (from ca. 50 ms to 600 - 1200 ms, depending on the task), the execution of task two is delayed until the first task has been processed for a given time. This is the case even if there is no competition concerning the motor output level. In some experiments [12] an additive effect has been observed (slope of the RT vs. time lag approaches -1). But also underadditive effects have been reported where this slope ranges between 0 and -1 ([13], p. 563).

In the third type considered here, the attentional blink (AB) experiment, a stimulus does not become consciously aware if it follows another stimulus too closely, thus showing forward masking. In the standard attentional blink experiment, a stimulus, for example a letter (e.g. 'X'), is given which, after a variable number of irrelevant patterns (e.g. digits), is followed by a second letter (e.g. 'Y'). Each pattern is presented for 100 ms. The first stimulus is always subjectively experienced and can therefore be reported. If the delay between the presentation of both relevant stimuli, the letters, is short enough (< 500–800 ms), subjects may however not be able to report the second stimulus, a phenomenon called the attentional blink ([14], p. 8).

3   The Model

In this section we will explain a model shown in Fig. 1b that at least qualitatively can simulate these effects. The model represents an extension of the structure shown in Fig. 1a. The essential difference concerns the processing of stimulus information. In the original model as shown in Fig. 1a, the appropriate stimulus is directly transformed into a signal that activates the corresponding motivation unit. The motivation units—there only two—inhibit each other forming a local winner-take-all (WTA) net. This principle can be found in the expanded version, too (Fig. 1b, MU1, MU2). However, the signal elicited by the stimulus undergoes a processing that also shows properties of a short term memory. The stimulus, being detected by a stimulus-specific filter (filt1, filt2), first triggers a unit showing low-pass filter properties (LPF). This information is then given to the motivation units (MU1, MU2). The motivation units now show the property of an integrator. Therefore, the information provided by the—in general short—stimulus (of e.g. pulse-like shape) is transformed into a sigmoid time function that continues to increase for some time after the stimulus is finished. The latter property is due to the, yet decreasing, activity of the low-pass filter after the stimulus is finished (this effect can be made stronger by application of a nonlinear LPF with a larger time constant for decreasing activity compared to increasing activity). At the same time the integrator is leaky, which means that the integrator behaves like a low-pass filter with a large time constant (much larger compared to that of the first LPF). The leak of the integrator is, however, closed as long as the excitatory input to the integrator is increasing. To add another nonlinear property, the activation of the integrator is bounded by a nonlinear saturation characteristic which is done to avoid unlimited growth in case the stimulus duration is very long.

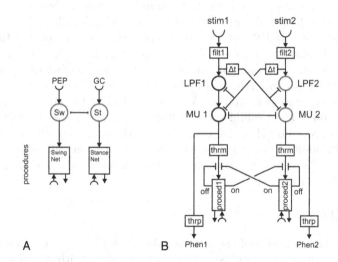

Fig. 1. (a) A section of Walknet, showing the controller of one leg. The motivation units Sw (swing) and St (stance) receive sensory input, PEP, concerning the leg position, and GC, monitoring ground contact, respectively. (b) A hypothetical network that is capable of dealing with some dual task experiments, for example the backward masking experiment. Stimulation of one of the procedures 1 or 2 activates a low-pass filter (LPF1, LPF2) followed by a motivation unit (MU1, MU2) and inhibits the corresponding units of the other procedure for a limited time (Δ t). As in (a), the motivation units are coupled via mutual inhibition. After the activation of one of the motivation units has reached threshold thr m the corresponding procedure is activated. If threshold thr_p is reached, the stimulus can be phenomenally experienced. As long as a procedure is active ("on"), the input to the competing procedure is inhibited. A feedback from the procedure provides an "off" signal to inhibit the input from its own motivation unit. Procedures are indicated by black rectangles, the motivation units are depicted in red or black circles. Excitatory connections are depicted by arrowheads, inhibitory influences by T-shaped connections. For further explanation see text.

A further difference to the simple treatment of sensory input as shown in Fig. 1a concerns the introduction of lateral inhibition arising from one sensor input that inhibits the LPF unit and the motivation unit of the other procedure. This influence is only active during a short time period (Fig. 1b, Δt) qualitatively corresponding to a high-pass filter. As mentioned, there is a mutual recurrent inhibition between the motivation units. Each motivation unit has two outputs characterized by two different thresholds, thr_m (related to motor response) and thr_p (related to phenomenal experience). The output using the lower threshold, thr_m, is sufficient to trigger the corresponding procedure and may thereby elicit a motor response. The higher threshold value, thr_p, when reached, leads to subjective, or phenomenal, experience of the stimulus (however this is realized by the neuronal system, for discussion of this matter including possible functional aspects see [8]). Reaching this threshold represents the condition for making reportability possible, i.e. it can activate a memory element containing the corresponding verbal expression (not shown in Fig. 1b, but see [8]).

In this form the model is able to show the properties of the backward masking experiment as explained above and the attentional blink as will be shown in Sect. 4. To comply with the PRP effect, too, a last expansion is required that, in a similar way, has also been applied for procedures used in Navinet. The procedure contains some kind of clock that represents the time needed to perform the behavior of the procedure (the clock might be represented endogenuously or be realized by using sensory feedback). As long as the clock has not yet approached its end, an "on" signal inhibits the connection between the motivation unit MU2 and its procedure (Fig. 1b, on). Thereby the execution of procedure two is delayed. In addition, after the procedure is finished, an "off" signal inhibits the input from its own motivation unit. This feedback avoids "stuttering" if the motivation unit is still active.

4   Results

To illustrate the behavior of the network shown in Fig. 1b, Fig. 2a shows the simple case where two stimuli S1 and S2 are presented with a temporal delay large enough so that no interaction between both procedures can be observed. In Fig. 2a as in the following figures, the activations of procedure one, provided on the left hand side of Fig. 1b, and which is stimulated first, is depicted by blue lines, whereas the activations of the second procedure are depicted in red lines. The result figures each show, from top to bottom, the time courses of the stimuli, the activation of the motivation units, the signal exciting the procedure (i.e. when the value of the motivation unit is above threshold thr_m), the time when the procedure is active, and finally the time periods when the motivation unit is above threshold thr_p, thus leading to phenomenal experience of the stimulus.

Fig. 2b shows the time courses of a backward masking experiment as studied by Fehrer and Raab [9], Neumann and colleagues (e.g. [10]) and Schmidt and Vorberg (e.g. [11]). As can be seen in this figure, the short stimulus S1 is responded by an activation of the procedure triggering a motor output, but not

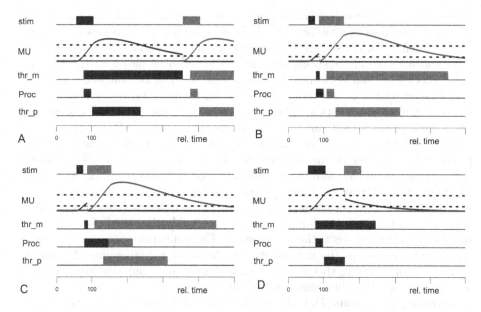

Fig. 2. Temporal development of the activation of some units (procedure 1, blue, procedure 2, red). Abscissa is relative time. Activation of a motivation unit may reach its motor threshold thr_m (lower dashed line) and/or the threshold thr_p (upper dashed line) for eliciting the phenomenal experience. (a) shows two stimuli given with a long time delay. (b) simulation of the backward masking effect. (c) Simulation of the PRP experiment, (d) Simulation of the attentional blink paradigm.

accompanied by phenomenal experience which is in agreement with the reported first result of the original study [9]. When the second stimulus follows with an appropriate time delay and S2 is presented long enough, the stimulus triggers both responses, i.e. it becomes phenomenally aware. There is no action produced because in the Fehrer-Raab paradigm, the motor response to the second stimulus (two square pattern) is characterized by a non-action, whereas in the experiment of Schmidt et al., both stimuli trigger an overt motor response.

Fig. 2b shows the results of the masking experiment which are in agreement with the findings of [9], i.e. while the first stimulus is not phenomenally recognized, it still triggers an action. How can this be explained? The first stimulus does not inhibit unit MU2, the representation of the input signal of procedure two, because in this experiment the latter is not stimulated as long as stimulus S1 is active. In contrast, the representation of the input signal of procedure one, given by unit MU1, can be suppressed, when stimulus two is given. Unit MU1 activated first may reach the level of thr_m, sufficient to activate the motor system, but not thr_p. Only the second stimulus has enough time to activate MU2 beyond thr_p and to reach the state of subjective experience, which allows the two-square pattern to become subjectively experienced. With the parameters used here, masking can be observed up to a lag of 220 simulation time steps between the beginning of stimulus 1 and stimulus 2. If this lag is zero and both stimuli are of the same duration, there is no masking, in agreement with the experimental findings. Activation of the behaviors, in this case, requires a duration and/or intensity of the stimuli being large enough (not shown).

In another case of backward masking, an experiment based on the Fehrer-Raab effect has been performed by Ansorge et al. [15]: the subject is given a visual target, either a square or a diamond (for 90 ms). When the subject sees the square, it should press the right button as fast as possible (RT measurement) and should press the left button, when it sees the diamond. Before the target is given, a (smaller) square or diamond, called prime, is projected at the same position, but only for 30 ms followed by a blank screen for 45 ms. The results show that RT is shorter than the control (a neutral stimulus, e.g. a circle is given as prime), when the same shape is given as prime and the target (congruent), whereas the RT is longer when prime and target are different (incongruent). As, in our simulation, only two inputs are considered, only the comparison between congruent and incongruent stimuli is possible. The simulation shows qualitatively comparable results: in the congruent case the onset for triggering the motor signal, i.e. for reaching threshold thr_m, is 5 units of simulation time, in the incongruent case this time amounts to 22 units.

Interestingly, Enns and Di Lollo [16] show a specific masking effect, the "common onset" paradigm, where both stimuli start at the same time, but stimulus two is active longer than stimulus one. In this case there is also a masking effect, which gets the stronger the longer stimulus two remains active after stimulus one is switched off. Both effects, according to these authors, cannot be explained by models available at that time. Both effects can however described by our model (not shown). This result is due to the inhibitory effect being active for a limited time only (Fig. 1b, Δt).

Enns and Di Lollo [16] have shown another masking effect, where stimulus one consists of four dots that do not superimpose or contact the pattern of stimulus two. This is critical because overlap or contact between both stimulus patterns is considered a crucial condition for the standard theories to explain masking. Therefore, according to these authors, no masking should have been found in this experiment. For our approach this result does not pose a problem, because we do not make any assumption concerning the geometrical arrangement of the stimuli. Therefore, in principle the four-dot masking experiment could be described on the level of our model, too, thereby, of course, disregarding the questions concerning the representation of spatial information.

To simulate the effect of the psychological refractory period (PRP), we apply the same stimuli as for the masking experiment, but assume that the execution of procedure one requires more time than in the masking experiment (Fig. 2c). As a consequence, the "on" signal of procedure one inhibits directly the influence from the competing motivation unit MU2 to procedure two. After procedure one has been finished, this inhibition is finished, too. Only then the behavior of procedure two can be started as observed in experiments showing the psychological refractory period.

To simulate the attentional blink experiment, we assume that the irrelevant stimuli (distractors) cannot pass the filter elements. Therefore, in the model we consider only the relevant stimuli. Both stimuli are of the same duration. The phenomenon of attentional blink can be observed in the simulation if the second relevant stimulus is given as long as the first stimulus elicits a high activation of the motivation unit (Fig. 2d, MU1). If stimulus two is not too strong (as in the experiment, both stimuli, one and two, had the same duration in the simulation), stimulus two does not lead to a motor response nor to phenomenal experience, although, when given without preceding stimulus one, both a motor response and a phenomenal experience would have been elicited (see Fig. 2a).

Thus, forward masking as found in the basic AB experiment can be explained by our model. However, our simple model does not explain more sophisticated experiments as are lag-one sparing, where no AB is found when no distractor is given between two consecutive relevant stimuli, i.e. without a distractor given between both relevant stimuli. In a similar experiment, the "blank between T1 and T2", no AB is found when there are no distractors, but a blank of about 100–150 ms between the first and the second relevant stimulus. This is indeed the case in the simulation, too, because we do not distinguish between irrelevant stimuli and blank signals.

5   Discussion

In this paper, we studied the expansion of a network that is based on a decentralized architecture consisting of procedural elements, and characterized by a heterarchical structure. The reactive network allows for selection of different behaviors able to control a hexapod robot. Within this network, responding to stimuli occurring within small time windows does not produce a problem, if the sensorimotor modules do not show overlapping elements and belong to the same context, i.e., are in accordance with an attractor state being allowed by the motivation unit network. When sensors and/or motor elements contributing to the behaviors are overlapping or are stored in different context areas, sufficient time is required between two subsequent stimuli to be answered properly. Problems can however occur if the time window becomes small. As such situations have been studied in so called dual-task experiments, we take these results as inspiration to implement a corresponding structure that can be integrated into our framework, the motivation unit based columnar architecture (MUBCA, [3]). To this end, we focus on essential aspects of the basic phenomena observed in human subjects as are forward masking, backward masking and PRP, but are not heading for a detailed simulation of the many psychological findings.

As depicted in Fig. 2, our model replicates some basic properties found in different (1) backward masking experiments, (2) experiments concerning PRP, and (3) forward masking experiments, the AB paradigm. The model is of extremely simple structure, which is possible basically because it abstracts from specific properties of specific sensors, in particular the spatial arrangement of stimuli within the visual field. This simplification allows to focus on the essential properties of the model as are lateral feedforward inhibition on the sensory level, mutual recurrent inhibition on the level of the short term memory for the stimuli, the motivation units, as well as recurrent inhibition on the level of the procedures. These inhibitory connections form bottlenecks at different levels.

Our model is based on an admittedly coarse level. In particular, no effort is made to compare the simulation times with time measured in the experiments (one reason is that the latter can also depend on the tasks selected). To give a rough estimate, one unit of simulation time may compare to about 2 ms real time. In addition, the time constant of the leaky integrator should be multiplied by about a factor in the order of 5. It should further be noted that the onset of a motivation unit activity does not allow to determine a value corresponding to the RT measured in human subjects, because in the simulation it is left open how much time is required to translate the activation of the motivation unit into a measurable motor response. Concerning the simulation of the AB effect, using the currently applied parameters, the pause between both relevant stimuli could be tripled compared to the case shown in Fig. 2c and the AB effect can still be observed. However, to simulate the effect for larger pauses (up to five times the duration of a stimulus has been used in the classical experiments), in the simulation the size of the leak of the integrator had to be smaller (or, in other words, the time constant of the low-pass filter had to be larger). This would easily be possible without consequences for the simulation of backward masking or the PRP experiment. Concerning the PRP effect, the timing does only depend on the mechanisms of inhibitory feedback from the procedure to its motivation unit, but not on the other parameters. The parameters used in the current simulation could therefore easily be changed to adopt different values.

When the "on" signal is a strong yes/no signal, an additive effect is observed in the simulation corresponding to results of Carrier and Pashler [17] and Jolicoeur (see review [13]). 'Additive effect' means that the temporal shift of the beginning of activation of MU2 is linearly related to the time between the beginning of stimulus one and stimulus two. Underadditive effects [13] have sometimes been observed and are discussed to result from binding problems occurring when the two tasks have to refer to a feature used by both tasks [18]. If in the simulation the inhibitory "on" signal would decrease slowly already during the time course of the execution of behavior one, the influence form the motivation unit MU2 may already activate procedure two before procedure one is finished. This assumption could lead to underadditive effects.

In the literature there are much more detailed models concentrating on specific paradigms, but they are either not implemented on a neuronal level and/or do not cover all three types of paradigms. In the following we will very briefly address more detailed approaches dealing with the simulation of specific findings of dual-task experiments. Zylberberg et al. [19] provide a complex internal neuronal structure forming a realistic simulation of mammalian brain properties. As input, simulated visual or auditory signals are applied whereas motor outputs are represented by simple go-nogo signals. The model comprises a large-scale network with about 20000 spiking neurons modeled to include the level of synapses. The model is able to describe in quantitative detail many psychological results concerning the PRP paradigm as well as aspects of the attentional blink paradigm. Due to the detailed simulation of neuronal properties the results can even be compared with neurophysiological data. The model does not deal with experiments on backward masking. In the Zylberberg et al. [19] approach separation between behavioral responses and phenomenal experience is not explicitly addressed.

Enns and Di Lollo [16] concentrate on the simulation of backward masking effects. Much more specific than our approach, these authors focus on spatial arrangement of shape and position of the stimuli, allowing for description of phenomena not possible for our abstract version. Parts of visual processing in our model is hidden in the filter function. As an interesting aside it should be mentioned that the approach of Enns and Di Lolla [16] includes phenomena like change blindness and inattentional blindness.

Probably the most advanced analysis of dual task experiments is given by the theory of task-driven visual attention and working memory (TRAM, [14]). This approach concentrates on processing of visual stimuli and covers a huge amount of experimental findings. Although based on earlier neuronal theories [20], TRAM is not yet realized as a detailed computational model. However it explains a number of effects found in backward and forward masking experiments with a special focus on the attentional blink paradigm. The PRP experiments are not addressed.

Thus, the three paradigms selected here are studied and simulated in much more detail by other authors compared to our approach, but none of these approaches covers all three paradigms addressed. Although of much coarser granularity, our model might therefore serve as a scaffold to be used for a broader approach. Future work will focus on how the two-procedure model studied here can be integrated in the complete network eraCog for the control of the hexapod robot Hector.

Acknowledgements. This work has been supported by the Center of Excellence Cognitive Interaction Technology (EXC 277), by the EC-IST EMICAB project # FP7–270182.

References

1. Schneider, A., Paskarbeit, J., Schäffersmann, M., Schmitz, J.: Biomechatronics for embodied intelligence of an insectoid robot. In: Jeschke, S., Liu, H., Schilberg, D. (eds.) ICIRA 2011, Part II. LNCS, vol. 7102, pp. 1–11. Springer, Heidelberg (2011)
2. Maes, P.: A bottom-up mechanism for behavior selection in an artificial creature. In: Proceedings of the First International Conference on Simulation of Adaptive Behavior on From Animals to Animats, pp. 238–246 (1991)
3. Schilling, M., Paskarbeit, J., Hoinville, T., Hüffmeier, A., Schneider, A., Schmitz, J., Cruse, H.: A hexapod walker using a heterarchical architecture for action selection. Frontiers in Computational Neuroscience 7 (2013)
4. Dürr, V., Schmitz, J., Cruse, H.: Behaviour-based modelling of hexapod locomotion: Linking biology and technical application. Arthropod Structure & Development 33(3), 237–250 (2004)

5. Schilling, M., Hoinville, T., Schmitz, J., Cruse, H.: Walknet, a bio-inspired controller for hexapod walking. Biological Cybernetics 107(4), 397–419 (2013)
6. Cruse, H., Wehner, R.: No need for a cognitive map: Decentralized memory for insect navigation. PLoS Computational Biology 7(3) (2011)
7. Hoinville, T., Wehner, R., Cruse, H.: Learning and retrieval of memory elements in a navigation task. In: Prescott, T.J., Lepora, N.F., Mura, A., Verschure, P.F.M.J. (eds.) Living Machines 2012. LNCS, vol. 7375, pp. 120–131. Springer, Heidelberg (2012)
8. Cruse, H., Schilling, M.: How and to what end may consciousness contribute to action? attributing properties of consciousness to an embodied, minimally cognitive artificial neural network. Frontiers in Psychology 4(324) (2013)
9. Fehrer, E., Raab, D.: Reaction time to stimuli masked by metacontrast. J. of Experimental Psychology 62, 143–147 (1962)
10. Neumann, O., Klotz, W.: Motor responses to non-reportable, masked stimuli: Where is the limit of direct parameter specification? In: Umiltà, C., Moscovitch, M. (eds.) Attention and Performance XV, pp. 123–150. MIT Press, Cambridge (1994)
11. Schmidt, T., Niehaus, S., Nagel, A.: Primes and targets in rapid chases: Tracing sequential waves of motor activation. Behav. Neurosc. 120, 1005–1016 (2006)
12. Pashler, H.: Dual-task interference in simple tasks: Data and theory. Psychological Bulletin 116, 220–244 (1994)
13. Jolicoeur, P., Tombu, M., Oriet, C., Stevanovski, B.: From perception to action: Making the connection. In: Prinz, W., Hommel, B. (eds.) Attention and Performance. Common mechanisms in perception and action, vol. XIX, pp. 558–586. Oxford University Press, Oxford (2002)
14. Schneider, W.: Selective visual processing across competition episodes: a theory of task-driven visual attention and working memory. Phil. Trans. R. Soc. B 368(20130060) (2013)
15. Ansorge, U., Klotz, W., Neumann, O.: Manual and verbal responses to completely masked (unreportable) stimuli: exploring some conditions for the metacontrast dissociation. Perception 27, 1177–1189 (1998)
16. Enns, J., Di Lollo, V.: What's new in visual masking? Trends in Cognitive Sciences 4, 345–352 (2000)
17. Carrier, L.M., Pashler, H.: Attentional limits in memory retrieval. Journal of Experimental Psychology: Learning, Memory, and Cognition 21, 1339–1348 (1995)
18. Hommel, B.: Automatic stimulus-response translation in dual-task performance. J. Exp. Psychol. Hum. Percept. Perform. 24, 1368–1384 (1998)
19. Zylberberg, A., Dehaene, S., Roelfsema, P.R., Sigman, M.: The human turing machine: a neural framework for mental programs. Trends Cogn. Sci. 15, 293–300 (2011)
20. Bundesen, C., Habekost, T., Kyllingsbæk, S.: A neural theory of visual attention and short-term memory (NTVA). Neuropsychologia 49, 1446–1457 (2011)

Modelling Legged Robot Multi-Body Dynamics Using Hierarchical Virtual Prototype Design

Mariapaola D'Imperio[1], Ferdinando Cannella[1], Fei Chen[1], Daniele Catelani[2], Claudio Semini[1], and Darwin G. Caldwell[1]

[1] Avanced Robotics Department
Istituto Italiano di Tecnologia, via Morego 30, Genova, Italy
{mariapaola.dimperio,ferdinando.cannella,fei.chen,
claudio.semini,darwin.caldwell}@iit.it
[2] MSC.Software srl
via Santa Teresa 12, Torino, Italy
daniele.catelani@mscsoftware.com

Abstract. Legged robots represent the bio-inspired family of robotic devices which has to perform the most complex dynamic tasks. It is essential for them to walk in unstructured terrains, carry heavy loads, climb hills and run up to a certain speed. A complete understanding of these performances and their optimization should involve both the control and the mechanics which has been ignored by robotic researchers for years. The solution we propose is a tradeoff between control and mechanics based on the Virtual Prototype Design Method. We build a simplified numerical model of a quadruped leg based on a hierarchial architecture. The proposed model is validated by comparing the numerical solution and the physical results coming from an extended campaign of experimental tests.

Keywords: Biomimetic, Legged Robots, Dynamics, Hydraulic Actuators, Virtual Prototype Design.

1 Introduction

Biomimetics is the scientific field who takes inspiration from nature to develop high efficient physical and mathematical models for robots. Researchers have always been interested in natural phenomena such as birds flying, snakes crawling, fishes swimming, horses galloping, dogs trotting of and last but not least humans walking. In the last decades scientists have built robots for imitating such behaviours. *Inspection* [1], *rehabilitation* [2] [3] and *rescue* robots [4] [5] [6] are the main topics covered by biomimetic ones (Fig. 1).

Inspection robots are used in all the situations where a human cannot go inside a "tunnel". Generally speaking that one is the area of crawling robots, thanks to their flexibility they can adapt to a narrow environment like a pipeline or the intestine of the human body. *Rehabilitation* robots have been designed to support and expand the physical capabilities of its users, particularly people

A. Duff et al. (eds.): Living Machines 2014, LNAI 8608, pp. 59–71, 2014.

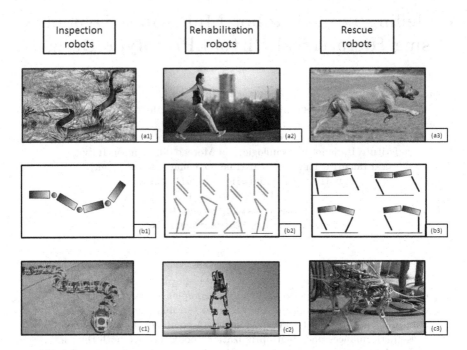

Fig. 1. (a1) (a2) (a3) Biological inspiration for robots. (b1) (b2) (b3) Dynamical concept modelling. (c1) Inspection (Crawling Robot) [7], (c2) Rehabilitation (Exoskeleton) [8], (c3) Rescue Robot (Legged Robot) [4].

with physical disabilities. The exoskeletons represent the biomimetic version of such family. *Rescue* robots are mainly designed to reach dangerous areas for the purpose of rescuing people. Common situations that employ rescue robots are earthquake, nuclear disasters, hostage situations and explosions. There are two main advantages in sending rescue robots in these scenarios, the first one is the personnel reduction while the second one is the ability to access to unreachable areas. *Legged robots* represent the biomimetic oriented family of such machines thanks to the capabilities they have to go through unstructured environments.

Legged robots can be classified in bipeds [9] [10], quadrupeds [4] [5] [6], hexapod [11] and octopods [12] ones. Generally speaking they are able to perform four different type of gait as walking, running, trotting and galloping. However, their motion abilities depend on the number of the legs they have. Quadrupeds are able to perform all the above described gaits while the others can perform only one or two type of motion.

Nowadays the panorama of quadrupeds is wide, it goes from robust to agile ones. Some example are LS3 [5], BigDog [13], HyQ [14], Cheetah [15] and Wildcat [16]. The Legged Squad Support System (LS3) is a very robust robot, it can carry up to 180 kg of squad equipment, sense and negotiate terrain and maneuver nimbly. BigDog is a dynamically stable quadruped able to pass through unstructured terrain, it can run at 6.4 km/h and climbing a 35 degree inclined

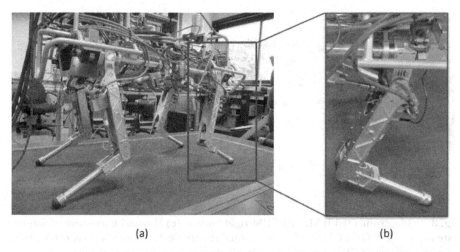

(a)   (b)

Fig. 2. (a) The Hydraulic Quadruped HyQ [4] [19], is designed to perform high dynamic task like jumping, running, climbing, etc. It is able to perform both indoor than outdoor operations like walking up to $2m/s$, jumping up to $0.5m$ and balancing the ground disturbance. HyQ weighs about $80kg$, is $1m$ long and $1m$ tall with fully stretched legs.; (b) The Hydraulically Actuated Leg (HAL) has 2 degrees of freedom (DOF) in the sagittal plane, the hip and knee flexion/extension permit the leg to move forward. The leg is built of a light-weight aerospace-grade aluminium alloy and stainless steel with two cylindrical hydraulic actuators. It allow to split the structure in two main groups of components, the hydraulic and the mechanical one.

terrain. HyQ is a hydraulic quadruped robot able to run up to $7.2\ km/h$ and it is also able to pass through unstructured terrain. Cheetah is the fastest legged robot in the World, it can run up to surpassing $46\ km/h$. Wildcat is a four-legged robot, it is able to run fast on all types of terrain using bounding and galloping gaits.

Modelling these dynamic features is a challenge. The traditional approach deals with this issues is based on control theory with the aim of managing the input of the system to obtain the desired output [17]. However there are some drawbacks in the traditional approach due to non-consideration of the structural aspects. First, the control is not always able to dealing with impulsive Ground Reaction Force (GRF); second, it does not take into account the joint flexibility; third it is not able to consider the structure flexibility. The Virtual Prototype Design (VPD) permits to find a tradeoff between control and mechanical aspects looking for an optimum solution.

VPD technique is well known since last five decades, it is widely used in industrial robotics where it allows saving time and money [18]. However to the best of our knowledge it is not broadly applied in the legged robot design. The main goal of the VPD is the creation of a Multi-Body Model (MBM) who is able to represent exactly the physical one. When the results of that MBM are satisfactory it is time to build the physical prototype.

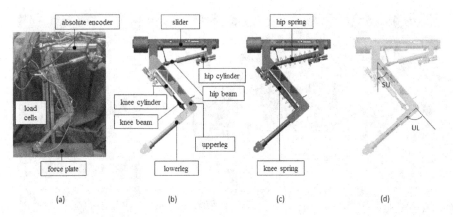

Fig. 3. (a) Instrumented HAL, (b) VPM rigid bodies, (c) Hip and knee spring-damper systems, (d) 2 DOF of HAL. SU represents the angle between the vertical to the upperleg axis while UL is the relative angle between upperleg and lowerleg.

However, building a good MBM could be very difficult due to the simultaneous presence both of complex physical phenomena and control laws. Modelling a high level model could be very risky. If not all the parameters involved in the simulation are well known, it easy to get wrong and misleading results. One of the most common scenario occurs at early stage of the design process, when not all the model details are available. For these reasons we propose a hierarchical solution to overpass that issue [20].

In this study we focus on building a simplified MBM for a quadrupedal leg. More in detail, the physical reference model we used is one of the Hydraulically actuated Quadruped (HyQ) leg (Fig. 2-(a)) called Hydraulically Actuated Leg (HAL), (Fig. 2-(b)). Since it is the first HAL virtual model, it is necessary to carry out an exhaustive campaign of experimental tests with the aim of estimating the MBM basic parameters. These tests are both statical and dynamical in order to have a description of a wide range of HAL motion. However, since it is a simple model we can only investigate the GRF impulsive propagation inside the structure and the influence of the joints flexibility, leaving to future works the modelling of the structural flexibility.

The rest of the paper is organized as follows. In Section II the HAL MBM building and testing are described, Section III contains the experimental tests, Section IV and Section V address results and conclusions respectively.

2   HAL Multy-Boldy Model

2.1   MBM Building

Each physical HAL contains around 450 parts made by different materials: plastic for the electronics, oil for the hydraulics, steel and alloy for the mechanics. It means that in this system there are more than 2700 DOF. The physical leg,

Table 1. Definition of the six lower pairs joints properties

Joint type	Relative displacement d_1	d_2	d_3	Relative rotation θ_1	θ_2	θ_3
Revolute	0	0	0	0	0	$\neq 0$
Prismatic	0	0	$\neq 0$	0	0	0
Screw	0	0	$p\theta_3$	0	0	$\neq 0$
Cylindrical	0	0	$\neq 0$	0	0	$\neq 0$
Planar	$\neq 0$	$\neq 0$	0	0	0	$\neq 0$
Spherical	0	0	0	$\neq 0$	$\neq 0$	$\neq 0$

instead has only two hydraulic DOF in the sagittal plane, it means that all the different components can be merged in several rigid bodies: one for each moving part of the structure. That choice allows the building of the simplest model of the hierarchical process.

A rigid body is treated as an assembly of components without relative movements, whose behaviour can be described referring to its center of the mass (*c.m.*). The nature of each single component must be taken in account during the merging operation, because the rigid assembly *c.m.* position depends on the inertia of every single sub element. The building process for the MBM starts with merging operation and it is developed using MSC Adams software.

The merging process results for the HAL mechanic components are three rigid bodies namely slider, upperleg and lowerleg (Fig. 3-(b)). The merging process results for the hydraulic components, instead, are rigid bodies and spring-damper actuators: hip cylinder, hip beam, knee cylinder and knee beam belong to the first group, while hip spring and knee spring belong to the second (Fig. 3-(b), 3-(c)). Each spring-damper is characterized by its own sitffness k and damping c coefficients. Two different HAL joints models are built starting from the aforementioned bodies. The first one is the Rigid Connected Model (RCM) while the second is the Flexible Connected Model (FCM).

RCM. The connections in the RCM were based on the surface contact, they are usually defined as *lower pairs*. Considering two bodies A and B and the vectors that describe the position and rotation of each of them, the constraint law is represented by Eq. 1 both for translations and for rotations.

$$\mathbf{x}_i^A \cdot (\mathbf{u}^A - \mathbf{u}^B) - d_i = 0 \qquad (1)$$
$$cos(\theta_i(\mathbf{x}_i^A \cdot \mathbf{x}_k^B)) - sin(\theta_i(\mathbf{x}_k^A \cdot \mathbf{x}_k^B)) = 0$$

where $\mathbf{x}^A$ represents A position and orientation; $\mathbf{x}^B$ is B position and orientation; d_i is the relative displacement between two bodies along i direction while θ_i describes the relative rotation between two bodies around i axes.

The scenario of rigid connections is composed by six different joints, whose properties are summarized in Tab. 1. These are mostly ideal connections that serve to evaluate the kinematics of the mechanism.

The HAL RCM internal constraints are represented by *cylindrical* connections, while its external joints are *translational* and *planar*. The first one ensures the leg vertical movement, the second one simulates the ground.

FCM. The connections in the FCM were modelled by using the *bushing* element which represents one of the flexible connections choices offered by MSC Adams. The bushings apply an action force on the reaction body that could be expressed by Eq.2.

$$F_j = -F_j \qquad (2)$$
$$T_j = -Ti - \delta \cdot F_i$$

where F_i is the translational force components acting on A; F_k is the translational force components acting on B; T_i is the torque component acting on A; T_k is the torque component acting on B; K_{ij} is a stiffness matrix elements; δ is the F_i arm while C_{ij} is a damping matrix elements.

The constitutive law for each connection is described by Eq. 3.

$$
\begin{bmatrix} F_x^A \\ F_y^A \\ F_z^A \\ T_x^A \\ T_y^A \\ T_z^A \end{bmatrix} = -
\begin{bmatrix}
K_{11} & 0 & 0 & 0 & 0 & 0 \\
0 & K_{22} & 0 & 0 & 0 & 0 \\
0 & 0 & K_{33} & 0 & 0 & 0 \\
0 & 0 & 0 & K_{44} & 0 & 0 \\
0 & 0 & 0 & 0 & K_{55} & 0 \\
0 & 0 & 0 & 0 & 0 & K_{66}
\end{bmatrix}
\begin{bmatrix} x \\ y \\ z \\ a \\ b \\ c \end{bmatrix}
$$
$$
+
\begin{bmatrix}
C_{11} & 0 & 0 & 0 & 0 & 0 \\
0 & C_{22} & 0 & 0 & 0 & 0 \\
0 & 0 & C_{33} & 0 & 0 & 0 \\
0 & 0 & 0 & C_{44} & 0 & 0 \\
0 & 0 & 0 & 0 & C_{55} & 0 \\
0 & 0 & 0 & 0 & 0 & C_{66}
\end{bmatrix}
\begin{bmatrix} V_x \\ V_y \\ V_z \\ \omega_x \\ \omega_y \\ \omega_z \end{bmatrix}
+
\begin{bmatrix} V_x \\ V_y \\ V_z \\ \omega_x \\ \omega_y \\ \omega_z \end{bmatrix}
\qquad (3)
$$

For a complex modelling it is possible to use a *field* connection that has all the terms K_{ij} and $C_{ij} \neq 0$. That choice involves the deep knowledge of the bearings this aspect is beyond the scope of paper.

2.2   MBM Testing

Three different groups of analysis are carried out using HAL MBM. In the first two RCM and FCM models are simulated without control laws, while in the third one FCM is tested with control law. For each of them the MBM performs three different tests namely static, quasi static and drop test. The bushing mechanical characteristics are taken from bearing commercial datasheets. The control law used in the presented model is a closed loop PID control with the same gain values applied during the experimental tests (described in the following paragraph).

The input of the first two groups of simulations are the measured displacement laws of SU and UL joints shown in Fig. 3-(d) as well as the estimated coefficient k

and c (described later). The aim of both of these comparisons is the validation of the aforementioned coefficients, the joints sensitivity analysis and the estimation of the GRF internal propagation. The main outputs are the force registered by both the hip and knee springs.

The third group of simulations, instead, is the closest to the physical tests. In this case, the joints rotations are governed by control laws and not by ideal motions like for the other simulations. The input data are the vertical movements of the slider and once the k and c coefficients for the springs. The output are the SU and UL numerical angles.

3   Experimental Tests

The experimental tests that are carried out on the HAL have the aim to get results useful both for the modelling and for the validation stage of the MBM. Three different tests are carried out using the HAL. They are **static, quasi static** and **drop test**. **Static test** results are useful to check the mass distribution of MBM, which can influence the dynamic response of the model. **Quasi static test** and **Drop tests** results are useful both for modelling the actuators and for validate the MBM. More in detail, the first one describes the low speed conditions while the second one describes the high speed ones.

The HAL motion behaviour is governed by a PID control system. It has the aim to modify the stiffness and the damping of the whole hydraulic system by adjusting the P and D gain values. All the aforementioned tests are carried out in two different conditions: *low stiffness* and *high stiffness*. In the first case $P = 150 Nmrad^{-1}$ while in the second one $P = 300 Nmrad^{-1}$. The D value is kept constant all the time $D = 6 Nmrad^{-1}s^{-1}$.

The HAL is instrumented with two angular encoders (*Avago AEDA3300 BE1, up to 80000 counts per revolution, resolution 0.0045deg*) place on hip and knee joints; one displacement sensor (*Absolute Encoder austriamicrosystem AS5045, signal 12Bit, resolution 0.0879deg*) installed on the slider; two load cells (*Burster 8417, force range 0-5 kN, accuracy 0.5 %*) installed on the actuators strokes and a force plate (*KISTLER 9260AA6, force range 2.5kN for Fx, Fy and 0-5kN for Fz*), which act like the ground where the leg stands or drops on. The vertical movement of the structure during all the tests is ensured by a sleeve attached to the slider and moving on a vertical bar (Fig.3 -(a)).

Static Test. The static tests are carried out starting from the initial position of the HAL leg, that depends on the chosen gain value. Afterward a group of three different payloads are applied on it: *0 kg, 3 kg* and *7 kg*. The results of the first stage of the test, when no load is applied, are useful to measure the internal reaction forces due to the robot own weight. The measured quantities are the force on the hip and knee pistons, the joint rotations and the GRF.

Quasi Static Test. The quasi static experimental tests starts with the HAL leg standing in its equilibrium position on the force plate. Afterwards it is pulled up and pushed down several times, following a sinusoidal movement. The measured

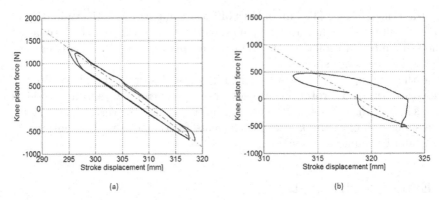

Fig. 4. Force-stroke displacement hysteresis drawn with a continuous line, Root Mean Square RMS drawn with a dash dot line for both Quasi Static Test (a) and (b) Drop Test

quantities are the same of the static test. Using the forces and actuator stroke displacement data it is possible to estimate the values for k and c coefficient respectively for both the actuators in low speed conditions. These data are used in input during the building process of MBM actuators for the numerical static simulation.

Drop Test. During the drop tests the HAL leg is left up to reach $5cm$ above the force plate, then it is released and dropped. The outputs were the same of the previous tests. Using the forces and stroke displacement data we got the c and k values for both the actuators in high speed conditions. These data are used in input during the building process of MBM actuators for the numerical dynamic simulation.

4   Results

This section addresses the results both of the experimental and the numerical tests. The first part is mainly focused on the experimental tests results used in input during the MBM building process while the second one addresses the MBM validation.

4.1   Experimental Tests Results

The experimental tests results used in the MBM actuators came from the quasi static and drop tests as already said before. More in detail the k and c coefficients are estimated once the force and stroke displacement for each actuator are known. The first of these last physical quantities derives from the load cell reasults while the second one can be calculated by the HAL kinematics.

Generally speaking the *stiffness* is proportional to the force ($F(t)$) applied on a body and to its related displacement (δ) as explained in Eq. 4.

$$F = k\delta \tag{4}$$

According to the Eq. 4, k coefficients are estimated from the the slope of the Root Mean Square (RMS) of the *force-displacement* curves obtained both for the low and the high speed tests as shown in Fig. 4.

The *damping* represents the predisposition of a body to the vibrations absorption. The c coefficients for the low speed tests come from the hysteretic diagram plotted in the force-stroke displacement work space (Fig. 4). Using the Eq. 5 [21] which describes the energy losses (ΔE_{cyc}) by a Single Degree of Freedom system, it is possible to estimate the aforementioned coefficients.

$$\Delta E_{cyc} = \int_{cyc} F(t)dx = \int_0^{2\pi/\omega} F(t)\dot{x}dt$$
$$= m\omega_n^2 A^2 |H(\omega)|\pi \sin(\varphi) = c\pi\omega X^2 \tag{5}$$

where $F(t)$ is the force that moves the piston, c is the damping coefficient, ω represents the force frequency and X is the maximum displacement amplitude.

The c coefficients for the high speed tests are estimated starting from the same measurement of the low speed tests by analyzing the momentum conservation as in Eq.6 [22]. That choice is due to the impossibility of having a closed hysteretic diagram using the high speed tests results (Fig. 4-(b))

$$mv_i^2 = c\frac{2\pi^2}{\Delta T}\Delta l_{max}^2 \tag{6}$$

where mv_i^2 represents the kinetic energy of the body during the dropping, ΔT is the time of the drop, Δl_{max} is the compression of the element that absorbs the impact energy and in our case is represented by the piston length. The kinetic energy has to consider the $F(t)_{imp}$ on the ground and the potential energy due to the weight of the leg ($mg\Delta T$), as shown in Eq. 7.

$$mv_i = -mg\Delta T + \int_0^{\Delta T} F(t)_{imp}dt \tag{7}$$

The coefficient estimated for both the low and high speed conditions are resumed in Tab.2. A statistical analysis is carried out on the experimental data. This show that the estimated c and k coefficients have a Gaussian distribution. 95% confidence intervals on the means are performed for these parameters in each condition, using the t-distribution with four degrees of freedom. Results show that each measured value is included in its relative interval, in this way measurement reliability is assured. For all the statistical analyses R software was used.

4.2   MBM Validation

The results of the first two group of simulations have the aim to validate the estimation of the k and c coefficients used for modelling the numerical springs and also, perform a joints sensitivity analysis and study the GRF propagation inside the HAL. In Fig. 5, Fig. 6, Fig. 7 is it possible to verify that the numerical

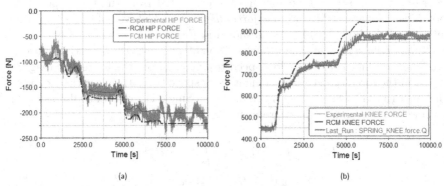

Fig. 5. Static Test Results with proportional gain P=300 $Nmrad^{-1}$. (a) Hip load cell vs Hip spring (b) Knee load cell vs Knee spring.

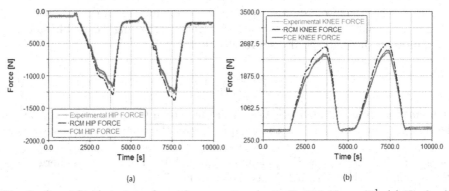

Fig. 6. Quasi Static test results with proportional gain P=300 $Nmrad^{-1}$. (a) Hip load cell vs Hip spring (b) Knee load cell vs Knee spring.

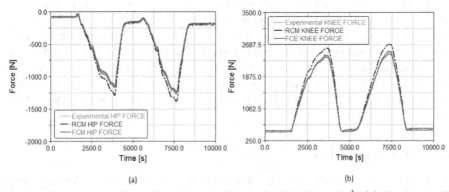

Fig. 7. Drop test results with proportional gain P=300 $Nmrad^{-1}$. (a) Hip load cell vs Hip spring (b) Knee load cell vs Knee spring.

Table 2. Stiffness and Damping Estimated Coefficients

		Hip Stiffness k [N/m]	Hip Damping c [Ns/m]	Knee Stiffness k [N/m]	Knee Damping c [Ns/m]
	Low	8.30E4	4.50E5	4.40E4	1.93E5
QUASI	σ	1.0%	22%	8.0%	22%
STATIC	High	1.84E5	2.1E6	1.16E5	3.8E5
TEST	σ	8.0%	22%	4.0%	19%
	Low	8.81E4	2.2E6	4.77E4	5.70E5
DROP	σ	0.0%	21%	2.0%	25%
TEST	High	1.75E5	2.55E5	8.75E4	1.50E5
	σ	8.0%	22%	4.0%	19%

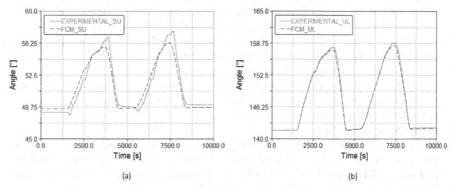

(a)   (b)

Fig. 8. FCM with control law Quasi Static test results with proportional gain P=300 $Nmrad^{-1}$. (a) Hip angle experimental vs Hip numerical angle (b) Knee angle experimental vs Knee numerical angle.

curves are both qualitatively and quantitatively near to the experimental ones. As expected, the FCM results are closer than the RCM ones. Figure 8 addresses the results of the third group of simulations. We decided to present here only the FCM Quasi Static numerical results, because they represent as well the other simulations. These tests in fact involve a wide range for angle variation respect to the static and drop tests. The agreement between experimental and numerical results shows that the MBM is able to predict the behaviour of the physical leg.

5 Conclusions

In this paper we propose a simplified MBM for one leg of a quadrupedal robot. That model represents the first stage of the hierarchial building process for a complete MBM with distributed flexibility. Even if it is a simplified model, it is a reliable representation of the physical one. We demonstrate the efficiency of the MBM owed to the agreement between the numerical and experimental results. The model could be used by the designer during the process of the bushings selection and also from the Control Systems Engineer to have an overall scheme before going into detail of the control design. Future investigations will consider several extensions of this work for refine the characteristics of the model both from the numerical and from the control point of view.

References

1. Sugiyama, Y., Hirai, S.: Crawling and jumping by a deformable robot. The International Journal of Robotics Research 25(5-6), 603–620 (2006)
2. Mikołajewska, E., Mikołajewski, D.: Exoskeletons in neurological diseases-current and potential future applications. Adv. Clin. Exp. Med. 20(2), 227–233 (2011)
3. Aach, M., Meindl, R., Hayashi, T., Lange, I., Geßmann, J., Sander, A., Nicolas, V., Schwenkreis, P., Tegenthoff, M., Sankai, Y., et al.: Exoskeletal neuro-rehabilitation in chronic paraplegic patients–initial results. In: Converging Clinical and Engineering Research on Neurorehabilitation, pp. 233–236. Springer (2013)
4. Semini, C.: Hyq-design and development of a hydraulically actuated quadruped robot. PD Thesis, University of Genoa, Italy (2010)
5. Dynamics, B.: Ls3: Legged squad support system (2012)
6. Lee, D.V., Biewener, A.A.: Bigdog-inspired studies in the locomotion of goats and dogs. Integrative and Comparative Biology 51(1), 190–202 (2011)
7. http://www.snakerobots.com/S7.html
8. http://www.eksobionics.com/
9. Moro, F.L., Tsagarakis, N.G., Caldwell, D.G.: A human-like walking for the compliant humanoid coman based on com trajectory reconstruction from kinematic motion primitives. In: 2011 11th IEEE-RAS International Conference on Humanoid Robots (Humanoids), pp. 364–370. IEEE (2011)
10. Edwards, L.: Petman robot to closely simulate soldiers (2010)
11. Cham, J.G., Karpick, J.K., Cutkosky, M.R.: Stride period adaptation of a biomimetic running hexapod. The International Journal of Robotics Research 23(2), 141–153 (2004)
12. Klaassen, B., Linnemann, R., Spenneberg, D., Kirchner, F.: Biologically inspired robot design and modeling. In: Proceedings of the ICAR 2003–11th International Conference on Advanced Robotics, pp. 576–581 (2003)
13. Raibert, M., Blankespoor, K., Nelson, G., Playter, R., et al.: Bigdog, the rough-terrain quadruped robot. In: Proceedings of the 17th World Congress, pp. 10823–10825 (2008)
14. Semini, C., Tsagarakis, N.G., Guglielmino, E., Focchi, M., Cannella, F., Caldwell, D.G.: Design of hyq–a hydraulically and electrically actuated quadruped robot. Proceedings of the Institution of Mechanical Engineers, Part I: Journal of Systems and Control Engineering 225(6), 831–849 (2011)
15. Lewis, M.A., Bunting, M.R., Salemi, B., Hoffmann, H.: Toward ultra high speed locomotors: design and test of a cheetah robot hind limb. In: 2011 IEEE International Conference on Robotics and Automation (ICRA), pp. 1990–1996. IEEE (2011)
16. http://spectrum.ieee.org/automaton/robotics/military-robots/whoa-boston-dynamics-announces-new-wildcatquadruped
17. Hogan, N.: Impedance control: An approach to manipulation. In: American Control Conference, pp. 304–313. IEEE (1984)

18. Rampalli, R., Ferrarotti, G., Hoffmann, M.: Why Do Multi-Body System Simulation? (2012)
19. Guglielmino, E., Cannella, F., Semini, C., Caldwell, D.G., Rodríguez, N.E.N., Vidal, G.: A vibration study of a hydraulically-actuated legged machine. In: ASME 2010 (2010)
20. Bucalem, M.L., Bathe, K.: The mechanics of solids and structures-hierarchical modeling and the finite element solutions. Springer (2011)
21. Priestley, M., Grant, D.: Viscous damping in seismic design and analysis. Journal of Earthquake Engineering 9(spec02), 229–255 (2005)
22. Meirovitch, L.: Analytical Methods in Vibrations. Macmillan (1967)

How Cockroaches Employ Wall-Following for Exploration

Kathryn A. Daltorio, Brian T. Mirletz, Andrea Sterenstein, Jui Chun Cheng,
Adam Watson, Malavika Kesavan, John A. Bender,
Roy E. Ritzmann, and Roger D. Quinn

Case Western Reserve University,
11111 Euclid Avenue, Cleveland, Ohio, USA 44106

Abstract. Animals such as cockroaches depend on exploration of unknown environments, and the complexity of their strategies may inspire robotic approaches. We have previously shown that cockroach behavior with respect to shelters and the walls of an otherwise empty arena can be captured with a stochastic state-based algorithm. We call this algorithm RAMBLER, Randomized Algorithm Mimicking Biased Lone Exploration in Roaches. In this work, we verified and extended this model by adding a barrier to our cockroach experiments. From these experiments, we have generalized RAMBLER to address an arbitrarily large maze. For biology, this is a model of the decision-making process in the cockroach brain. For robotics, this is a strategy that may improve exploration for goals in certain environments. Generally, the cockroach behavior seems to recommend variability in the absence of planning, and following paths defined by the walls.

1 Introduction

Robotic explorers have great potential, but in an unknown environment, the best goal-seeking strategy cannot be determined a priori. Gradient descent methods arrive at some targets (local minima) quickly, while missing others. Adding randomness can help, for example (Azuma et al., 2010). Randomness is used internally in high-level planning algorithms, for example in (LaValle and Kuffner, 2001). However, any randomness can add even more inefficiency, so algorithms are tuned for each application to balance the inherent trade-off between greedy goal tracking (direct but incomplete) and stochastic exploration (finds even hidden goals eventually). Improved stochastic searches could make high level planning algorithms faster.

The tactile boundaries are often under-exploited. Generally walls and obstacles are skirted just enough to get to the goal (Daltorio et al., 2010). Tactile boundaries could be a guide, natural landmark, or leverage point for pushing off or for climbing up. Resources like charging stations, targets like dirt to be vacuumed, and other agents in swarms may be more likely to be found along walls. In some cases, wall-following approximates information-optimizing (Fox 2013). Since following walls with tactile sensors is feasible (Lamperski et al. 2005), these tactile features need not be obstacles to be avoided; rather, the walls define uniquely important paths.

A. Duff et al. (eds.): Living Machines 2014, LNAI 8608, pp. 72–83, 2014.
© Springer International Publishing Switzerland 2014

Insect navigation has inspired engineers to consider strategies for moving with limited information and computation, such as "Bug" algorithms (Lumelski and Stepanov, 1986; Taylor and LaValle, 2009), or gaits generated with artificial neurons (Beer, 2003). A criticism of some of these works is that these imagined strategies were not related to experimentally observed animal behaviors (Webb, 2009). Cockroaches have been widely studied responding to isolated elements of their environment (Ritzmann et al., 2012; Jeanson et al., 2003; Harley et al., 2012; Harvey et al., 2009; Canonge et al., 2009) so we hope to learn more about the animal and about exploration by putting those behaviors into context of their overall multi-sensory goal-exploration strategy.

Our previous model (Daltorio et al. 2013) demonstrates that little mapping or memory is required to capture cockroach strategy for finding a shelter in an unknown arena. The cockroaches appear to be using a random walk in the arena and a wall-following state for the boundaries, however the transition probabilities are biased by the visual location of the goal. We called this finite state machine algorithm RAMBLER, Randomized Algorithm Mimicking Biased Lone Exploration in Roaches. The state transition probabilities were measured from the data, but the metric we used to validate the model was the overall path length to the goal. In this work, we validate that model for cockroaches in a more complex arena and we extend RAMBLER to a more general form to deal with obstacles.

Most mobile robots need to deal with obstacles in their desired path. Attempts to use the RAMBLER algorithm on a robot in an arena with a barrier between the robot and the goal resulted in problematic behaviors (Tietz 2012), since the behavior for obstacle edges was previously unspecified given the concave wall boundary. Thus in this work, we performed additional trials with the cockroaches, placing a clear acrylic barrier in the arena. This gives us an opportunity to compare the cockroach's interaction with clear walls vs. opaque walls, and determine what happens when the goal is on the other side of a wall.

2   Cockroach Behavior around Clear Barrier

The cockroach experiments were similar to our previous work (Daltorio et al. 2013), except that we added a clear plastic barrier to the arena. For that work, we had built a walled arena to contain the cockroaches in order to study how they behaved with respect to a goal in the environment. After testing several potential goal features, we chose a shelter that consisted of a roof of red plastic film to be a goal. However, as we discovered, the animals were also attracted to the walls and corners of the arena, perhaps because the walls provide another sort of shelter. For clarity, we will refer to the feature with the film roof as the "shelter", the enclosing vertical plastic sides as "walls" which define the boundary of the arena. The "corners" are intersections of the walls, and the clear obstructions added within the walls are "barriers". The arena is 914mm square with a small starting chamber, shown in Fig. 1. Before each trial, an intact female cockroach, *Blaberus discoidalis*, was separated from the lab colony and the arena was cleaned to eliminate traces from previous trials. The cockroach was placed in the starting chamber and allowed to settle before the plastic gate was lifted. The cockroaches entered the arena, which was evenly lit by overhead lights

(1500 lux) except under the shelter (300 lux). Above the walls, black fabric shrouded the arena because otherwise there was asymmetry in the time spent on either side of the arena due to room colors and lighting. The first 60 seconds of exploration behavior was filmed at 20 Hz by an overhead camera using the Motmot image acquisition package (Straw and Dickinson, 2009). Each animal was used only once. The video was post-processed to find the position of the cockroach's visual center and its body orientation in each frame using the Caltech Multiple Fly Tracker (http://ctrax.sourceforge.net/) and the associated FixErrors toolbox for MATLAB (MathWorks, Inc., Natick, MA, USA) (Branson et al., 2009). Because there were still occasional inversions of the tracked angle, we also corrected any instantaneous 180° flips. All further analysis was done in MATLAB.

In this work, we added a barrier between the start-chamber and rear wall where the shelter was located. The L-shaped barrier is 450 mm long. The short side of the L is 40mm long and keeps the barrier upright. The primary difference between the barrier and the boundary walls is opacity: the barrier is clear acrylic whereas the walls are acrylic, spray painted white. Note that the top surface of the barrier is painted black to aid in analysis. Also, the walls of the arena need not be encountered in the shortest path from the starting chamber to the shelter. In each of 188 trials, Fig. 1, the long side of the barrier was placed in the arena parallel to the wall with the entrance chamber. Of these, in 53 of the trials the barrier was centered within the arena such that the cockroach could pass between the barrier and the wall on either side of the barrier. In the remaining 135 trials, the barrier abutted the center of the left wall. Of these, sometimes the small side of the L-shaped barrier was against the wall (70 trials) and sometimes the small side of the L-shaped barrier was in the center of the arena extending away from the starting chamber. Most of these trials included a shelter, but 21 of the 53 centered barrier trials, 4 of the 70 trials with the flat edge in the center of the arena and 7 of the 65 trials with the small edge of the L in the arena center did not have a shelter.

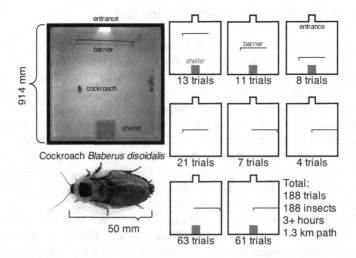

Fig. 1. Cockroaches were tested in eight different arena configurations

These trials result in 320 distinct encounters with the barrier, where each encounter begins when the cockroach approaches within 50mm (the antenna length) and concludes when the cockroach is more than 100 mm away from the barrier. Only 34 of these encounters occurred in trials with no shelter. Thus when a shelter is present, there are an average of 1.8 barrier encounters per trial whereas without a shelter, there is an average of 1.0 barrier encounters per trial.

2.1   Characterization of Barrier-Following as Wall-Following

From these data, we looked for evidence that the cockroaches turned away from the barrier from the beginning of the trial. This was not the case, in fact we found no indication that the cockroach could see the barrier until the cockroach was in antenna range. Thus we will assume that the barrier is irrelevant to navigation before a tactile encounter.

Next, we looked at how the cockroach followed the barrier. "Wall-following" appears to be a different state than exploring the area away from walls. The escape response directions are different (Ritzmann, 1993) and sharp turns can be tracked even at high speeds (Camhi and Johnson, 1999). Even their speed distribution is different (Bender et al 2011): away from the wall there is a non-zero mode speed while close to the wall the distribution is dominated by stops and slow walking. Barrier following has this same distribution, so we treat barrier following like wall following.

Next, the RAMBLER parameters associated with the wall (departure rate, turnaround rate and corner passage) were measured with respect to the barrier in Fig. 2. For the best comparison, only the data away from the ends of the barrier were considered so that the barrier would be present along the entire length of the body, as when the cockroach was along the wall. Overall from these data, we find that we can treat the barrier-following as a special case of wall-following in which the angle to the shelter can be negative, *i.e.* through the wall.

First, along the positive angle axis, the barrier departure rate is similar to the wall departure rate. This is a validation of our wall departure rate results since these are a different set of animals in experiments with a different arena configuration performed by different people and analyzed by an independent block of code. Also, it suggests that when the tactile stimulus is on one side and the visual stimulus is on the opposite side of the body (positive angle in Fig. 2), the opacity of the wall does not matter. Second, the negative angles in this plot provide an extension to RAMBLER: what to do if the tactile stimulus is between the animal and the visual attractive stimulus? The data suggest that the insect is more reluctant to leave the barrier when the shelter is visible through the barrier, and especially reluctant if the shelter is visible in more forward directions. Note that the confidence intervals on the negative angles are tighter because there are more encounters on the entrance side of the barrier than there are on the shelter side of the barrier. An interesting question for future work is whether this would be true if the cockroach had observed the shelter at the beginning of the trial and then an opaque barrier appeared. Would they remember the shelter location or behave as if it had never existed?

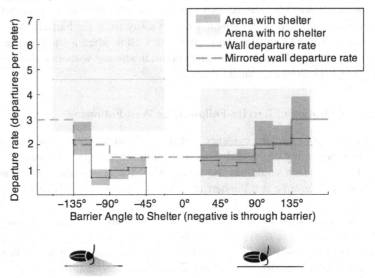

Fig. 2. The barrier departure rates are similar to the wall departure rates. The barrier departure rate was calculated by dividing the number of departures by the sum of the distance walked at each timestep along the long parts of the L-shaped barrier. To get a confidence interval on that value, the mean was taken of a sample of the Boolean variable that is true if and only if the point is the last point of an encounter that is within 50 mm of the barrier. This sampling was weighted by the distance walked in that timestep. 1000 such means were found and the 15th and 85th percentiles give a 70% bootstrap confidence interval. The barrier data on the left side of the figure is a new situation for which have no comparable boundary wall data, but instead we compared to the mirrored line (dashes). Sectors in the diagrams indicate the relative direction of the shelter.

Finally, we compared the frequency of changing direction along the barrier. We identified turnarounds as points in which the direction around the barrier changed and also the direction of walked distance for at least 50mm each side of the point were in opposite lateral directions. The average turnaround rate was 0.7 turnarounds per meter, which is approximately the wall turnaround rate when facing the shelter (Daltorio et al. 2013). Most of these 44 turnarounds were on the side of the barrier closer to the entrance chamber. Finally, when the cockroach walked into the wall from the barrier, it turned back 30% of the time. This agrees well with the behavior in corners of two walls (Daltorio et al. 2013).

2.2   Characterizing Edge Behavior

In the empty arena, a cockroach encounters only corners when following the walls. Corners may cause additional antenna contacts and often a stop to investigate the feasibility of climbing out of the arena. The new trials create a new situation in which

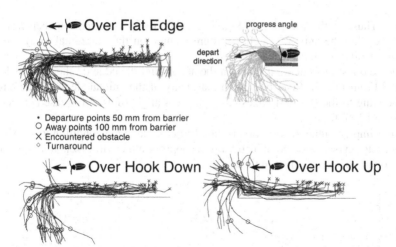

Fig. 3. The tracks of cockroaches aligned and plotted together, rotated such that the long part of the L-shaped barrier was oriented at zero degrees and mirrored horizontally and/or vertically and plotted so that the tracks move to the left and down over the edge. There were one to three cases for each of the three types in which the cockroach followed the whole edge from the front to back of the barrier. These obscured the results and are not plotted here but are included in the quantitative analysis in Fig. 4. The diagram in the upper right shows how we will quantify these paths as a departure at a certain progress angle from a constant radius arc and at a certain departure direction.

the cockroach is following a wall that ends. Here the options are: turn back, locally search for a new tactile contact surface, keep walking in the same direction as it was headed before, or depart the barrier at some other angle.

To analyze the behavior of cockroaches as they walked past the edge of the barrier, only encounters in which the animal walked along the barrier between 100mm and 50mm from an edge were included. There are four categories of edges considered: (1) "over flat edge" - around the simple straight edge (around the top of a letter L) (2) "over hook down" - over the outer part of the hook of the L (counterclockwise around the bottom of a letter L) (3) "over hook up" - around the inner park of the hook (clockwise around the bottom of a letter L) or (4) coming to the edge of the barrier and encountering a wall. In case (4), we did not distinguish between situations where the cockroach encountered the small part of the L along the wall and situations in which the small part of the L was in the center of the arena. In case (4) there are generally only two outcomes: following the wall upwards or turning around back onto the barrier: just like a corner between two walls. For cases (1), (2), and (3), the typical tracks for a cockroach rounding an edge are shown in Fig. 3. After an edge, the paths seem to follow a searching spiral, gradually leading away from the barrier.

The departure points form an arc in Fig. 3 because they are defined by proximity to the edge. Further, there is a visible pattern in the way the cockroaches come away from the barrier. The walking angle of the animal depends on the degree about the circle that it progressed around the edge. Regardless of whether the edge is a straight edge or a hooked edge, this relationship between how far around the edge the animal walked and the final orientation of the animal as it walks away is similar. Note that this suggests we can approximate this behavior much like the departure from a wall, only using an arc instead of straight path followed by a pivot away from the tactile stimulus. Thus, in the top plot of Fig. 4, a relationship is found between how far around the edge the animal searches (progress angle on the horizontal axis) and the resulting heading (on the vertical axis).

Then the question is at what point on the arc around the edge do the animals leave the wall? From Fig. 4, it can been seen that many of the animals leave at the angles corresponding to the sharpest edges (so a progress angle of 180° for case (1), 90° for case (3) and 270° for case (2)).

For our implementation we took the data from the flat edge condition to see if it fit the other edge types when combined with the appropriate environmental stimuli (Fig. 6). We found no shelter effect in these data.

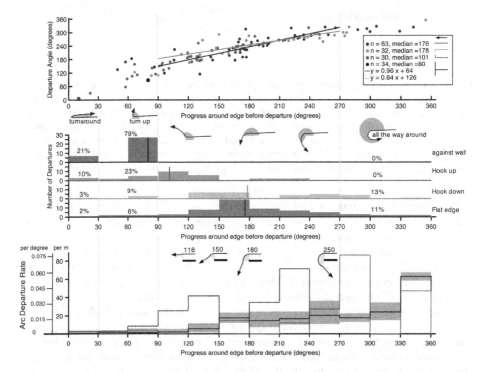

Fig. 4. How far around the edge the cockroach walks. In the top plot, if we define an "away point" at the first 100mm from the wall and use the line segment between the departure point and away point to define the departure angle, as in defining wall departures in (Daltorio et al. 2013), we can see a relationship between the angle of the departure point around the edge and the departure angle. In the second plot of Fig. 4, we binned the departures by progress angle. Turnarounds have the smallest progress angle. Continuing all the way around the barrier corresponds to the largest progress angle. In the third plot, we converted the bins into departure rates. We approximated each path as an arc around the edge with radius 50mm. The rates are determined by comparing the departures at that bin with the departures at bins of greater angle, which passed that bin without departure. For the flat edge, where we have the most data, the departure rate is higher when it begins to arc around the edge than along the straight walls or barriers. The departure rate further increases by an order of magnitude when the body becomes perpendicular to the edge. Similar trends seem to happen in the hook up and hook down condition but with departure rate increases at different angles corresponding to edges of the hook.

3   Generalizing RAMBLER

Since cockroaches treat barriers and walls similarly and edge departures can be quantified by departure rates and departure angles, we can extend our RAMBLER algorithm as suggested by Fig. 5. This is more general than our previous paper's state diagram, because we have clarified not just how to handle the boundaries of the space but also the obstacles within the space. Another difference between our state diagram in the previous paper is that this one only includes states that are sustained for indeterminate timesteps. For the details of intermediate actions such as how to depart the wall by turning away see the pseudocode in the appendix of (Daltorio, 2013).

RAMBLER
Random Algorithm Mimicking Biased Lone Exploration in Roaches

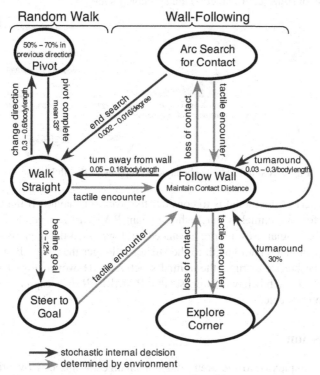

Fig. 5. Our proposed controller to navigate unknown environments

This model is tested at two levels. First, we will look at the edge behaviors and compare to Fig. 4. The edge state in Fig. 3 is built with the characterization from walking off the flat side of the barrier. Hook edge types are experienced as combinations. For example, going around the "over hook down" configuration will lose contact from wall following and search for a new contact and then, if the agent continues without departing the arc, it will find the short segment of the barrier and treat it as a wall which will then end in searching edge behavior and possibly continue until it reaches the back side of the barrier for another wall-following behavior state. The "over hook up" condition is first a corner, then a wall, then an edge, etc. The simulation behavior has less variability built into it, and generally follows the wall a little closer than the cockroaches in corners. As a result, sometimes the cockroach does not treat the inside of the hook as a corner, perhaps because it detects the edge of the short segment. The arcs tend to increase radius slightly as the agent turns so sometimes the measured "departure point" of the simulated data is a little different than the point at which simulation decided to depart which explains some differences in the top and bottom subplot. Nonetheless, the simulated results in Fig. 6 are similar to the cockroach data. To improve it further, we would need a better virtual antenna design combined with data about how cockroaches use their antennae.

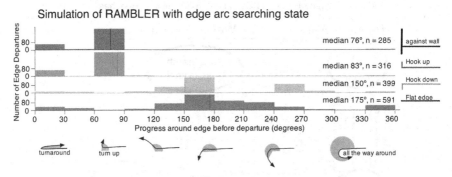

Fig. 6. A simulated edge behavior compares well with Fig. 4

The second test of the model is to consider the overall behavior. If the path length to the shelter for the animal is much shorter than RAMBLER predicts for barriers, then perhaps the animal uses a more advanced strategy when the arena is more complicated. If the animal's path length to the shelter is longer than RAMBLER predicts then perhaps the barrier disrupts the animal's behavior. However, Fig. 7 shows that the simulated cockroach behavior suggests that RAMBLER does capture naïve exploration at this length scale.

4   Discussion

Adding more complexity to the arena did not change the insect's overall strategy. When a tactile guide is lost, the cockroach searches for tactile contact. Our results suggest that this does not require the animal to characterize the edge (how long it is, whether there is a hook, etc.) but that a combination of walls can be treated as a series of wall-following (constant contact surfaces), edge searching (unexpected losses of contact surfaces) and corner exploration (unexpected gain of a second contact surface). We expect that closer observations of cockroach antennae during these key conditions will help us to better model the task-level control of these actions.

We hope that this characterization of the overall strategy will corroborate future theories of cockroach brain function and will be useful as robotic search heuristics.

As an example of an environment that might benefit from such heuristics, we modified an obstacle titled "Bug Zapper" (Choset et al., 2010) that is a simple example of a situation that might foil a naïve wall-following and gradient descent strategy. In this environment, we compared simulations of 3 stochastic algorithms: our RAMBLER algorithm if it can sense the goal, our RAMBLER algorithm if it cannot sense the goal, and finally a random bounce algorithm. The random bounce algorithm goes straight until encountering an obstacle and then chooses a random pivot direction from a uniform distribution of angles, similar to what robotic vacuums might do.

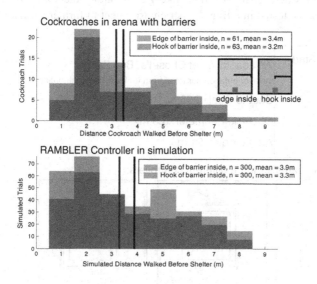

Fig. 7. A comparison of cockroach and simulation path length from the entrance to shelter. The two most common arena conditions for the barrier cockroach trials were when there was a shelter and a barrier centered against the side wall with either the hook of the L at the center of the inside of the arena pointing down ("hook inside") or with the flat edge of the L at the inside of the arena ("edge inside"). The distance walked before the shelter (or the total trial path length if the animal never reached the shelter) was measured. The means are plotted as vertical lines on the distributions. The mean path length is less when the hook leads the animal toward the shelter. Even though this difference is not statistically significant in the animal data ($p = 0.5$ MATLAB kstest2), it is captured in the controller implementation. The simulation samples the initial conditions from 100mm into the arena. This indicates that RAMBLER is a good model of our experiments.

A naïve algorithm (Choset et al., 2010) of following encountered walls until back on the path to the goal would do unusually poorly in this environment. If the agent has a policy of turning to the right at obstacles it has to follow the whole spiral around to the goal. If the agent has a policy of turning to the left at the encountered obstacle, it will loop around the outside of the spiral forever, never reaching the goal. Bug algorithms would solve this labyrinth by remembering a few key parameters as it walks around the edge (Choset et al., 2010). For example, Bug1 would walk around the whole perimeter and then go back to the best departure point. RAMBLER doesn't require such long term memory and it doesn't require precision deterministic movements. In Fig. 8, RAMBLER performs better than random bouncing in this arena and adding the shelter bias helps the robot get there even more often. Note that we widened the labyrinth paths slightly from (Choset et al, 2010) and increased the trial time to 2 min (from the cockroach trial times of 1 min) because otherwise even fewer of the random bounce trials got into the goal. This demonstrates the value of wall-following, especially in cases with complex obstacles, thread-the-needle like passageways, or where goals are more likely to be along the walls. In future work, cockroaches can be tested in a Bug Zapper environment to further validate this model.

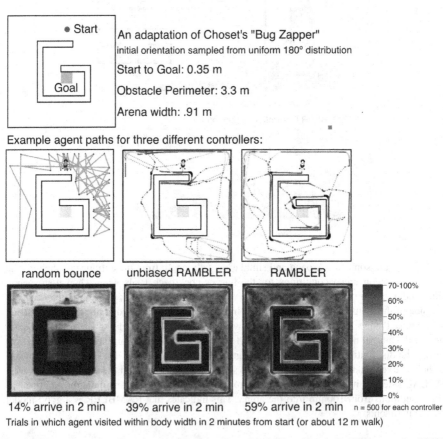

Start to Goal: 0.35 m

Obstacle Perimeter: 3.3 m

Arena width: .91 m

An adaptation of Choset's "Bug Zapper"
initial orientation sampled from uniform 180° distribution

Example agent paths for three different controllers:

random bounce   unbiased RAMBLER   RAMBLER

14% arrive in 2 min   39% arrive in 2 min   59% arrive in 2 min   n = 500 for each controller

Trials in which agent visited within body width in 2 minutes from start (or about 12 m walk)

Fig. 8. Three algorithms in an environment author titled "Bug Zapper" (Choset et al., 2010). The agents are shown for scale at the starting position and the paths through the arena are shown. RAMBLER's wall following brings the agent toward the goal more often. If the shelter can be seen, the agent follows the wall even more and thus finds the red square goal even more often. 500 simulated trials for each condition.

Acknowledgements. We would like to thank the students of the BioRobotics Team Research, including Nicholas Szczecinski, Victoria Webster, Douglas Porr, John Richards, Amy Brown. This work was supported by NSF research Grant No. IIS-1065489 and NSF grant IOS-1120305.

References

1. Azuma, S., Sakar, M.S., Pappas, G.J.: 49th IEEE Conference on Decision and Control, pp. 6337–6342 (2010)
2. Beer, R.D.: Adaptive Behavior 11, 209–243 (2003)
3. Bender, J.A., Simpson, E.M., Tietz, B.R., Daltorio, K.A., Quinn, R.D., Ritzmann, R.E.: Journal of Experimental Biology 214, 2057–2064 (2011)
4. Branson, K., et al.: Nature Methods 6, 451–457 (2009)
5. Camhi, J.M., Johnson, E.N.: Journal of Experimental Biology 202, 631–643 (1999)
6. Canonge, S., et al.: Journal of Insect Physiology 55, 976–982 (2009)
7. Choset, H., Hager, G.D., Dodds, Z.: Robotic motion planning: Bug algorithms. Lecture Notes, Carnegie Melon University,
   http://www.cs.cmu.edu/~motionplanning/lecture/
   Chap2-Bug-Alg_howie.pdf
8. Daltorio, K.A., et al.: Proceedings of the 2010 IEEE/ION Position Location and Navigation Symposium (2010 ION/IEEE PLANS), Indian Wells, CA (2010)
9. Daltorio, K.A.: Ph.D. Thesis. Case Western Reserve University (2013)
10. Daltorio, K.A., et al.: Adaptive Behavior 21(5), 404–420 (2013)
11. Fox. Living Machines 2013, London, UK, pp. 108–118 (2013)
12. Harley, C.M., Ritzmann, R.E.: J. Exp. Bio. 213, 2851–2864 (2010)
13. Harley, C.M., English, B.A., Ritzmann, R.E.: J. Exp. Bio. 212(3) (2009)
14. Harvey, C.D., Coen, P., Tank, D.W.: Nature 484, 62–68 (2012)
15. Jeanson, R., et al.: Journal of Theoretical Biology 225, 443–451 (2003)
16. Lamperski, A.G., Loh, O.Y., Kutscher, B.L., Cowan, N.J.: IEEE ICRA 2005, pp. 3838–3843 (2005)
17. LaValle, S.M., Kuffner, J.J.: In: Donald, B.R., Lynch, K.M., Rus, D. (eds.) Algorithmic and Computational Robotics: New Directions, pp. 293–308. A K Peters, Wellesley (2001)
18. Lumelski, V.J., Stepanov, A.A.: IEEE Transactions on Automatic Control AC-31(1) (1986)
19. Ritzmann, R.E.: In: Beer, R., Ritzmann, R.E., Mckenna, T. (eds.) Biological Neural Networks in Invertebrate Neuroethology and Robotics, ch. VI. Academic Press (1993)
20. Ritzmann, R.E., et al.: Frontiers of Neurosciences 6(97) (2012)
21. Straw, A.D., Dickinson, M.H.: Source Code for Biology and Medicine 4(9) (2009)
22. Taylor, K., LaValle, S.M.: IEEE ICRA, pp. 3981–3986 (2009)
23. Tietz, B.R.: Masters Thesis. Case Western Reserve University (2012)
24. Webb, B.: Adaptive Behavior 17(4), 269–286 (2009)

Machines Learning - Towards a New Synthetic Autobiographical Memory

Mathew H. Evans, Charles W. Fox, and Tony J. Prescott

Sheffield Centre for Robotics (SCentRo), University of Sheffield,
Sheffield, S10 2TN, U.K.
{mat.evans,charles.fox,t.j.prescott}@shef.ac.uk

Abstract. Autobiographical memory is the organisation of episodes and contextual information from an individual's experiences into a coherent narrative, which is key to our sense of self. Formation and recall of autobiographical memories is essential for effective, adaptive behaviour in the world, providing contextual information necessary for planning actions and memory functions such as event reconstruction. A synthetic autobiographical memory system would endow intelligent robotic agents with many essential components of cognition through active compression and storage of historical sensorimotor data in an easily addressable manner. Current approaches neither fulfil these functional requirements, nor build upon recent understanding of predictive coding, deep learning, nor the neurobiology of memory. This position paper highlights desiderata for a modern implementation of synthetic autobiographical memory based on human episodic memory, and proposes that a recently developed model of hippocampal memory could be extended as a generalised model of autobiographical memory. Initial implementation will be targeted at social interaction, where current synthetic autobiographical memory systems have had success.

Keywords: Synthetic, Autobiographical, Memory, Episodic, Hippocampus, Robotics, Predictive, Coding, Deep, Learning.

1 Introduction

We receive a continual and very high band-width stream of sensory data during our waking hours. Our autobiographical systems process this data in a highly specific and adaptive fashion so as to provide quick access (within seconds) to relevant information experienced from hours to decades earlier. Autobiographical memory (AM) is defined as the recollection of events from one's life. Though similar in conception to Tulving's episodic memory [52], AM goes beyond simple declarative facts of an event to recall of rich contextual details of a scene [13]. AM is a prerequisite for developing the narrative self [32], related to and conceptually similar to the temporally extended self [43]. This version of the self is the individual's own life story, developed through experience, remembered, and projected into the future.

A. Duff et al. (eds.): Living Machines 2014, LNAI 8608, pp. 84–96, 2014.

For robots to behave in a flexible and adaptable manner, and succeed in complex sensorimotor tasks, it is essential that they store their experience appropriately and use this information during online processing. For example, in human-robot interaction a memory of a familiar person or game would be intuitively advantageous especially in developing trust or attachment. This concept is familiar in probabilistic robotics, where an informative prior is essential for accurate inference. At a purely computational level, it has been shown that learning through 'episodic control' is more efficient than building a forward model or developing habits when experience is limited and tasks are complex, as is often the case in real-world robotics applications [24].

Innovative sensor designs, miniaturised HD cameras and affordable hard drives have endowed modern robots with an impressive capacity for gathering and storing information from the world. This information can be pooled across a range of modalities (e.g. vision, audition, touch, LIDAR, depth) at high bandwidth. However, robots remain poor at extracting or retrieving task-relevant information when needed, either offline from their vast archives, or online during streaming [54].

In practice, data streams are filtered through feature detectors, processed by machine learning black boxes, or passively compressed into annotated histories where no such tools are available. As a result, despite having the ability to encode virtually everything that happens to them and access to vast stores of additional information online, robots are poor at determining which aspects of their history are important for making decisions and performing actions, or for framing engagement with people.

How does the brain solve this problem? By forming autobiographical memories that evolve over the life-time of an agent that places events in the context of the self and its goals. In this position paper we outline common approaches to autobiographical memory modelling in robotics, and contrast this with more modern understanding of memory function and organisation (see Figure 1). Four main principles are identified: compression, pattern completion, pattern separation and unitary coherent perception. To address each of these principles in a single architecture we outline a modelling framework based on predictive coding: deep learning for hierarchical representation and compression of sensory inputs (modality specific hierarchies in Figure 1, thought to reside in the sensory processing circuitry of the brain), and episodic memory formation through a particle filter and Boltzmann machine hippocampus model (Following Fox and Prescott [16], see Figures 2 and 3). We close with a discussion of our specific implementation goals, centring our episodic memory system within a rich sensorimotor system to aid ongoing processing.

2   Current Approaches to Synthetic Autobiographical Memory

What is the state of the art in this field? Synthetic autobiographical memory (SAM) broadly has two flavours; (i) storage of unedited data streams (e.g. for

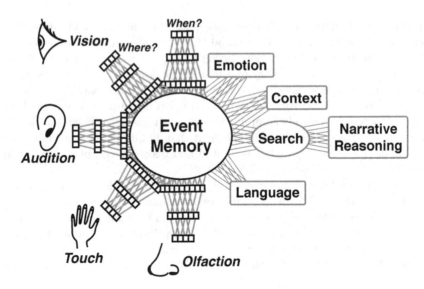

Fig. 1. An overview of our proposed model of biological autobiographical memory. Based on the basic systems model of Rubin [44], event memories are formed from convergent activity across disparate brain systems. Algorithmically we turn to deep networks for sensory processing [3], and predictive coding to frame learning [11]. Storage and recall is modulated by emotional, contextual and linguistic information. Narrative reasoning seeds searches, recognitions, reconstructions and predictions (the core of pattern separation and completion) by re-activating modality-specific schemas [21].

maximum likelihood estimation [25]), or (ii) annotated high-level sequences (e.g. of social interactions [36]). The initial problem with approach (i) - storage capacity - is being alleviated to a degree by improvements in hard-drive capacity. However, storing a complete sensory history becomes un-wieldy quickly and the ability to search and utilise this vast store of information to inform ongoing processing becomes seriously problematic.

Recent success has been had in demonstrating the effectiveness of approach (ii) such as in cooperative tasks [36], or learning through social interaction [37]. These models are largely symbolic, similar in spirit to classic models of cognition like ACT-R [1] and others (reviewed in [54]), with hand-set higher level representations and action scripts coordinated into event memories [38]. As in early cognitive science, these models of memory have appealed to the computer metaphor for the mind, though thorough critiques of this approach have been articulated [10]. Pointedly, short-term and long-term storage of homogenous information is an idea from the 1960s but continues today in many models of memory [29, 5].

There remain fundamental gaps between cognition, memory and learning in these models. The compression of experience into memories is not handled in an adaptive way as it is in nature, and these models neither capture the rich feature-set of biological memory nor exploit advances in machine learning to increase the power and flexibility of stored information. Our position is that memory, and as a result our model of AM, is central to action and formed adaptively through experience with the world.

3   Characteristics of Biological Autobiographical Memory

As noted in Wood et al. [54], the cognitive science and neurobiology of memory has changed drastically over recent years, but synthetic approaches have not kept pace. In particular, models of episodic and autobiographical memory are not compatible with theories of active, distributed memory systems [44] or ideas such as the predictive brain [11]. Forming new memories is not only about data compression, though this is important, but about selectivity and efficiency in the mechanisms of memory coding and retrieval. Cognitive scientists and neurobiologists assert that long-term memory formation is not a passive process of logging data into generic storage [2][44], but a highly active process depending on factors such as the depth to which an event is processed, or the wider context of the current task [26][42].

Biological memory is highly distributed and tightly-woven into ongoing cognition [54]. The neural underpinnings of different kinds of memory and imagery cannot be separated from the circuitry of perception and action. For example, when a person is asked to imagine rotating an object, neural activity is elevated in the visual and somatosensory cortices [12].

In order to match the function of human AM, a SAM system should have the following properties.

- **Compression.** Human memory systems are vast, and the detail to which events can be recalled is extensive. However, we certainly do not remember everything that happens in our lives, and items are not stored with equal detail. Reducing the volume of data to be stored through active data compression involves attenuating redundant information (efficient coding [48]), and prioritising attention and storage resources to information most critical for achieving goals.
- **Pattern Completion.** Reconstructing an event from brief exposure to part of that event, or from an impoverished, noisy or degraded version of the full scene e.g. when experiencing a familiar environment in the dark. This operation follows the idea of a *schema* [21][40]. Schemas can be thought of as generative models for a particular basic system [44]. Such schemas could be seeded with a small piece of information - e.g. you were at a children's party - and a rich scene can then be filled in - i.e. there was a cake, gifts, guests, and games were played.
- **Pattern Separation.** The process of transforming similar representations or memories into highly dissimilar, nonoverlapping representations [31]. This is important as we have many experiences that are similar to each other but nonetheless must be remembered as distinct. For example in discriminating edible from inedible plants. This function also relates to chunking of distinct event sequences from longer ongoing experiences, a core feature of episodic

memory (see Figure 2). During later retrieval it is important only to re-call relevant information for a particular task, thus separating relevant from irrelevant memories in a given context.

- **Unitary Coherent Perception.** Intuition attests that we experience a sin-gle *unitary* version of the world, not a probability distribution or blur over possible world states. Neither do we experience a world where independent percepts in a scene conflict with or contradict one another; our experience is *coherent*. These seemingly obvious features of experience, while having some recent experimental support [18], are in opposition to optimal decision mak-ing theories [4] and some forms of the Bayesian brain hypothesis (for a recent review see [39]). At the least, a maximum a posteriori approximation must be made at the 'percept' stage moment-to-moment even if the underlying computations are Bayesian.

  Helpfully, as a heuristic, unitary coherent (UC) perception may avoid the NP-hard computational complexity of full Bayesian inference of having to consider every possible interpretation of a current scene [18]. The existence of immediate perception as UC places strong constraints on models of AM: if AM is to store aspects of immediate percepts, then the recall and storage of memories would also be UC rather than fully Bayesian. This in turn will place constraints on theories of reconceptualisation of past episodes, which may otherwise require wider storage of probabilities.

3.1 The Role of the Hippocampal Network in Autobiographical Memory

What is known about the neural circuits thought to underly AM in the brain? The extended hippocampal network, and its coordinated interaction with neo-cortex and subcortical structures, is thought to be the neurobiological sub-strate of the four AM functions outlined above. At a gross level, damage to the hippocampal-entorhinal cortex network famously causes anterograde amne-sia [30]. At a cellular level the microstructure of the CA3 region suggests it likely serves as a 'convergence zone enabling information from different sensory modalities to be associated together (an idea originated by Marr [27]).

Hippocampal neural activity shows attractor dynamics [53] enabling content-addressable re-activation of the entire stored representation through activation of a particular memory by the dentate gyrus (DG) [28]. This role of older DG granule cells in associative pattern completion is complemented by the finding that neurogenesis supports pattern separation by young DG granule cells [31]. Critically, for pattern completion in AM, it has been shown that patients with hippocampal lesions cannot recall nor imagine spatially coherent events [22], and this reconstructive process can be modelled at the level of neurons [9]. Together these results highlight the importance of this structure in the combination and coordination of memory sequences, tying together activity across distributed circuits in the brain. We base our model of episodic memory on the hippocampal system.

These findings from the study of biological memory suggest that to create effective SAM we could adopt the following strategy. Compression and pattern completion can be accounted for by an implementation of the predictive brain hypothesis [11]. This allows both efficient and accurate recognition of events from partial information, as well as affording decision making, planning and reconstruction through inference in the model. High level unitary representations of a scene or event are then coordinated into coherent sequences by iterative pattern completion and separation operations in a model based on the hippocampus.

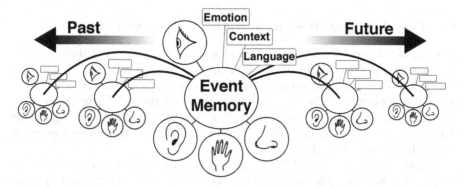

Fig. 2. Episodic memory in our model functions by combining disparate information across modalities into a single coherent percept from moment to moment. These transient percepts are monitored and coordinated over time into event sequences and episodic memories through iterative pattern completion and separation operations.

4 Capturing the Characteristics of Biological Autobiographical Memory in a Single Modelling Framework

4.1 Compression and Pattern Completion: The Predictive Brain Hypothesis

In distributed episodic memory the individual sensor or motor systems encode their own memories in a domain-appropriate format [44]. Encoding of sensory input has been shown to be very efficient in biological systems, with neural populations being tuned to the scene statistics of the world [48]. A further efficiency is not to simply encode the world as it is, but to develop a generative model of the world and encode only events that could not be predicted.

So called Predictive Brain hypotheses [11][19]have become popular in recent years, claiming to provide a unified computational account of a range of cognitive capacities and seeking to explain an increasing number of neuroscience phenomena e.g. [20][41][50].

Reconstructing a past event can be understood as one of the functions of such a prediction engine, using past experience to anticipate and make sense of events as they happen. Seeding such a mechanism with appropriate clues will allow retrieval of a past episode. The same system, operating continuously, can also serve to fill-in and enrich the representation of the current situation, preparing the platform for more informed and appropriate action. It is well known that people can fill in sensory scenes with expected information, and the reconstruction of memories from experienced or suggested fragments can lead to false memories even in individuals with 'highly superior' AM [33].

This powerful and efficient learning strategy would be highly appropriate for SAM and has not previously been applied to that domain.

4.2   Deep Learning: Towards a Synthetic Predictive Brain

A recent development in the field of machine learning is the arrival of practical deep learning systems [3] for abstracting information from data in an unsupervised way. Though neural networks have been investigated for decades [7, 45], recent developments of efficient training methods for deep (i.e. more than three layer) neural networks (DNNs) finally provide existence proofs of algorithms that can reconstruct (predict) complex sensory scenes in the manner required by predictive brain hypotheses.

Current excitement around DNNs can be ascribed to two prominent features. Firstly, they are now able, for the first time, to match or exceed human performance in certain benchmark pattern recognition tasks due to large-scale implementation on GPUs (see [3] for a review). Secondly, the kinds of invariant higher order representations developed by neural networks bare striking similarity to the tuning curves of neurons in higher order sensory cortex [55].

The unique processing architectures that result from training a DNN on data from a particular sensory domain could be described as a schema [21], an important concept in memory as discussed earlier. We propose that it is precisely the interaction of a compact episodic memory, and the sequential re-activation of an appropriate schema that underlies the compression and pattern completion capabilities of autobiographical memory. Therefore, it is to these DNN methods that we turn to compress sensory data streams and provide inputs to the episodic memory system.

4.3   Pattern Separation and Unitary Coherent Perception: Hippocampus as a Unitary Coherent Particle Filter

The vast majority of invasive neuroscience research uses rodent models. As a result, spatial reasoning or navigation tasks are often used to probe the biological foundations of memory. Many models of hippocampal function that are faithful to circuit-level details are therefore based on navigation or spatial reasoning [6]. Our model [16] has been developed as an effective algorithm for spatial inference (an idea recently expanded upon by Penny et al. [35]), while at the same time accounting for other known hippocampus-circuit phenomena such as the existence of sensory-pattern-specific and object-specific ('grandmother') cells. In robotics too, navigation has historically been a widely studied problem leading to widely-cited solutions to the Simultaneous Localisation And Mapping (SLAM). Our model may be viewed as a particular bio-inspired implementation of the SLAM algorithm, with EM algorithm steps mapping to neural activation and Hebbian learning respectively.

Our approach combines and extends research in both robotics and computational neuroscience fields: mapping well studied algorithms that have been shown to perform spatial reasoning and memory in navigation onto a neural circuit known to perform this function. Here we propose an update of this navigation model to autobiographical memory, echoing the well-articulated arguments of Buzsáki and Moser [8], based on the theory that the underlying computations for spatial and semantic memory are fundamentally the same.

The problem of locating oneself in an environment involves associating a given percept with a particular *place* (localisation) and encoding the transitions from one place to another by monitoring sensory differences, odometry and memories of previous traversals of the same environment. Autobiographical memory can be seen as essentially the same computation, but associating a given percept to an *event*, while transitions from event to event can be handled by recursive pattern separation and completion operations. The analogy between space and AM is seen clearly in the well-known mnemonic technique of the 'memory palace' which explicitly uses spatial memory to organise sequences of object percepts [49].

It had been proposed before that the behaviour of animals during learning could be modelled as a particle filter [15], and elsewhere that sequential learning in the hippocampus could be modelled with a Temporal Restricted Boltzmann Machine (TRBM) [51], a machine learning algorithm. Fox and Prescott [16] took a TRBM model for navigation and extended it to include a mapping to hippocampal-entorhinal cortex circuitry that included: unitary coherence; sensor/odometry inputs from entorhinal cortex to DG; CA3 associative learning of input-place mappings; CA1 decoded localisation posteriors (place cells); and a subiculum - septum 'lostness detection' loop (see Figure 3). In a series of papers the model was extended to include online learning with a biomimetic sub-theta cycle after-depolarisation [17], and scaled up to complete a real-world navigation task by processing visual inputs through SURF feature extraction and K-Nearest Neighbour clustering [46].

We consider decoding localisation posteriors (DG-CA3-CA1 in the model) to be a pattern completion loop, whereas lostness detection and correction (Sub-Sep) to be pattern separation. By compressing an agent's sensory history through deep generative networks instead of SURF features, and by expanding the range of inputs in the model to the full gamut of sensory modalities, we aim to develop a more complete account of synthetic autobiographical memory.

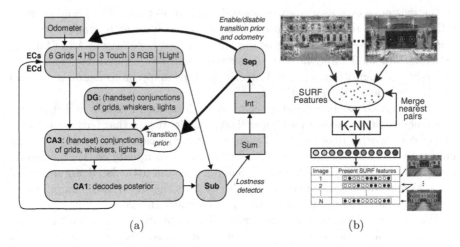

(a)   (b)

Fig. 3. (a) The Bayes filter hippocampus model, adapted from Fox and Prescott [16]. See main text for a synopsys of components. (b) A schematic of the SURF visual feature extraction step, constructing feature vector inputs to the Bayes filter hippocampus model, adapted from Saul et al. [46]. We aim to replace this feature extraction step with a range of modality-specific deep generative networks.

5 Conclusions and Future Directions

5.1 Advantages of Our Approach

The Deep Learning and UC hippocampus architecture presented here, based on basic systems and predictive coding theories of brain function, has a number of advantages over previous approaches. Using DNNs to learn predictive models of sensory scenes allows the agent to capture the full explanatory power of the data. Current approaches constrain how memories are stored at design time, limiting the practical utility of that information for new tasks in the future.

Our approach leverages state of the art machine learning approaches, bringing benchmark performance in speech and image processing to new modalities and new tasks. Code for many of these methods have been optimised for execution on GPUs, allowing the exploitation of cheap high-performance computing for real-time operation on robots. In addition, the UC hippocampus model has already been shown to work on real robot platforms with real-world sensory data, which is encouraging going forward.

5.2 Implementation Objectives: Social Interaction with an iCub Robot

An ideal testbed of our approach to SAM would be to embed the model on a mobile or humanoid robot to compress, parse and store rich streams of incoming data to improve ongoing processing performance. Our particular interest is in developing robots for social interaction, both for the inherent technical challenges and strong history of SAM in this task. Building on benchmarks established with existing SAM approaches [38][37], we hope to develop a system that can recognise people, remember interactions, and adapt interactions to specific people. Initial results of this work will be presented at the conference. Extensions to other problems in autonomous robots, and task domains such as life-long learning (as in [34]), will follow in due course.

Potential difficulties of implementing our approach are many and include: securing sufficient computational resources to run the algorithms (especially the DNNs) in real time; storage considerations during development of the data compression algorithms; interfacing the output of the SAM system with motor planning and narrative reasoning, which would ultimately involve encoding the memories themselves in terms of motor-control consequences; interfacing the UC episodic memory hippocampus model with rich predictive DNNs is non trivial; and event separation which would ultimately require a hierarchical implementation - a large problem in itself but one that is an interesting current area of research e.g. [47].

An additional active area of research is to advance machine learning algorithms to better capture the rich statistical structure of the world. Standard Deep Learning approaches are based on Restricted Boltzmann machines and their variations [3]. The latent representational structure in these models are relatively simple. The relationships between representations and latent variables within the learnt hierarchy, relationships that exist in the world and the data, are not captured by these models. Deep Gaussian processes [14], a recently proposed extension of Gaussian Process Latent Variable models [23], potentially provide a method for capturing rich hierarchical statistical relationships between latent variables. Integrating a Deep Gaussian Process with our UC hippocampus could substantially increase the power of this model.

Acknowledgements. The authors would like to thank the EU taxpayer for funding this research (EU grant no. 612139 WYSIWYD - "What You Say Is What You Did"), and our colleagues at the Sheffield Centre for Robotics for interesting discussions that stimulated this work.

References

[1] Anderson, J.R.: Act: A simple theory of complex cognition. American Psychologist 51(4), 355 (1996)

[2] Baddeley, A.: Essentials of Human Memory (Classic Edition). Psychology Press (2013)

[3] Bengio, Y.: Learning deep architectures for ai. Foundations and Trends in Machine Learning 2(1), 1–127 (2009)

[4] Bernardo, J.M., Smith, A.F.: Bayesian theory, vol. 405. John Wiley & Sons (2009)

[5] Berntsen, D., Rubin, D.C.: Understanding autobiographical memory: Theories and approaches. Cambridge University Press (2012)

[6] Bird, C.M., Burgess, N.: The hippocampus and memory: insights from spatial processing. Nature Reviews Neuroscience 9(3), 182–194 (2008)
[7] Bryson, A.E., Denham, W.F., Dreyfus, S.E.: Optimal programming problems with inequality constraints. AIAA Journal 1(11), 2544–2550 (1963)
[8] Buzsáki, G., Moser, E.I.: Memory, navigation and theta rhythm in the hippocampal-entorhinal system. Nature Neuroscience 16(2), 130–138 (2013)
[9] Byrne, P., Becker, S., Burgess, N.: Remembering the past and imagining the future: a neural model of spatial memory and imagery. Psychological Review 114(2), 340 (2007)
[10] Carello, C., Turvey, M.T., Kugler, P.N., Shaw, R.E.: Inadequacies of the computer metaphor. In: Handbook of Cognitive Neuroscience, pp. 229–248 (1984)
[11] Clark, A.: Whatever next? predictive brains, situated agents, and the future of cognitive science. Behavioral and Brain Sciences 36(3), 181–204 (2013)
[12] Cohen, M.S., Kosslyn, S.M., Breiter, H.C., DiGirolamo, G.J., Thompson, W.L., Anderson, A., Bookheimer, S., Rosen, B.R., Belliveau, J.: Changes in cortical activity during mental rotation a mapping study using functional mri. Brain 119(1), 89–100 (1996)
[13] Conway, M.A.: Autobiographical memory: An introduction. Open University Press (1990)
[14] Damianou, A.C., Lawrence, N.D.: Deep gaussian processes. arXiv preprint arXiv:1211.0358 (2012)
[15] Daw, N., Courville, A.: The pigeon as particle filter. In: Advances in Neural Information Processing Systems, vol. 20, pp. 369–376 (2008)
[16] Fox, C., Prescott, T.: Hippocampus as unitary coherent particle filter. In: The 2010 International Joint Conference on Neural Networks (IJCNN), pp. 1–8. IEEE (2010)
[17] Fox, C., Prescott, T.: Learning in a unitary coherent hippocampus. In: Diamantaras, K., Duch, W., Iliadis, L.S. (eds.) ICANN 2010, Part I. LNCS, vol. 6352, pp. 388–394. Springer, Heidelberg (2010)
[18] Fox, C., Stafford, T.: Maximum utility unitary coherent perception vs. the bayesian brain. In: Proceedings of the 34th Annual Conference of the Cognitive Science Society (2012)
[19] Friston, K.: The free-energy principle: a unified brain theory? Nature Reviews Neuroscience 11(2), 127–138 (2010)
[20] Friston, K., Kiebel, S.: Predictive coding under the free-energy principle. Philosophical Transactions of the Royal Society B: Biological Sciences 364(1521), 1211–1221 (2009)
[21] Ghosh, V.E., Gilboa, A.: What is a memory schema? a historical perspective on current neuroscience literature. Neuropsychologia 53, 104–114 (2014)
[22] Hassabis, D., Maguire, E.A.: The construction system of the brain. Philosophical Transactions of the Royal Society B: Biological Sciences 364(1521), 1263–1271 (2009)
[23] Lawrence, N.: Probabilistic non-linear principal component analysis with gaussian process latent variable models. The Journal of Machine Learning Research 6, 1783–1816 (2005)
[24] Lengyel, M., Dayan, P.: Hippocampal contributions to control: The third way. In: Advances in Neural Information Processing Systems, pp. 889–896 (2007)
[25] Lepora, N., Fox, C., Evans, M., Diamond, M., Gurney, K., Prescott, T.: Optimal decision-making in mammals: insights from a robot study of rodent texture discrimination. Journal of The Royal Society Interface 9(72), 1517–1528 (2012)

[26] Lisman, J., Grace, A.A., Duzel, E.: A neohebbian framework for episodic memory; role of dopamine-dependent late ltp. Trends in Neurosciences 34(10), 536–547 (2011)

[27] Marr, D.: Simple memory: a theory for archicortex. Philosophical Transactions of the Royal Society B 262, 23–81 (1971)

[28] McClelland, J.L., McNaughton, B.L., O'Reilly, R.C.: Why there are complementary learning systems in the hippocampus and neocortex: insights from the successes and failures of connectionist models of learning and memory. Psychological Review 102(3), 419 (1995)

[29] Miller, G.A.: The cognitive revolution: a historical perspective. Trends in Cognitive Sciences 7(3), 141–144 (2003)

[30] Milner, B., Corkin, S., Teuber, H.L.: Further analysis of the hippocampal amnesic syndrome: 14-year follow-up study of hm. Neuropsychologia 6(3), 215–234 (1968)

[31] Nakashiba, T., Cushman, J.D., Pelkey, K.A., Renaudineau, S., Buhl, D.L., McHugh, T.J., Barrera, V.R., Chittajallu, R., Iwamoto, K.S., McBain, C.J., et al.: Young dentate granule cells mediate pattern separation, whereas old granule cells facilitate pattern completion. Cell 149(1), 188–201 (2012)

[32] Neisser, U., Fivush, R.: The remembering self: Construction and accuracy in the self-narrative, vol. (6). Cambridge University Press (1994)

[33] Patihis, L., Frenda, S.J., LePort, A.K., Petersen, N., Nichols, R.M., Stark, C.E., McGaugh, J.L., Loftus, E.F.: False memories in highly superior autobiographical memory individuals. Proceedings of the National Academy of Sciences 110(52), 20947–20952 (2013)

[34] Paul, R., Rus, D., Newman, P.: How was your day? online visual workspace summaries using incremental clustering in topic space. In: 2012 IEEE International Conference on Robotics and Automation (ICRA), pp. 4058–4065. IEEE (2012)

[35] Penny, W.D., Zeidman, P., Burgess, N.: Forward and backward inference in spatial cognition. PLoS Computational Biology 9(12) 9(12), e1003383 (2013)

[36] Petit, M., Lallée, S., Boucher, J.D., Pointeau, G., Cheminade, P., Ognibene, D., Chinellato, E., Pattacini, U., Gori, I., Martinez-Hernandez, U., et al.: The coordinating role of language in real-time multimodal learning of cooperative tasks. IEEE Transactions on Autonomous Mental Development 5(1), 3–17 (2013)

[37] Pointeau, G., Petit, M., Dominey, P.F.: Successive developmental levels of autobiographical memory for learning through social interaction. IEEE Transactions on Autonomous Mental Development (forthcoming)

[38] Pointeau, G., Petit, M., Dominey, P.F.: Embodied simulation based on autobiographical memory. In: Lepora, N.F., Mura, A., Krapp, H.G., Verschure, P.F.M.J., Prescott, T.J. (eds.) Living Machines 2013. LNCS, vol. 8064, pp. 240–250. Springer, Heidelberg (2013)

[39] Pouget, A., Beck, J.M., Ma, W.J., Latham, P.E.: Probabilistic brains: knowns and unknowns. Nature Neuroscience 16(9), 1170–1178 (2013)

[40] Preston, A.R., Eichenbaum, H.: Interplay of hippocampus and prefrontal cortex in memory. Current Biology 23(17), R764–R773 (2013)

[41] Rao, R.P., Ballard, D.H.: Predictive coding in the visual cortex: a functional interpretation of some extra-classical receptive-field effects. Nature Neuroscience 2(1), 79–87 (1999)

[42] Rennó-Costa, C., Lisman, J.E., Verschure, P.F.: The mechanism of rate remapping in the dentate gyrus. Neuron 68(6), 1051–1058 (2010)

[43] Rochat, P.: Criteria for an ecological self. The Self in Infancy: Theory and Research 112, 17 (1995)

[44] Rubin, D.C.: The basic-systems model of episodic memory. Perspectives on Psychological Science 1(4), 277–311 (2006)
[45] Rumelhart, D.E., Hinton, G.E., Williams, R.J.: Learning representations by back-propagating errors. MIT Press, Cambridge (1988)
[46] Saul, A., Prescott, T., Fox, C.: Scaling up a boltzmann machine model of hippocampus with visual features for mobile robots. In: 2011 IEEE International Conference on Robotics and Biomimetics (ROBIO), pp. 835–840. IEEE (2011)
[47] Schapiro, A.C., Rogers, T.T., Cordova, N.I., Turk-Browne, N.B., Botvinick, M.M.: Neural representations of events arise from temporal community structure. Nature Neuroscience (2013)
[48] Simoncelli, E.P.: Vision and the statistics of the visual environment. Current Opinion in Neurobiology 13(2), 144–149 (2003)
[49] Spence, J.D.: The memory palace of Matteo Ricci. Penguin Books Harmondsworth (1985)
[50] Srinivasan, M.V., Laughlin, S.B., Dubs, A.: Predictive coding: a fresh view of inhibition in the retina. Proceedings of the Royal Society of London. Series B. Biological Sciences 216(1205), 427–459 (1982)
[51] Taylor, G., Hinton, G., Roweis, S.: Modeling human motion using binary latent variables. In: Schölkopf, B., Platt, J., Hoffman, T. (eds.) Advances in Neural Information Processing Systems, vol. 19 (2007)
[52] Tulving, E.: Elements of episodic memory. Oxford Psychology Series (1985)
[53] Wills, T.J., Lever, C., Cacucci, F., Burgess, N., O'Keefe, J.: Attractor dynamics in the hippocampal representation of the local environment. Science 308(5723), 873–876 (2005)
[54] Wood, R., Baxter, P., Belpaeme, T.: A review of long-term memory in natural and synthetic systems. Adaptive Behavior 20(2), 81–103 (2012)
[55] Yamins, D.L., Hong, H., Cadieu, C., DiCarlo, J.J.: Hierarchical modular optimization of convolutional networks achieves representations similar to macaque it and human ventral stream. In: Advances in Neural Information Processing Systems, pp. 3093–3101 (2013)

A Phase-Shifting Double-Wheg-Module
for Realization of Wheg-Driven Robots[*]

Max Fremerey[1], Sebastian Köhring[1], Omar Nassar[1], Manuel Schöne[1],
Karl Weinmeister[1], Felix Becker[2], Goran S. Đorđević[3], and Hartmut Witte[1]

[1] Technische Universität Ilmenau, Chair of Biomechatronics, 98693 Ilmenau, Germany
{maximilian-otto.fremerey,sebastian.koehring,omar.nassar,
manuel.schoene,karl.weinmeister,hartmut.witte}@tu-ilmenau.de
[2] Technische Universität Ilmenau, Chair of Technical Mechanics,
98693 Ilmenau, Germany
felix.becker@tu-ilmenau.de
[3] University of Niš, Control Engineering Department and Robotic Lab,
18000 Nis, Serbia
goran.s.djordjevic@elfak.ni.ac.rs

Abstract. Following mechatronic design methodology this paper introduces a
phase-shifting double-wheg-module which forms an alternative approach for
wheg-driven robots. During construction focus was placed on a smooth locomotion of the wheg-mechanism over flat terrain (low alternation of the CoM in
vertical y-direction) as well as the ability to overcome obstacles. Simulations
using the multi-body simulation tool ADAMS View® were executed in order to
prove estimations done. Using the results of simulation and calculation a first
prototype was designed, manufactured and tested by experiment.

Keywords: whegs, bio-inspired robotics, locomotion.

1   Introduction and Motivation

Stable and robust walking over unstructured and unknown terrain is still a challenging
task for a robot or an autonomously acting machine. Considering the issue the robot
or the autonomously acting machine has additionally to deal with different kinds of
obstacles, the implementation of (bio-inspired) legs for locomotion purposes displays
a possible approach.

TEKKEN II [6], CHEETAH-CUB robot [14] or BIGDOG [11] are formidable examples
for successful bio-inspired walking machines. Thereby it is quite interesting that
(while neglecting other design criteria) robustness can be achieved either by compli-

---

[*] This work is supported by NATO grant EAP.SFP 984560 and the 'Thüringer Innovationszen-
trum Mobilität (ThIMo)' in line with the project 'Silvermobility - Nahfeldmobilitätskonzepte
für die Altersgruppe 50+' (no. 2011FGR0127) funded by the European Social Fond (ESF),
and the Thuringian ministry of economy, work and technology TMWAT via 'Thüringer Auf-
baubank' (TAB).

A. Duff et al. (eds.): Living Machines 2014, LNAI 8608, pp. 97–107, 2014.
© Springer International Publishing Switzerland 2014

cated software algorithms (BIGDOG), embodiment (well-designed, even compliant mechanics like in CHEETAH-CUB), or a balanced mixture of both.

However, legged robots exhibits some disadvantages due to their kind of locomotion. One major issue is the high number of actuators used. BIGDOG requires 16 [11] and even CHEETAH-CUB featuring a bio-inspired pantograph-like mechanism (c.f. Witte & Fischer [4]) has two actuators per leg and therewith eight in sum [14]. In addition, the proper synchronization of the legs still needs effort in control.

Thus the authors want to highlight another possible, yet established approach for robust robot locomotion: the concept of whegs (like shown in section 2), and their improvement. Fig.1 illustrates the principle of a wheg (wheel + leg) and its derivation from a wheel. Thereby a wheg consists of a center (hub) and a different number of fixed spokes. According to application and aim, the spokes might be bendy or stiff.

Fig. 1. Illustration of a wheel and two whegs consisting of a centre and a different number n of spokes

Whegs as a combination of the wheel and leg principles provide advantages of both. They achieve a high horizontal velocity due to the use of a continuously rotating actuator, while keeping control effort low. Aside the spokes give the wheg the ability to overcome obstacles in a leg like manner: 1) A stance phase where the spoke is in contact with the substrate and 2) the swing phase where the spoke is not in contact with the substrate.

By controlling the speed and phasing of a wheg relative to the other whegs, a wheg-equipped robot simply adapts to different environments. Therefore whegs impose efficient dynamic stability and robustness in interacting with unexpected obstacles and terrain.

2   State of the Art of Selected Wheg-Driven Robots

The use of whegs, or rimless wheels respectively, is no new idea. A brief study of a rimless spoked wheel in 2D with a single degree of freedom was carried out by McGeer. In 1997 Coleman et al. [2] carried out a 3D mathematical analysis including 3D stability of a rimless wheel with rigid spokes rolling on an inclined ramp, considering the effect of gravity.

One of the first robots that used rimless wheels (or whegs) was PROLERO (PROtotype of LEgged ROver) by Alvarez et al. in 1996 [1]. Featuring six simple spokes, driven by a single actuator each, PROLERO preceded the WHEGS[TM] series as well as

RHEX [15]. RHEX, a robot having six rimless wheels with one passive compliant spoke each, was developed in 2001 in Bühler's group at McGill University by Saranli et al. [12]. Each wheg was driven by a single actuator allowing a change of rotational speed of phase of each wheg individually. Being redesigned several times and equipped with differently shaped whegs (e.g. a c-shaped semi-circle), RHEX is able to overcome rough terrain in a stable manner and even human-sized steps with up to 42 % slope [8].

An alternative approach was introduced in 2002 by Quinn et al. [9]. They developed WHEGS^{TM} I; a cockroach inspired robot featuring six whegs, each wheg having three instead of one spokes. Aside the required driving torque is delivered by a single drive and separated towards the whegs by gear mechanisms. Therewith the drive motor is able to run at constant speed instead of accelerating and decelerating like in RHEX during each walking cycle. In addition WHEGS^{TM} I has compliant axles which enable the robot to overcome a large variety of obstacles, stairs and barriers without changing parts of the design. Based on WHEGS^{TM} I Quinn and coworkers developed several other wheg driven robots like WHEGS^{TM} II, DAGSI WHEGS, MINIWHEGS, CLIMBING MINI WHEGS, or SEADOG [15].

Following the idea of having a robust robot for unknown environments DFKI Bremen introduced ASGUARD robot [3]. Its purpose is the use of whegs for autonomous outdoor missions on various harsh substrates. Therewith focus is placed on a proper communication and sensor strategy.

Another small sized wheg-driven robot is WARMOR [5]. Equipped with only four whegs each featuring four compliant spokes (Material: glass-fiber reinforced plastics, length: 75 mm, thickness 0.3 mm) the robot is able to deal with unstructured environments covered e.g. with gravel or debris. Due to a balanced mechanical design the robot uses the energy stored in the spokes during touch down for reducing the amount of energy required during next lift off. For robustness purposes the robot is also able to flip without any lack in performance.

3   Advantages and Disadvantages of Wheg Driven Systems

In summary whegs are an applicable variant if the environment the robot/technical system has to deal with consists of planes with different orientations and obstacles unknown in size and orientation. Therefore these obstacles might be of random type like stones or debris as well as structured like steps or platforms.

However due to the discontinuous spoke-to-ground contact of the wheg the center of mass (CoM) of the wheg (and therefore the robot) is forced to an alternating movement perpendicular to the substrate. This alternation is considered as smoothness of the wheg-driven system. It is best when the number of spokes n is infinite; the wheg becomes a wheel. The ability to overcome obstacles is poor. The smoothness is worst having whegs with n = 1, which results in a high vertical movement of the center of mass, but is the best way to deal with obstacles (compare fig. 3). However all previous wheg-driven systems show limitations either concerning the smoothness or the range of feasible obstacle height. Therefore modifications of hitherto common and widely spread wheg robot design are necessary.

For overcoming higher obstacles than a common wheg does, Hong et al. in 2006 developed a new concept called IMPASS (Intelligent Mobility Platform with Active spoke system) [7]. Here the focus was placed into individual foot placement to overcome even extreme irregular terrain. Therefore Hong et al designed a wheg where the length of each spoke could be actively changed (independent of each other).

To improve this smoothness of wheg-driven robots without losing the ability to overcome obstacles Shen et al. in 2009 introduced WHEEL-LEG HYBRID ROBOT [13]; a design where an additional actuator changes the shape of the wheg between a circle and a semi-circle (by two c-shaped spokes). So the WHEEL-LEG HYBRID ROBOT is able to use a wheel-like configuration while traveling over flat terrain and a wheg-like configuration when dealing with unstructured environment. This principle, but designed as a triple wheel, also was shown by Quaglia et al. in 2013 with their EPI.Q robot family [10].

4   Influence of Number of Spokes on Wheg Kinematics

Finding a proper ratio between an acceptable smoothness (means less alternation of CoM in y-direction) and good obstacle dealing seems obvious for an adequate wheg design. Therefore the calculation of both design criteria is introduced successively. Fig. 2 illustrates all relevant elements of a single wheg like spoke length r, the number of spokes n the angle between two spokes α (= 360°/n), the height y_{min} using to calculate the alternation y and (theoretically) feasible obstacle height h.

Due to fig. 2 y_{min} is used to calculate the alternation y. It is described by

$$y_{min} = r \cdot \cos(\alpha) \tag{1}$$

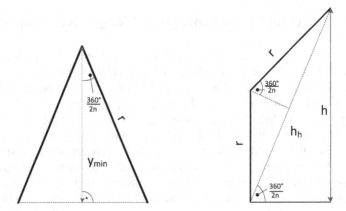

Fig. 2. Calculation of alteration (= smoothness) in y-direction using y_{min} (left) and (theoretically) feasible obstacle height h subject to number of spokes n (right)

$$y = r - y_{min} \qquad (2)$$

$$\alpha = \frac{360°}{(2 \cdot n)} \qquad (3)$$

For the theoretically feasible obstacle height h eq. 4 is used:

$$h = \sin(\alpha) \cdot sqrt\ (2r† \cdot (1 - \cos(\alpha)) \qquad (4)$$

The results of the alternation of the CoM in y-direction as well as calculation of theoretically feasible obstacle height subject only to number of spokes n is shown in fig 3. Here the percental results eliminating the need for spoke length r are shown.

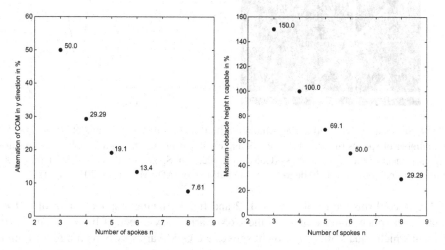

Fig. 3. Results of percental alteration (= smoothness) of CoM in y-direction (left) and theoretically feasible obstacle height h both subject to number of spokes n (right)

With increasing number of spokes the alternation of the CoM of the observed single wheg decreases which results in an improved smoothness and a smooth locomotion. But in addition the ability to overcome obstacles decreases, too. For example, having eight whegs, the CoM only alternates 7.61 % of spoke length in y-direction. However the feasible obstacle height amounts 30 % of spoke length only.

To prove these results and for having a model for further investigations (especially for torque required subject to the shape of the robot) a calculation-based simulation was done. Software used is the multi body simulation tool ADAMS View® 2010. Fig. 4 illustrates the investigated model of a single wheg featuring four and eight spokes. Focus of analyze is placed on the vertical movement in y-direction of the CoM of the wheg again. Therefore the wheg is forced to roll on a flat plane. The interaction between ground and each spoke is described by a common Coulomb friction force.

The investigation of the theoretically feasible obstacle height was not done due to the fact that during simulation the foot-down position becomes also important when dealing with feasible heights. Neglecting this influence in aid of an easy first estimation like shown in eq. (4), a comparison to simulation offers no further insights.

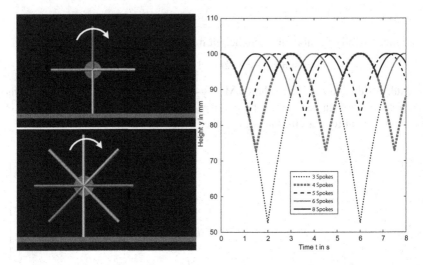

Fig. 4. Selected models used during simulation (left) and results of wheg models with a different number of spokes (three, four, five, six and eight spokes) rolling on a flat plane and resulting movement of the center of mass (CoM) in y-direction. Spoke length during simulation: 100 mm, rotational joint speed: 30 degree/sec, simulation tool: ADAMS View® 2010 (right).

Fig. 4 confirms the results of eq. 1, 2 and fig. 3. Having a spoke length of   100 mm and a spoke number of four, the CoM varies in y-direction by a maximum of 50 mm which equals 50 %. For eight spokes the CoM varies by a maximum of 8 mm (≈ 8 %).

By following eq. 1 to eq. 4 an acceptable ratio between smoothness and possible obstacle height is favored. In addition to still existing solutions described in section 2 and 3 the authors successively introduce the design approach of the phase-shifting double-wheg-module which offers the ability to change the number of spokes during real-time motion in an active way.

5   The Phase-Shifting Double-Wheg-Module

5.1   Mechanics

In order to change the ratio between smoothness of the robot and possible obstacle height the design of the phase-shifting double-wheg-module is introduced. It consists of a first wheg (outer wheg) which is rigidly attached to a shaft. Around this shaft a hallow axle turns a second wheg (inner wheg). Each shaft is driven by a distinct actuator. Following the results of the simulation each single wheg of the

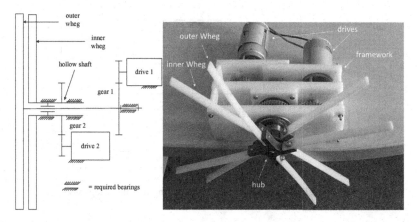

Fig. 5. Technical principle of the phase shifting double wheg-module (left) and assembled version (right)

double-wheg-module consists of four spokes. Therefore a change between a wheel-like eight spoke variant for smooth locomotion over flat terrain as well as a four spoke variant for crossing obstacles is feasible. Fig. 5 left illustrates the technical principle.

Having a light-weighted wheg module for later robotic purposes, the drives are connected to the shafts via a gear (here a 1:1 gear drive). For a quick change of material and geometries of spokes during experiments each wheg hub (material: steel) features a defined interface. During first experiments spokes are considered to be rectangular flat bars. Material used for spokes and framework is POM. Fig. 5 right illustrates the assembled module.

5.2   Electronics and Control

Due to the use of two motors (MAXON Inc., no. 222049) with planetary gear boxes the two whegs are able to turn independently of each other. Both motors are driven by a self-made motor-driver board and an ARDUINO UNO microcontroller board. Sensors used are rotary encoders from MAXON with 32 pulses per revolution. The control loop for a smooth setting of a phase difference between inner and outer wheg is displayed in fig. 6.

It consists of two parts. The first part is a common speed control for the outer wheg (wheg a). It is realized by PI control. The inner wheg (wheg b) is controlled by a cascaded control where the outer loop is the position control. Set value of control is the desired phase shift between the inner and the outer wheg which is compared to the present phase shift: the difference of positions of inner and outer wheg.

Therefore the position control consists of a P-part only while the inner loop of the outer wheg features the same speed control parameters like the inner wheg but feed-forward.

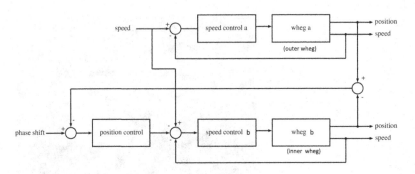

Fig. 6. Closed loop control of the phase-shifting double-wheg-module

5.3   Experiments

To test the phase-shifting double-wheg-module, several basic experiments are executed. During all experiments the wheg module was jacked up, so the spokes of the whegs did not contact the ground. The speed was set to 8 rounds per minute.

In fig. 7 the results of maintaining a predetermined phase-shift, here 0°, over the time ($\approx$ 30 s) are shown. All snapshots were taken from KINOVEA motion analysis.

Fig. 7. Closed loop control of the phase-shifting double-wheg-module

Like shown in fig. 7, the measured present phase-shift (successive named offset) between inner and outer wheg (inside the KINOVEA motion analyzing software) ranges up to 7° due to backlash of the gear and inaccuracy of manufaction (set value of phase-shift: 0°). This range is kept for almost 20 seconds (mid-image of fig. 7). After $\approx$ 30 seconds, offset amounts up to 10°. However, the offset does not further increase beyond 30 seconds.

For analyzing the ability of the wheg-module to handle disturbances, the wheg module was driven with a set value of 0° again. Although being disturbed (here selected spokes of a wheg were slowed down manually, cp. fig. 8) the offset between inner and outer wheg amounts up to 9°. The time for compensating the disturbance is about 2 s.

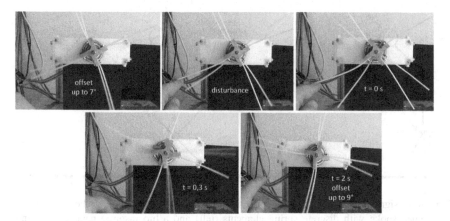

Fig. 8. Disturbance of the phase-shifting double-wheg-module

Changing between different set values was investigated in addition. Fig. 9 illustrates the performance of the phase-shifting double-wheg-module when changing the set value from 0° to 45° (time required: ≈ 0.8 s).

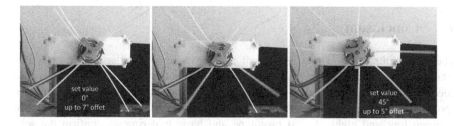

Fig. 9. Closed loop control of the phase-shifting double-wheg-module

Therefore the change of set values causes the desired change between a four-spoke and an eight spoke configuration of the phase-shifting double-wheg-module. While having a set value of 0°, offset amounts up to 7° again. A set value of 45° causes an offset up to 5° (measured angle: 50°) between inner and outer wheg.

6   Future Work

As revealed in section 5 the phase-shifting double-wheg-module needs improvement. Although showing the desired behaviour, the current control loop parameters require fine-tuning for reducing present offset. Therefore a model in MATLAB Simulink® will be introduced which represents the latest mechanical setup.

Aside the authors develop an experimental setup for the determination of the alternation of the CoM of the phase-shifting double-wheg-module as well as resulting ground reaction forces. In addition the authors currently investigate a different spoke design by simulation: Instead of having a flat rod a compliant spoke element is

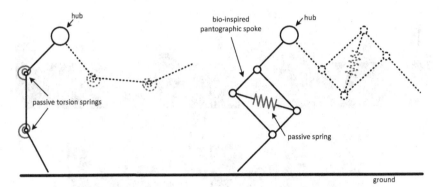

Fig. 10. Design variants of compliant spokes for the phase-shifting double-wheg-module, here a c-shaped spoke with discrete spring elements (left) and a bio inspired pantographic spoke (right) (only two spokes per wheg are shown)

desired. Thereby design variants range from a slightly c-shaped spoke with discrete spring elements to a bio-inspired pantographic spoke in order to reduce shocks and to make use of energy recuperation during locomotion.

7   Conclusion

The article introduces a new variant of a wheg design, the phase-shifting double-wheg-module. This module enables a smooth locomotion on flat ground as well as adequate obstacle dealing as a result of the use of an online changeable number of spokes. Due to executed calculations and simulation, the number of spokes switches between four and eight. A first prototype and basic experiments confirm the scheduled operation mode of the phase-shifting double-wheg-module.

Acknowledgement. The authors want to thank Danja Voges for her contribution in terms of layout of figures and plots.

References

[1] Alvarez, M., de Peuter, W., Hillebrand, J., Putz, P., Matthyssen, A., de Weerd, J.F.: Walking robots for planetary exploration missions. In: WAC 1996, Montpellier, France (1996)

[2] Coleman, M.J., Chatterjee, A., Ruina, A.: Motions of a rimless spoked wheel: a simple three-dimensional system with impacts. Dynamics and Stability of Systems 12(3), 139–159 (1997)

[3] Eich, M., Grimminger, F., Bosse, S., Spenneberg, D., Kirchner, F.A.: A Hybrid-Wheel Security and SAR-Robot Using Bio-Inspired Locomotion for Rough Terrain. In: International Workshop on Robotics for Risky Interventions & Surveillance of Environment (2008)

[4] Fischer, M.S., Witte, H.: Legs evolved only at the end! Philosophical Transactions of the Royal Society A: Mathematical. Physical and Engineering Sciences 365(1850), 185–198 (2007)

[5] Fremerey, M., Djordjevic, G.S., Witte, H.: WARMOR: Whegs Adaptation and Reconfiguration of MOdular Robot with Tunable Compliance. In: Prescott, T.J., Lepora, N.F., Mura, A., Verschure, P.F.M.J. (eds.) Living Machines 2012. LNCS, vol. 7375, pp. 345–346. Springer, Heidelberg (2012)

[6] Fukuoka, Y., Kimura, H., Cohen, A.H.: Adaptive dynamic walking of a quadruped robot on irregular terrain based on biological concepts. The International Journal of Robotics Research 22(3-4), 187–202 (2003)

[7] Hong, D., Laney, D.: Preliminary Design and Kinematic Analysis of a Mobility Platform with Two Actuated Spoke Wheels. In: Proceedings of the 2005 IEEE/RSJ Conference on Intelligent Robots and Systems (2006)

[8] Moore, E.Z., Campbell, D., Grimminger, F., Buehler, M.: Reliable stair climbing in the simple hex-apod'RHex'. In: Proceedings of the IEEE International Conference on Robotics and Automation, ICRA 2002, vol. 3, pp. 2222–2227. IEEE (2002)

[9] Quinn, R.D., Offi, J.T., Kingsley, D.A., Ritzmann, R.E.: Improved mobility through abstracted biological principles. In: IEEE/RSJ International Conference on Intelligent Robots and Systems, vol. 3, pp. 2652–2657. IEEE (2002)

[10] Quaglia, G., Oderio, R., Bruzzone, L., Razzoli, R.: A Modular Approach for a Family of Ground Mobile Robots. International Journal of Advanced Robotic Systems 10 (2013)

[11] Raibert, M., Blankespoor, K., Nelson, G., Playter, R.: Bigdog, the rough-terrain quadruped robot. In: Proceedings of the 17th World Congress, pp. 10823–10825 (2008)

[12] Saranli, U., Buehler, M., Koditschek, D.E.: RHex: A simple and highly mobile hexapod robot. The International Journal of Robotics Research 20(7), 616–631 (2001)

[13] Shen, S., Li, C., Cheng, C., Lu, J., Wang, S., Lin, P.: Design of a leg-wheel hybrid mobile platform. In: Proceedings of the IEEE/RSJ International Conference on Intelligent Robots and Systems, pp. 4682–4687 (2009)

[14] Spröwitz, A., Tuleu, A., Vespignani, M., Ajallooeian, M., Badri, E., Ijspeert, A.J.: Towards Dynamic Trot Gait Locomotion: Design, Control, and Experiments with Cheetah-cub, a Compliant Quadruped Robot. The International Journal of Robotics Research, 1–19 (2013)

[15] Biologically Inspired Robotics, Case Western Reserve University, http://biorobots.case.edu/projects/whegs/ (March 13, 2014)

Design Principles for Cooperative Robots with Uncertainty-Aware and Resource-Wise Adaptive Behavior

Carlos García-Saura, Francisco de Borja Rodríguez, and Pablo Varona

Grupo de Neurocomputación Biológica, Escuela Politécnica Superior
Universidad Autónoma de Madrid, 28049 Madrid, Spain
carlos.garciasaura@estudiante.uam.es, {f.rodriguez,pablo.varona}@uam.es
http://www.ii.uam.es/~gnb/

Abstract. In this paper we describe several principles for designing and implementing bio-inspired robotic collaborative search strategies. The design approach is particularly oriented for algorithms that can tackle search problems that deal with uncertainty, such as locating odor sources that have spatial and temporal variance. These kind of problems can be solved more efficiently by a reasonable amount of collaborative robots, and thus we propose a low-cost platform based on the open-source philosophy. The platform allows to evaluate different collective strategies that emerge from the interaction among robots that are aware of the uncertainty and make a wise use of all available sensors and resources. This includes an adaptive use of sensor signals and actuators, and a good communication strategy. We introduce GNBot, a flexible open-source robotic platform, and a virtual communication network topology approach to validate uncertainty-aware and resource wise bio-inspired search strategies.

Keywords: Cooperative robots, search algorithms, smart sensing, adaptive behavior, low-cost robotic platform, 3D printing, open-source, machine olfaction.

1 Introduction

Environmental exploration and monitoring, surveillance, target search and rescue, etc. are tasks that can be undertaken by cooperative robots (for a recent review on this topic see [1]). When the environment is fully characterized and the monitoring/search targets are stationary, straightforward rule-based heuristics including brute force approaches lead to efficient cooperative search. Bio-inspired approaches can come into play when search strategies have to deal with a large degree of unavoidable uncertainty.

The problem of odor source detection, localization and characterization using cooperative swarm robots has been studied for more than a decade considering mainly stationary odor sources [2,3,4,5,6]. Finding multiple odor sources that vary on time using cooperative search strategies has attracted less attention [7].

A. Duff et al. (eds.): Living Machines 2014, LNAI 8608, pp. 108–117, 2014.

This problem requires the use of search algorithm solutions that ensure that all sources are located in minimum time, avoiding obstacles, overlapping, and redundant interactions among robots which have limited resources to perform the search. Bio-inspired strategies that consider knowledge about the different range limitations of the employed sensory modalities, adaptive modality information integration, and feedback from the resource status can contribute to the design of algorithms that deal with these severe restrictions.

This is particularly relevant in the cases where there is uncertainty in the definition of the search region, in the effective range of the sensors, the efficient use of available energy resources (see Fig. 1), and on the latency and specific distribution of the odor sources [8,9,10]. Examples of such strategies are Lévy walks that combine clusters of short length steps with longer movements between them. In several animal species this strategy leads to optimal search [11]. Of course, Lévy walk strategies are changed or at least modulated when relevant information from different modalities becomes available during the search, including behavioral information from the same or other species. In this paper we propose a set of hardware and software design principles for open-source and low-cost collaborative robots to implement and validate a wide range of such bio-inspired adaptive strategies useful for uncertainty-aware robotic search.

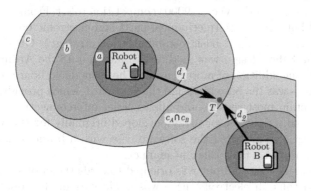

Fig. 1. *Handling uncertainty based on knowledge from the different sensor ranges and energy resources.* The figure illustrates different sensor ranges for two robots. The a-range represents the proximity sensor; b-range is the effective odor detection range for the electronic nose; and c-range represents each robot's estimation of its effective search area. In this example, the swarm must ensure that target T is sensed for the presence of odor sources. In the figure, robot B is closer to the target $(d_2 < d_1)$, but its battery is lower than robot A's. The uncertainty of robot B to get close enough so that point T is covered by the sensory range, combined with the cooperative strategy among both robots, could tip the balance towards the motion of robot A. In general, movement decisions could be modulated by the different range limitations of the employed sensory modalities, and by the limited resources of the robots at a given time.

2   Flexible Robotic Platform

For the robot design we used as a base existing open-source projects in which most mechanical parts are intended to be manufactured with a low-cost 3D printer. These are called *printbots* [12], and one example is the MiniSkybot[1]. Printbots provide an accelerated design path not only for creating basic educational robots [13] but also for making robots that can be used for research, such as the platform presented in this paper. With this technique it is possible to easily reuse, develop and incorporate the most useful parts of previous designs in order to fulfill the requirements of a robot needed for a given task.

Regarding swarm robotics, printbots have been preferred over other solutions for both the reduced manufacture cost and, most importantly, for the high adaptability of the designs to the search problems. For instance, the usage of 3D printed structures opens the possibility of easily testing the performance of various spatial distributions of the odor sensors in the robot, by manufacturing custom parts that can accommodate the different configurations.

2.1   Specifications of the GNBot

The proposed robot design (see Fig. 2) is a derivative of the ArduSkybot[2], an educational printbot based on the Arduino UNO[3] electronic board, and it also incorporates work from the Vector-9000[4] competition robot. From these projects, we integrated both hardware (mechanical parts and electronic boards) and software (robot firmware and a basic communication strategy). On the hardware side, the ArduSkybot design was modified to incorporate the Arduino MEGA[5] board to allow the addition of more sensors. The Printshield board -designed for the ArduSkybot- was the base to design the GNBoard, which provides a compact solution to contain most sensors and allows easier interconnection. The GNBot also incorporated a light sensor array developed originally for the Vector-9000, and this project also provided the software base with a framework to interface with the computer in a fault-tolerant manner.

The selection of sensory input was oriented towards the odor source localization task, and thus an electronic nose was made part of the robot. The odor sensor works with a heater element whose temperature can be controlled electronically by a power transistor placed for this purpose on the GNBoard. Sensor temperature control allows the usage of different modulation strategies to enhance sensor performance and to allow adaptation to the intensity of odor sources [14]. The mounted sensor is the *TGS-2600* gas detector from Figaro[6], but any sensor with a similar polarization scheme could be used.

---

[1] http://www.iearobotics.com/wiki/index.php?title=Miniskybot_2
[2] https://github.com/carlosgs/ArduSkybot
[3] http://arduino.cc/en/Main/ArduinoBoardUno
[4] https://github.com/carlosgs/carlosgs-designs/tree/master/
   Vector-9000-a-fast-line-follower-robot
[5] http://arduino.cc/en/Main/ArduinoBoardMega
[6] http://figarosensor.com/

Next, the other key element that had to be taken into account was the monitorization of battery voltage in order to give robots knowledge of the status of their energy resources. The implementation of this sensing capability was done by placing a resistive divider that adapts battery voltage to the $[0, 5]V$ range that can be measured with the Arduino board. Robots also incorporate an IR analog rangefinder (ref: *GP2Y0A21*) to avoid collision with obstacles and other robots within the search environment, as well as a light sensor array intended to be used to identify light landmarks. Finally, an electronic compass (ref: *HMC5883L*) provides the orientation knowledge. This way, each robot and the swarm can have a basic multi-modal perception of their status and location in the search area.

Fig. 2. *The GNBot and GNBoard.* List of sensors present in the robot: *a)* TGS-2600 odor sensor, *b)* IR rangefinder, *c)* LDR light sensor array, and *d)* electronic compass module. Battery voltage is also monitored by the power supply *(e)*. The base actuators are two-servo motors and the wireless interface is based on a ZigBee module (*XB24-BWIT-004*).

The motion of the robot is achieved using two continuous rotation servo-motors (*SM-S4303R*) as the main actuators. Servo-motors provide a compact and low-cost solution to achieve the digital speed control needed, and they are frequently used in bio-inspired robot designs [15,16]. The main downside of this kind of actuator is the high current demand -particularly during transient motions- which requires the use of an adequate power supply. For our design we decided to use a switching power supply rather than a linear regulator, provided the much higher efficiency. Switching power regulators also have a broad input voltage range, allowing to make better use of the full capacity of the batteries since they can be connected in series without negative effect in the performance. The power supply of the GNBoard uses an *LM2576-5* switching regulator that is capable of delivering up to 3A, which is more than sufficient to cover the energy demand of the entire robot.

The development of this platform is kept open-source and accessible via GitHub[7] to allow re-use and community evolution of the project.

2.2   Adaptability of Sensors and Actuators

Collaborative search algorithms may require adaptive control of the sensory input in order to maximize sensitivity, and the proper management of actuator elements to reduce energy demand during the exploration. A first approach can include monitoring events on the battery performance (see Fig. 3) to implement basic decision making on the search strategy.

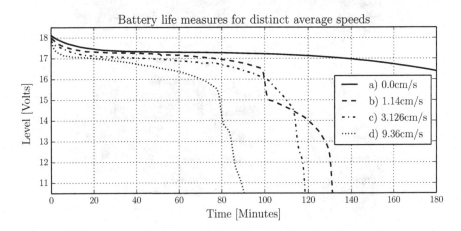

Fig. 3. *Battery life measures for distinct average speeds.* One GNBot powered by two fully-charged 9V 300mAh rechargeable batteries in series configuration was the setup used for this study. A simple obstacle-avoidance routine was used to ensure the continuous motion of the robot. *Operating time: a) 255min b) 131min c) 119min d) 91min. Distance range estimation: a) 0m b) 90.29m c) 222.26m d) 505.44m.*

The measurements in Fig. 3 show that servo-motors do not have a linear *power/performance* relationship, as they maintain a very high power consumption even at very low speeds. Using this information the search strategy can be adapted to maximize search range. This can be done, for instance, by disabling

---

[7] https://github.com/carlosgs/GNBot

the motion of the robot when waiting for the olfactory sensor measurement to be completed. The knowledge of various differentiated stages on the battery life is useful information that can be incorporated to trigger a decision-making event during search, to optimize the use of energy resources. For example, abrupt changes on battery level can trigger a speed change, shutting down high consumption sensors such as the electronic nose, or altering the decision of which robot approaches a given target (cf. Fig. 1). In the case of search strategies based on Lévy walks, the motion decision extracted from the Lévy distribution can also be modulated in real time with the battery life estimation.

Apart from bio-inspired energy management, sensing often requires to implement gain-control in different sensory modalities to better adapt to variability in each environmental context. Modulation of the temperature of single odor sensors can serve as a virtualization of sensor arrays and it also allows for spatio-temporal encoding which can be needed for more advanced detection and classification tasks. The GNBoard is conveniently provided with the circuitry necessary to achieve this sort of adaptive modulation in odor sensing.

2.3 Communication among Robots

In the case of collaborative swarm robots, one of the key design decisions that needs to be made is the selection of a proper communication method. Not only there is a trade-off between the *working range, maximum information throughput and cost*, but there are some other facts that must also be taken into account:

- *Working environment:* Radio-Frequency (RF) communications generally provide a robust system for most applications, but sometimes other solutions can provide a better balance between performance and cost. For instance, for underwater applications RF signal attenuation may become an issue, and the use of sound waves, light pulses, or even tethering with a cable become reasonable options.
- *Power requirements, adaptivity and remote-end sensing:* Since mobile robots have very limited energy resources, an efficient system should be generally preferred in order to maximize the operation time. In the same line, some interfaces provide a way to switch among different power schemes in real time, as well as having the capability to measure the signal intensity received from the other end. These should be preferred since the adaptivity in communications is key towards an optimal usage of energy resources.
- *Networking capability:* Not all communication systems offer the possibility to address data to different end nodes, and this is a must to allow scaling up the size of the swarm. This issue is further discussed in the next section.

ZigBee[8] (*IEEE 802.15.4*) has been chosen as an integrated solution, since it is a highly configurable platform that is compatible with most of the points detailed above. It provides an excellent networking layer, and since it is designed for low power applications, ZigBee has the possibility of adapting RF energy usage

---

[8] http://www.digi.com/technology/rf-articles/wireless-zigbee

in real time. The main downsides of this solution are the maximum throughput rates and the timing constrains, that must be taken into account when considering the use of more complex sensory input such as real time video streams.

Communication with the robots through these modules is asynchronous via User Datagram Protocol (UDP), and thus real time event handling is critical. The links must be fault tolerant, which can be achieved by using redundancy to ensure that messages are received and processed correctly by each node, and soft-state should be preferred to avoid deadlocks and allow fast recovery. Reliability in communications is a key element to implement the virtualized network topology explained in next section.

2.4   Virtualization of the Network Topology

Uncertainty-aware and resource-wise collaborative search algorithms may rely on very different network topologies to maximize the chances of successful search and minimize time or energy consumption while, at the same time, dealing with context-specific communication range restrictions. In particular, the spatial scale of the search problem and the actual detection range of the odor sources are important factors to design the network topology, that could be changed for instance according to the energy level information. As shown in Fig. 4, a base tree topology makes it possible to emulate and test many different architectures.

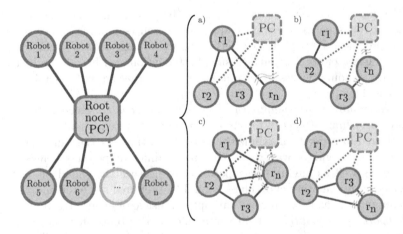

Fig. 4. *Network topology.* The star architecture shown on the left serves as the base to emulate a wide range of topologies. While underlying communications are centralized, the implemented algorithms can have very different requirements. As all information flows through the central computer, virtualized connectivities between robots can be defined in order to achieve various architectures such as *a) tree/star, b) line, c) fully interconnected*, or *d) mesh*, among others. The emulation of many other characteristics of physical links (*variable delays, jitter, data corruption, packet loss...*) can be used to test the resilience of the implemented search algorithms in a controlled manner.

The approach used is to abstract all the calculations to a root computer, effectively using each robot as a peripheral. A centralized infrastructure is used to command each autonomous robot independently, but a convenient layer of abstraction allows testing algorithms that are decentralized (see panels *a-d* in Fig. 4). Main advantages of this approach are:

- First, it is able to reduce costs since robots are kept simple, with reduced computational ability. For a fixed funding, cutting down the cost of each robot makes it possible to create more of them and thus have a bigger swarm.
- Second, as all data flows through the root node, it can be logged and analyzed in order to evaluate the performance of each algorithm and allows easier debugging. Having all information in one place is particularly convenient when testing distributed algorithms.
- Third, it provides a layer of abstraction. The code that specifies the behavior of each robot runs on the central computer, and thus a high-level programming language (such as Python[9]) can be used. This way it is possible to focus on developing the algorithms rather than dealing with the limitations of memory and power of the micro-controller on board each robot.
- Finally, it is important to emphasize that the centralized architecture supports a large dynamic range of complexity of the algorithms, that can be kept simple (i.e. chaotic search [17]) or complex (i.e. particle filtering [4]).

The emulated topologies could be dynamically reconfigured in real time to adapt to the search requirements (i.e. groups of robots may establish separate sub-networks when getting far from others to do local search, and later share the search results with the rest of the group). The chosen ZigBee communication protocol natively supports the deployment of such architectures in the real world, which makes it very convenient towards the actual implementation of the virtually-optimized topologies.

3   Discussion

In the case of odor searching tasks, collaborative robot strategies have to face uncertainty arising from the definition of the search area, the latency of the targets, the actual range and efficiency of the different sensory modalities, the available resources and expected duration, etc. Bio-inspired strategies that use adaptive context-dependent sensory integration and decision making can help to implement collaborative search under severe uncertainty conditions.

Many behavioral and neuroethological studies emphasize the capacity of different animals to switch between different navigation and search strategies (e.g. see for review [18,19]). Animals have an extremely flexible and efficient sensory-motor integration that results in successful search strategies that guarantee survival in situations where uncertainty is large such as the localization of odor sources in order to find nutriment. The incorporation of relevant information

---

[9] http://www.python.org/

regarding various sensor modalities into Lévy walk strategies, as in some animal species [11], provides inspiration for the design of novel search strategies for collaborative robots that work environments with such kind of uncertainty restrictions.

The wide variety of possibilities for designing and implementing bio-inspired searches which rely on different sensory information integration and motor decision making, calls for novel flexible robotic platforms that can meet the requirements arising from handling uncertainty and resource availability within these paradigms. The approach discussed in this paper addressed this issue from the perspective of a solution that allows maximum flexibility, scalability and reuse. We have proposed several hardware and software design principles and an open-source robotic platform, the GNBot, which allow the usage of low-cost robots for the design and validation of collaborative and adaptive search algorithms.

Acknowledgments. Authors acknowledge support from MINECO IPT-2011-0727-020000, TIN2012-30883 and TIN2010-19607, as well as from a 2013/14 Collaboration Grant from the Spanish Ministry of Education, Culture and Sports.

References

1. Hu, J., Xu, J., Xie, L.: Cooperative Search and Exploration in Robotic Networks. Unmanned Systems 1(1), 121–142 (2013)
2. Dunbabin, M., Marques, L.: Robots for Environmental Monitoring: Significant Advancements and Applications. IEEE Robotics Automation Magazine 19(1), 24–39 (2012)
3. Hayes, A., Martinoli, A., Goodman, R.: Swarm robotic odor localization. In: Proceedings of the 2001 IEEE/RSJ International Conference on Intelligent Robots and Systems, vol. 2, pp. 1073–1078 (2001)
4. Marques, L., Nunes, U.: Particle swarm-based olfactory guided search. Autonomous Robots 20, 277–287 (2006)
5. Marjovi, A., Marques, L.: Multi-robot olfactory search in structured environments. Robotics and Autonomous Systems 59(11), 867–881 (2011)
6. Marjovi, A., Marques, L.: Optimal spatial formation of swarm robotic gas sensors in odor plume finding. Autonomous Robots 35(2-3), 93–109 (2013)
7. Mcgill, K., Taylor, S.: Robot algorithms for localization of multiple emission sources. ACM Computing Surveys 43(3), 1–25 (2011)
8. Edwards, A.M., Phillips, R.A., Watkins, N.W., Freeman, M.P., Murphy, E.J., Afanasyev, V., Buldyrev, S.V., Da Luz, M.G.E., Raposo, E.P., Stanley, H.E., Viswanathan, G.M.: Revisiting Lévy flight search patterns of wandering albatrosses, bumblebees and deer. Nature 449(7165), 1044–1048 (2007)
9. Torney, C., Neufeld, Z., Couzin, I.D.: Context-dependent interaction leads to emergent search behavior in social aggregates. Proceedings of the National Academy of Sciences of the United States of America 106(52), 22055–22060 (2009)
10. Hein, A.M., McKinley, S.A.: Sensing and decision-making in random search. Proceedings of the National Academy of Sciences 109(30), 12070–12074 (2012)
11. Reynolds, A.M.: Effective leadership in animal groups when no individual has pertinent information about resource locations: How interactions between leaders and followers can result in Lévy walk movement patterns. EPL 102(1), 18001 (2013)

12. Gonzalez-Gomez, J., Valero-Gomez, A., Prieto-Moreno, A., Abderrahim, M.: A new open source 3d-printable mobile robotic platform for education. In: Rückert, U., Joaquin, S., Felix, W. (eds.) Advances in Autonomous Mini Robots, pp. 49–62. Springer, Heidelberg (2012)
13. García-Saura, C., González-Gómez, J.: Low cost educational platform for robotics, using open-source 3d printers and open-source hardware. In: ICERI 2012 Proceedings of the 5th International Conference of Education, Research and Innovation, November 19-21, pp. 2699–2706. IATED (2012)
14. Yáñez, D.J., Toledano, A., Serrano, E., Martin de Rosales, A.M., de Rodriguez, F.B., Varona, P.: Characterization of a clinical olfactory test with an artificial nose. Frontiers in Neuroengineering 5(1) (2012)
15. Meyer, F., Sproewitz, A., Berthouze, L.: Passive compliance for an (RC) servo-controlled bouncing robot. Advanced Robotics 20(8), 953–961 (2006)
16. Herrero-Carrón, F., Rodríguez, F.B., Varona, P.: Bio-inspired design strategies for central pattern generator control in modular robotics. Bioinspiration & Biomimetics 6(1), 016006 (2011)
17. Zhu, K., Jiang, M.: An improved artificial fish swarm algorithm based on chaotic search and feedback strategy. In: International Conference on Computational Intelligence and Software Engineering, CiSE 2009, pp. 1–4 (December 2009)
18. Dollé, L., Sheynikhovich, D., Girard, B., Chavarriaga, R., Guillot, A.: Path planning versus cue responding: a bio-inspired model of switching between navigation strategies. Biological Cybernetics 103(4), 299–317 (2010)
19. Arleo, A., Rondi-Reig, L.: Multimodal sensory integration and concurrent navigation strategies for spatial cognition in real and artificial organisms. Journal of Integrative Neuroscience 6(3), 327–366 (2007)

Insect-Inspired Tactile Contour Sampling Using Vibration-Based Robotic Antennae

Thierry Hoinville[1], Nalin Harischandra[1], André F. Krause[1,2], and Volker Dürr[1]

[1] Department of Biological Cybernetics, University of Bielefeld, Germany
[2] Cognitive Interaction Technology – Center of Excellence (CITEC),
University of Bielefeld, Germany
thierry.hoinville@uni-bielefeld.de

Abstract. Compared to vision, active tactile sensing enables animals and robots to perform unambiguous object localization, segmentation and shape recognition. Recently, we proposed a bio-inspired, CPG-based, active antennal control model, so-called Contour-net, which captures essential characteristics of antennal behavior in climbing stick insects. In simulation, this model provides a robust and effective way to trace contours and classify various 3D shapes. Here, we propose a physical robotic implementation of Contour-net using vibration-based active antennae. We show that combining tactile contour tracing with vibration-based distance estimation yields fairly accurate localization of contact events in 3D space.

Keywords: Tactile contour tracing, Tactile sampling, Tactile localization, Antenna, Hopf oscillator, Stick insect.

1 Introduction

Tactile sensing enables humans and animals to actively perceive extensive properties of surrounding objects that are hard, if not impossible, to obtain through vision [10,18,8,13]. Notable examples are surface texture, chemical properties, temperature, compliance and humidity. However, even regarding properties that are known to be routinely processed from visual input, like object contour and distance, direct physical contact yields unambiguous robust estimates. In fact, the sense of touch is independent of light conditions and directly provides reliable three-dimensional information useful for object localization, segmentation and shape recognition.

Insect antennae and mammal whiskers have inspired robotic research in the area of tactile sensors. Early work by Kaneko et al. [3] describes an artificial antenna using a flexible beam capable of detecting 3D contact locations and surface properties. Russel and Wijaya [15] applied an array of whiskers that passively scan over an object to recognize its shape using advanced pre-processing contact-points and decision trees. Robotic whisker arrays were used by Solomon and Hartmann [17] to generate 3D spatial representations of the environment and extract object shapes. Related work done by Kim and Möller [4] used a vertical whisker array to detect the vertical shape and curvature of objects. Sullivan

A. Duff et al. (eds.): Living Machines 2014, LNAI 8608, pp. 118–129, 2014.

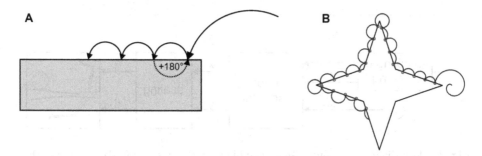

Fig. 1. **A**: Tactile contour-tracing template. Each contact induces a 180° phase forwarding of a circular movement. Combined with velocity control of a tactile probe, the resulting change in movement velocity and direction causes the probe to "bounce off" the surface at every contact event, resulting in a successive scan along the object's shape. **B**: 2D simulation of contact-triggered contour tracing. *Black*: star-shaped object. *Blue*: trajectory of the antennal tip. *Red dots*: contact locations.

et al. [19] performed Bayesian classification of surface texture using an array of actuated hair-like elements. Drawing inspiration from the American cockroach, Lee et al. [9] developped an artificial antenna and a control model for high-speed wall following of a wheeled robot. Recently, Mongeau et al. [11] added hairs to this antenna and studied implications for mechanics and sensing.

Recently, we proposed a bio-inspired, CPG-based, active antennal control model called "Contour-net" [6]. This template model [9] captures essential characteristics of antennal behavior in walking and climbing stick insects: rhythmic searching movements using an antenna with two revolute joints with strong coupling and a contact-triggered switch to local sampling [16,5,7]. Specifically, the antennal velocity is driven by a CPG modulated by a single, binary sensor signal. Upon the first contact with an obstacle (Fig. 1A), the behavior switches from a broad antennal searching pattern to a local contour sampling pattern with higher frequency and lower amplitude. Each contact event triggers a 180° phase shift in the CPG, which leads the antenna to "bounce off" the surface of the object repeatedly. In [6], we have shown in simulation that this simple model provides a robust and effective way to trace contours of 2D and 3D objects of various shapes (Fig. 1B). Furthermore, sampled contact points may directly be used for neural network-based shape classification [6].

In this paper, we extend the previous work by implementating Contour-net on a physical robotic platform equipped with two vibration-based active antennae as developped in [12]. To assess the effectiveness of our implementation, we report its performance relating to 3D localization.

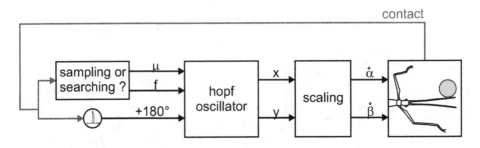

Fig. 2. Block diagram of Contour-net. The binary contact signal (red) determines the frequency and amplitude of the Hopf Oscillator. The discrete impulse block guarantees that the oscillator phase is forwarded once by 180°, resulting in a movement away from the object surface. The output of the oscillator is then scaled and drives a simulated or real robot antenna using velocity control of antennal joints.

2   Methods

2.1   Contour-Net

Figure 2 shows a block diagram of the Contour-net model. Like in [6], the antennal rhythmic pattern is generated by a Hopf oscillator defined in Cartesian space such that

$$\dot{x} = \gamma \left(1 - x^2 - y^2\right) x - 2\pi f y \tag{1}$$
$$\dot{y} = \gamma \left(1 - x^2 - y^2\right) y + 2\pi f x \tag{2}$$

with γ defining the speed of convergence to the limit cycle, i.e. unit circle of frequency f.

Angular velocity commands applied to the antenna, $\dot{\alpha}$ and $\dot{\beta}$, are simply determined from scaling the Hopf oscillator state variables x and y as follows:

$$\dot{\alpha} = s_1 x \tag{3}$$
$$\dot{\beta} = s_2 y \tag{4}$$

where s_1 and s_2 are the maximum velocity commands for the antennal pan and tilt joints, respectively.

After a contact is detected by the antenna, the parameters f, s_1, s_2 are instantly changed such that antennal movement switches from a broad elliptic "searching pattern" to a local circular "sampling pattern" of higher frequency and lower amplitude. Once a given time span has passed without encountering a further contact event, the parameters are changed back to the "searching pattern". Each contact event results in a 180° phase shift implemented by inverting both state variables of the Hopf oscillator, i.e. $x \to -x$ and $y \to -y$.

2.2   Robotic Hardware and Software

We implemented Contour-net on an Erratic mobile robot (Videre Design LLC). This differential wheeled platform measures 40 cm (L) × 37 cm (W) × 18 cm (H)

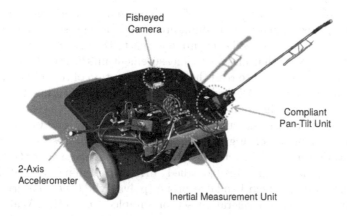

Fig. 3. The insectoid robot used to implement Contour-net

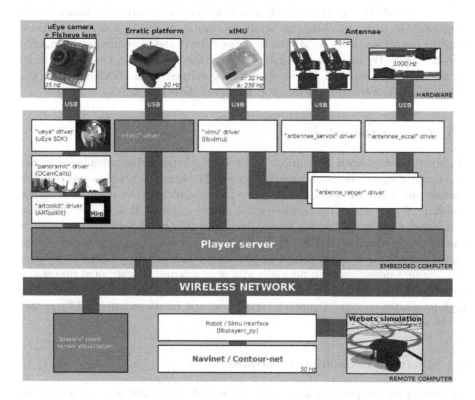

Fig. 4. Robotic software architecture based on Player server. Six drivers (i.e. Player plugins) had been developed, including the image processing pipeline and the "antenna_ranger" handling tactile processing for contact detection and localization. On the client side, a common Python interface has been developed for the Player architecture and the Webots simulator. Contour-net and Navinet controllers can therefore be transferred seamlessly between the simulation and reality.

and weighs 15 kg. For the purpose of studying multisensory integration, this robot (Fig. 3) is equipped with an omnidirectional vision system and two active tactile antennae, but only the latters are relevant to the present work and will be described below. Note that an inertial measurement unit x-IMU (x-io Technologies Ltd.) may also be used to read the global orientation of the robot, or for self-motion compensation [2]. All devices are connected via serial bus USB2 to the embedded computer. This dual-core 1.6 GHz system handles the communication with the various devices and supports the necessary image and tactile low-level processing. As depicted on Fig. 4, all low-level sensory processes are organized within Player's server architecture (http://playerstage.sourceforge.net), as a stack of connected plug-ins (so-called "drivers") running in parallel.

Each antenna is a 40 cm bendable probe (polyacrylic tube) sensorized at its tip only, by a two-axis acceleration sensor sampled at 1 kHz (Analog device ADXL321). The two antennae are mounted onto two pan-tilt-units (PTU), each allowing active rotation around two orthogonal axes, controlled via two servo motors (HiTEC HS-755MG) driven by a Polulu mini maestro, 12-channel, USB controller. In order to obtain an adaptive compliant tactile system, a current limiting circuit was incorporated to the power line to the servo motors driving the pan-tilt-units. By limiting the current, maximum torque that can be produced by the servos was limited, effectively avoiding damage to the flexible polyacrylic tubes in an emergency situation (e.g. if contact detection fails).

2.3   Contact Detection

Antennal contact with any object is detected whenever the magnitude of the acceleration signal reaches a given threshold. The accelerometer measures acceleration along two orthogonal axes, therefore considering the magnitude of the resultant vector allows contact detection at any angle of attack. The threshold was tuned empirically to 15 g which gives relatively good performance, even in case of gentle contacts (Fig. 5). This simple method relies to a large extent on the motor compliance we introduced. In fact, in preliminary experiments, we found that contact detection with normally stiff servos required additional processing to achieve reasonable performance. In particular, we had to use a Chebyshev high-pass filter to extract contact events from the interfering natural frequency of the servos.

2.4   Distance Estimation Procedure

Depending on the position of contact, the length of the free oscillating segment of the antenna differs. This results in damped oscillations of different frequency composition which can be retrieved from the acceleration signal. Either the two predominant frequency components (due to harmonics) or the entire frequency spectrum of the damped oscillation can be taken into account for estimating the position of contact along the antenna [12]. In this study, we used a subsampled version of the entire frequency spectrum.

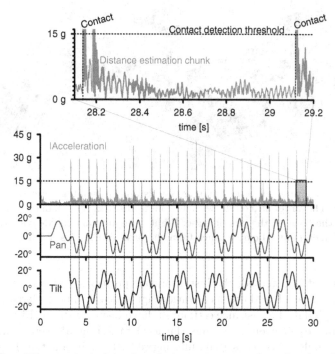

Fig. 5. Repetitive contact detection during contour-tracing. Acceleration measured at the antennal tip (grey and green), pan and tilt angles of the servo-motors (red and blue) during contour-tracing. Antennal contacts are detected whenever the acceleration magnitude reaches a threshold (dashed line at 15 g). As soon as a contact has been detected, the pan and tilt angles are recorded (see insert on top), and a chunk of acceleration data begin to be buffered for later processing, such as distance estimation or material recognition. The buffer is reset whenever the acceleration signal crosses the threshold, even if on contact is followed by multiple threshold crossings (as in the example shown in the insert). This way, only the damped oscillations occurring after contact are captured. Each complete chunk of acceleration data is then sent to a 3D contact localisation routine, together with the corresponding pair of pan and tilt values (black dots on pan-tilt traces).

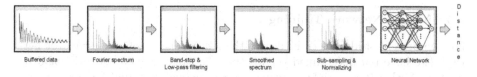

Fig. 6. The flow chart illustrating the sensor signal processing architecture. The buffered raw signal is used to compute the Fourier spectrum. Then the frequencies above 250 Hz are set to zero to remove high-frequency noise. The frequency band of 38-58 Hz is set to zero, too, removing 50 Hz cross-talk from the servo-motors. Then the spectrum is smoothed and sub-sampled eightfold, yielding 64 data points. Finally, these are normalised before being fed into the 4-layered neural network.

Fig. 7. Data acquisition setup for training the distance estimator network of the left antenna. Left: The antenna collide with the large disc (25 cm diameter) at 20 cm distance from the basis of the plastic tube. Right: The antenna collide with the small disc (10 cm diameter) at 10 cm distance from the basis of the plastic tube.

Once a contact is detected, the instantaneous positions and velocities of the pan-tilt-units are saved, and the acceleration signal starts to be buffered in memory until enough data for distance estimation have been captured (Fig. 5). Since a single antennal touch event may cause the acceleration signal to cross the detection threshold multiple times, signal buffering restarts each time the threshold is reached. This insures that the memorized chunk contains only data corresponding to after-contact damped oscillations. The distance estimation procedure is called once the registered chunk of data contains 512 values.

The complete flow of information during distance estimation is shown in Fig. 6. The frequency spectrum of the buffered chunk is computed using fast Fourier transform (FFTW library v.3.3.3, [1]). After filtering, interpolation and normalization, the frequency spectrum is subsampled in order to be fed to a feed-forward multi-layer perceptron which computes the distance estimate. We based our implementation on the Fast Artificial Neural Network Library (FANN v.2.2.0, http://leenissen.dk/fann/wp/) for the flexibility it provides in choosing the neural network properties and learning algorithms.

2.5   Neural Network Training

For the present study, we trained only the left antenna distance estimator network (we have not determined yet whether the same trained network could be used for both left and right antennae or two networks should be trained separatly to cope with individual disparity in antennal mechanics). The training was performed off-line using the RPROP algorithm [14]. Training data were obtained from several trials of contour-tracing performed sequentially on two distinct wooden discs of 10 and 25 cm diameters, as shown in Fig. 7. Prior to each trial, the target disc was manually placed in front of the left antenna, in such a way that the antenna could touch its circumference at a given constant distance. Consequently, discs were all approximatively aligned parallely to the robot frontal plane with their axis passing throught the joint center of the pan-tilt-unit (see Fig. 7). Distances were measured along the antenna from the basis of the flexible tube (i.e. 8.5 cm away from the joint center). The small disc was placed at six proximal distances ranging from 8 to 18 cm, whereas the large disc was used for six distal distances ranging from 20 to 30 cm. Data were logged with the tactile system running the contour-tracing algorithm described above. Upon first contact, the controller switched from the searching mode to the sampling mode, successfully tracing the entire circumference of the target. Two minutes of sampling per trial yielded around 100 contact events. Thus, the complete data set included 1200 contact events (100×12 distances).

3 Results

3.1 Distance Estimation

For an initial investigation of the most suitable set of distance estimator network parameters, i.e., the number of layers, number of units, or choice of activation functions per layer, we used the Matlab neural network toolbox (http://www.mathworks.com). Networks with two or three hidden layers were tried, and the number of units per layer was varied from 5 to 80 in steps of 10 or 15. The network showing best performance in distance estimation was a 4-layered network with 20, 10 and 5 neurons for the first, second and third hidden layer, respectively. While the hidden layers had neurons with a sigmoid activation function, the output layer was composed of neurons with linear activation function.

At first, we trained the neural network with the whole dataset and found that distance estimation within the distal range was considerably more unreliable than within the proximal range (see Fig. 8A). Across the entire working-range, only 77% of the total variance was captured by linear regression from measured distance to target distance. Closer inspection revealed that the frequency spectra for distal contact events hugely varied with regard to the peak location of the high-frequency band. As a consequence, we limited the distance estimation task to a proximal range (8 to 22 cm), including 800 contact events from the original data set. For this reduced data set, the ANN was able to estimate the distances reliably (Fig. 8B), a linear regression line explaining 93% of the variance of the measured distances.

3.2 Tactile 3D Localization

For each touch event, logged data include the pan and tilt angles at contact, and the acceleration time courses measured at the antennal tip (Fig. 5). The combination of antennal orientation angles and the distance estimated from the acceleration data (Fig. 8) enables localizing the contact point in 3D space.

Fig. 9 shows the 3D point clouds corresponding to the reduced data set of contour-sampled discs. For comparison, Fig. 9A and B show both the calculated

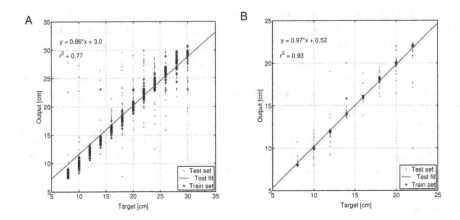

Fig. 8. Confusion plots, showing the distance output of the ANN compared to the true distance of the object. **A**: Plot for the total data set including the most distal contacts where the clear deviation visible. **B**: Plot for the proximal to middle data set. The equation is the linear fit to the estimated data.

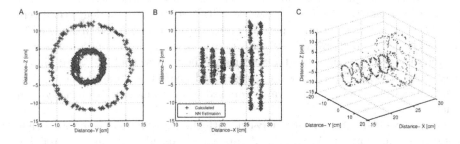

Fig. 9. Tactile 3D localization. Frontal and lateral projections of the contact locations at different distances are shown in **A** and **B**. Locations were calculated using the known distance (blue crosses) or the distance estimated by the ANN (red dots). **C**: 3D view of the ANN-estimated locations. Origin of coordinates is the joint center of the PTU.

Table 1. The root-mean-squared error for the diameter estimation. For the comparison, the errors for the calculated values are also shown. All distances are in mm. N=100 per diameter and distance.

Diameter	Distance	Training Set		Testing Set	
		Calculated	By ANN	Calculated	By ANN
100	80	7.3	7.5	7.9	7.8
100	100	8.6	8.7	8.5	8.5
100	120	9.7	9.9	11.4	11.4
100	140	7.7	7.7	8.3	8.4
100	160	8.8	8.9	9.4	9.5
110	180	10.3	10.3	10.3	10.4
250	200	8.2	8.4	8.6	8.8
250	220	8.3	8.7	8.5	13.4

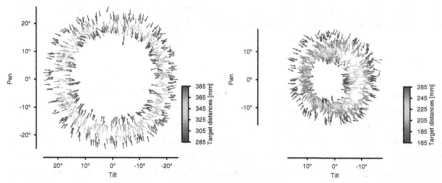

Fig. 10. Pan and tilt angles and instantaneous velocity at the moment of contact with
the 25 cm (left) or 10 cm (right) discs. Arrows of the same color belong to the same
trial, where the distance of the disc was varied between trials. Notice how the arrows
are aligned nearly radially, i.e., locally perpendicular to the contour of the disc. Local
variability of direction and global variability of instantaneous velocity are low.

contact positions (using the known target distance in combination with measured
orientation angles) and the corresponding estimated positions (using the output
of the trained neural network). Both clouds of points are reasonably similar,
without an obvious bias or a strong difference in variability. The variability
within the calculated 3D locations is due to the fact that calculation relies upon
the measured pan and tilt data, subject to slight misalignment of the target in
front of the antennal pan-tilt-unit.

Table 1 lists the root mean squared error of the radius (size) estimation at
different distances from the robot, separated into testing and training data sets.
The error is calculated by relating the estimated positions to the true circumference of the discs. Since the errors for the large disc are in the same range as those
for the small disc, the tactile localisation error appears to be size-independent.
Moreover, the error is only weakly dependent on contact distance.

Owing to a slight bending of the antenna while in contact with the disc, distance estimation is biased towards larger values, and antennal orientation angles
are biased towards smaller values. Thus, the errors listed in Table 1 contain both
a systematic bias towards smaller object size, and a noise component.

3.3   Surface Normal Estimation

In [6], our simulations provided evidence that contact data sampled by Contour-
net could be sufficient to recognize properly 3D object shapes using neural networks or other classifiers. However, the classification performance depends in
a significant manner to the regularity of the scan-path. One approach we suggested to improve the scan-path regularity (for further detail see [6]) relies on
estimating the surface normal of the sampled object at each contact point.

Fig. 10 shows the angular velocity vectors of the antenna at the instants of contact for each disc placed at different distances. Overall, those vector sets give a fairly good approximation of what is expected for circular shapes, the directional variability being larger for the smaller disc than for the larger disc.

4   Discussion

In this paper, we have shown that our antennal sampling template model, so-called Contour-net, can be applied to control a physical robotic antenna. This implementation was greatly facilitated by the limited assumption made on the tactile signal, namely only a binary contact information is required. Virtually any type of tactile probe matches this requirement. Here we have used the antenna developed in [12] and have shown that combining Contour-net with vibration-based distance estimation is effective to localize touch events in 3D space.

Yet, our approach needs to be validated on more complex 3D shapes than discs and presented from multiple perspectives. In particular, it is unknown how surface normal estimation, which shape classification depends on [6], would deteriorate with more angular contours, like those of cubes or pyramids. The number of contact events required for proper classification of 3D shapes becomes also a critical constraint in physical world as the time dedicated to sampling is necessarily limited.

Another open question concerns the applicability of our approach while the robot is in motion, exploring the environment. Indeed, vibration-based antennae are prone to pick up irrelevant perturbations generated for instance by the robot rolling on uneven floor. Cancelling of self-induced stimulation via forward models has been already explored [12,2], but not in the context of tactile contour sampling and localization.

Shape classification appears of limited relevance concerning insect biology. For example, stick insects rather use tactile sampling and localization for climbing over obstacles [18]. Antennal contact commonly induces re-targeting of a front leg after a delay of 40 ms [16]. The ratio to the average swing duration (200 ms) is one fifth. Given that our distance estimation procedure is delayed by 500 ms (it requires a data chunk of 512 values sampled at 1 kHz), implementing similar re-targeting on a legged robot equipped with our robotic antenna would constrain the swing duration to a lower limit of 2.5 seconds, limiting significantly the achievable top speed of the robot. It is not clear to which extent the data chunk can be shorten while keeping fairly accurate distance estimates.

Finally, an interesting way to improve our robotic antennae would be to measure their bending. This additional information could be used to improve localization performance by correcting for the bias towards smaller object size and larger distance estimates. It could also complement the acceleration-based contact detection in the cases it often fails, e.g. when the antenna gets trapped in a corner or slides along a surface after hitting it with a small angle of attack.

Acknowledgments. This work has been supported by the European project EMICAB FP7 – 270182.

References

1. Frigo, M., Johnson, S.G.: The design and implementation of FFTW3. Proceedings of the IEEE Special Issue on Program Generation, Optimization, and Platform Adaptation 93(2), 216–231 (2005)
2. Harischandra, N., Dürr, V.: Predictions of self-induced mechanoreceptive sensor readings in an insect-inspired active tactile sensing system. In: International Symposium on Adaptive Motion of Animals and Machines (AMAM), pp. 59–62 (2013)
3. Kaneko, M., Kanayma, N., Tsuji, T.: Active antenna for contact sensing. IEEE Transactions on Robotics and Automation 14(2), 278–291 (1998)
4. Kim, D.E., Möller, R.: Biomimetic whiskers for shape recognition. Robotics and Autonomous Systems 55(3), 229–243 (2007)
5. Krause, A.F., Dürr, V.: Active tactile sampling by an insect in a step-climbing paradigm. Frontiers in Behavioural Neuroscience 6(30), 1–17 (2012)
6. Krause, A.F., Hoinville, T., Harischandra, N., Dürr, V.: Contour-net - a model for tactile contour-tracing and shape-recognition. In: Proceedings of the 6th International Conference on Agents and Artificial Intelligence, pp. 92–101 (2014)
7. Krause, A.F., Winkler, A., Dürr, V.: Central drive and proprioceptive control of antennal movements in the walking stick insect. Journal of Physiology-Paris 107(1-2), 116–129 (2013)
8. Lederman, S., Klatzky, R.: Haptic perception: A tutorial. Attention, Perception, & Psychophysics 71(7), 1439–1459 (2009), doi:10.3758/APP.71.7.1439
9. Lee, J., Sponberg, S.N., Loh, O.Y., Lamperski, A.G., Full, R.J., Cowan, N.J.: Templates and anchors for antenna-based wall following in cockroaches and robots. IEEE Transactions on Robotics 24(1), 130–143 (2008)
10. Lee, M., Nicholls, H.: Review article tactile sensing for mechatronics - a state of the art survey. Mechatronics 9(1), 1–31 (1999)
11. Mongeau, J.M., Demir, A., Lee, J., Cowan, N.J., Full, R.J.: Locomotion- and mechanics-mediated tactile sensing: antenna reconfiguration simplifies control during high-speed navigation in cockroaches. J. Exp. Biol. 216, 4530–4541 (2013)
12. Patanè, L., Hellbach, S., Krause, A.F., Arena, P., Dürr, V.: An insect-inspired bionic sensor for tactile localization and material classification with state-dependent modulation. Frontiers in Neurorobotics 6(8), 1–18 (2012)
13. Prescott, T.J., Diamond, M.E., Wing, A.M.: Active touch sensing. Philos. Trans. R. Soc. Lond. B Biol. Sci. 366, 2989–2995 (2011)
14. Riedmiller, M., Braun, H.: A direct adaptive method for faster backpropagation learning: The RPROP algorithm. In: Proceedings of the IEEE International Conference on Neural Networks, pp. 586–591 (1993)
15. Russell, R.A., Wijaya, J.A.: Object location and recognition using whisker sensors. In: Australasian Conference on Robotics and Automation, pp. 761–768 (2003)
16. Schütz, C., Dürr, V.: Active tactile exploration for adaptive locomotion in the stick insect. Philosophical Transactions of the Royal Society B: Biological Sciences 366(1581), 2996–3005 (2011)
17. Solomon, J.H., Hartmann, M.J.: Biomechanics: Robotic whiskers used to sense features. Nature 443(7111), 525 (2006)
18. Staudacher, E., Gebhardt, M.J., Dürr, V.: Antennal movements and mechanoreception: neurobiology of active tactile sensors. Adv. Insect Physiol. 32, 49–205 (2005)
19. Sullivan, J.C., Mitchinson, B., Pearson, M.J., Evans, M., Lepora, N.F., Fox, C.W., Melhuish, C., Prescott, T.J.: Tactile discrimination using active whisker sensors. IEEE Sensors Journal 12(2), 350–362 (2012)

A Predictive Model for Closed-Loop Collision Avoidance in a Fly-Robotic Interface

Jiaqi V. Huang and Holger G. Krapp

Department of Bioengineering,
Imperial College London, London SW7 2AZ, UK
h.g.krapp@imperial.ac.uk

Abstract. Here we propose a control design for a calibrated fly-brain-robotic interface. The interface uses the spiking activity of an identified visual interneuron in the fly brain, the H1-cell, to control the trajectory of a 2-wheeled robot such that it avoids collision with objects in the environment. Control signals will be based on a comparison between predicted responses – derived from the known robot dynamics and the H1-cell responses to visual motion in an isotropic distance distribution – and the actually observed spike rate measured during movements of the robot. The suggested design combines two fundamental concepts in biological sensorimotor control to extract task-specific information: active sensing and the use of efference copies (forward models). In future studies we will use the fly-robot interface to investigate multisensory integration.

Keywords: motion vision, brain machine interface, blowfly, closed-loop control, predictive model.

1   Introduction

The blowfly, *Calliphora vicina*, is one of the most agile fliers amongst airborne insects. In several laboratories, flies have been used successfully as model systems for studying the neural mechanisms underlying visually guided behaviour [1] including collision avoidance [2].

The final goal of this project is to investigate how the blowfly combine information about state changes from multiple sensor systems for flight and gaze control under closed-loop condition. At the current stage, the main interest lies in developing control architecture for collision avoidance based on the signals of an identified interneuron, the H1-cell.

The H1-cell is located in the third optic lobe, the lobula plate, where blowflies employ about 60 motion-sensitive directional-selective interneurons called lobula plate tangential cells (LPTCs) [3]. These cells are part of ipsilateral and contralateral networks, the output elements of which connect to the various motor systems supporting gaze and flight stabilization. Some LPTCs were also thought to be involved in collision avoidance [2, 4]. The H1-cell is a spiking cell which, along its axonal connection, transmits visual motion information from the lobula plate in one brain

A. Duff et al. (eds.): Living Machines 2014, LNAI 8608, pp. 130–141, 2014.
© Springer International Publishing Switzerland 2014

hemisphere to the contralateral lobula plate in form of action potentials. It's excited, i.e. it increases its spike rate, when experiencing back-to-front horizontal optic flow over the ipsilateral eye and inhibited during motion in the opposite direction [5]. If presented with no motion at all, it still generates a comparatively low spontaneous spike rate. The response properties of the H1-cell have been studied for decades [6], and the neuron is easily accessible by means of extracellular recording.

Collision avoidance is a common behavioural feature observed in basically all flying insects, including: locust [7], fruit fly [8] and blowfly [4]. Previous approaches to study collision avoidance range from behavioural experiments by using induction coils mounted on the head and thorax of a fly [9] to capturing the flight trajectory by means of high speed videography [10]. In those studies, information such as flight velocities and accelerations as well as body and gaze orientations were collected, to analyse the relationships between optic flows fields the flies experience and the movements they generate as a response. In replay experiments optic flow reconstructed from known flight trajectories was presented to stationary flies using an LED-based virtual environment wide-field display, FliMax, while recording the neural activity of LPTCs [4]. Recently a different approach was developed to study visuo-motor control in flies under closed-loop conditions using a robotic platform [11]. The robot was equipped with high speed video cameras and mounted on a rotation turntable. The video cameras monitored rotation-induced optic flow which was presented to a stationary fly using high speed CRT monitors while the response of the H1-cell were recorded. The resulting spiking activity of the cell was then used in a closed-loop stabilization task to reduce relative motion between the robot on the turntable and the visual environment [11].

We have now developed a miniaturized platform [12] to be mounted on a small two-wheeled robot specially designed for on-board recordings from the blowfly H1-cell. Such a platform will enable studies on multisensory integration where the fly is actually moving in space and both the mechanosensory as well as the visual system are stimulated under closed-loop conditions. This will allow us to compare the responses of specific visual interneurons – which reflect multisensory signal integration [13–15] – under experimental conditions where the different sensor systems are selectively disabled or manipulated. The difference to previous studies would be that closed-loop conditions are combined with the simultaneous stimulation of several sensor systems.

The on-board recording system is designed to support two major objectives: (i) The ability to record signals from motion-sensitive interneurons and using them for the control of the mobile robot under closed-loop conditions. (ii) To monitor neuronal activity over an extended period of time, which requires the implementation of a feedback control architecture where the activity of the H1-cell may be used to avoid collision of the robot with any obstacles in the surroundings.

Most of the technical challenges associated with objective number (i) have been resolved [12]. Achieving objective number (ii), however, requires a strategy because of the H1-cell's preference for back-to-front motion. Forward movement of the robot induces front-to-back horizontal optic flow that would result in a strong inhibition of H1-cell activity drastically reducing the signal range available for visually guided collision avoidance.

A potential solution of this problem would be the implementation of a preprogrammed sinusoidal trajectory of the robot which combines translational and rotational self-motion components. Alternating clockwise and counter-clockwise rotational self-motion components of the robot should produce alternating front-to-back and back-to-front patterns of optic flow resulting in consecutive inhibition and excitation of the two H1-cells. Such temporally alternating activity pattern is expected to provide a sufficiently large output signal range of the cell suitable to modulate the preprogrammed robot trajectory.

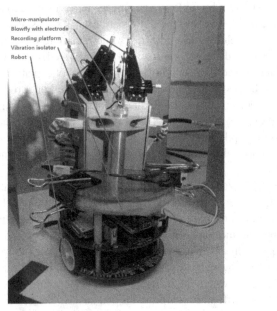 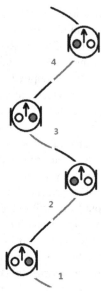

Fig. 1. Blowfly-robot assembly and the pre-programmed self-motion trajectory of the robot. (**Left**) The assembly contains: robot, sponge for vibration damping, recording platform and blowfly. (**Right**) The robot sinusoidal trajectory, the right and left H1-cells are excited during rotational self-motion components in section 1 & 3 and 2 & 4, respectively.

In this paper we propose a control architecture that avoids collision of the blowfly-robot platform with obstacles in the environment. In the next sections, a feedback control with predictive models is introduced, as well as the methods of how the predictive models are built.

2   Predictive Control Architecture

2.1   Interface

We recently established a miniaturized recording platform designed to be mounted on a small two-wheeled robot (Pololu© m3pi) [12]. As an extension of that previous work we have now implemented a simple brain-machine interface that includes the following functionalities: the measurement of H1-cell action potentials, the conversion of the action potential into a time-continuous spike rate, and the transformation of the spike rate into a motor control voltage.

These tasks are supported by an ARM processor (NXP© LPC1768) mounted on top of the robot that receives input from an ADC module at a rate of 5KS/s. Individual H1-cell action potentials are detected if the recorded signal exceeds a pre-programmed threshold voltage. The detected action potentials are used to calculate the time-continuous spike rate within an integration time interval of 50ms. We developed a simple control law based on an inverted model of the H1-cell's velocity tuning. It generates a PWM (pulse width modulation) signal as a function of the time-continuous H1-cell spike rate which is used to controls the speed of the motors of the robot wheels.

2.2   Predictive Feedback Control

If the H1-cell did prefer horizontal front-to-back optic flow, a simple and direct solution to achieve collision avoidance would be to compare the spike rates from both cells in either side of the brain while the robot was moving forward on a straight trajectory. In such a case the cell sensing optic flow induced by objects close by would generate a higher spike rate than its contralateral counterpart processing optic flow generated by more distant objects. The difference in terms of spike rates produced by the two cells could then be used directly to steer clear of potential obstacles. As mentioned above, however, the H1-cell prefers back-to-front movements and a preprogrammed sinusoidal trajectory will be needed to induce a sufficiently large output signals range in the H1-cells. The drawback of this strategy is that the excitation of the two cells during a sinusoidal trajectory will not be simultaneously available as the neurons would be driven one after the other. Therefore, the cells' outputs cannot be compared continuously but only after two consecutive clockwise and counter-clockwise semi-rotations have been completed (cf. Fig. 1, Right). This would incur intolerable delays in a closed-loop feedback controller.

To overcome excessive delays we instead propose a predictive feedback control algorithm. In this algorithm a continuous difference signal is not compute between the activities of both H1-cells but between the actually measured spike rates of the cells and the spike rate predicted by a forward model. The forward model would be based on the relationship between the input voltage to the motors driving the wheels and the robot velocity as well as the velocity dependence of the H1-cell responses assuming an average distance distribution of objects within the environment. The resulting difference signals could then be used as a continuous feedback signals that modifies the preprogrammed trajectory to avoid collisions.

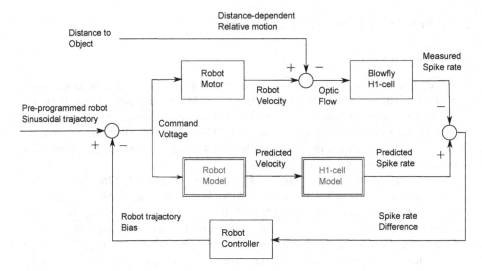

Fig. 2. Control loop for collision avoidance of blowfly-robot with predictive model. For explanation see text.

The block diagram of such a *predictive feedback controller* is shown in figure 2. A pre-programmed sinusoidal signal generates the command voltage that drives the DC motors. The resulting rotational self-motion components produce optic flow sensed by the H1-cells, the spike rates of which encode angular velocity. At the same time the command signal is run through a model of the robot dynamics relating input voltage to angular velocity. The output of that operation is passed on to a model of the H1-cell's angular velocity dependence which generates a predicted spike rate given the robots momentary angular velocity. As the model of the H1-cell assumes an average distance distribution, the output will differ from the actually measured spike rate in case the robot approaches any objects close by. The resulting difference will then be added to the preprogrammed sinusoidal command voltage.

This control architecture requires a specification of the forward model components, i.e. the command voltage-angular velocity relationship of the robot and the velocity tuning of the H1-cell, given an average distance distribution. With respect to the distance distribution we consider the 3d-layout of the lab environment.

3   Predictive Models

3.1   Robot Model

To obtain the command voltage-angular velocity relationship of the robot, we systematically varied the command voltage while monitoring the changes of the robot's angular position during rotations around its vertical axis using high-speed videography at a temporal resolution of 512 frames/second.

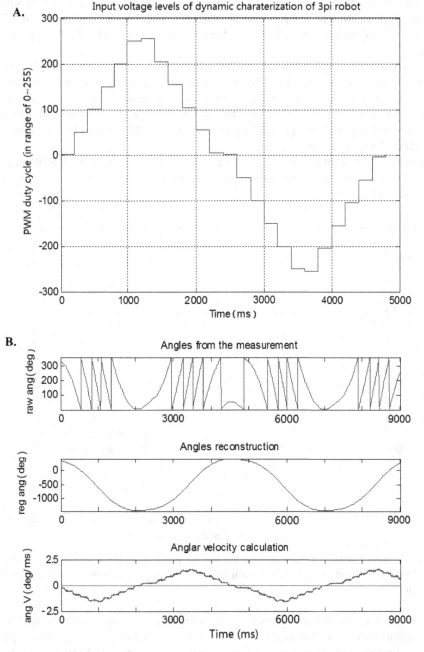

Fig. 3. Data processing from angular position to angular velocity of the robot. (**A**) The input command voltage profile for characterizing the robot angular velocity. (**B**) The measurements of (Top) the raw angular position data extracted from high speed camera. (Middle) the reconstructed angular position data (without 360 degree limitation). (Bottom) the angular velocity plot, derived from the angular position data.

8-bit PWM voltage was used to control the robot wheel motor, the PWM duty cycles can be interpreted to equivalent DC voltages. The duty cycle of the PWM voltage has 255 difference levels. And the robot motors were driven by 9.25V (regulated power).

An asymmetric angular velocity profiles (Fig. 3A) were pre-programmed where robot accelerations and decelerations were approximated by asymmetric step functions (Fig. 3B Bottom). The command voltage was increased from 0 to 9.07V (duty cycle: 250/255) with 1.81V (50/255) steps during acceleration sections. The initial command voltage of 9.25V (255/255) was reduced to 0.18V (5/255) at -1.81V steps during deceleration sections.

Angular positions of the robot in each frame were analyzed using the Matlab image processing toolbox. From the angular positions we derived the angular velocity as illustrated in figure 3B.

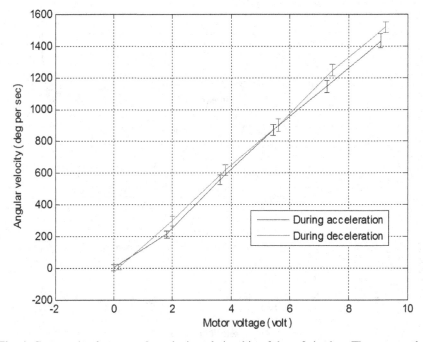

Fig. 4. Command voltage-angular velocity relationship of the m3pi robot. The mean and standard deviations of angular velocities of the m3pi robot during acceleration and deceleration in steps. (n=100)

After the angular position values were extracted from the images taken by the high speed camera, the profile of angular positions was obtained (Fig. 3B Top). The angular position values of the original data were limited to the range of 0-360 degrees, but were then transformed into a continuous angular position value (Fig. 3B Middle). The angular velocity was obtained by temporal differentiation of the robot's angular orientation (Fig. 3B Bottom).

To plot the velocity curve (Fig. 4), 100 data samples of robot angular velocity were taken from each step of the angular velocity (Fig. 3B Bottom), from which the mean and standard deviation were calculated. The results (Fig. 4) show, a linear relationship between input voltage (PWM equivalent DC voltage) and output angular velocity of the m3pi robot. The slight hysteresis effect noticeable is not expected to have a detrimental effect on the performance of the robot under closed-loop conditions. After the gears were properly lubricated, the maximum angular velocity of the robot reached about 1500 deg/s.

3.2   H1-Cell Model

The H1-cell velocity tuning has been obtained using a setup described in an earlier account [12] where the visual motion stimulus consisted of a vertical black-and-white grating with a fixed spatial frequency. In the context of the control architecture proposed here we determined the velocity tuning in the laboratory which includes a wide range of spatial frequencies and will eventually be the environment within which the robot is supposed to operate.

Experiments were carried out on female blowflies, *Calliphora vicina*, from 4-11 days. To reduce the amount of movements that could degrade the quality of the recordings the fly's legs and proboscis were removed and the resulting wounds were sealed with bee wax. Small droplets of bee wax were applied to immobilize the wing hinges. The head of the fly was fixed in between two pins of a custom-made fly holder, after adjusting its orientation according to the 'pseudopupil methods' [16]. The thorax was bent downwards and fixed to the holder to expose the back of the head capsule. A small hole was cut into the back of the head capsule under optical magnification using a stereo microscope (Stemi 2000, Zeiss©). Fat and muscle tissue were removed and physiological Ringer solution (for recipe see Karmeier et al. [17]) was applied to prevent desiccation of the brain.

We used tungsten electrodes (3 MΩ tungsten electrodes, FHC Inc., Bowdoin, ME, USA) for extracellular recording from H1-cell. The electrode was positioned next to the H1-cell using a mini micromanipulator so that the amplitude of the recorded H1-cell spikes was at least 2 times higher than the largest amplitudes of the background noise (SNR > 2:1). The signals were amplified by a nominal gain of 10^4 and sampled by a data acquisition board (NI USB-6215, National Instruments Corporation, Austin, TX, USA) at a rate of 20KS/s.

To generate wide-field optic flow stimuli the fly on the recording platform, which was attached to the axel of a stepper motor, was rotated in the lab environment. The experiments were performed under controlled room light conditions. A single stimulus trail of the blowfly H1-cell lasted 10 seconds. During the first half a second the fly was kept stationary, followed by a rotation for 3 seconds in the cell's null direction (H1-cell inhibited) and then 3 seconds in the preferred direction (H1-cell excited) at a constant angular velocity, after which the fly remains stationary again for 3.5 seconds. We rotated the fly at 10 angular velocities: 3 deg/s, 15 deg/s, 45 deg/s, 60 deg/s, 75 deg/s, 120 deg/s, 165 deg/s, 210 deg/s, 255 deg/s and 300 deg/s. These velocities were applied in a shuffled sequence, which was repeated in a mirrored order for another 9 times.

We calculated the spike rates within the three-second time intervals during pre-ferred direction and null direction rotation. The mean and standard deviations were calculated from the spike rates obtained over 10 repetitions. For recordings from sin-gle cell upon preferred direction motion, the tuning curves show a logarithm inclina-tion with increasing angular velocity (Fig. 5 Left). During null direction motion the activity of the cells is strongly inhibited throughout the entire range of angular veloci-ties applied. In one of the experiments both H1-cells were recorded simultaneously. The responses of both cells to motion in the preferred direction showed a similar de-pendence on angular velocity (Fig. 5 Left, black lines).

Altogether, the results (Fig. 5 Right) obtained in the laboratory environment sug-gest a monotonic relationship between angular velocities and spike rate of H1-cell over a range from 0 to 300 deg/s that may be described by a logarithmic function. This experimentally derived input-output characteristic may inform the H1-model in our predictive feedback control architecture.

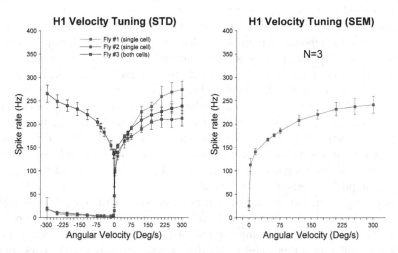

Fig. 5. Velocity tuning of blowfly H1-cell in lab environment. (**Left**) Mean and standard devia-tion of spike rates. The positive and negative angular velocities indicate motion in the preferred direction and anti-preferred direction of the recording cell, respectively. In recordings from both cells, null direction simulation of one cell is the preferred direction of the other. (n=10) (**Right**) average velocity tuning of 3 H1-cells. Mean and standard error of spike rates. (N=3, n=10).

4   Discussion

4.1   H1-Cell and Robot Velocity Tuning

The velocity tunings of both H1-cell and robot show monotonic input-output relation-ships. We are confident that these experimental results will be sufficient for the specification of forward models in our predictive control architecture illustrated in figure 2.

The monotonic but non-linear velocity tuning of the H1-cell obtained in the laboratory environment may be best described by a logarithmic function (Fig. 5). Similar results were previously reported by Lewen et al [18]. Lewen and colleagues recorded H1-cell activity outdoors under natural light conditions over a range of angular velocities from 0.3 deg/s up to more than 4000 deg/s. The H1-cell responses to angular velocities between 0.3 deg/s and 300 deg/s we presented here are almost identical to those Lewen's et al. [18]. Despite its non-linear shape, the tuning curve still provides a unique relationship between angular velocity and the spike rate which is a favorable feature in the context of a closed-loop control system.   It is worth mentioning that the temporal frequency tuning of the H1-cell, i.e. angular velocity over the spatial wavelength of a grating pattern, assumes a bell-shaped form where temporal frequencies - left and right of the optimal one - result in identical response [12].

The characterization of the H1-cell spiking activity we present here was performed during pure rotations of the fly. In future experiments when the animal will be mounted on the robot it will experience optic flow due to rotation and translation. The translation-induced contribution to the local flow vectors will depend on distance where close-by objects elicit greater deviation from the predicted responses. In an asymmetric environment with respect to the distance distribution the resulting difference between the prediction errors obtained for the spike rates of the left and the right H1-cell will be used to steer the robot clear of any obstacles.

The linear relationship between command voltage and angular velocity (Fig. 4) of the m3pi robot simplifies the transfer function of the forward robot model to a single gain. In addition, the maximum angular velocity of the robot almost reaches the highest angular rates observed in freely flying blowflies' 1700 deg/s [10].

The characterization of the robot dynamics was performed without any payload on the robot. Additional instrumentation such as the miniaturized recording platform will certainly increase the inertia of the robot and reduce its dynamic properties. This potential issue may be overcome by a series of DC motors with different power ratings available for the m3pi robot.

4.2   The Control Law with the Predictive Model

Interestingly, a flight trajectory that combines translational and rotational self-motion components has been observed in previous studies. Kern et al. [19] reported blowflies in free flight to perform an alternating pattern consisting of fast body saccades followed by phases of almost pure translator side-slip. This pattern flies generate when flying in confined spaces was interpreted as a strategy to separate rotatory from translator optic flow in time. During the translation phases specific LPTCs were suggested to signal relative distance to surrounding objects, information that could be used for collision avoidance [2, 4, 20].

Our predictive control architecture is inspired by the generation of efference copies in the nervous system. The use of efference copies (forward models) is general principle in many cases of sensory information processing. It is an efficient way of removing any known (predictable) components from the sensory input and thus enables the application of comparatively high feedback gains to the difference signal without the risk of output saturation when controlling motor tasks. Efference copies are likely to play another important role in the integration of reflex and voluntary behavior. They may be used to change the sensitivity of sensor systems during voluntary motor actions [21] which would otherwise be immediately counteracted by inner-loop reflex control systems.

The control design we propose may also be compared to a strategy known as 'active sensing' – specifically, active vision – where known voluntary movements are used to "scan" the visual environment [22]. One of the examples is peering: some insects move their body from side to side along a translatory trajectory. The result is the induction of motion parallax which allows the system to assess relative distance information. This is used by praying mantis, for instance, to work out the distance to potential prey before striking [23].

The question as to how different modalities contribute to the control of multimodal guidance, navigation, and control tasks will eventually be subject to future studies using our fly-brain machine interface.

Acknowledgments. We'd like to thank Kit Longden, Martina Wicklein, Ben Hardcastle and Peter Swart for all the helps, experience sharing and discussion on the work presented. This work was partially supported by US AFOSR/EOARD grant FA8655-09-1-3083 to HGK.

References

1. Taylor, G.K., Krapp, H.G.: Sensory Systems and Flight Stability: What do Insects Measure and Why? In: Casas, J., Simpson, S.J. (eds.) Advances in Insect Physiology, pp. 231–316. Academic Press (2007)
2. Lindemann, J.P., Weiss, H., Möller, R., Egelhaaf, M.: Saccadic flight strategy facilitates collision avoidance: closed-loop performance of a cyberfly. Biol. Cybern. 98, 213–227 (2008)
3. Borst, A., Haag, J.: Neural networks in the cockpit of the fly. J. Comp. Physiol. A 188, 419–437 (2002)
4. Lindemann, J.P., Kern, R., van Hateren, J.H., Ritter, H., Egelhaaf, M.: On the Computations Analyzing Natural Optic Flow: Quantitative Model Analysis of the Blowfly Motion Vision Pathway. J. Neurosci. 25, 6435–6448 (2005)
5. Krapp, H.G., Hengstenberg, R., Egelhaaf, M.: Binocular Contributions to Optic Flow Processing in the Fly Visual System. J. Neurophysiol. 85, 724–734 (2001)
6. Hausen, K.: Functional characterization and anatomical identification of motion sensitive neurons in the lobula plate of the blowfly Calliphora erythrocephala. Z Naturforsch, 629–633 (1976)
7. Blanchard, M., Rind, F.C., Verschure, P.F.M.J.: Collision avoidance using a mod-el of the locust LGMD neuron. Robot. Auton. Syst. 30, 17–38 (2000)
8. Tammero, L.F., Dickinson, M.H.: Collision-avoidance and landing responses are mediated by separate pathways in the fruit fly, Drosophila melanogaster. J. Exp. Biol. 205, 2785–2798 (2002)

9. Schilstra, C., van Hateren, J.H.: Using miniature sensor coils for simultaneous measurement of orientation and position of small, fast-moving animals. J. Neurosci. Methods 83, 125–131 (1998)
10. Bomphrey, R.J., Walker, S.M., Taylor, G.K.: The Typical Flight Performance of Blowflies: Measuring the Normal Performance Envelope of Calliphora vicina Using a Novel Corner-Cube Arena. PLoS ONE 4, e7852 (2009)
11. Ejaz, N., Peterson, K.D., Krapp, H.G.: An Experimental Platform to Study the Closed-loop Performance of Brain-machine Interfaces. J. Vis. Exp. (2011)
12. Huang, J.V., Krapp, H.G.: Miniaturized Electrophysiology Platform for Fly-Robot Interface to Study Multisensory Integration. In: Lepora, N.F., Mura, A., Krapp, H.G., Verschure, P.F.M.J., Prescott, T.J. (eds.) Living Machines 2013. LNCS, vol. 8064, pp. 119–130. Springer, Heidelberg (2013)
13. Parsons, M.M., Krapp, H.G., Laughlin, S.B.: A motion-sensitive neurone re-sponds to signals from the two visual systems of the blowfly, the compound eyes and ocelli. J. Exp. Biol. 209, 4464–4474 (2006)
14. Parsons, M.M., Krapp, H.G., Laughlin, S.B.: Sensor Fusion in Identified Visual Interneurons. Curr. Biol. 20, 624–628 (2010)
15. Huston, S.J., Krapp, H.G.: Nonlinear Integration of Visual and Haltere Inputs in Fly Neck Motor Neurons. J. Neurosci. 29, 13097–13105 (2009)
16. Franceschini, N.: Pupil and Pseudopupil in the Compound Eye of Drosophila. In: Wehner, R. (ed.) Information Processing in the Visual Systems of Anthropods, pp. 75–82. Springer, Heidelberg (1972)
17. Karmeier, K., Tabor, R., Egelhaaf, M., Krapp, H.G.: Early visual experience and the receptive-field organization of optic flow processing interneurons in the fly motion pathway. Vis. Neurosci. 18, 1–8 (2001)
18. Lewen, G.D., Bialek, W., de Ruyter van Steveninck, R.R.: Neural coding of natu-ralistic motion stimuli. Netw. Bristol. Engl. 12, 317–329 (2001)
19. Kern, R., Boeddeker, N., Dittmar, L., Egelhaaf, M.: Blowfly flight characteristics are shaped by environmental features and controlled by optic flow information. J. Exp. Biol. 215, 2501–2514 (2012)
20. Lindemann, J.P., Egelhaaf, M.: Texture dependence of motion sensing and free flight behavior in blowflies. Front. Behav. Neurosci. 6, 92 (2013)
21. Chen, Y., McPeek, R.M., Intriligator, J., Holzman, P.S., Nakayama, K.: Smooth Pursuit to a Movement Flow and Associated Perceptual Judgments. In: Becker, W., Deubel, H., Mergner, T. (eds.) Current Oculomotor Research, pp. 125–128. Springer US (1999)
22. Egelhaaf, M., Kern, R., Lindemann, J.P., Braun, E., Geurten, B.: Active Vision in Blowflies: Strategies and Mechanisms of Spatial Orientation. In: Floreano, D., Zufferey, J.-C., Srinivasan, M.V., Ellington, C. (eds.) Flying Insects and Robots, pp. 51–61. Springer, Heidelberg (2010)
23. Rossel, S.: Binocular Spatial Localization in the Praying Mantis. J. Exp. Biol. 120, 265–281 (1986)

Neuromechanical Simulation of an Inter-leg Controller for Tetrapod Coordination*

Alexander Hunt[1], Manuela Schmidt[2], Martin Fischer[2], and Roger D. Quinn[1]

[1] Case Western Reserve University, Cleveland OH 44106, USA
[2] Friedrich-Schiller-Universität Jena 07743 Jena, Germany

Abstract. A biologically inspired control system has been developed for coordinating a tetrapod walking gait in the sagittal plane. The controller is built with biologically based neurons and synapses, and connections are based on data from literature where available. It is applied to a simplified, planar biomechanical model of a rat with 14 joints with an antagonistic pair of Hill muscle models per joint. The controller generates tension in the muscles through activation of simulated motoneurons. Though significant portions of the controller are based on cat research, this model is capable of reproducing hind leg behavior observed in walking rats. Additionally, the applied inter-leg coordination pathways between fore and hind legs are capable of creating and maintaining coordination in this rat model. Ablation tests of the different connections involved in coordination indicate the role of each connection in providing coordination with low variability.

Keywords: Neural Controller, Rat, Mammal, Central Pattern Generator, Inter-leg Coordination.

1 Introduction

Legged robots are unable to walk over uneven ground as well as animals. Animals dynamically adapt to varying terrain by changing gaits, adjusting footfall positions, and responding to perturbations. The neural circuits which accomplish these tasks reside in complex hierarchies, oscillating and interacting with each other at different time scales based on feedback from sensory information, making them difficult to understand. The prevailing theory in neurobiology is that hierarchies in the central nervous system sub-divide complex tasks into subtasks. With the brain at the top and motoneurons at the bottom, each level relies on level-appropriate sensory information to act on the levels below. Neurological experimentation suggests that a large set of neurons involved in steady state walking are located in the thoracic ganglia for arthropods [1], and the spinal cord for vertebrates [2].

Many of these neurological measurements suggest hypothetical circuits, and modeling is a useful tool to test that controllers based on these circuits are sufficient to produce the desired behavior [3]. Modeling typically starts at lower,

---

* This work was supported by DARPA M3 grant DI-MISC-81612A

A. Duff et al. (eds.): Living Machines 2014, LNAI 8608, pp. 142–153, 2014.
© Springer International Publishing Switzerland 2014

basic levels and builds on findings to develop more complex models. Insect modeling has progressed significantly in this way. Early hypotheses concerning insect walking began with a set of leg coordination rules based on years of observing stick insects [4]. Implementation of these rules on a robot helped to test how they could produce reliable, coordinated walking over even ground [5]. Further models of insect walking have focused on coordination methods in individual legs, beginning with finite state rules [6], and how descending commands can be used to affect leg states and transitions in behavior [7,8]. More detailed modeling of these systems has examined specific sensory pathways and how central pattern generators (CPGs) can be coordinated to produce different stepping behaviors with more detailed biological neuron models [9,10].

Vertebrate systems are also being modeled. Knowledge of stance to swing transitions in stepping cats [11] has resulted in a model that demonstrates the necessity of both joint position feedback and limb loading feedback on replicating the behavior [12]. Further research on this model has shown that the replicated stepping patterns are inherently stable [13]. Research on salamanders has demonstrated how CPGs can be organized for the generation of different gaits and smoothly transition between swimming and walking [14].

Non-detailed evidence of CPGs in mammalian systems has resulted in many modeling theories. Early theories hypothesize the existence of a single CPG per leg, driving transitions between stance and swing. However, more recent models utilize multiple oscillating circuits at multiple hierarchical levels [15] which is supported by recent neurological data [16], indicating the CPG systems may look more similar to those in insects than previously hypothesized [17]. A model of a CPG for every joint coordinated through sensory feedback pathways has been shown to successfully replicate many behaviors seen in CPG coordination [18].

The work presented in this paper expands upon a state controller for tetrapod hind leg stepping by implementing it in a neurologically based CPG controller similar to the one produced by Szczecinski [10]. The neural system controls a dynamic model of a rat, and has been implemented on both hind and fore legs. This work also incorporates inter-leg pathways discovered in previous animal research and coordinates all four legs of the model into a stable symmetrical gait.

2   Biomechanical Model

The simulation is conducted in the Vortex physics engine (CM Labs, Montreal, Quebec) implemented by Animatlab [19]. Rigid body dynamics, ground contact, body collisions, and friction are all simulated. The inertia of a solid body is determined by the shape of a triangulated mesh and a given density. The biomechanical model was constructed through scans of rat bones and then digital matching of these bones to a rat walking up a rope using 3D x-ray high speed video. This reconstruction was limited to motion in the sagittal plane, resulting in a total of 14 degrees of freedom, four for each front leg (scapula, shoulder, elbow, and wrist) and three for each hind leg (hip, knee, and ankle). The model uses the linear Hill muscle model to generate forces within the rat model [19]. Tension, T, is developed in the muscle according to:

$$\frac{dT}{dt} = \frac{k_{se}}{c} \left(k_{pe} x + c\dot{x} - \left(1 + \frac{k_{pe}}{k_{se}} \right) \cdot T + A \right) \tag{1}$$

where x is the muscle length, k_{se}, k_{pe}, and c are the series, parallel, and damping elements, and A is the activation level of the muscle, which has a length-tension component. Muscles are chosen as the activation method for two main reasons. First, it is likely that some aspects of the control system are tuned directly to the mechanical properties inherent in muscles. Second, actuator compliance is an important part of achieving robust motions that can reject perturbations passively. To simplify the control, each joint is controlled by two antagonistic muscles. Muscle properties were set to produce the desired torques calculated by Witte et al. while maintaining stability of motion [20].

Sensory feedback in the model is similar to types Ia, Ib, and II, and although simplified representations of nature, they contain important elements.

$$\text{Ia} = k_a x_{series} \qquad \text{Ib} = k_b T \qquad \text{II} = k_c x_{parallel} \tag{2}$$

where k_a, k_b, and k_c are gain parameters set by the user to act as injected current into a neuron. Ia feedback is dependent on both length and velocity of muscle movement. Ib is dependent on the tension developed in the muscle, and II is dependent on length through the length of the parallel elastic element x_2. The gains were experimentally adjusted to produce an injected current which gives usable information through a sensory neuron across the full range of muscle lengths and contractions.

2.1   Neural Model

This work is built on the dynamics of a leaky integrator, and is capable of modeling individual non-spiking interneurons, the firing rate of a population of neurons, or a single spiking neuron after the addition of a spiking threshold. We are currently not concerned with the specifics on how action potentials are generated and have left out Hodgkin-Huxley sodium and potassium currents. Instead, we are concerned with how signals propagate through the network, and how individual neurons and populations of neurons activate, deactivate, and contribute to network behavior. Each neuron is governed by the equation:

$$C_m \frac{dV}{dt} = g_L \cdot (E_L - V) + I_{ext} + I_{ions} \tag{3}$$

where V is the membrane potential, C_m is the membrane capacitance, E_L is the leak potential, g_L is the leak conductance, I_{ext} are the external synaptic and injected current inputs and I_{ions} are additional ionic currents that can be added to increase the dynamics of particular neurons. Constants such as C_m, g_L, and E_L are based on typical spinal cord interneuron values, but may be more directly implemented with known values when knowledge of them is made available. This limits the number of free parameters in the network and allows us to focus on the strength and types of connections between neurons. The synapse model is:

$$I_{syn} = g_{syn} \cdot (E_{syn} - V_{post}) \tag{4}$$

where E_{syn} is the potential of the synapse, V_{post} is the postsynaptic membrane potential, and g_{syn} is piecewise linear activation function defined as

$$g_{syn} = g_{max} \cdot min\left(max\left(\frac{V_{pre} - E_{lo}}{E_{hi} - E_{lo}}, 0\right), 1\right) \tag{5}$$

where g_{max} is a user defined maximum conductance, V_{pre} is the presynaptic membrane potential, and E_{lo} and E_{hi} are the two user-defined voltage threshold values that define the piecewise-linear function where $E_{lo} < E_{hi}$. The above model is also capable of representing the activation of a population of neurons with the same type of synapse where the population activity varies between 0 when $V_{pre} < E_{lo}$ and g_{max} when $V_{pre} = E_{hi}$. Most neurons used in this model represent average voltage values of a population, similar to those of other networks that model mammalian walking behaviors [18,9].

This neural model offers many advantages. Because the model is biologically based it is expandable and not limited to the currently modeled system. Known neural pathways, ionic currents, and synapse plasticity which influence locomotor behavior can be implemented directly into the model, several of which are described in 3.1 and 3.2. These additional currents and properties that present a significant impact on the behavior of a particular area of the neural system may be added to individual neurons appropriately without further increasing the complexity of the rest of the system. One particular area of relevance to our study is how a set of neurons can produce a pattern generating behavior. The ion channels we have included in the model that contribute to active bursting phases are calcium channels, where the calcium current, I_{Ca}, is defined as:

$$I_{Ca} = g_{Ca} \cdot m(V) \cdot h(V) \cdot (E_{Ca} - V) \tag{6}$$

$$\frac{dm}{dt} = \frac{m_\infty(V) - m}{\tau_m(V)} \qquad \frac{dh}{dt} = \frac{h_\infty(V) - h}{\tau_h(V)} \tag{7}$$

where E_{Ca} is the equilibrium calcium channel voltage, g_{Ca} is the calcium conductance and $\tau_m(V) = \phi_m\sqrt{\epsilon_m(V)}$ and $\tau_h(V) = \phi_h\sqrt{\epsilon_h(V)}$. The functions $m_\infty(V) = \frac{1}{1+\epsilon_m(V)}$ and $h_\infty(V) = \frac{1}{1+\epsilon_h(V)}$ are sigmoids, where $\epsilon_m(V) = exp(-S_m \cdot (V_m - V))$ and $\epsilon_h(V) = exp(-S_h \cdot (V_h - V))$. $\phi_m, \phi_h, S_m, S_h, V_m$ and V_h are constants. By starting with a relatively simple model, we are able to lay out the basic structure of the network in a tractable manner. Through the use of biologically based neurons we are able to increase complexity where necessary. The higher level of complexity can lead to the emergence of more capable and dynamic behaviors.[21].

3   Control Network

3.1   Intra-leg Network

The exact nature of how the nervous system controls muscle activation is not known [22]. For simplicity, the muscle control system in our network is tuned for position based control, and leaves out the effect of Renshaw cells, Ia feedback, and other known motoneuron influences. This control develops tension in the muscle based on type II afferent feedback (stretch receptors) modulated by a constant force inhibiting neuron and is illustrated in Fig. 1. As the muscle gets closer to its desired length, excitation of the motor neuron decreases. We use kinematic data [23] and recently collected unpublished data to determine what these positions are and tune the system to achieve this motion. Exact position control is not yet achieved however, due to complex dynamics of the muscles and momentum of the limbs.

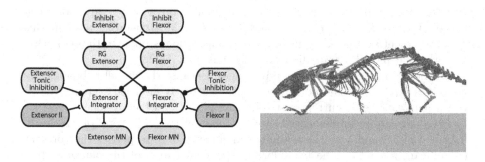

Fig. 1. Left: Single leg joint controller. Circular ends represent inhibitory synapses while triangular ends represent excitatory synapses. Each motoneuron is excited by type II (red) feedback unless inhibited by a CPG half-center (blue). Each motoneuron is also inhibited through an adjustable tonic inhibition (green). Right: Animatlab model of rat. Model reconstruction was done using bone matching of high speed, 3D x-ray video.

Most intra-leg network pathways in our model are developed directly from proposed mechanisms in mammalian literature, which is focused almost exclusively on the hind leg. Stance-to-swing transition is the most studied phenomenon, and it occurs from both reduced firing in Ib Golgi tendon afferents, and increased firing from hip flexor stretching [11]. This integration of signals is shown in Fig. 2 and can be visually followed with the purple dashed synapses; it is indicated by the Swing-Ready neuron (far left), which is excited by Flexor Ia and depressed by Calf Ib. These afferents also contribute to the initiation of swing by initiating contraction of the thigh and shin muscles. Positive force feedback [24] also plays a role in the network. This load feedback can initiate stance if the leg is not already in this state (blue pathways), as well as increase muscle activation (not shown). Experiments have shown that stance is initiated by reduced firing of the hip flexor type II afferents [25]. This indicates that the hip is forward and causes contraction of the extensor and calf muscles and can be followed by the brown pathways. Using only these theories has proven difficult in developing a walking model that is able to step stably, and one notably absent key to the puzzle is the initiation of contraction in the quadriceps muscles. We hypothesize that quadricep contraction is initiated by stretching of the extensor muscle during the swing phase. This causes the quadriceps to start contracting part way through swing, as can be determined through analysis of the kinetic data [20].

The foreleg network is not shown, but it directly parallels connections in the hind-leg network. Scapula movement mimics that of the femur for stance and swing, and the hand mimics that of the foot. However, the inclusion of a fourth joint did require the network to be modified. Movements of the tibia are paralleled by movements of the humerus and the ulna operating in opposite (one extends while the other flexes) to extend and contract the leg during stance and swing.

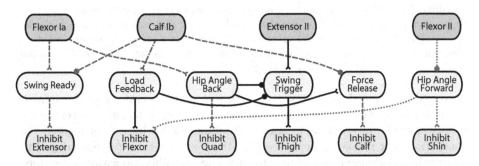

Fig. 2. Intra-leg coordination for the rear rat leg model. Three joints are controlled in this diagram: Hip (flexor and extensor muscles), knee (quad and thigh muscles), and ankle (calf and shin muscles). All blocks are integrator neurons. Sensory information is transduced to current and injected into the sensory neurons (top). Feedback is filtered by a layer of interneurons (middle) and is used to coordinate the CPGs (bottom). Coordinating pathways are inhibitory (circle end) or excitatory (triangle end). Pathways inspired through biological research can be followed as dashed purple [11], dot-dashed blue [24], and dotted brown [25] lines.

3.2   Inter-leg Network

Ipsilateral inter-leg connections are also based on mammalian literature as well as recent, unpublished work in the Fischer lab. They are an order of magnitude weaker than intra-leg connections [4]. Akay et al. [26] postulate three main coordinating influences between fore and hind legs. 1) Foreleg extension reduces onset of the hind limb flexor. 2) The end of activation of the hind limb flexor advances activation of the forelimb flexors. 3) Activation of the forelimb extensors contributes to inhibiting hind limb flexors. These can be followed with the green pathways and labels in Fig. 3. Coordination between the two front legs is

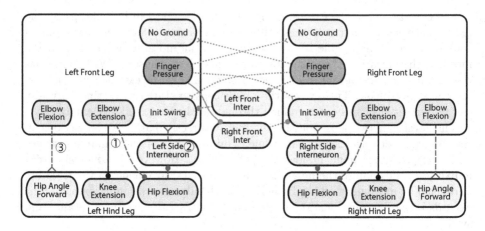

Fig. 3. Inter-leg coordination network. These connections are weaker than connections coordinating intra-limb movement. Contralateral connections encourage only one foot to be on the ground at a time, are developed from [4] and can be followed with the orange dotted lines. Ipsilateral legs are coordinated based on research performed on cats, can be followed with the green dashed lines, and are numbered according to the mechanisms determined by Akay et al. [26].

borrowed from knowledge of insect literature [4] and observations of the animal. If one leg is in swing, the contralateral leg is encouraged to either enter stance if it is in swing, or continue in stance. Additionally, we have implemented the connection that if a leg is in stance then the contralateral leg is encouraged to enter swing. Parameters were tuned to encourage coordination within a few steps of model initiation, but not impose coordination immediately.

3.3   Model Testing

Several tests were run on the model to determine its effectiveness. In all tests, the model was dropped from a short height with limbs in a walking configuration similar to that at the beginning of stance. The hip joints were constrained to a minimum of 5.83 cm from the ground and uniform random noise of 0.1 mV was inserted into the CPG-RG level to encourage different starting neural configurations (which joints begin in flexion, and which in extension) and to keep the system from finding locally minimum solutions during testing. The tests varied the inter-leg connections between fore and hind legs. Inter-leg connection tests consisted of 1) fully connecting system as described above (C123), 2) removal of connection 1 (C23) 3) removal of connection 2 (C13) and 4) removal of connections 1 & 2 (C3). Attempts without connection 3 did not result in a coordinated walking gait and are therefore not presented. Tests were recorded for 5s and analyzed from 1.5-5s (after the model had time to initialize and develop coordination). Each configuration was run 40 times. Kinematic data were collected for fore and hind legs during C123. Mean and standard deviation of step timing was taken for all tests that gained symmetrical coordination within the first 1.5s.

4 Results

4.1 Kinematic Results

The developed network is capable of producing stepping motions for both front and hind legs. Due to the greater available knowledge of hind limb connections and limitations in the modeling environment, the hind leg data fits more closely than the foreleg data. The mean and one standard deviation for animal and simulation data can be seen in Fig. 4. The hip joint of the model has smaller amplitude than the animal, though the phase matches closely. The knee joint closely matches the double extension flexion phases visible in the animal. The ankle joint is the most poorly matched of the hind leg joints. It is in extension during the beginning of swing whereas the animal is clearly in flexion. The joint also takes longer to extend than the animal pre-stance.

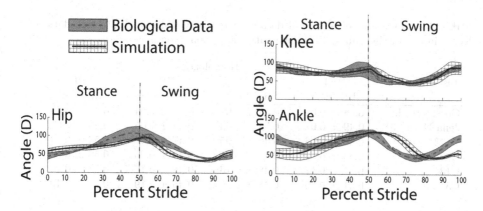

Fig. 4. Comparison of averaged biological recordings with results of the simulation for a single stride over multiple trials. Mean data for each is shown with the dashed and solid lines and one standard deviation is shown shaded or hashed around it. Stride is broken into stance from 0%-50% and swing 50%-100%.

4.2 Inter-leg Coordination Results

Not all of the trials resulted in coordination by the 1.5s mark. C123 produced coordination 80% of the time. C23 produced coordination 53%, C13 produced coordination 75%, and C3 produced coordination 65% of the time.

The resulting coordination is similar to that recorded by Górska et al. [27], however it contains some distinct differences when compared to recent data taken in the Fischer lab. In all instances, homologous limbs (both front, or both hind) are directly out of phase, touching down halfway through the stride of the opposing leg. For ipsilateral legs, the simulation matches data recorded by Górska et al., hind limbs touch down 40% of the way through the stride of the ipsilateral forelimb. This is about 20% out of phase compared to the data collected in the Fischer lab, in which the hind limbs touch down 60% through

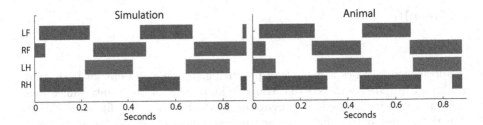

Fig. 5. Footfall diagram of simulation compared with animal data taken at FSU Jena. Solid bars indicate stance of a particular foot. A single step cycle is approximately 450ms long for both the live animal and simulation. For the symmetric gait, front legs alternate stance and swing and are approximately 50% out of phase. When compared to the live animal, the hind legs' phase of the simulation is shifted forward by 20%, which could be corrected by adding a neural delay in connection 3 of the inter-leg network.

Table 1. Comparison of biological data collected by Górska et al. [27] and data collected in the Fischer lab at FSU Jena with simulation experiments

	Homologous			
Test	HF1	HF2	HH1	HH2
	LF→RF	RF→LF	RH→LH	LH→RH
Górska et al.[27]	0.48 ± 0.05	0.52 ± 0.05	0.51 ± 0.04	0.49 ± 0.05
Jena Data	0.49 ± 0.07	0.53 ± 0.03	0.53 ± 0.05	0.49 ± 0.04
C123	0.53 ± 0.08	0.46 ± 0.05	0.47 ± 0.04	0.53 ± 0.04
C23	0.54 ± 0.04	0.46 ± 0.06	0.45 ± 0.05	0.54 ± 0.05
C13	0.51 ± 0.17	0.48 ± 0.16	0.48 ± 0.17	0.51 ± 0.16
C3	0.51 ± 0.20	0.48 ± 0.19	0.45 ± 0.19	0.55 ± 0.19

	Diagonal		Homolateral	
Test	D1	D2	L1	L2
	RH→LF	LH→RF	LF→LH	RF→RH
Górska et al.[27]	0.11 ± 0.05	0.08 ± 0.05	0.40 ± 0.05	0.41 ± 0.06
Jena Data	-0.10 ± 0.08	-0.13 ± 0.04	0.65 ± 0.06	0.60 ± 0.05
C123	0.04 ± 0.06	0.10 ± 0.04	0.43 ± 0.03	0.43 ± 0.04
C23	0.04 ± 0.07	0.11 ± 0.04	0.42 ± 0.03	0.43 ± 0.04
C13	-0.01 ± 0.17	0.08 ± 0.15	0.41 ± 0.12	0.42 ± 0.10
C3	-0.02 ± 0.16	0.07 ± 0.15	0.43 ± 0.11	0.40 ± 0.09

the phase of the preceding limb. Coordination is extremely tight for the walking animals, with standard deviations of 3-8% of the stride. Simulation coordination with the C123 network is close to that of the original coordination, with standard deviation from previous coordination of 3-8%. C23 is also close with standard deviations of 3-7%. C13 still results in coordination, however standard deviations increase to 10-17%, and C3 results in ranges from 9-20%.

5 Discussion and Conclusions

5.1 Kinematics

The presented neural system is capable of producing forward walking motions of the rat simulation with several kinematic movements similar to that of the rat. The hip and knee joints are almost completely within one standard deviation of the biological data while most other joints follow the movements closely. Many joints that do not match well do not because the simulation environment limits the available range of motions for several joints, and torque output of muscles changes significantly based on the changing moment arms of the point-to-point muscle actuators. Though more detailed network construction may lead to better fitting of the biological data, these results support the validity of this initial construction of the network and model design. Timing between joints may become more tightly coupled with the addition of biarticular muscles, and larger, more consistent ranges of motion can be achieved with muscles wrapping around joints.

5.2 Inter-leg Coordination

The data for coordination suggests several conclusions. The developed inter-leg network has no connection between hind legs, and coordination is maintained by connections between the fore and hind legs. Using the mechanisms postulated by Akay et al. the developed network is able to produce a stable symmetrical gait. The data of this gait matches data collected by Górska et al. closely, and supports this as a potential means of coordination for mammals. Not only is the timing between foot touchdowns very similar, but the network keeps the timing as closely coupled as that of the animal. These data contrast with recently collected data on locomoting rats and previously collected data on locomoting cats [28], in which stance of the hind leg occurs after swing of the foreleg begins. This could be accomplished with the current network by adding more neural delay to connection 3, and further research may prove this to be sufficient.

In the model network, an additional pathway was added to connection 1 to slow down movement of knee extension in addition to hip flexion. This connection could be further tested in the animal to determine if knee extension is indeed affected by elbow flexion during the hind limb swing phase.

Performing ablation tests of the inter-leg network indicates the roles of each connection in producing tight, coordinated movement. Connection 3 is sufficient and necessary to produce coordination in the proposed scheme. Connection 2 only acts rostrally and is not sufficient for producing coordination because the hind legs are not coupled contralaterally. Although connection 2 is not sufficient for producing coordination, removal of the connection from the network (C13) reduces the likelihood of walking coordination. Although connection 1 influences hind leg behavior, it does not directly influence phase and only affects motoneuron activation. Like connection 2, connection 1 is not capable of producing symmetrical coordination by itself. In fact, it often causes front and hind legs to fall into phase with each other, producing the opposite of the desired effect. However, connection 1 appears to be important in reducing deviations in walking coordination.

The developed neural system coupled with a biomechanical simulation allows for different theories and network connections to be added and subtracted to study the effects that these systems have on behavior of the animal. It was successfully used to produce symmetrical coordination in a walking rat model, and will be further used to test additional pathways hypothesized to be of importance in coordinating joints and limbs in walking mammals.

References

1. Büschges, A., Schmitz, J., Bässler, U.: Rhythmic patterns in the thoracic nerve cord of the stick insect induced by pilocarpine. Journal of Experimental Biology 198(Pt. 2), 435–456 (1995)
2. Brown, T.G.: On the Nature of the Fundamental Activity of the Nergous Centres; Together with an Analysis of the Conditioning of Rhythmic Activity in Progression, and a theory of the evolution of function in the Nervous System. Journal of Physiology - London (48), 18–46 (1914)
3. Pearson, K.G., Ekeberg, O., Büschges, A.: Assessing sensory function in locomotor systems using neuro-mechanical simulations. Trends in Neurosciences 29(11), 625–631 (2006)
4. Cruse, H.: What mechanisms coordinate leg movement in walking arthropods? Trends in Neurosciences 13(1990), 15–21 (1990)
5. Espenschied, K.S., Quinn, R.D., Chiel, H.J., Beer, R.: Leg Coordination Mechanisms in the Stick Inspect Applied to Hexapod Robot Locomotion. Adaptive Behavior 1(4), 455–468 (1993)
6. Ekeberg, O., Blümel, M., Büschges, A.: Dynamic simulation of insect walking. Arthropod Structure & Development 33(3), 287–300 (2004)
7. Schilling, M., Hoinville, T., Schmitz, J., Cruse, H.: Walknet, a bio-inspired controller for hexapod walking. Biological Cybernetics 107(4), 397–419 (2013)
8. Rutter, B.L., Taylor, B.K., Bender, J.A., Blümel, M., Lewinger, W.A., Ritzmann, R.E., Quinn, R.D.: Descending commands to an insect leg controller network cause smooth behavioral transitions. In: Intelligent Robots and Systems (IROS 2011) (2011)
9. Daun-Gruhn, S.: A mathematical modeling study of inter-segmental coordination during stick insect walking. Journal of Computational Neuroscience, 255–278 (2010)
10. Szczecinski, N.S., Brown, A.E., Bender, J.A., Quinn, R.D., Ritzmann, R.E.: A Neuromechanical Simulation of Insect Walking and Transition to Turning of the Cockroach Blaberus discoidalis. Biological Cybernetics (2013)
11. Pearson, K.G.: Role of sensory feedback in the control of stance duration in walking cats. Brain Research Reviews 57(1), 222–227 (2008)
12. Ekeberg, O., Pearson, K.G.: Computer simulation of stepping in the hind legs of the cat: an examination of mechanisms regulating the stance-to-swing transition. Journal of Neurophysiology 94(6), 4256–4268 (2005)
13. Harischandra, N., Ekeberg, O.: System identification of muscle-joint interactions of the cat hind limb during locomotion. Biological Cybernetics 99(2), 125–138 (2008)

14. Bicanski, A., Ryczko, D., Knuesel, J., Harischandra, N., Charrier, V., Ekeberg, O., Cabelguen, J.M., Ijspeert, A.J.: Decoding the mechanisms of gait generation in salamanders by combining neurobiology, modeling and robotics. Biological Cybernetics 107(5), 545–564 (2013)
15. McCrea, D.A., Rybak, I.A.: Organization of mammalian locomotor rhythm and pattern generation. Brain Research Reviews 57(1), 134–146 (2008)
16. Zhong, G., Shevtsova, N.A., Rybak, I.A., Harris-Warrick, R.M.: Neuronal activity in the isolated mouse spinal cord during spontaneous deletions in fictive locomotion: insights into locomotor central pattern generator organization. The Journal of Physiology 590(Pt. 19), 4735–4759 (2012)
17. Büschges, A., Borgmann, A.: Network modularity: back to the future in motor control. Current Biology: CB 23(20), R936–R938 (2013)
18. Markin, S.N., Klishko, A.N., Shevtsova, N.A., Lemay, M.A., Prilutsky, B.I., Rybak, I.A.: Afferent control of locomotor CPG: insights from a simple neuromechanical model. Annals of the New York Academy of Sciences 1198, 21–34 (2010)
19. Cofer, D., Cymbalyuk, G., Reid, J., Zhu, Y., Heitler, W.J., Edwards, D.H.: AnimatLab: a 3D graphics environment for neuromechanical simulations. Journal of Neuroscience Methods 187(2), 280–288 (2010)
20. Witte, H., Biltzinger, J., Hackert, R., Schilling, N., Schmidt, M., Reich, C., Fischer, M.S.: Torque patterns of the limbs of small therian mammals during locomotion on flat ground. The Journal of Experimental Biology 205(Pt. 9), 1339–1353 (2002)
21. Daun-Gruhn, S., Büschges, A.: From neuron to behavior: dynamic equation-based prediction of biological processes in motor control. Biological Cybernetics, 71–88 (July 2011)
22. Brownstone, R.M., Bui, T.V.: Spinal interneurons providing input to the final common path during locomotion. Progress in Brain Research (902), 81–95 (2010)
23. Fischer, M.S., Schilling, N., Schmidt, M., Haarhaus, D., Witte, H.: Basic limb kinematics of small therian mammals. The Journal of Experimental Biology 205(Pt. 9), 1315–1338 (2002)
24. Prochazka, A., Gillard, D., Bennett, D.J.: Positive force feedback control of muscles. Journal of Neurophysiology 77(6), 3226–3236 (1997)
25. McVea, D.A., Donelan, J.M., Tachibana, A., Pearson, K.G.: A role for hip position in initiating the swing-to-stance transition in walking cats. Journal of Neurophysiology 94(5), 3497–3508 (2005)
26. Akay, T., McVea, D.A., Tachibana, A., Pearson, K.G.: Coordination of fore and hind leg stepping in cats on a transversely-split treadmill. Experimental Brain Research. Experimentelle Hirnforschung. Expérimentation Cérébrale 175(2), 211–222 (2006)
27. Górska, T., Zmyslowski, W., Majczynski, H.: Overground locomotion in intact rats: interlimb coordination, support patterns and support phases duration. Acta Neurobiologiae Experimentalis 59(2), 131–144 (1999)
28. Cruse, H., Warnecke, H.: Coordination of the legs of a slow-walking cat. Experimental Brain Research, 147–156 (1992)

A Minimum Attention Control Law
for Ball Catching

Cheongjae Jang[1], Jee-eun Lee[1], Sohee Lee[2], and Frank C. Park[1]

[1] Robotics Laboratory, Seoul National University, Seoul, South Korea
[2] Department of Electrical Engineering and Information Techonology,
Technische Universität München, Munich, Germany
fcp@snu.ac.kr
http://robotics.snu.ac.kr/fcp

Abstract. We present an attention-minimizing LQR-based feedback control law for vision-based ball catching. Taking Brockett's control attention functional as our performance criterion, and under the simplifying assumption that the optimal control law is the sum of a linear time-varying feedback term and a time-varying feedforward term, we show that our control law is stable, and easily obtained as the solution to a finite-dimensional optimization problem over the set of symmetric positive-definite matrices. We perform numerical experiments for robotic ball-catching and compare our control law with a discretized version obtained as the solution to a mathematical programming problem. Like human ball catching, our results also exhibit the familiar transition from open-loop to closed-loop control during the catching movement, and also show improved robustness to spatiotemporal quantization. Our approach is applicable to more general control settings in which multi-tasking must be performed under limited computation and communication resources.

Keywords: Minimum attention, spatiotemporal quantization, optimal feedback control, ball catching.

1 Introduction

Programming and controlling a robot to catch a ball of known shape and size is by now regarded as a fairly routine task—using vision to track a ball in flight, and applying an appropriate visual servo control law to guide the robot arm to the ball, for example, are no longer viewed as technically challenging. On the other hand, if the robot were now asked to juggle ten or more balls, using low-performance, noisy vision sensors, and able to draw upon only a limited portion of the available computational and communication resources, the problem then becomes much more challenging. Today, robots are becoming increasingly complex in structure—a typical humanoid robot has on the order of 30 to 50 degrees of freedom, for example, with a corresponding number of sensors and actuators—and are being asked to perform increasingly complex tasks, many of them simultaneously. Despite this trend, current robot motion planning and

A. Duff et al. (eds.): Living Machines 2014, LNAI 8608, pp. 154–165, 2014.
© Springer International Publishing Switzerland 2014

control laws for the most part do not take into account physical limitations on computation, communication, and memory.

Humans, in contrast, are remarkably adept at multitasking, and with modest effort can learn how to juggle two or more balls. Beginning with the seminal work of [4], how humans catch objects has been the subject of considerable research in both the human motor control [5], [6] and computer animation literature [7]. Some of the main findings include: (i) Human eye movements consist of rapid catch-up saccades that try to quickly get a rough estimate of the object location, and smooth pursuit movements that continuously track an object. Because the object's flight duration is quite short, and human vision is relatively slow, the type and sequence of eye movements are critical to successful catching. (ii) The hand begins to move early when only crude estimates of the object position are available, and under open-loop control. Later in the ball's trajectory the open loop arm motion is followed by a closed-loop homing phase.

The aim of this paper is to develop a control strategy for vision-based robot ball catching that attempts to minimize the required computation and communication resources. There are a number of ways to quantitatively formulate such a criterion: (i) Measuring the robustness of the control law to both space and time quantization (this usually involves a discretization of the system with respect to both state and time), and (ii) Minimizing the control attention functional of Brockett, which is a cost functional on control law that quantify the cost on control implementation from the change of the control w.r.t. state and time change [1], [2], [3]. Although the two formulations are different manifestations of the same qualitative feature, in the former case the discretization of the system often leads to a high-dimensional nonlinear programming problem that is highly sensitive to parameters, fraught with numerous local minima, and subject to discontinuous gradients. The latter formulation, in contrast, leads to an intractable infinite-dimensional optimization problem.

The main contribution of this paper is an attention minimizing control law based on the linear quadratic regulator (LQR). We make the simplifying assumption that the optimal control law is the sum of a linear time-varying feedback term and a time-varying feedforward term; from an implementation perspective this assumption is quite reasonable, as most practical control laws are of this structure (feedback to compensate error along with feedforward to optimize certain performance criterion). The resulting control law is stable and easily obtained as the solution to a finite-dimensional optimization problem over the set of symmetric positive-definite matrices.

We perform numerical experiments for robotic ball-catching that compare our minimum attention LQR control law with a discretized version obtained as the solution to a mathematical programming problem. We show that, remarkably, our optimal LQR-based minimum attention control law shares many of the main features of human ball tracking and catching, e.g., the transition from open-loop to closed-loop control during the movement, and improved robustness to spatiotemporal quantization. Our approach is in fact applicable to much more general control settings, in which multi-tasking must be performed by the system under limited computation and communication resources.

The paper is organized as follows. In Section 2 we examine some of the formulations that have been developed to measure controller robustness to spatiotemporal quantization, focusing in particular on Brockett's attention functional. Section 3 describes our main result, the minimum attention LQR tracking controller. Results of numerical ball-catching experiments are reported in Section 4.

2   Robustness to Quantization and Control Attention

Practical implementations of control laws ultimately require spatiotemporal quantization of the control and measurement signals at some stage. One way in which control performance can be measured is with respect to the resolution of the underlying quantization. There is in fact a growing body of literature on the control of quantized systems, and more generally, on the control of systems subject to data rate constraints (see, e.g., [9], [10], [11], [12]). Most of these works make various simplifying assumptions about the underlying system that are often too restrictive for typical robots (e.g., all systems are linear), or focus on a particular aspect of control (e.g., feedback stabilization). For robots, one must consider not only the control implementation costs, but also the quality of the trajectories.

Another way to measure a controller's attention that makes contact with this body of work is as follows. Given the system $\dot{x} = f(x) + G(x)u$ with an associated performance criterion

$$J(u) = \phi(x(t_f), t_f) + \int_{t_0}^{t_f} L(x, u, t)\, dt, \tag{1}$$

we consider its discretized version: the state equation and cost function are of the form

$$x_{k+1} = \Phi(x_k, u_k, t_k) \tag{2}$$

$$J(u_0, \ldots, u_{N-1}) = \sum_{i=0}^{N} \Psi(x_k, u_k, t_k). \tag{3}$$

The control can only be updated at a finite set of N ordered times $\{t_0, \ldots, t_{N-1}\}$, and over each interval $[t_i, t_{i+1}]$, the control u_i is assumed constant. We then consider the following finite-dimensional optimization problem: Find the update times $\{t_i\}$ and controls $\{u_i\}$ that minimize the cost J subject to the discrete-time state dynamics (2).

By converting the original continuous-time problem into a finite-dimensional optimization problem, it would seem that any number of gradient-based optimization algorithms can be used. It has been pointed out in [8], however, that the gradients can become discontinuous at the update times. More problematically, the resulting optimization problem can become highly sensitive to parameters and initial conditions, with numerous local minima caused by the inherent non-linearity and discontinuity of the problem. Our later experiments on ball catching illustrate some of the practical difficulties with this approach.

In this paper we shall instead exploit Brockett's minimum attention functional [1], [2], [3]. Given a system with state equation $\dot{x} = f(x, u, t)$, where t denotes time, $x \in X \subseteq \mathbb{R}^n$ is the state vector (X denotes the state space), and $u(x, t) \in \mathbb{R}^m$ is the control vector, the attention functional $J(u)$ is defined as

$$J(u) = \int_0^{t_f} \int_X \alpha \left\| \frac{\partial u}{\partial x} \right\|^2 + (1 - \alpha) \left\| \frac{\partial u}{\partial t} \right\|^2 \, dx \, dt, \tag{4}$$

where $\alpha \in [0, 1]$ is a scalar weighting term. The basic premise behind this attention functional is that the simplest control law to implement is a constant input; the more frequently a control changes, the more effort is required to implement it. Control laws typically depend on both the state and time, so that the cost of implementation can be linked with the rate at which the control changes with respect to changes in state (the first term of (4)) and changes in time (the second term of (4)). The parameter α continuously adjusts the ratio of the cost placed on the closed-loop term relative to that of the open-loop term.

While the attention functional is intuitively appealing, solutions to the variational problem are very difficult to come by—existence of solutions is not ensured, and the multidimensional integral over both the state space X and time usually renders the problem intractable. One way to make this problem more tractable is to restrict our focus to systems of the form $\dot{x} = f(x) + G(x)u$, with $u = K(t)x + v(t)$, where $K(t)$ is a time-varying feedback gain matrix and $v(t)$ is a feedforward term. Though this assumption ignores any non-linear relation of the input to state and the optimization result in this form may be suboptimal, it is reasonable to implement and most practical control laws are in this structure. In this case the integral over the state space X can be ignored, and the attention functional reduces to

$$J(u) = \int_{t_0}^{t_f} \alpha \left\| K(t) \right\|^2 + (1 - \alpha) \left\| \dot{u}(t) - K(t)\dot{x} \right\|^2 \, dt, \tag{5}$$

Assuming this class of controls, some limited results on minimum attention trajectories have been obtained for simple kinematic mobile robot models [16]. In the next section, we show how to merge the LQR control framework with the minimum attention problem, and derive a minimum attention feedforward-feedback control law that exploits the many advantages of LQR control, while minimizing attention in the sense of Brockett.

3   Minimum Attention LQR Tracking

In this section we consider the problem of deriving an LQR tracking control law that minimizes attention in the sense of Brockett. Suppose we are given a reference trajectory $r(t)$, which is obtained via, e.g., measurement-based estimation of an object's trajectory. Consider the following tracking problem (here Q, R, P_f are all assumed symmetric positive-definite):

$$\min_u J(u) = \frac{1}{2}(x(t_f) - r(t_f))^T P_f(x(t_f) - r(t_f))$$
$$+ \frac{1}{2} \int_{t_0}^{t_f} (x - r)^T Q(t)(x - r) + u^T R(t)\, u\, dt \tag{6}$$

subject to $\dot{x} = A(t)x(t) + B(t)u(t)$, with $x(0)$ given. The optimal control is [13]:

$$u(x, t) = -R(t)^{-1}B(t)^T(P(t)x + s(t)) \tag{7}$$
$$-\dot{P} = PA + A^T P - PBR^{-1}B^T P + Q, \quad P(t_f) = P_f \tag{8}$$
$$-\dot{s} = -PBR^{-1}B^T s + A^T s - Qr, \quad s(t_f) = -P_f r(t_f). \tag{9}$$

If the state equation is nonlinear (i.e., $\dot{x} = f(x, u, t)$), a sequential linear quadratic (SLQ) algorithm [14] can be used to obtain the optimal solution for the linearized system with $A(t) = \frac{\partial f}{\partial x}$, $B(t) = \frac{\partial f}{\partial u}$. Provided the SLQ algorithm converges, the optimal solution and control are obtained as trajectories $(x^*(t), u^*(t))$, but the optimal control can still be expressed as the sum of a feedback and feedforward term as follows:

$$u(x, t) = u^*(t) - R(t)^{-1}B(t)^T P(t)(x - x^*(t)) \tag{10}$$
$$-\dot{P} = PA + A^T P - PBR^{-1}B^T P + Q, \quad P(t_f) = P_f \tag{11}$$
$$\text{where} \quad A(t) = \frac{\partial f}{\partial x}|_{(x^*(t), u^*(t))}, \; B(t) = \frac{\partial f}{\partial u}|_{(x^*(t), u^*(t))}.$$

This control is the same as $u^*(t)$ for the optimal trajectory, but when the system is disturbed by noise, the presence of the feedback term assures stability near the optimal solution $x^*(t)$. Lyapunov stability is satisfied for this control if $Q > 0$, $P(t) > 0$ for all $t \in [0, t_f]$, and the Lipschitz condition holds for $\frac{\partial F}{\partial x}(x, t)$ (i.e., $\|\frac{\partial F}{\partial x}(x_1, t) - \frac{\partial F}{\partial x}(x_2, t)\| \le L\|x_1 - x_2\|$) where $F(x, t) = f(x, u(x, t))$ [15]. For the 2-DOF robot arm that we use later in our case studies, the Lipschitz condition holds for bounded $\dot{q}_1, \dot{q}_2$, and bounded solutions $P(t)$ of Riccati equation. Here, Q, R, P_f are chosen to satisfy the above stability conditions.

Based on this we now derive our minimum attention LQR tracking controller. As can be seen from Equations (7),(10), the LQR control contains both a feedforward term and a linear feedback term obtained as the solution to a Riccati equation. We now consider the problem of determining the form of the LQR controller that minimizes attention for general nonlinear systems as specified by (10). The attention functional for this controller is given by

$$\min_{P_f, Q, R} J_{att} = \int_{t_0}^{t_f} \alpha \left\| \frac{\partial u}{\partial x} \right\|^2 + (1 - \alpha) \left\| \frac{\partial u}{\partial t} \right\|^2 dt \tag{12}$$
$$= \int_{t_0}^{t_f} \alpha \|K(t)\|^2 + (1 - \alpha) \|\dot{u}(t) - K(t)\dot{x}\|^2 dt, \tag{13}$$

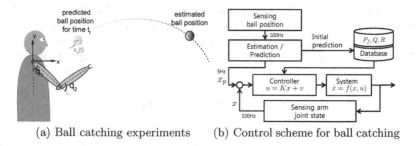

(a) Ball catching experiments   (b) Control scheme for ball catching

Fig. 1. Ball catching experiments: Initial arm posture is $q_1 = 0°, q_2 = 0°$. Ball position is estimated by sensors (e.g., eyes) at 100Hz with noise, and the ball position at t_f is predicted by a Kalman filter.

subject to $u = K(t)x + v(t)$, where $K(t) = -R(t)^{-1}B(t)^T P(t)$, $v(t) = u^*(t) + R(t)^{-1}B(t)^T P(t)x^*(t)$ with the solution to Equation (6), $x^*(t), u^*(t)$, and $P(t)$ satisfying Equation (11). Here, the integration over the state space can be ignored for small neighborhoods around the optimal solution, $x^*(t)$. J_{att} can be adjusted by changing the values of Q, R and P_f, and the attention functional can now be minimized with respect to symmetric positive-definite Q, R, P_f. The result is a stable trajectory tracking controller that incurs minimal control attention costs, and is robust to small errors in the state and time.

4   Ball Catching Experiments

4.1   Problem Formulation and Experiment Design

For the experiment, we consider a simplified two-dimensional version of the ball catching problem. Referring to Figure 1(a), the ball is assumed to be thrown to a person, who catches the ball at a designated time t_f on the $x - y$ plane. The state of the ball $z \in \mathbb{R}^4$ is defined to be the position (x, y) and velocity $(\dot{x}, \dot{y})$ of the ball. Suppose the sensor measuring the complete state of the ball is subject to Gaussian noise. The discretized state equations with N time intervals is given as follows ($\Delta t = t_f/N = 10$ ms, $k = 1, \ldots, N$):

$$z_{k+1} = Az_k + B + w_k \tag{14}$$

$$s_k = Hz_k + v_k \tag{15}$$

where $A = \begin{bmatrix} I_{2\times2} & I_{2\times2}\Delta t \\ O_{2\times2} & I_{2\times2} \end{bmatrix}, B = \begin{bmatrix} O_{3\times1} \\ g\Delta t \end{bmatrix}, H = I_{4\times4}$, g is the gravitational acceleration, z_k is the state of the ball and s_k is the measurement at time index k, $w_k \sim (0, Q_k)$ and $v_k \sim (0, R_k)$ are the Gaussian noise entering into the state equations for the ball dynamics and sensor measurements, respectively. Here, $Q_k = \sigma_1 diag\{0, 0, 1, 1\}$ and $R_k = \sigma_2 I_{4\times4}$ are constants.

With this dynamic model, Kalman filter is used to estimate the current state and predict the state of the ball at time t_f. The joint reference trajectory is obtained by solving the inverse kinematics of the arm for the predicted ball position. Initially the prediction of the final ball state is inaccurate due to noisy measurements, but with time, the prediction becomes more accurate.

To model the arm dynamics, we use a two-link planar model with dynamic equations of the form

$$\tau = M(q)\ddot{q} + C(q,\dot{q})\dot{q} + N(q,\dot{q}), \tag{16}$$

where $q \in \mathbb{R}^2$ denotes the joint angles and $\tau \in \mathbb{R}^2$ represents the input joint torques. The kinematic and inertial parameter values are taken from [17] and chosen to closely match those of a typical human arm.

The state of the arm is set to $x = (q_1, q_2, \dot{q}_1, \dot{q}_2)^T$, with state equations $\dot{x} = f(x, u)$, where the input u is the joint torque vector in Equation (16). The reference state, $x_p(t)$, is set to the joint angle obtained by solving the inverse kinematics for the predicted ball position (assuming the joint velocity is zero).

4.2   Preprocessing of LQR Cost Functions

To obtain a minimal attention tracking controller, we consider a tracking problem of Equation (6) for the reference trajectory $r(t) = x_p(t)$, subject to $\dot{x}(t) = f(x(t), u(t))$. Note that the optimal Q, R, and P_f need to be determined in real-time in order to apply to ball catching. As a practical measure, Q, R, and P_f are optimized a priori for specific regions where grasping is likely to occur. We partition the robot workspace into several regions, and assume a set-point as the reference trajectory for each region. We then find and store the optimal Q, R, P_f, and optimal input u. The reference points are set as $(x, y) = r(\cos\theta, \sin\theta)$, with $r \in \{0.2, 0.3, 0.4, 0.5\}$ and $\theta \in \{0°, 15°, 30°, 45°, 60°, 75°, 90°\}$ where shoulder joint position is set as the origin, and r is the distance from shoulder joint to the reference point. For each of these fixed set points SLQ algorithm is applied to obtain the optimal solution. Since the most important objective is to catch the ball, we add to the attention functional an additional term reflecting the final position error, to ensure that the hand can reach the predicted ball position at t_f.

Our optimization problem then becomes

$$\min_{P_f, Q, R} J = \mu J_{error} + J_{att} \tag{17}$$

where $J_{error} = \|x(t_f) - x_p(t_f)\|^2_{pos}$ is the final positional error in the state space and J_{att} is the attention functional of Equation (13). For this set of numerical experiments, we set $P_f = diag(\rho_1, \rho_2, 0.01\rho_1, 0.01\rho_2)$, $Q = diag(1, 1, 0.01, 0.01)$, $R = diag(\rho_3, \rho_3)$. The reason each component of P_f and Q is set to 1 and 0.01, respectively, is that for our catching problem, the state related to the final position is more important than velocities. Also, we set appropriate lower bounds for ρ_1 and ρ_2 as 1, since if they become too small, the SLQ algorithm may become overly sensitive to the initial guess.

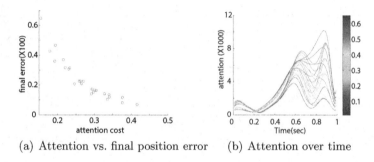

(a) Attention vs. final position error   (b) Attention over time

Fig. 2. The relationship between attention functional cost and final position error when $\alpha = 0.99$. (a) Plot of the attention functional cost vs. final position error for the LQR ball catching controller (b) The color of each graph represents the level of final position error of the LQR ball catching control. Blue indicates a small position error.

Observe from Figure 2(a) that there exists a tradeoff between attention and the final position error for a specified range of values for P_f, Q, R. A large feedback gain reduces the final error, but it also increases the cost of the attention functional. Figure 2(b) shows the attention functional as a function of time, with the final position error values indicated by color. The figure shows that the overall control attention increases as the final error decreases. This tendency becomes more apparent near the final time, i.e., more attention is needed immediately prior to catching the ball in order to reduce the final error.

4.3   LQR Tracking Control with Updating References

In a real ball catching task, when the first prediction of the final ball position is given, we determine the Q, R, P_f, and u from the stored values that best fits the initial estimate. Since it is not possible to solve for the LQR control with the future reference trajectory $x_p(t)$, we use the latest prediction of the ball as a reference point. During the task, corrective movements must be calculated by solving for the LQR control sequentially as the prediction is updated. The control scheme is described in Figure 1(b). For our simulation, the update rate of the reference is set as 5Hz.

Figure 3(a), illustrates our simulation system for ball catching and the corresponding optimal trajectory of the hand. For this experiment $\rho_1 = 21.69, \rho_2 = 17.48, \rho_3 = 1.64$ is used from the optimization result of the previous section. As shown in Figure 3(b), the feedback gain is initially small, but has large values near the final time. A large feedback gain near the final time can be explained by the need of feedback controller to catch the ball by minimizing the final positional error in Equation (17). The feedback gain in the middle is needed to maintain the angle of the elbow joint, which is important to reduce the torque to lift the forearm toward the ball. Also, the feedforward inputs tend to decrease while feedback gains tend to increase near t_f; this is particularly true at the elbow joint. These results are consistent with the way humans catch balls,

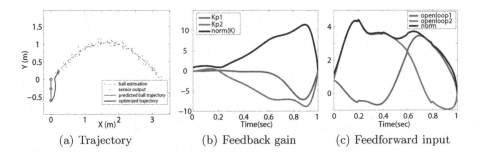

(a) Trajectory   (b) Feedback gain   (c) Feedforward input

Fig. 3. Ball catching simulation. The distance between end-effector and ball prediction position is 0.63cm at t_f, which can be thought as a successful catch.

i.e., initially the arm moves under feedforward control, while toward the end, feedback gains are increased so as to accurately catch the ball.

For the two LQR controllers—one of which takes into account attention—we now conduct the following set of numerical experiments. Time is uniformly discretized over the movement duration, with $100Hz$ or $10Hz$ update rate. Joint position and velocity measurements are quantized to $0.001°, 0.001°/s$ or $5°, 5°/s$ increments, respectively. Then control under quantization can be obtained as follows:

1. **Time quantization.** If $u = u(x, t)$ (i.e., both the state x and time t appear explicitly in the control), over the interval $[t_{i-1}, t_i]$ the actual control input is set to the following average value of $u(x,t)$: $u_i = \frac{1}{t_i - t_{i-1}} \int_{t_{i-1}}^{t_i} u(x_{i-1}, t) \, dt$.

2. **Space quantization.** If the actual state at time t_{i-1} is $x(t_{i-1})$, then the state x_{i-1} for calculating the control law under quantization is measured with limited resolution γ: $x_{i-1} = floor(\frac{x(t_{i-1})}{\gamma}) * \gamma + \frac{\gamma}{2}$.

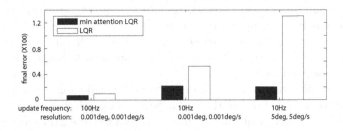

Fig. 4. Final joint position error under time and space quantization

Figure 4 shows the final joint position error(J_{error} in Equation (17)) under time and space quantization. The minimum attention LQR controller seems to be more robust to quantization over both time and space. These results confirm that considering the control attention costs make the control be robust to spatiotemporal quantization.

4.4 Ball Catching Subject to Attention Limits

To compare the results of our minimum attention LQR tracking controller with existing approaches, we assume that there exists limits on the overall control attention rather than minimizing controller attention, and assume feedback gains and feedforward inputs that can be updated only a finite number of times. Also a constraint is imposed on the norm of feedback gains, to limit attention related to feedback. The goal is to catch the ball at a designated target location while minimizing input torques. For this problem we assume a control input of the form $\tau = -K_p(t)(q - q_d(t)) - K_v(t)(\dot{q} - \dot{q}_d(t)) + v(t)$. It is further assumed that over the interval $[t_{i-1}, t_i]$, the feedback gains $K_p(t)$, $K_v(t)$, and feedforward term $v(t)$ are respectively discretized to constant values K_{p_i}, K_{v_i}, and v_i, where K_{p_i} and K_{v_i} are diagonal and positive-definite. The robot state $x(t) = \begin{bmatrix} q^T & \dot{q}^T \end{bmatrix}^T$ is assumed to be estimated perfectly, and the desired state of the arm $x_p(t)$ is obtained as done in the previous section. The control objective is to minimize

$$\min_{t_i, K_i, v_i} J = (x(t_f) - x_p(t_f))^T Q(x(t_f) - x_p(t_f)) + \int_{t_0}^{t_f} \|u(t)\|^2 dt, \qquad (18)$$

subject to $\int_{t_0}^{t_f} \|K(t)\|^2 \, dt \leq A$, where $x(t_f)$ and $x_p(t_f)$ are the final state and desired state of the arm, respectively, $u(t)$ is the control input trajectory, and A is a constant to limit feedback gains. This problem then becomes a finite-

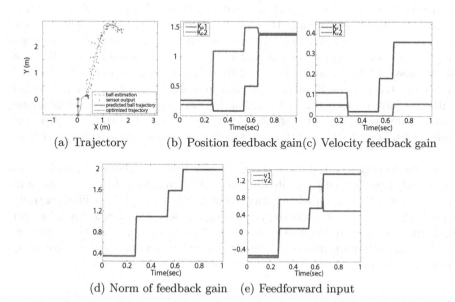

(a) Trajectory   (b) Position feedback gain(c) Velocity feedback gain

(d) Norm of feedback gain   (e) Feedforward input

Fig. 5. For a given goal position of the ball, the optimal feedback gains and feedforward inputs are shown. The optimal update time is given as (0.28s, 0.54s, 0.66s). The result included is the optimum among several local minima from different initial conditions.

dimensional optimization problem over the set $\{t_{i-1}, K_{p_i}, K_{v_i}, v_i\}_{i=1,...,N}$. The arm movement duration is set to one second.

Results of the optimization for the case $N = 4$ are shown in Figure 5. As shown in Figure 5(d), the norm of the feedback gains tend to increase with time. This implies that more attention is needed to catch the ball as time goes by with limitations on attention. The trajectory of the hand as well as the tendency of the gain are consistent with the results of the previous section. Note that the values of the positional feedback gain K_p are larger than those of the velocity feedback gain K_v, implying that for ball catching movements with varying predicted position, position information is more useful than velocity information. Despite the limitations on control inputs, the main results are consistent with the way humans catch balls.

It should be noted, however, that this problem is highly nonlinear, and the objective function is also nonconvex. The problem is highly sensitive to initial conditions, with the likelihood of numerous local minima. Also the optimization procedure takes considerably longer as the dimension of the problem increases. (The number of optimization variables $= 7 \times N - 1$).

5   Conclusions

In this paper we have examined the ways in which attention can be accounted for in the generation of robot trajectories and their associated control laws. With robots being asked to perform increasingly complex tasks, often simultaneously, it is timely to ask whether and how robot motion planning and control laws should consider these limitations on the available computation, communication, and memory resources. We have examined in some detail the attention functional of [1] as a tool for quantifying a control law's implementation costs. We have shown how to forge it into a practical tool, by obtaining a minimum attention trajectory tracking LQR control law that guarantees stability near the desired path. We have developed and applied such a framework to the two dimensional ball-catching problem. Quantization experiments show that the controller optimized with respect to the attention functional is more robust to quantization error.

Alternate notions for quantized control with limited attention have also been examined, together with a finite-dimensional optimization framework for finding such laws. We compared this framework to the optimized tracking controller from the perspective of attention. Further experimental studies and a more comprehensive analysis are required to draw any firm conclusions, but the framework for bringing attention into the discussion on robot planning and control seems to be a useful one.

Acknowledgments. The research was supported by a grant to Bio-Mimetic Robot Research Center funded by Defense Acquisition Program Administration, SNU-IAMD, and the BK21+ program in mechanical and aerospace engineering at Seoul National University.

References

1. Brockett, R.W.: Minimum attention control. In: Proc. IEEE Int. Conf. Decision and Control, pp. 2628–2632 (1997)
2. Brockett, R.W.: Minimizing attention in a motion control context. In: Proc. IEEE Int. Conf. Decision and Control, pp. 3349–3352 (2003)
3. Brockett, R.W.: Notes on the control of the Liouville equation. In: Control of Partial Differential Equations. Lectures Notes in Mathematics, vol. 2048, Springer, Berlin
4. Woodworth, R.S.: Accuracy of voluntary movement. The Psychological Review: Monograph Supplements 3(3) (1899)
5. Zago, M., McIntyre, J., Senot, P., Lacquaniti, F.: Visuo-motor coordination and internal models for object interception. Experimental Brain Research 192(4), 571–604 (2009)
6. Dessing, J.C., Peper, C.L.E., Bullock, D., Beek, P.J.: How position, velocity, and temporal information combine in the prospective control of catching: data and model. Journal of Cognitive Neuroscience 17(4), 668–686 (2005)
7. Yeo, S.H., Lesmana, M., Neog, D.R., Pai, D.K.: Eyecatch: simulating visuomotor coordination for object interception. ACM Transactions on Graphics (TOG) 31(4) (2012)
8. Flasskamp, K., Murphey, T., Ober-Blobaum, S.: Switching time optimization in discretized hybrid dynamical systems. In: Proc. IEEE Conf. Decision and Control, pp. 707–712 (2012)
9. Nair, G., Fagnani, F., Zampieri, S., Evans, R.J.: Feedback control under data rate constraints: an overview. Proc. IEEE 95(1), 108–137 (2007)
10. Hristu, D.: Optimal Control with Limited Communication. Ph.D. dissertation, Harvard University, Cambridge Massachusetts (1999)
11. Liberzon, D.: Hybrid feedback stabilization of systems with quantized signals. Automatica 39, 1543–1554 (2003)
12. Li, K., Baillieul, J.: Robust and efficient quantization and coding for control of multidimensional linear systems under data rate constraints. Int. J. Robust Nonlinear Control 17, 898–920 (2007)
13. Kirk, D.E.: Optimal control theory: an introduction. Courier Dover Publications (2012)
14. Sideris, A., Bobrow, J.E.: An efficient sequential linear quadratic algorithm for solving nonlinear optimal control problems. IEEE Transactions on Automatic Control 50(12), 2043–2047 (2005)
15. Khalil, H.K., Grizzle, J.W.: Nonlinear systems, vol. 3. Prentice Hall, Upper Saddle River (2002)
16. Lee, S., Park, F.C.: Mobile robot motion primitives that take into account the cost of control. In: Latest Advances in Robot Kinematics, pp. 429–436. Springer, Netherlands (2012)
17. Nakano, E., Imamizu, H., Osu, R., Uno, Y., Gomi, H., Yoshioka, T., Kawato, M.: Quantitative examinations of internal representations for arm trajectory planning: minimum commanded torque change model. J. Neurophysiology 81, 2140–2155 (1999)

Simulated Neural Dynamics Produces Adaptive Stepping and Stable Transitions in a Robotic Leg

Matthew A. Klein[1], Nicholas S. Szczecinski[1],
Roy E. Ritzmann[2], and Roger D. Quinn[1]

[1] Dept. of Mechanical and Aerospace Engineering,
[2] Dept. of Biology,
Case Western Reserve University,
10900 Euclid Ave., Cleveland, OH, 44106, USA
{matthew.a.klein,nicholas.szczecinski,
roy.ritzmann,roger.quinn}@case.edu
http://biorobots.case.edu/

Abstract. Animals exhibit flexible and adaptive behavior. They can change between modes of locomotion or modify the details of a step to better suit their environment. Insects have massively distributed control architectures in which each joint has its own central pattern generator (CPG), which is coordinated with its neighbors only through sensory information. Different modes of walking (forward, turning, etc.) can be produced by changing which CPGs are affected by which sensory information, called a reflex reversal. The presented robotic leg is controlled by a computational neuroscience model of part of the nervous system of the cockroach *Blaberus discoidalis*. It steps adaptively to correct for unexpected obstacles and can reverse reflexes to produce turning motions.

Keywords: Legged-locomotion, robot control, computational neuroscience.

1 Introduction

The control of legged locomotion has long been an area of interest in robotics. Highly articulated legs provide many advantages over wheels and tracks when navigating uneven terrain in both natural and man-made environments. The problem lies in coordinating many degrees of freedom and complex geometries to generate propulsion and support while simultaneously adapting to a changing terrain. Some success has been found using classic control strategies. However, new strategies, based on our knowledge of insect leg control, show great promise in this area[19].

Much is known about the control of walking in insects, particularly the stick insect [1][3][6][8][15] and cockroaches [5][18][25][26]. Experiments on the nervous systems of insects have revealed a reflex-based, distributed architecture to be responsible for controlling joint movement in stick insects and cockroaches, as opposed to a centralized system that controls each joint. These experiments have

A. Duff et al. (eds.): Living Machines 2014, LNAI 8608, pp. 166–177, 2014.

also identified many of the sensory signals that coordinate walking in a single stick insect leg[3]. It has also been shown that changing leg motion, for instance from forward walking to turning, often involves making minor changes to the way sensory input is handled at one or two joints while maintaining the same coordination elsewhere [2][4][14][18].

We present a robotic leg modeled after the cockroach *Blaberus discoidalis*, controlled by a computational model of the cockroach locomotor system. The model builds on the success of the Ekeberg controller [13] by adapting their known stick insect reflexes to a cockroach morphology[24]. Our model was inspired largely by [11][12] and implements a similar network of Hodgkin-Huxley-like dynamic neurons that describe the electrical currents modifying each neuron's membrane voltage. This network is based on our previous simulation work, with some changes to address requirements specific to a robot[24].

This work serves both as a trial for a robotic control scheme and also as a test bed for biological hypotheses. We present results from experiments that show a stepping leg that addresses many of the issues raised by previous work[16][21], particularly the smoothness and stability of motion, both while walking and transitioning to turning.

2   Methods

The robotic leg and the biological neural network (BNN) that controls its motion is shown in Fig. 1. The figure illustrates the flow of information from sensory input (top) to motor output (bottom). The sensory neurons (Fig. 1(a)) transduce joint angle and load information into a simulated membrane voltage. The output of these sensory neurons coordinates endogenously oscillating central pattern generators (CPGs) (Fig. 1(c) and (d)). The voltages of the CPG half-centers are compared (Fig. 1(e)), and the servos are commanded to flex or extend based on which is more excited. As the joints are actuated, joint angle and load information is updated, thereby closing the feedback loop. A layer of interneurons (Fig. 1(b)) routes the coupling between sensory neurons and CPGs to alter the reflexes caused by incoming sensory information, which can result in transitions between different behaviors.

The BNN was originally simulated in [24] using AnimatLab, a neuromechanical simulation tool [9]. We utilized the equations for its built-in neuron and synapse models to model our network. The details of the models used are described below.

2.1   Neuron and Synapse Models

Neuron and Synapse Models. The building block for the BNN described above is the neuron, which we have chosen to model as a single compartment leaky neuron model with optional ion channel dynamics similar to those in the Hodgkin-Huxley model. Detailed spiking dynamics are not modeled, instead letting the membrane voltage of a neuron represent either one nonspiking neuron,

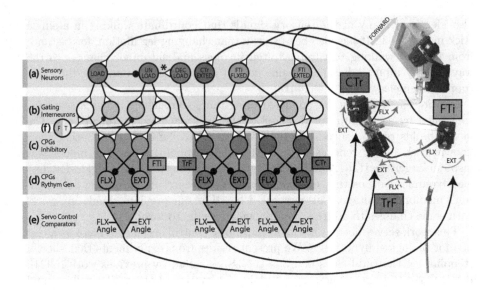

Fig. 1. Sensor data activates sensory neurons (a) which are routed through a layer of sensory interneurons (b) to CPG units (c) and (d) which control servos (e). Sensory interneurons are modulated by a context neurons (f) to change path of synaptic pathways.

a population of nonspiking neurons, or the spiking rate of a population of spiking neurons. Using this model is justified because insect locomotor systems are known to possess nonspiking neurons, which process sensory information and drive motor activity [7]. Also, the global behavior of a large population of synchronized spiking or nonspiking neurons is functionally similar to the output of a single nonspiking neuron, which is a simpler and computationally less expensive alternative to simulating the individuals within the population [17][22].

The change in a neuron's membrane potential, v_i, is described by the following differential equation:

$$C_m \frac{d}{dt}(V_i) = -I_L(V_i) - \sum_{j \neq i} I_{syn}(V_i, V_j) + I_{app} \, , \tag{1}$$

where C_m is the membrane capacitance of the neuron's membrane, $I_L(V_i) = g_L(V_i - E_L)$ is the leak current, I_{app} is the sum of all intrinsically or externally applied currents, and $I_{syn}(V_i, V_j) = g_{syn}s_\infty(V_j)(V_i - E_{syn,j})$ is the synaptic current from neuron j to neuron i. $s_\infty(V_j)$ is a sigmoid function that increases monotonically from 0 to 1 over the conductance range of the synapse, V_l to V_h,

$$s_\infty(V_j) = \begin{cases} 0, & V_j < V_l \\ \dfrac{V_j - V_l}{V_h - V_l}, & V_l \leq V_j \leq V_h \\ 1, & V_j > V_h \, . \end{cases} \tag{2}$$

Synaptic currents, $I_{syn}(V_i, V_j)$, become active when presynaptic voltage, V_j, enters the conductance range and $s_\infty(V_j)$ becomes nonzero, thereby moving the stable point of Equation (1) closer to $E_{syn,j}$. In this way, synaptic currents can be made excitatory or inhibitory by setting $E_{syn,j}$ above or below the voltage range of the postsynaptic neuron.

We also added a decaying synapse to replicate data seen in [28] and [27] where load sensors discharge for a short duration at the at the end of stance. This synapse follows the model of an excitatory synapse with an added multiplier that decays exponentially after the presynaptic neuron voltage passes V_l. Note that this is merely an attempt to replicate the data and not an attempt to describe the underlying neural mechanism. Examples of an excitatory, inhibitory, and decaying synapse can be seen in Fig. 2(a)-(b).

Context and Gating Interneurons. It is possible to set the synaptic conductance range of a neuron, such that a presynaptic neuron can cause it to operate within or outside of that range. In this way, the ability of a neuron to communicate with subsequent neurons can be turned on or off creating a synaptic gating effect. To a certain degree, this effect mimics primary afferent depolarization (PAD) described in [23]. Fig. 2(c) and (d) show an example of this phenomenon.

Central Pattern Generators. We have adapted CPGs found in [11] and [12] to work within the modeling framework of AnimatLab[9]. CPG units are made up of two pairs of neurons, each forming one half of a half-center oscillator. Each half-center pair consists of a neuron following equation 1 that is excited by a neuron with an additional term

$$C_m \frac{d}{dt} V_i = -I_{Ca}(V_i, m_i, h_i) - I_L(V_i) - \sum_{j \neq i} I_{syn}(V_i, V_j) + I_{app} . \qquad (3)$$

where $I_{Ca} = g_{Ca} m_i h_i (V_i - E_{Ca})$ is an additional nonlinear current term that models the influx of Ca^{2+} ions through voltage gated ion channels. The conductance of these ion channels is modulated by the fast opening and slow closing of gating variables m and h whose dynamics are described by

$$h_i' = (h_\infty(v_i) - h_i)/\tau_h(v_i) , \qquad (4)$$
$$m_i' = (m_\infty(v_i) - m_i)/\tau_m(v_i) \qquad (5)$$

where $m_\infty(V_i)$ and $h_\infty(V_i)$ are respectively increasing and decreasing sigmoid functions in the range $[0, 1]$ and $\tau_m(V_i)$ and $\tau_h(V_i)$ determine the time scale at which m and h change. Additionally, for this group of neurons I_{app} provides enough tonic current to naturally depolarize the neuron.

The oscillator is completed by inhibiting each calcium dependent neuron by the interneuron from the opposing half-center, forming the CPG unit seen in Fig. 3(a). The phase plane plot in Fig. 3(d) illustrates how the CPG exhibits

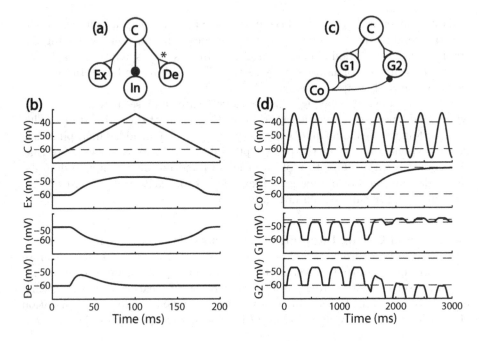

Fig. 2. (a)-(b) As the control neuron, C, moves through its synaptic conductance range, it excites Ex, inhibits a self-excited In, and temporarily excites De when it increases above V_l. An "*" denotes a decaying synapse. (c)-(d) When the context neuron, Co, is inactive, G2 oscillates within its conductance range. However, as the context neuron excites G1 and inhibits G2, G1 begins to conduct while G2 is silenced.

endogenous oscillations even in the absence of sensory input through a shift in the V-nullcline. This differs slightly from [11] and [12] where escape from the V-nullcline results in a relatively constant period of hyperpolarization for each half-center.

The total period of the CPG as well as the time each half-center is excited during the total period can be modulated by descending currents (Fig. 3(c)). Alternatively, phasic descending currents can reset the phase of the CPG by forcing one side to switch states or remain in a state for an extended duration.

2.2   Hardware Implementation

The robotic leg is modeled after the right middle leg of the cockroach *Blaberus discoidalis* at a scale of 25.4:1. It is constructed of aluminum leg segments and the five joints are actuated by Dynamixel AX-12+ smart servos. The servos also act as sensors providing joint angle, velocity, and torque feedback. A strain gauge is placed on the trochanter segment which is sensitive to out-of-plane bending. The strain gauge was integrated into a quarter Wheatstone bridge circuit and the signal amplified to interface with an analog to digital converter (ADC).

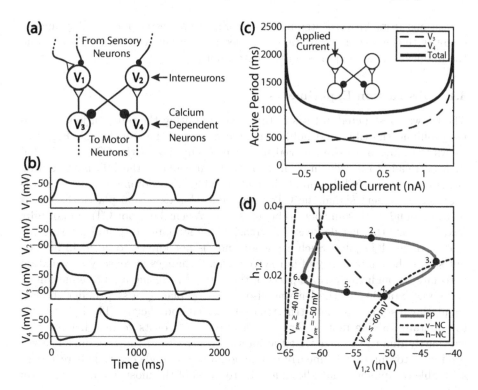

Fig. 3. (a) CPG structure. (b) Time course. (c) The active period of either half-center can be lengthened or shortened by injecting a current into one side. The total period remains relatively constant over a small range. (d) The phase plot illustrates the generation of oscillations.

To control the robot, the network was reconstructed in LabView (National Instruments, Austin, TX) and simulated on a laptop (2.0 GHz Intel Core2Duo). The neural dynamics were numerically integrated using Euler one-step integration with a time step of 1 ms. The laptop interfaced with a NI CompactRIO-9074 over a wireless network via User Datagram Protocol (UDP), which used a built in FPGA to read from the ADC connected to the strain gauge and to create a TTL level half duplex UART for communication with the servos.

While the single strain gauge is a simplification of the cockroach's complex groups of campaniform sensilla, sense organs sensitive to compressive strains in the exoskeleton, it provides the load feedback necessary for the function of the BNN.

3   Results

Experiments were performed to validate that the leg can step adaptively when faced with obstacles and that CPGs make stepping more stable. In addition, transitions from forward stepping to turning were tested for stability and to quantify the change in leg motion. This section presents the data collected, and the implications are explored in the Discussion.

3.1   Obstruction Experiments

The leg was obstructed in tests in three ways: limiting the range of joints, manually pulling the tibia and holding it, and changing the step height. Fig. 4(a) shows the kinematics from a trial in which the motion of the CTr was limited by a barricade to prevent its full flexion. When walking, the FTi flexes during swing and reaches full flexion at the end of swing. Chordotonal organs monitoring that movement signal the CTr to extend. This extension places the tarsus on the ground, with some lag, to begin stance. When the robot CTr cannot fully flex, however, this lag decreases, increasing the stepping rate and altering the motion of the FTi joint. Such a phenomenon is not known to occur in insects, but shows the adaptability of our system; small changes in sensory information percolate through the various parts of the network, causing adaptive motion.

The adaptability of the leg was also tested by placing a step under the leg, simulating a step onto an obstruction or a lowered body height. The leg was able to maintain walking motion for obstacles up to 65% of its vertical height. This was partly due to the fact that the internal controller of the Dynamixel servos produce a torque that is linearly proportional to the angle error. This results in a tunable compliance that allows for the rotation of the most distal thorax coxa joint.

Further trials were run in which walking was interrupted by forcibly grabbing and pulling the tibia to a fully extended position. Fig. 4(b) shows output from this trial. Restraining the leg not only extended the FTi joint, but also loaded the leg, stopping motion in all three joints (highlighted in orange). As soon as the leg was released, it resumed its rhythm with its first step. Allowing sensory information to halt CPG activity in extreme situations is not only biologically accurate[6], but also would prevent a robot from damaging itself should it get stuck in the field.

A similar experiment was performed with the system after its CPGs had been removed. This was done by removing the neurons in Fig. 1(d) (whose calcium currents drive CPG rhythm) and instead using those in Fig. 1(c) to determine which position the servo should approach. Fig. 4(c) shows the results of a trial in which the leg was perturbed while walking. It was able to walk, but differed from the CPG version in two key ways. The first is that its motion was qualitatively less smooth. When positioning was driven directly by sensory information, the noise in the sensory feedback caused rapid threshold crossing in the servo comparators (Fig. 1(e)), producing jerking motion. This issue has been addressed in past work with an FSM controller by simulating muscle dynamics [20].

The second difference is that the version without CPGs could be halted indefinitely by sensory input. Fig. 4(c) shows that the leg was eventually forced into an equilibrium position from which it could not escape. This is different than

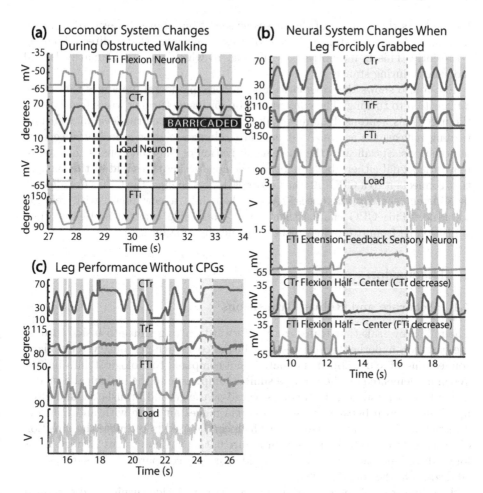

Fig. 4. (a) A barricade prevents the tarsus from moving to its extreme posterior position. (b) The tibia is forcibly held in place preventing sensory cues from being met. (c) The tibia is held with the CPGs disabled. This trial fails unlike the previous version.

the results in Fig. 4(b); when CPGs were present, the rhythm may be forced to pause. However, oscillations return immediately when the obstruction is removed. Without CPGs, specific instances of sensory information may indefinitely halt all motion, even after the obstruction ceases.

3.2   Transition Experiments

The system employed here can transition smoothly and stably from walking to turning. Reflex reversals are enacted by a context neuron (Fig. 1(f)) that changes the pathways between sensory neurons and CPGs. This simple redirection of sensory information results in functional changes in the behavior of the leg.

Fig. 5 shows the position of the tarsus as seen from above. In forward walking, movement during the stance phase produces a movement that is directed mostly posteriorly. This is in contrast to turning, in which there is primarily lateral movement during stance.

Fig. 5 also shows the kinematics and underlying neural dynamics that cause a transition to turning. The context neuron for inside turning was stimulated by the user. This neurons dynamics make it slowly approach the level at which all turning reflexes are active. As it activates, changes can be seen in the sensory interneurons. Specifically, the baseline activation of these neurons is changed to move the signal either above or below the conducting threshold of the synapse that connects it to a CPG. As the active pathways change, the CPGs are affected by different sensory influences, forcing them into a new phase with relation to stance. This CPG change then causes noticeable changes in FTi motion, transitioning from extending in stance to flexing in stance.

4   Discussion

4.1   Accurate and Robust Transitions

The reflex reversals that transform the forward walking gait into inside turning are based on hypothesized connections in the cockroach nervous system. These connections were developed, simulated, and compared to biological data in [24]. When implemented on the robot, a similar comparison to the biological data can be made. During stance, the tarsus is placed on the ground, translated, and lifted up. A line drawn between these two points makes an angle with an anteriorly directed line in the transverse plane which can be used to estimate the direction of thrust created by a single step. In Figure 6, this angle averages 23 degrees for forward walking and 50 degrees for inside turning, compared to 10 degrees and 40 degrees in the cockroach[24].

The time scale of the transition from walking to inside turning is determined by the membrane capacitance of the context neuron. In Figure 6 this value is five thousand times larger than the other neurons in the network, resulting in a transition that takes 3 or 4 steps, which is consistent with observations from *Blaberus*[24]. By significantly lowering the membrane capacitance, stable transitions can be performed instantaneously. For comparison with biological data, a slower transition is important to reveal similarities between the two. However, for a walking machine, faster transitions will result in quicker response times.

Also, CPGs help prevent the leg from getting stuck during transitions. Previous work with a FSM, controlled only by sensory input, required more precise timing and special cases to handle transitions between walking and turning[21]. During these behaviors, a specific sequence of sensory events is expected. However, when the leg finds itself outside of its expected sequence due to external

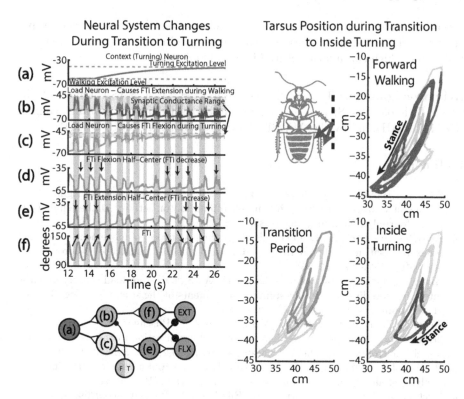

Fig. 5. Left: Plots showing kinematic and neural changes in the system when commanded to turn. Gray indicates stance. Line (a) shows how the TURN context neuron slowly activates as the user stimulates it. (b) and (c) show how the bias of the sensory interneurons changes as the activation of the context neuron (A) changes. The changes in (b) and (c) cause the CPG half-centers in (d) and (e) to change phase in relation to stance, causing the associated joint to follow suit (f). Right: The plot shows tarsus position during the transition from walking to inside turning.

perturbations or a sudden reflex reversal, it can get stuck in a steady state position. A CPG, acting in parallel with the sensory reflexes, is able to set the leg back in motion. Similar results could be achieved with any non-sensory method of switching the active motoneuron group.

5 Conclusion

A robotic model of a single cockroach middle leg produced stepping, inside turning, and transitions between those behaviors via a simulated nervous system based on insect neurobiology. It benefits from Hodgkin-Huxley-like neuron and synapse models, a biologically informed network organization, compliant actuation, and emulation of actual insect load sensors. It adapts to several types of obstructions and perturbations, sometimes producing unintended adaptive behaviors, which can inform future biological experiments.

Acknowledgments. This work was supported by a NASA Office of the Chief Technologists Space Technology Research Fellowship (Grant Number NNX12AN24H) and by the AFOSR (Grant Number FA9550-10-1-0054).

References

1. Akay, T., Bässler, U., Gerharz, P., Büschges, A.: The role of sensory signals from the insect coxa-trochanteral joint in controlling motor activity of the femur-tibia joint. Journal of Neurophysiology 85(2), 594 (2001)
2. Akay, T., Büschges, A.: Load signals assist the generation of movement-dependent reflex reversal in the femur-tibia joint of stick insects. Journal of Neurophysiology 96(6), 3532–3537 (2006)
3. Akay, T., Haehn, S., Schmitz, J., Büschges, A.: Signals from load sensors underlie interjoint coordination during stepping movements of the stick insect leg. Journal of Neurophysiology 92(1), 42–51 (2004)
4. Akay, T., Ludwar, B.C., Göritz, M.L., Schmitz, J., Büschges, A.: Segment specificity of load signal processing depends on walking direction in the stick insect leg muscle control system. The Journal of Neuroscience 27(12), 3285–3294 (2007)
5. Bender, J.A., Simpson, E.M., Ritzmann, R.E.: Computer-assisted 3D kinematic analysis of all leg joints in walking insects. PloS One 5(10) (2010)
6. Bucher, D., Akay, T., DiCaprio, R.A., Büschges, A.: Interjoint coordination in the stick insect leg-control system: the role of positional signaling. Journal of Neurophysiology 89(3), 1245–1255 (2003)
7. Büschges, A., Kittmann, R., Schmitz, J.: Identified nonspiking interneurons in leg reflexes and during walking in the stick insect. Journal of Comparative Physiology A, 685–700 (1994)
8. Büschges, A., Schmitz, J., Bässler, U.: Rhythmic patterns in the thoracic nerve cord of the stick insect induced by pilocarpine. Journal of Experimental Biology 198(Pt. 2), 435–456 (1995)
9. Cofer, D., Cymbalyuk, G., Reid, J., Zhu, Y., Heitler, W.J., Edwards, D.H.: AnimatLab: a 3D graphics environment for neuromechanical simulations. Journal of Neuroscience Methods 187(2), 280–288 (2010)
10. Cruse, H.: What mechanisms coordinate leg movement in walking arthropods? Trends in Neurosciences 13, 15–21 (1990)
11. Daun-Gruhn, S., Rubin, J.E., Rybak, I.A.: Control of oscillation periods and phase durations in half-center central pattern generators: a comparative mechanistic analysis. Journal of Computational Neuroscience 27(1), 3–36 (2009)
12. Daun-Gruhn, S., Tóth, T.I.: An inter-segmental network model and its use in elucidating gait-switches in the stick insect. Journal of Computational Neuroscience (2010), doi:10.1007/s10827-010-0300-1
13. Ekeberg, O., Blümel, M., Büschges, A.: Dynamic simulation of insect walking. Arthropod Structure & Development 33(3), 287–300 (2004)
14. Hellekes, K., Blincow, E., Hoffmann, J., Büschges, A.: Control of reflex reversal in stick insect walking: effects of intersegmental signals, changes in direction and optomotor induced turning. Journal of Neurophysiology 107(1), 239–249 (2011)

15. Hess, D., Büschges, A.: Role of Proprioceptive Signals From an Insect Femur-Tibia Joint in Patterning Motoneuronal Activity of an Adjacent Leg Joint. Journal of Neurophysiology 81(4), 1856–1865 (1999)
16. Lewinger, W.A., Rutter, B.L.: Sensory Coupled Action Switching Modules (SCASM) generate robust, adaptive stepping in legged robots. Climbing and Walking Robots (September 2006)
17. Markin, S.N., Klishko, A.N., Shevtsova, N.A., Lemay, M.A., Prilutsky, B.I., Rybak, I.A.: Afferent control of locomotor CPG: insights from a simple neuromechanical model. Annals of the New York Academy of Sciences 1198, 21–34 (2010)
18. Mu, L., Ritzmann, R.E.: Kinematics and motor activity during tethered walking and turning in the cockroach, Blaberus discoidalis. Journal of Comparative Physiology. A, Neuroethology, Sensory, Neural, and Behavioral Physiology 191(11), 1037–1054 (2005)
19. Raibert, M.H.: Legged Robots That Balance. MIT Press (1986)
20. Rutter, B.L., Lewinger, W.A., Blümel, M., Büschges, A., Quinn, R.D.: Simple Muscle Models Regularize Motion in a Robotic Leg with Neurally-Based Step Generation. In: IEEE ICRA, pp. 10–14 (April 2007)
21. Rutter, B.L., Taylor, B.K., Bender, J.A., Blumel, M., Lewinger, W.A., Ritzmann, R.E., Quinn, R.D.: Descending commands to an insect leg controller network cause smooth behavioral transitions. In: Intelligent Robots and Systems (IROS 2011) (2011)
22. Spardy, L.E., Markin, S.N., Shevtsova, N.A., Prilutsky, B.I., Rybak, I.A., Rubin, J.E.: A dynamical systems analysis of afferent control in a neuromechanical model of locomotion: I. Rhythm generation. Journal of Neural Engineering 8(6), 065003 (2011)
23. Stein, W., Schmitz, J.: Multimodal convergence of presynaptic afferent inhibition in insect proprioceptors. Journal of Neurophysiology 82, 512–514 (1999)
24. Szczecinski, N.S., Brown, A.E., Bender, J.A., Quinn, R.D., Ritzmann, R.E.: A Neuromechanical Simulation of Insect Walking and Transition to Turning of the Cockroach Blaberus discoidalis. Biological Cybernetics (2013)
25. Watson, J.T., Ritzmann, R.E.: Leg kinematics and muscle activity during treadmill running in the cockroach, Blaberus discoidalis: I. Slow running. Journal of Comparative Physiology. A, Sensory, Neural, and Behavioral Physiology 182(1), 11–22 (1998)
26. Zill, S.N., Büschges, A., Schmitz, J.: Encoding of force increases and decreases by tibial campaniform sensilla in the stick insect, Carausius morosus. Journal of Comparative Physiology. A, Neuroethology, Sensory, Neural, and Behavioral Physiology 197(8), 851–867 (2011)
27. Zill, S.N., Keller, B.R., Duke, E.R.: Sensory signals of unloading in one leg follow stance onset in another leg: transfer of load and emergent coordination in cockroach walking. Journal of Neurophysiology 101(5), 2297–2304 (2009)
28. Zill, S.N., Schmitz, J., Büschges, A.: Load sensing and control of posture and locomotion. Arthropod Structure & Development 33(3), 273–286 (2004a)

Blending in with the Shoal: Robotic Fish Swarms for Investigating Strategies of Group Formation in Guppies

Tim Landgraf[1], Hai Nguyen[1], Joseph Schröer[1], Angelika Szengel[1],
Romain J.G. Clément[2], David Bierbach[2], and Jens Krause[2]

[1] Freie Universität Berlin, FB Mathematik u. Informatik
Arnimallee 7, 14195 Berlin, Germany
[2] Leibniz-Institute of Freshwater Ecology & Inland Fisheries
Müggelseedamm 310, 12587 Berlin, Germany
tim.landgraf@fu-berlin.de
http://biorobotics.mi.fu-berlin.de

Abstract. Robotic fish that dynamically interact with live fish shoals dramatically augment the toolset of behavioral biologists. We have developed a system of biomimetic fish for the investigation of collective behavior in Guppies and similarly small fish. This contribution presents full implementation details of the system and promising experimental results. Over long durations our robots are able to integrate themselves into shoals or recruit the group to exposed locations that are usually avoided. This system is the first open-source project for both software and hardware components and is supposed to facilitate research in the emerging field of bio-hybrid societies.

Keywords: biomimetic robots, biomimetics, swarm intelligence, social behavior, social networks, swarm tracking.

1 Introduction

The study of collective animal behavior can benefit substantially from the use of biomimetic robots. Once accepted by the animal group as a conspecific, the robots can be used to test theoretical models of group formation, leadership, mate choice and other biological functions and mechanisms. In recent years this approach has been shown in various animal models, such as in cockroach shelter seeking [1], honeybee dance communication [2], bowerbird courtship behavior [3] and decision making in fish [4]. While the complexity of those robotic systems is still low in number of actuators and behavioral repertoire (for a review see [5] or [6]), a few interactive, closed-loop systems for robotic fish have been proposed only recently [7–9]. Those systems all use tracking devices that visually recognize the individuals of the shoal. This information is fed back to the control of the robots and therewith allows a dynamic interplay between robots and animals. However, significant biological findings, obtained with such interactive systems, have yet to be put forward. In interactive systems, the robots have to

A. Duff et al. (eds.): Living Machines 2014, LNAI 8608, pp. 178–189, 2014.

reach group acceptance with proper appearance but also on the level of social behavior. The integration of artificial agents therefore is a fragile process and can only be sustained with robust tracking systems, a finely tuned set of motion controllers and carefully composed robotic behaviors. In this contribution we present a detailed description of our interactive system for robotic fish, soon be published as open-source software and hardware specifications. We hope that this will facilitate research in this exciting new field of bio-hybrid systems. This paper is divided into five parts: 1) general system description, 2) computer vision implementation for tracking the robots and live fish, 3) description of interactive behaviors 4) experimental validation and results and 5) conclusions.

2 General System Description

Small shoals of up to 20 full-grown *Guppies* are kept in a water tank of 1 m^2 floor area filled with only 15 cm of water. The tank is positioned at about 1.40 m above the ground. Two-wheeled differential drive robots are moving below the tank on a transparent platform (Figure 3). Each of the robots holds up a neodymium magnet to the bottom side of the tank in which fish replicas, magnetically coupled, follow the robots' movements (see Figure 1). The artificial fish are moulded using dead template animals, painted and finished for a realistic morphology and appearance. The robots carry two infrared LEDs on their bottom sides. On the ground, a camera with an IR-pass filter glass is facing upwards to track the movements of the robots. A second camera is affixed above the tank to track both real and artificial animals within the tank. One computer is running a program for tracking the robots and sending motion commands to each individual robot over a wireless channel. A second computer evaluates the video feed of the shoal camera and sends tracking data to the first computer via a local area network connection.

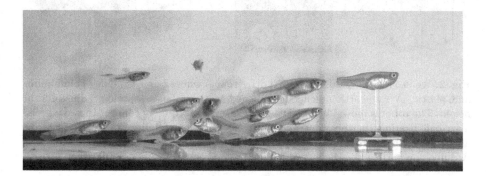

Fig. 1. The replica is attached to the magnetic base with a thin transparent plastic stick of 1 mm width. The photograph shows a Guppy replica ahead of a swarm of 12 animals.

2.1   Robot Design and Control

The two-wheeled robot's aluminium frame and the wheels are custom built with a base area of 7 *cm* × 7 *cm* as depicted in Figure 2. The main electronics are three Arduino-compatible boards (so called "shields") for main processing (Arduino Uno), WiFi communication (Copperhead WifiShield) and motor control (DFRobot Motor Shield). We use a two-cell LiPo battery pack (7.4 *V* nominal output) that connects to a voltage regulator (Fairchild Semiconductor LM350T) and a voltage divider.

The former provides a constant voltage of 5 V with currents of up to 3 A irrespective of the battery level. The regulator's output is fed to the Arduino main board and the motor shield which generates PWM signals for each motor from this constant input voltage. The voltage divider scales down the battery output which then can be measured on an analog pin of the Arduino board with respect to an internal reference (1.1 V). Two Faulhaber DC gear motors are affixed to the side plates and directly connect to the wheels that we produce from circular plates into which V-shaped notches are turned. Rubber gasket rings fitted in the notches serve as treads for traction. Each robot carries a strong rare-earth magnet at the end of a plastic rod held up against the tank's floor without physical contact. The tank itself is made of glass that due to the mass of the water column bends slightly downwards. The distance of the magnet to the glass is therefor less than 1 mm near the center of the tank and approx. 2 mm in the periphery. The magnet's poles are aligned in parallel to the running surface.

Fig. 2. We use two-wheeled differential drive robots whose frame is made of aluminium and plastic. A rare-earth magnet is attached to the robot's top side, adjustable in height. The robot is moved by two DC gear-motors that directly connect to two wheels. Two ball bearings in the front and back are used for stability. Three IR-LEDs in alignment with the robots forward direction are inset into the base plate. A stack of Arduino-compatible boards is used for motor control and wifi communication.

The control program for the agents is executed on a personal computer. The main control loop is run at a frequency of 30 Hz as determined by the bottom camera's frame rate. In each time step, a command packet is issued and sent to the individual robots via WiFi (UDP). Each robot has a unique IP-address and only receives its respective packets. We use a fixed length protocol with a two bytes header, 12 bytes data and two bytes checksum. The motion command consists of two motor speeds as computed by a PID controller based on the current robot's position and future target points. Targets can be defined as a static sequence or, dynamically, by integrating the output of the shoal tracking. The proportional component of the controller is defined by two sigmoidal functions, that define the forward and turning speeds over the Euclidean and angular distances to the target point, respectively. Once a robot has reached its next target position, the next location in the sequence is selected as the new target. In interactive behaviors, properties of the shoal or its members are used to define target points on the fly for every new time step. Depending on the starting conditions and the shoal's behavior, this might result in very different trajectories. In section 4 we describe the interactive behaviors in more detail. The firmware on the robots receives WiFi packets and translates the data content. Currently, the robots implement the following sub-routines:

1. Sending status packets back to the control computer at 1 Hz, including the robot's unique identifier and battery level
2. Toggling of infra-red LEDs for the identification of robot
3. Generating PWM signals according to the received motor speed values
4. Toggling the so-called burst mode, in which slow motor speeds are translated to PWM signals that are preceded by short bursts of a higher frequency to overcome initial friction and inertia.

3   Computer Vision

We propose using two sensory channels: one for controlling the robotic agents and one for measuring social interactions within the tank. Especially in dense groups, robots might get confused with real fish when using just one tracking system from above. This is a major difference to the systems proposed by [8] and [7].

3.1   Robot Tracking

We propose attaching three IR-LEDs to the bottom sides of the robots. When operated on a transparent platform, each robot can be easily localized in the video feed of a camera on the ground. For this task, we utilize a standard webcam (Logitech Pro 9000) with the IR-block cover glass replaced by an IR-pass filter. Hence, the infrared LEDs produce very bright spots in the otherwise dark camera image. Hence, the computer vision pipeline is designed minimalistically:

First, a global threshold is applied to binarize the image, which is then denoised by an erosion operation. Possible remaining gaps in the blobs are filled by applying a dilation. Subsequently, we seek connected components that likely represent IR-LEDs. Since the distance of the LEDs on each robot is known and constant, the distance of the blobs in the image is known and constant as well.

Fig. 3. The wheeled robots are moving on a transparent plastic plate in a space below the water tank. A camera on the floor is used to track the individual robots that each have three infrared LEDs inset in their baseplate.

Two of the three LEDs are placed in close vicinity, such that their resulting blobs merge to a significantly larger object than the one resulting from the single LED. Using the distance constraint we identify pairs of heterogeneously sized blobs that then define the location and orientation of each robot. Since robots might come close to each other we additionally check whether the candidate blob pair is not too far away from previous detections. Before the robots are operated, the program initially broadcasts a request to all available robots to report their IDs. Then, sequentially, each robot is requested to blink its LEDs. The vision system detects this event and assigns the respective ID to the formerly unidentified virtual object in the tracking arena. For all following frames new object detections are assigned the ID of the nearest previously known robot locations. All coordinates are rectified and translated to the world coordinate system using a user-defined rectangle that matches the arena's outline in the camera projection.

3.2 Fish Tracking

All interactions of robots and live fish are observed from above the tank by a second camera. In order to detect all individuals we use a background subtraction procedure that models foreground and background pixel distributions with a mixture of Gaussians [10]. Since the tank's bottom and walls are laminated with white plastic the fish appear as clear dark objects in the camera image. Once converged, the background model shows an empty tank such that

Fig. 4. A (cropped) screenshot of the shoal tracking system. Each individual of the shoal is identified by a number. An ellipse and a triangular tail is depicted to mark each fish's position. The orange tail marks the only robot in this recording. The green dots signify that those animals have been found to be in one subgroup, whose center is denoted by the white dot. The circle around the dot denotes one standard deviation.

we get clear positive peaks in the difference image of background and current live frame. The individual fish are detected by applying a global threshold to the difference image. All regions having above-threshold values are used for seeding a customized multi-agent particle filter (unpublished, based on [11]). Each of the 500 particles, representing a point in state space, is scored using the difference image: A particle is defined by its world coordinates, an orientation angle and length and width. A given parameter combination defines an ellipse in image space. The match to the image is computed as the difference of the sum of all pixels within the ellipse and the sum of pixels in a ring around the ellipse.

The particle filter uses importance resampling to select a number of particles that by this process iteratively converge to the optimal locations, as defined by the matching function. Per frame, after a number of resampling iterations, clusters of particles are averaged to obtain a robust location estimate for each individual. The fishs' orientation angles cannot be robustly obtained on still images. Since fishes usually swim forwards, the direction is disambiguated by integrating the motion vector over a fixed time window. The system assigns an ID to every fish object, i.e. a cluster of particles in state space, and tracks them using a simple model of fish motion. For a given cluster center, its motion speed in three-dimensions (planar position and orientation) defines an expected position for future frames. We distribute new hypotheses along this motion axis using Gaussians centered at the measured velocities. Since the number of fishes in the tank is known and constant we keep an according number of cluster representatives and update their locations iteratively over time. Figure 4 depicts a sample shoal with overlaid tracking information. The shoal tracking system was validated in two regards: first, the positional error of static objects was calculated by placing fish-sized metal blocks on known positions and comparing the system's output to the reference positions. The average error is below 1 mm (std: 2 mm) but might be larger when objects are moving or the polygon that defines the homography is set erroneously. Secondly, the tracking error was determined by counting the number of individuals that were either lost by the system or assigned the wrong ID (e.g. after swimming close to another individual). If an individual gets lost and found again, we would only count the loss. Five different video sequences of the same duration (1000 frames, 40 seconds) and varying number of individuals were subject to the tracker. In average 1.2 errors per minute occur - most of them produced in one sequence with many fishes overlapping.

4   Interactive Behaviors

Interactive behaviors use the continuous stream of sensory feedback from the fish tracking system. For each frame a TCP-packet containing individual positions and cluster information to the robot control computer. There, a callback routine is triggered that collects the data and makes it available in a respective data structure that holds the robots' positions as well as provided by the robot tracking component. We have defined atomic behaviors, such as following a certain individual, going to a location in the tank, waiting, exploring, wall following, alignment with individuals, and so on. Atomic behaviors are preconfigured with appropriate values for parameters such as motion velocities, distance constraints with regard to other individuals and temporal parameters such as typical waiting durations. Each interactive behavior is implemented as an exchangeable module that logically assembles atomic behaviors dependent on the actual situation. All interactive behavior modules exhibit a generic structure. They can access the current set of world information (positions and orientations of all fishes and robots) and produce a target point or motor speeds directly for every time step. For each robotic agent a different behavior can be defined. For example, one agent can be set to follow another robot for all times and this one can be configured to follow a live fish. We have implemented a number of complex behaviors:

Follow Swarm Center. This behavior finds clusters of fish and chooses either the largest or closest cluster to follow. A target point is produced that either lies in the middle of the swarm, or, if the swarm is very dense, on the cluster bounds. If no clusters can be found, the nearest neighbor is selected to be followed. Collision avoidance is implemented rudimentarily: If individuals are likely being hit by the replica, the robot slows down and continues its way to the target leaving the potential evasive maneuver to the fish.

Swarm Integration Behaviors. We have implemented a behavior to mimic schooling inspired by the model of Couzin and coworkers [12] that describes fish schooling behavior as a simple sequence of attraction, alignment and

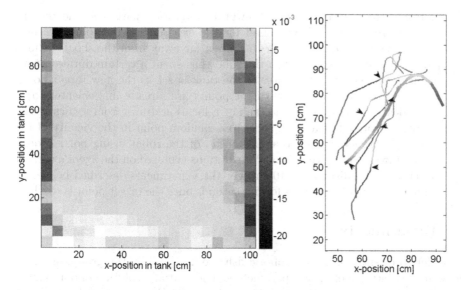

Fig. 5. Left: Difference image of animals' probability of presence in interactive recruitment experiments and reference scenario. All red and orange squares denote locations that were visited less often in the recruitment scenario. White and yellow squares, predominantly in the center of the tank, denote regions with higher occurence compared to a no-robot scenario. Right: Animals often copy the trajectory of the robot. The figure shows the movements of five live individuals and the path described by the robot (thick line). Time is color-coded and ranges from red over blue to violet for the duration of 7 seconds. The arrows denote the positions of all individuals at the time point the robot stopped near the target point. All animals follow after the robot took the lead towards the center of the tank. Intriguingly, most individuals seem to copy the direction of the robot more than half a second after the robot initiated its move.

repulsion behaviors. Therefore, three concentric zones are defined around the robot. If e.g. the agent measures another individual within its repulsion zone, a target point away from that animal is produced. If a fish is swimming in the alignment zone, the robot is rotated in order to match the orientation of that animal. If an animal is in the attraction zone, it is approached. The rejection rule has priority over the other rules; if there are any fish present in the rejection zone, only the rejection maneuver will be executed. If there are no individuals within the rejection zone, but both the alignment and attraction zones contain individuals, the robot will use the average of the motion vectors resulting from the alignment and attraction rules.

Predator Behavior. The predator behavior is a variant of *Follow Swarm Center*. The motion speeds are higher and the target point is always the center of the swarm. Minimal distances to individuals are ignored.

Recruitment Behaviors. There are many combinations of atomic behaviors and parameters that make up recruitment behaviors. All basically consist

of two stages, an approach and swarm integration part, and the recruitment phase. In the behavior used for the experiments the robot executes
the behavior *Follow Swarm Center* when far away from the shoal (defined
parametrically). Second, when the robot has spent a certain duration with
or inside the swarm, the center of the tank is set as the new target point.
Following further options for target points are already implemented: a) a
fixed location anywhere in the tank or a location drawn from a normal distribution centered at a given spot, b) a random point in the vicinity of the
robot, c) a random point that lies ahead of the robot, using polar coordinates drawn from a two normal distributions centered on the agent's current
direction and a distance of 10 cm. For the experiments described below, we
configured the robot to wait for five seconds once the target point is reached.

5   Experiments

In order to quantify how well robotic fish are accepted by the group, one may
look at a variety of behavioral parameters. The spatial resolution of our camera
system yields approximately 10 pixels per fish only. Hence, our analysis focuses
on body positions and derived parameters only. In this paper we compare how
fish are distributed in different scenarios. As a reference, a natural shoal, without
robots was kept in the water tank for 30 minutes and was video recorded. In
the second experiment, the robot described a static square-shaped trajectory for
the same time. In the third experiment, the robot was configured to follow the
shoal's center and, in the last experiment, executed the recruitment behavior
for luring the animals to the center of the water tank, a location that is usually
avoided. The positions and orientations were extracted and the tracking data was
evaluated subsequently in MATLAB. We generated a 2-dimensional histogram
of the fishs' positions which is normalized by the total number of occurences.
This results in a map of presence probability for any fish over the entire tank
area. Furthermore, we calculate the average inter-individual distance over all N
animals for each time step t as:

$$d_{II}(t) = \frac{2}{N(N-1)} \sum_{j=1}^{N} \sum_{\substack{i=j+1, \\ i \neq j}}^{N} \|\boldsymbol{p}_i(t) - \boldsymbol{p}_j(t)\|_2$$

where $\boldsymbol{p}_i(t)$ is the planar position of the i-th animal at time t. This measure
grows larger when the shoal spreads, even when similarly dense sub-shoals are
formed. To capture the general groups cohesion we calculate the average nearest-
neighbor distance as:

$$d_{NN}(t) = \frac{2}{N(N-1)} \sum_{j=1}^{N} \min_{\substack{i=1..N, \\ i \neq j}} \|\boldsymbol{p}_i(t) - \boldsymbol{p}_j(t)\|_2$$

This is done for the no-robot scenario as well as for the mixed groups
scenarios, ignoring the positions of the robot in all cases. In the following we

characterize the natural, no-robot scenario and show how robots influence the shoal's behavior.

5.1   Results

Natural shoals tend to first explore the tank and then stay in one of the four corners most of the remaining time of the test. There are individual differences in the frequency of exploratory behaviors. In one no-robot experiment, two of the seven animals frequently left the group alone, individually. This results in a broader distribution of inter-individual and nearest-neighbor distances. The animals in the other experiments were less bold, reflected by a narrower distribution. The 2-dimensional distribution of the probability of presence exhibits high peaks at the corners and almost no fish near the center of the tank. In two independent trials of 30 minutes duration animals occupied a region of 50 cm x 50 cm centered on the middle of the tank in only 5 % of the time. A robot running on a static square-shaped trajectory does occasionally attract animals. However, the following behavior (Figure 5 depicts a sample run) is often interrupted after only a few seconds. In less than 8 % of the time animals are present in the inner square. Operating a robot with the interactive recruitment behavior described above yields a presence in the inner square in more than 15 % of the time. Animals that were recruited by the robot are staying for a short while in the inner

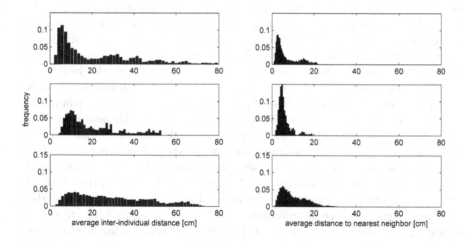

Fig. 6. Distributions of average inter-individual distance (left column) and average distance to nearest neighbor (right column) for three scenarios: the upper row shows the distributions of the reference case with no robots. The middle row depicts very similar distributions measured with one robot following the shoal's center. The recruitment behavior as shown in the last row exhibits a broader distribution. Note that in the two latter cases the robots have been excluded in the computation of the distance averages to reflect the shoal-only behavior.

square but then shortly swim off to the periphery, presumably because the robot stopped and stood still once it reached the tanks center. Example videos of the experiments can be found at [13]. Figure 5 depicts the difference of the distributions of the no-robot scenario and the recruitment behavior. Generally, each individual in the shoal seems anxious to stay with the group. Animals that explore the tank on their own can be observed only rarely. This is expressed in the distributions of average inter-individual and nearest-neighbor distances. Figure 6 shows the distributions of both measures over full experiments. Remarkably, a simple swarm following behavior produces a distributions resembling those of the reference case. The interactive recruitment behavior attracts single or a few individuals only, which is reflected in broader distributions, similar to the experiment with bold animals. However, Kolmogorov-Smirnov tests for equality of distributions yield no significant similarities for all pairs of distribution combinations.

6   Conclusions

We have developed a low-cost multi-agent platform for biomimetic robotic fish that resemble in appearance and behavior their live counterparts. The system allows dynamic interactions with single individuals or groups of fish using exchangeable modules of behavioral logic. The system is scalable for using many other agents under even larger tanks. However, due to the size of the wheeled robot under the tank, the density of robotic shoals is limited. The tracking of individual fish is robust, though it should be further improved in future research, e.g. by using cameras with higher spatial resolution. This would improve the quality of upcoming biological research by allowing smaller (i.e. younger) guppies to be used in experiments in which age distribution affects the focal behavior. The experiments indicate that closing the feedback loop enables the robot to display natural behavior expressed in similar distributions of inter-individual and nearest-neighbor distances. By applying a simple recruitment behavior, a single robot was able to lure away a few individuals from the shoal. However, the behavioral parameters, such as motions speeds and minimal distances to other individuals might still be chosen sub-optimally. In a few occasions, the robot seems to come too close to the fish or move too fast within the shoal. The affected individuals quickly evade the robot. In part, our results reflect these occasions in the distance distributions as well as in the spatial presence distribution. By opening our code base to the public, we invite other groups to join this specific research line and hope that this will facilitate research in bio-hybrid systems in general. Currently, we are preparing experiments with live *Guppies* and two robotic fish to investigate group decision making. For example, depending on the individuals' different experiences previously made with robots which were able or unable to find food. In addition, we plan to test this system with an implementation of a variety of theoretical models for schooling, mate choice and other biological functions in order to investigate whether the simulated results match our observations.

References

1. Halloy, J., Sempo, G., Caprari, G., Rivault, C., Asadpour, M., Tache, F., Said, I., Durier, V., Canonge, S., Ame, J., et al.: Social integration of robots into groups of cockroaches to control self-organized choices. Science 318(5853), 1155–1158 (2007)
2. Landgraf, T., Oertel, M., Rhiel, D., Rojas, R.: A biomimetic honeybee robot for the analysis of the honeybee dance communication system. In: IROS, pp. 3097–3102 (2010)
3. Patricelli, G.L., Uy, J.A.C., Walsh, G., Borgia, G.: Sexual selection: male displays adjusted to female's response. Nature 415(6869), 279–280 (2002)
4. Faria, J.J., Dyer, J.R., Clément, R.O., Couzin, I.D., Holt, N., Ward, A.J., Waters, D., Krause, J.: A novel method for investigating the collective behaviour of fish: introducing 'robofish'. Behavioral Ecology and Sociobiology 64(8), 1211–1218 (2010)
5. Krause, J., Winfield, A.F., Deneubourg, J.L.: Interactive robots in experimental biology. Trends in Ecology & Evolution 26(7), 369–375 (2011)
6. Mitri, S., Wischmann, S., Floreano, D., Keller, L.: Using robots to understand social behaviour. Biological Reviews 88(1), 31–39 (2013)
7. Bonnet, F., Rétornaz, P., Halloy, J., Gribovskiy, A., Mondada, F.: Development of a mobile robot to study the collective behavior of zebrafish. In: 2012 4th IEEE RAS & EMBS International Conference on Biomedical Robotics and Biomechatronics (BioRob), pp. 437–442. IEEE (2012)
8. Swain, D.T., Couzin, I.D., Leonard, N.E.: Real-time feedback-controlled robotic fish for behavioral experiments with fish schools. Proceedings of the IEEE 100(1), 150–163 (2012)
9. Landgraf, T., Nguyen, H., Forgo, S., Schneider, J., Schröer, J., Krüger, C., Matzke, H., Clément, R.O., Krause, J., Rojas, R.: Interactive robotic fish for the analysis of swarm behavior. In: Tan, Y., Shi, Y., Mo, H. (eds.) ICSI 2013, Part I. LNCS, vol. 7928, pp. 1–10. Springer, Heidelberg (2013)
10. Zivkovic, Z., van der Heijden, F.: Recursive unsupervised learning of finite mixture models. IEEE Transactions on Pattern Analysis and Machine Intelligence 26(5), 651–656 (2004)
11. Isard, M., Blake, A.: Condensation conditional density propagation for visual tracking. International Journal of Computer Vision 29(1), 5–28 (1998)
12. Couzin, I.D., Krause, J., James, R., Ruxton, G.D., Franks, N.R.: Collective memory and spatial sorting in animal groups. Journal of Theoretical Biology 218(1), 1–11 (2002)
13. Landgraf, T., Nguyen, H., Schröer, J., Szengel, A.: Robo Fish LM 2014 Videos (2014), http://robofish.mi.fu-berlin.de/videos/lm2014

Capturing Stochastic Insect Movements with Liquid State Machines[*]

Alexander Lonsberry, Kathryn Daltorio, and Roger D. Quinn

Case Western Reserve University
Department of Mechanical and Aerospace Engineering
10900 Euclid Ave. Cleveland, OH, USA

Abstract. A Liquid State Machine (LSM) is trained to model the stochastic behavior of a cockroach exploring an unknown environment. The LSM is a recurrent neural network of leaky-integrate-and-fire neurons interconnected by synapses with intrinsic dynamics and outputs to an Artificial Neural Network (ANN). The LSM is trained by a reinforcement approach to produce a probability distribution over a discrete control space which is then sampled by the controller to determine the next course of action. The LSM is able to capture several observed phenomenon of cockroach exploratory behavior including resting under shelters and wall following.

Keywords: Liquid State Machine, Artificial Neural Network, Stochastic Control, Biologically Inspired Robotics.

1 Introduction

There are many examples of complex machines today that rely on inspiration from nature. However, the behavioral strategies of animals are difficult to replicate. In prior work, we mimicked exploratory behavior of cockroaches seeking shelters in arenas by manually defining finite states and analyzing the data to characterize transition properties. The RAMBLER (Randomized Algorithm Mimicking Biased Lone Exploration in Roaches) algorithm [1, 2] replicates the highly variable animal movement while also capturing the slight shelter-seeking biases in the random walk. However RAMBLER does not model stops, curving paths, or continuous speed changes. In this work, the same source data used for RAMBLER is used in conjunction with Reservoir Computing techniques to generate a controller with neurological parallels.

Reservoir Computing techniques have been developed to overcome the difficulties in training recurrent neural networks directly and to mimic observed phenomenon in

---

[*] This work is based upon work support by [*] Defense Advanced Research Projects Agency (DARPA) Maximum Mobility and Manipulation (M3) research grant No. DI-MISC-81612A and by the National Science Foundation (NSF) under grant No. IIS-1065489. Any opinions, findings, and conclusions or recommendations expressed in this material are those of the authors and do not necessarily reflect the views of either DARPA or NSF.

A. Duff et al. (eds.): Living Machines 2014, LNAI 8608, pp. 190–201, 2014.
© Springer International Publishing Switzerland 2014

the brain. Recurrent neural networks are different than traditional artificial neural networks (ANNs) in that they have forward and backward connections. Reservoir Computing models incorporate a predetermined or "fixed" recurrent network, which is deemed the 'reservoir'. The outputs of neurons in the reservoir are combined with a trainable output layer. Thus the recurrent network itself requires no training. Instead, learning is handled solely in the output layer.

Some complex neurological functions have been observed to be akin to Reservoir Computing techniques. In monkeys, a weighted sum of activity from a sample of cortical neurons has been shown to predict hand trajectory [3]. These networks can also explain how visual signals are processed in the visual cortex [4], and offers a model for explaining sequential signal processing in primate brains [5, 6]. It has also been shown that reservoirs of neurons correspond to spatial coding of information in the brain [7].

Brain circuits may compute several computational tasks in parallel and at different time-scales. Time-varying dynamics can be achieved by building the network from dynamic, facilitating synapses that connect spiking neurons [8]. These systems exhibit rich dynamics, like waves on the surface of water after a perturbation. For this reason, reservoirs composed of spiking neurons connected with synapse that have internal dynamics are commonly referred to as "liquids". The system comprised of a liquid reservoir and trainable output layer is referred to as a Liquid State Machine (LSM) [9]. The LSM reservoir is a recurrent neural network constituted of leaky-integrate-and-fire (LIF) neurons connected together via synapses with phenomologically accurate dynamics [10]. LSMs have proven effective in many challenging nonlinear tasks; examples include controlling a planar two link arm [11], robotic object tracking and motion prediction [12], and   music classification [13].

Presented here an LSM with an Artificial Neural Network (ANN) output layer is trained using a reinforcement method to emulate the exploring behavior of a cockroach. The LSM inputs sensory data and produces a probability distribution over the space of possible actions. An action is sampled at random based on the probability distributions. The resulting model is shown to emulate some cockroach behavior.

2   Methods

2.1   Behavioral Data

Biological experiments were performed in which cockroaches were tracked via camera while exploring a walled arena. The arena is 914mm square with a small starting chamber. Before each trial, the arena was cleaned to eliminate traces from previous trials and an intact female cockroach, Blaberus discoidalis, was separated from the lab colony. The cockroach was placed in the starting chamber and allowed to settle before the opaque plastic gate was lifted. The cockroaches entered the arena, which was evenly lit by overhead lights (1500 lux) except under the shelter (300 lux). Above the walls, black fabric shrouded the arena to eliminate background light. The first 60 seconds of exploration behavior was filmed at 20 Hz by an overhead camera using the Motmot image acquisition package [14]. Each animal was used only once.

The video was post-processed to find the position of the cockroach's visual center and its body orientation in each frame using the Caltech Multiple Fly Tracker (http://ctrax.sourceforge.net/) and the associated FixErrors toolbox for MATLAB (MathWorks, Inc., Natick, MA, USA) [15]. Because there were still occasional inversions of the tracked angle, we corrected any instantaneous 180° flips in Matlab. Speed and direction of movement were recorded as the cockroaches explored the arena with and without a goal (shelter from light) in the environment. From the position and orientation, we approximated the cockroaches antenna and visual sense of the walls using virtual LIDAR-like free ranges. We calculated the intersection of rays from the head at 30° increments and walls, those ranges, after appropriate transformation, are inputs to the LSM.

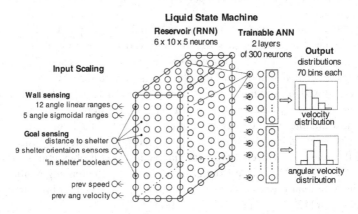

Fig. 1. Inputs to the reservoir of the LSM are converted into signals all normalized over [0,1]. These inputs are then connected through synapses that stimulate randomly selected neurons in the reservoir. Neurons in the reservoir are connected to each other at random. The output of each neuron in the reservoir is input to an ANN. The ANN is trained to output probability distributions of the animals' velocity and angular velocity.

2.2   Neural Controller

It is observed that the cockroach's behavior appears stochastic. To replicate the behavior a neurologically inspired controller, denoted as Π and partially depicted in Fig. 1, is trained to produce probability distributions over a discrete space of velocity controls and subsequently sample a control from the probability distributions.

LSM: Reservoir Details
The LSM which produces the probability distributions in Π, is comprised of models drawn from literature [8], Leaky-Integrate-and-Fire (LIF) neurons connected via dynamic synapses, where all the associated parameters are inspired from literature [16] and which have been modified based on experimentation. Neuron dynamics are described by the following:

$$\tau_m \frac{dV_m}{dt} = -\left(V_m - V_{Rest}\right) + R_m \left(I_{syn} + I_{injection} + I_{noise}\right) \tag{1}$$

where V_m is the neuron membrane potential measured in mV, τ_m is the neuron membrane time constant chosen from a Gaussian distribution with mean value of 25 ms and standard deviation of 3.5 ms. Here $V_{Rest} = 0$ mV is the resting potential of the neuron, $V_{Thresh} = 15$ mV is the spiking threshold where if this value is exceeded an action potential is emitted on the following time step. The membrane potential is reset to V_{Reset}, which is chosen from a uniform distribution over the interval [5, 8 mV], at the time-step following the spike and is held at this voltage level for a refractory period of 2 ms. Membrane resistance R_m is 1 MΩ , I_{syn} is current from synaptic connections, $I_{injection}$ is a tonic background injection drawn from a uniform distribution [11.0, 14.5 nA], and I_{noise} is background noise which is drawn from a Gaussian distribution with 0 mean and stand deviation of 4.5 nA.

Synaptic connections in the reservoir is based on the distance between the LIF neurons which are placed in a 3D lattice structure at integer positions along each axis, forming a cube-like structure as shown in Fig. 1. The reservoir used here is constituted of 300 neurons in a 5 x 5 x 12 arrangement. The probability of synaptic connection from neuron i to neuron j, is denoted as c_{ij} and is a function of the Euclidean distance between the neurons $D(n_i, n_j)$ in the lattice structure:

$$P\left(c_{ij} \mid D\left(n_i, n_j\right)\right) = C_{ij} \exp\left(-D^2 \left(n_i, n_j\right) / \lambda^2\right) \tag{2}$$

where $\lambda = 2.1$ controls the number of connections in the reservoir. In the reservoir 20% of the neurons are set to being inhibitory (I) and 80% are excitatory (E). The value of C_{ij} is dependent on the type of presynaptic and postsynaptic neuron, that is $C_{EE} = 0.8$ for a connection from an excitatory neuron to another excitatory neuron and subsequently $C_{EI} = 0.8$, $C_{IE} = 0.7$, and $C_{II} = 0.7$.

Synapses connecting the neurons within the reservoir incorporate short-term plasticity (STP) based on the model proposed in [17]:

$$\begin{aligned} A_k &= A u_k R_k \\ u_k &= U + u_{k-1} \left(1 - U\right) \exp\left(-\Delta_k / F\right) \\ R_k &= 1 + \left(R_{k-1} - u_{k-1} R_{k-1} - 1\right) \exp\left(-\Delta_k / D\right) \end{aligned} \tag{3}$$

where A_k is the amplitude of current injected into the postsynaptic neuron following the k-th presynaptic action potential, $u_k \in [0,1]$ and $R_k \in [0,1]$ model the effects of facilitation and depression respectively, F and D are time constants controlling facilitation and depression, and U is the probability of neurotransmitter release, where F, D, U are set as in [18]. The values of u_k and R_k are dependent on the time between presynaptic spikes Δ_k where $k \geq 2$. The value of the current flowing into the postsynaptic neuron at time t following the k-th action potential decays exponentially by $\exp(-t/\tau_s)$ where $\tau_s = 30$ ms.

LSM: Input and Data Representation
Insects are equipped with complex sensors that relay the state of the physical animal
and the environment to the nervous system. Here, we have chosen several simple
ways of encoding information about the walls and shelter relative to the insect. A
combination of signals, denoted compactly together as z, are input to the reservoir
along with the current velocity and angular velocity, v and ω respectively. All these
sensory inputs, which are normalized over [0, 1] at each time-step, are connected to
reservoir neurons in probabilistic fashion, where an input signal can be connected to
more than one neuron. The reservoir has five layers of neurons, and each layer is arranged in a 5×12 configuration. The probability of an input signal being connected to
a neuron in the reservoir is dependent on which layer the neuron is in: 0.1, 0.07, 0.05,
0.04, 0.04 to layer 1, 2, 3, 4, and 5, respectively. Inputs signals are connected by simple linear synapses with weights w, randomly chosen over uniform distribution [0,
50 nA]. If the i-th sensor in the set z has a value of z_i, the current I injected to the LIF
neuron j is $I = w_{ij}z_i$.

Wall Sensors
The animals' sensory perception of the bounding walls of the arena is modelled by
two sets of sensors. The first set is comprised of 12 distance sensors that are positioned on the head of the animal and radiate outwards at equally spaced angles on the
interval [0, 2π rads], similar to LIDAR sensors. The data sampled from these sensors
is $z_V = \{z_{V_1}, z_{V_2}, ..., z_{V_{12}}\}$ where z_{V_i}, given $i = 1,...,12$, are scalar distance measurements normalized over [0, 1] based on the size of the arena.

The second set of five sensor values is "antennae-like" and are similar to those
above in that they radiate out from the head (at orientations of $-\pi/2$, $-\pi/4$, 0, $\pi/4$, and
$\pi/2$) and measure the distance to the walls. The values are however filtered with a
sigmoidal function such that they are large when the animal is near to the walls. The
sigmoidal range sensors are modeled as $z_a = \{z_{a_1}, ..., z_{a_5}\}$ where z_{a_i}, given $i = 1,...5$, are the filtered distance values:

$$z_{a_i} = \frac{1}{1 + e^{\gamma\left(z_{a_i}^o - \eta\right)}} \qquad (4)$$

where $z_{a_i}^o$ is the pre-filtered value measured by a distance sensor, $\eta = 50mm$ is the
approximate length of cockroach antennae, and $\gamma = 10.0$ was found to be a suitable
value. Having range information encoded with sensitivity at close ranges may help the
model mimic antenna reactions.

Shelter
The position of the animal relative to the shelter is represented by z_{s_d} and
$z_s = \{z_{s_1}, ..., z_{s_9}\}$. The first value, z_{s_d}, is the distance between the animal and the shel-
ter normalized over [0, 1] based on the size of the arena. The values z_{s_i}, given

$i = 1,\ldots,9$, encode orientation of the animal relative to the shelter. These nine orientation values can be thought as being generated by an array of sensors spaced evenly over a circle on the interval $[-\pi, \pi]$, where each sensor's position on the circle is relative to the centerline of the animal. These sensors produce higher values when the orientation of shelter relative to the simulated agent is near to their assigned angular position value. Calculation of these nine sensor values is done as follows:

$$z_{s_i} = g(s_i) = \begin{cases} \dfrac{1}{\sqrt{2\pi\sigma}} e^{-\left(\frac{s_i - 2\pi}{2\sigma}\right)^2} & \pi \leq s_i < 2\pi \\[3mm] \dfrac{1}{\sqrt{2\pi\sigma}} e^{-\left(\frac{s_i}{2\sigma}\right)^2} & -\pi \leq s_i \leq \pi \\[3mm] \dfrac{1}{\sqrt{2\pi\sigma}} e^{-\left(\frac{s_i + 2\pi}{2\sigma}\right)^2} & -2\pi \leq s_i < -\pi \end{cases} \tag{5}$$

where the values s_i, given $i = 1,..,9$, are from the nine evenly spaced sensors:

$$s = \{-4n+\theta, -3n+\theta, -2n+\theta, -n+\theta, 0, n+\theta, 2n+\theta, 3n+\theta, 4n+\theta\} \tag{6}$$

where $n = 2\pi/9$ and $\theta \in [-\pi, \pi]$ is heading angle to the center of the shelter relative to the centerline of the animal. In this way, we encode the orientation of the shelter in a smooth form. When the shelter is directly in front of the simulated agent, z_{S_5}, which represents the space directly in front of the animal, is the largest and z_{S_1} and z_{S_9}, which represent the space behind the agent, are the smallest. Whereas, if the shelter is directly behind the animal, z_{S_5} is the smallest and z_{S_1} and z_{S_9} are the largest. Also, the simulated animal is informed if it is under the shelter. It receives a Boolean signal of either 0, if it is outside of the shelter, or 1, if it is underneath the shelter.

LSM Output Layer: Artificial Neural Network

The output of the reservoir neurons are fed into an artificial neural network (ANN). The ANN has two hidden layers each with 300 neurons modeled with sigmoidal activation functions. The LIF neurons in the reservoir produce spike trains that are not readily suitable as input into the ANN. To make the reservoir output more suitable the spike trains are filtered as in [19] via:

$$f_i(t) = \sum_{j=1}^{k} \exp\left(-\frac{t - {}^i t_j}{\tau_f}\right) \tag{7}$$

where ${}^i t_j$ is the time at which the i-th neuron in the reservoir emitted its j-th spike and $\tau_f = 30$ ms. These outputs are input into the first layer of the ANN. The ANN is trained to produce a probability distribution over a discrete set of velocities and angular velocities the animal can produce.

2.3   Learning a Discrete Action Space

The controller Π samples sensory data at 20 Hz, the sample rate during experimentation, and subsequently outputs a control u: $u(t) = \{v, \omega\}$, where v (forward velocity) and ω (angular velocity) are constrained to sets of allowable values $v \in \{v_1,..., v_n\}$ and $\omega \in \{\omega_1,..., \omega_m\}$, where n and m are the number of velocity and angular velocity values respectively. The actual insects have an internal controller we define as Π^*. The goal is to train the model Π to tend towards Π^*. It is observed that the movement of the animals is highly variable. Thus Π is trained to produce a probability distribution over the set of allowable controls. A control value is produced by sampling these probability distributions generated by the LSM. These probability distributions are not known apriori and therefore reinforcement learning is used to learn the probability distributions over the action space.

A reinforcement technique is used to train the ANN output layer (see Fig. 1). Reinforcement learning is iterative and begins with the LSM receiving state information $z(t)$. The LSM subsequently produces an estimate of the distributions $P(v)$ and $P(\omega)$. If this was a typical reinforcement scheme, the controller Π would subsequently choose an action $u(t)$ based on the probability distributions and allow the agent to interact with the environment (in this case move a simulated cockroach). A reward or penalty would then be generated based on the action and controller would be modified based on how "good" or "bad" the action was. This process is not used here. Rather than actually sampling the distributions to generate an action $u(t)$, the action made by one of the animals during experimentation, denoted as $u(t)^*$, is used instead. Essentially, the controller Π is being trained by forcing it to follow the same path(s) of animal(s) in the arena. While training, the simulated cockroach takes the same actions as the animals were observed to do. Though this seems to be akin to a supervised learning method, this is not quite the case as the probability distributions are not known. Rather, only the action $u(t)^*$ chosen by the animal is known. Thus at each time step of the recorded experiments, the LSM produces the probability distributions that are then modified such that the probability of the velocity and angular velocity $u(t)^*$ are increased while the probability of all other actions are decreased. These modified probability distributions are then used as the target distributions to train the ANN output using backpropagation [20].

The critical portion of this process is modifying the LSM-produced distributions $P(v)$ and $P(\omega)$ based on $u(t)^*$. The distributions, like the action spaces, are discrete. There is a probability associated for each velocity v_i, where $i = 1,...,n$, and each angular velocity ω_i, where $i = 1,...,m$. For simplicity, σ_i is defined as the probability that either v or ω take on the i-th value in their respective set of actions: $\sigma_i = p(\alpha_i / \Pi, z)$, where α can represent either v or ω. At each successive time step, the LSM generates the probability distribution $P(\alpha)$. The distribution $P(\alpha)$ is modified by positively reinforcing the probability of j-th action, the action taken by the animal, and penalizing the rest in the following manner:

$$\hat{\sigma}_i = \begin{cases} \sigma_i + (1 - \sigma_i)\mu & \text{if } i = j \\ \sigma_i - \sigma_i\mu & \text{if } i \neq j \end{cases} \tag{8}$$

where the hat refers to the modified distribution. Now that $\hat{P}(\alpha)$ is formed, it is used as target data for the ANN. It is hypothesized, and verified via testing, that over many training iterations Π converges toward the so far unknown Π^*, as will be shown in the results.

3   Results

3.1   Implementation

20 LSMs were generated and trained. As the reservoirs are created at random, not all the controllers performed equivalently. To gauge performance the trained LSMs were simulated with the training data and at each time-step the LSM would generate the probability output distributions over the action space. The difference between the centroid of the probability distributions and the actual action of the animal was recorded. The mean of the difference between the animal's velocity and the centroid of the LSM probability distribution was 50 ± 130 mm/sec and 5 ± 13 rad/sec for velocity and angular velocity respectively, which is different than the desired 0 mean value.

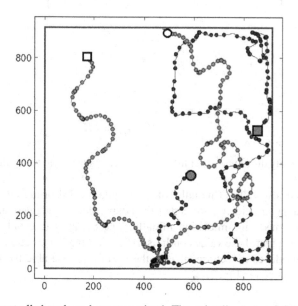

Fig. 2. LSM controlled cockroach versus animal. The animal's start and finish are marked by the large circle and large square both with white fill. The LSM controlled agent's start and finish is marked by the large circle and a large square both with grey fill. The trajectories have circular markers that are dropped onto the plot every 0.1 seconds.

The controllers are simulated in Python and are largely built using the Brian [21] neural simulator toolbox. Input to the LSM changes every 0.05s matching experimental data sampled at 20 Hz. The LSM itself is simulated with a time step of 0.0001s. A total of 170 animal trials are used as training data. During training, the controller was shown all the animal trials 5 times, in randomized order.

3.2   Convergence of Probability Distributions

Though the probability distributions over the action space is not known for individual states, the mean distributions over all time can be found and used as a baseline for comparison against the controller Π. These distributions denoted as $P^*(v)$ and $P^*(\omega)$ for velocity and angular velocity respectively are found by counting the frequency at which v_i and ω_j occur over all animal trials. By examination of Fig. 3, it can be seen that the distributions do tend to the actual distribution but do not fully converge. Near both zero velocity and zero angular velocity the LSM has the correct distribution shape but does not quite reach the same magnitude level. The LSM tends to produce higher probability values at the ends of the distributions as compared to the animal. Thus the LSM does move towards the correct distribution but does not reach the correct magnitudes.

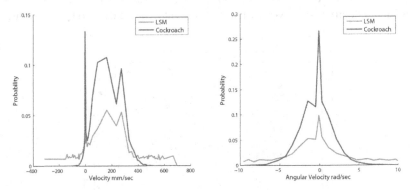

Fig. 3. Probability distribution over the control space of the animal versus the LSM

As the high and low-values (the tails) of the resultant LSM-produced distributions for velocity and angular velocity, are higher than expected, exaggerated and uncharacteristic turns were made in simulation initially. These values (the tails) have been discarded as we consider these values outliers. After limiting these actions we find the LSM controller produces trajectories that to the naked eye resemble the trajectory of the actual animal as seen in Fig. 2.

3.3   Occupancy

To compare the LSM based control versus that of the animal of RAMBLER [1] occupancy measurements are used and are displayed in Fig. 4. These occupancy measurements display the amount of time the animal, simulated or otherwise, spends in specific areas of the arena. The animal, RAMBER, and LSM alike all tend heavily towards the boundaries of the environment and particularly the corners. Interestingly, the LSM has a very strong preference for the walls, more so than RAMBLER or the animal itself. Also LSM based agent has some preference for the shelter region. When the shelter is on the bottom wall, the LSM controller tends to favor the bottom wall. When the shelter is in the top right corner the effect of shelter seeking is most pronounced. This is perhaps because the shelter is in the corner and the LSM and animal alike prefer the corners. When the shelter is on the rightmost boundary, again the LSM prefers that bounding wall.

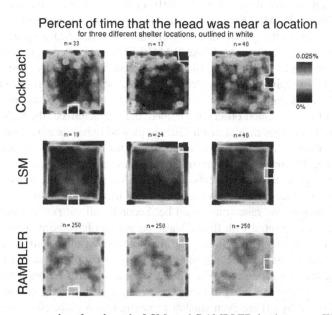

Fig. 4. Occupancy graphs of cockroach, LSM, and RAMBLER in the arena. White squares outline the shelter placement. The number *n* indicates the number of trails used to generate the graphs.

4   Discussion

Implementation of a neurologically inspired, stochastic LSM based controller produces qualitatively cockroach-like behavior in that the paths are stochastic and result in more time spent along the wall with a bias towards the shelters. Plots of the distribution over the control space tend to the correct modes, though outliers at the boundaries are overrepresented.

Examining wall-following might clarify the differences in occupancy grids between the animal and the LSM model. The simulated agents do not follow the walls as nicely as animals can with their sensitive antennae [22]. RAMBLER does not attempt to model wall-following at the velocity control level: there is a wall-following action state. It may be that LSM would be able to better capture wall-following behavior by checking for sensory input bias. For example, when following the wall closely, sensitivity to a few of the sensors on one side is essential. However, due to the random nature of the reservoir generation it may be that these input signals only have a small effect on the reservoir activity. Enforcing bilateral symmetry, biasing connections from key directions, or changing the way the wall proximity is encoded might allow the LSM to perform more like the animal.

There are several important things to note in comparing this LSM method with the finite state approach in RAMBLER. First, these two algorithms had different objectives. RAMBLER was never intended to model stops or sensory-modulated speed changes, which is why it is not particularly effective at matching the occupancy graphs. However, RAMBLER was intended to capture the shelter-seeking bias, so it does get the agent to the goal with a shorter path than the LSM approaches. Thus, we can see that a custom approach like RAMBLER can be used to hunt for specific trends. In contrast, an advantage to this LSM approach is that many different aspects of the behavior (stopping, shelter-preference, curves, wall-following) can be captured without explicitly defining those behaviors. Second, because these desired behaviors do not have to be explicated observed, defined and characterized, the LSM approach is much faster to implement, and much faster to expand to handle new data.

Implementation of the LSM as a neurologically inspired method for real-time control is feasible. The research demonstrates that these abstractions are very capable, however there are some impediments. Firstly, LSMs are complex tools, and to the author's best knowledge, they are without a general metric or heuristic to specify how the large or complex the reservoir should be. Secondly all the neurons and synapses have parameters associated with them. These parameters have been chosen based on literature and experimentation. It may be that optimizing them with respect to the application would combine the advantages of customizing and automating model creation. The training scheme for the output layer could also be improved. As it is presented here, when the ANN produces very small probability values, the change in those values made by (8) are even smaller or negligible. A more effective scheme to handle small values could potentially eliminate the problem with the tails of the distributions being too high.

References

1. Daltorio, K., Tietz, B., Bender, J., Webster, V., Szczecinski, N., Branicky, M., Ritzmann, R., Quinn, R.: A model of exploration and goal-searching in the cockroach, Blaberus discoidali. Adapt. Behav. 21, 404–420 (2013)
2. Daltorio, A., Tietz, B., Bender, J., Webster, V.: A stochastic algorithm for explorative goal seeking extracted from cockroach walking data. In: 2012 IEEE INternational Conference on Robootics and Automation (ICRA), pp. 2261–2268 (2012)
3. Wessberg, J., Stambaugh, C., Kralik, J., Beck, P., Laubach, M., Chapin, J., Kim, J., Biggs, J., Srinivasan, M., Nicolelis, M.: Real-time prediction of hand trajectory by ensembles of cortical neurons in primates. Nature 408, 361–365 (2000)

4. Nikolić, D., Haeusler, S., Singer, W., Maass, W.: Temporal dynamics of information content carried by neurons in the primary visual cortex. In: Adv. Neural Inf. Process. Syst., pp. 1041–1048 (2006)
5. Dominey, P., Hoen, M., Inui, T.: A neurolinguistic model of grammatical construction processing. J. Cogn. Neurosci. 18, 2088–2107 (2006)
6. Blanc, J., Dominey, P.: Identification of prosodic attitudes by a temporal recurrent network. Cogn. Brain Res. 17, 693–699 (2003)
7. Buonomano, D., Maass, W.: State-dependent computations: spatiotemporal processing in cortical networks. Nat. Rev. Neurosci. 10, 113–125 (2009)
8. Maass, W., Legenstein, R., Bertschinger, N.: Methods for estimating the computational power and generalization capability of neural microcircuits (2005)
9. Maass, W., Natschläger, T., Markram, H.: Computational models for generic cortical microcircuits. Comput. Neurosci. (2004)
10. Tsodyks, M., Markram, H.: The neural code between neocortical pyramidal neurons depends on neurotransmitter release probability. Proc. Natl. Acad. Sci. USA 94, 719–723 (1997)
11. Joshi, P., Maass, W.: Movement generation with circuits of spiking neurons. Neural Comput. (2005)
12. Burgsteiner, H.: On learning with recurrent spiking neural networks and their applications to robot control with real-world devices (2005)
13. Ju, H., Xu, J., VanDongen, A.: Classification of musical styles using liquid state machines. In: Neural Networks (IJCNN) (2010)
14. Straw, A., Dickinson, M.: Motmot, an open-source toolkit for real-time video acquisition and analysis. Source Code Biol. Med. 4 (2009)
15. Branson, K., Robie, A., Bender, J., Perone, P., Dickinson, M.: High-throughput ethomics in large groups of Drosophila. Nat. Methods. 6, 451–457 (2009)
16. Joshi, P., Maass, W.: Movement generation and control with generic neural microcircuits. In: Ijspeert, A.J., Murata, M., Wakamiya, N. (eds.) BioADIT 2004. LNCS, vol. 3141, pp. 258–273. Springer, Heidelberg (2004)
17. Markram, H., Lübke, J., Frotscher, M., Sakmann, B.: Regulation of synaptic efficacy by coincidence of postsynaptic APs and EPSPs. Science (80) (1997)
18. Joshi, P.: From memory-based decisoin to decision-based movements: A model of interval discrimination followed by action selection. Neural Networks 20, 298–311 (2007)
19. Ju, H., Xu, J., VanDongen, A.: Classification of musical styles using liquid state machines. In: The 2010 Internationa Joint Conference on Neural Networks, IJCNN, pp. 1–7. IEEE (2010)
20. Haykin, S.: Neural Networks and Learning Machines. Prentice Hall (2008)
21. Goodman, D., Brette, R.: Brian: a simulator for spiking neural networks in Python. Front. Neuroinform. (2008)
22. Chapman, T.P., Webb, B.: A model of antennal wall-following and escape in the cockroach. J. Comp. Physiol. 192, 949–969 (2006)

Acquisition of Synergistic Motor Responses through Cerebellar Learning in a Robotic Postural Task

Giovanni Maffei[1], Marti Sanchez-Fibla[1],
Ivan Herreros[1], and Paul F.M.J. Verschure[1,2]

[1] SPECS, Technology Department, Universitat Pompeu Fabra,
Carrer de Roc Boronat 138, 08018 Barcelona, Spain
[2] ICREA, Institucio Catalana de Recerca i Estudis Avan cats,
Passeig Llu s Companys 23, 08010 Barcelona
{giovanni.maffei,marti.sanchez,ivan.herreros,paul.verschure}@upf.edu

Abstract. Coordination of synergistic movements is a crucial aspect of goal oriented motor behavior in postural control. It has been proposed that the cerebellum could be involved in the acquisition of adaptive fine-tuned motor responses. However, it remains unclear whether motor patterns and action sequences can be learned as a result of recurrent connections among multiple cerebellar microcircuits. Within this study we hypothesize that such link could be found in the Nucleo-Pontine projection and we investigate the behavioral advantages of cerebellar driven synergistic motor responses in a robotic postural task. We devise a scenario where a double-joint cart-pole robot has to learn to stand and balance interconnected segments by issuing multiple actions in order to minimize the deviation from a state of equilibrium. Our results show that a cerebellum based architecture can efficiently learn to reduce errors through well-timed motor coordination. We also suggest that such strategy could reduce energy cost by progressively synchronizing multiple joints movements.

Keywords: postural adjustments, cerebellum, action sequence, bio-mimetic, robotics, simulation.

1 Introduction

Coordination of different movements is a crucial aspect of goal oriented motor control. When pinching an object with a hand, one needs to synchronize multiple muscles activity in order to synchronously press the opponent fingers on the surface of the object and firmly grasp it [1]. Similarly, when performing the action of standing up from a sitting position a set of muscles in different parts of the body is synchronously active. The contraction of muscles in the ankle and leg moves the center of mass of the body forward and allows the body to move away from the chair. The contraction of muscles in the hip and torso is therefore triggered in order to compensate for the postural disturbance, pushing

A. Duff et al. (eds.): Living Machines 2014, LNAI 8608, pp. 202–212, 2014.
© Springer International Publishing Switzerland 2014

the torso backwards and setting the center of mass to a new stable position [1]. Experimental studies on human healthy subjects show that in a quiet standing task, a reciprocal relationship exists between the angular accelerations of the hip and ankle joints. It appears that either positive or negative angular acceleration of ankle joint is compensated for by oppositely directed angular acceleration of the hip joint [2]. Similar reciprocal opposite responses in human quite standing are found also in [3], where a more detailed analysis on the magnitude of the responses shows the ankle torque and the hip torque peaking almost synchronously. Such results suggest that precise co-activation of multiple muscles and finely tuned synchronization of responses are beneficial in terms of accuracy in motor control. Such ability, however, is not innate but it is acquired and improved with time and experience.

Studies conducted on subjects affected by cerebellar lesions show that they perform poorly in tasks where well timed motor responses are required. For example it has been reported that lesions to the cerebellar cortex could reduce the ability to anticipate the effect of expected perturbation in a postural task [4]. It has also been suggested that similar lesions could impair the sense of rhythm and the reproduction of rhythmical patterns by finger tapping [5]. These results make the cerebellum an ideal candidate for the neural substrate responsible for acquisition and finely timed coordination of multiple motor responses.

Despite the temporal dynamics of postural control mechanisms underlying the acquisition of adaptive motor responses has been widely studied, it remains unclear how the synchronization of multiple responses is performed. It is possible that two or more independent motor responses are sequentially triggered by proprioceptive feedback, where the sensory outcome of the effects of an action would serve as trigger for a following action. However, pure feedback control driving two connected joints would be too slow and could not entirely account for synchronized responses as the ones shown in [2]. In order to account for such short time delays, it is possible that sequential motor responses are linked and sequentially triggered in a direct forward manner bypassing feedback control, at least in the early stages of the action course.

With this hypothesis in mind we extend the theoretical work of [6] and [7] proposing that a sequence of motor responses could be acquired by multiple cerebellar microcircuits linked together. It is indeed possible that a recurrent connection from the output stage (the cerebellar nuclei) to the input stage (pontine nuclei) of the cerebellum could be established by the Nucleo-Pontine Projections. Such projections could indeed, hypothetically, trigger a sequence of responses by feeding the output of one microcircuit back to cerebellar cortex, therefore allowing the co-activation of multiple muscles performing a synergistic motor sequence. Extending the work proposed in [8,7], we hypothesize that such neural substrate could be a crucial mechanism explaining synchronization of multiple motor responses with possible advantages in terms of accuracy and energy cost optimization.

In order to test our hypothesis we implement a cerebellar based adaptive control architecture for the acquisition of co-activated adaptive responses. The architecture is composed by two cerebellar microcircuits linked by a nucleo-pontine projection in a way that the output of one microcircuit will serve as an input to the second microcircuit, promoting the association of two sequential motor responses. Coherently with previous posture modeling studies [2], we devise a simulated two-joints cart-pole setup where a robot engaged in a postural task has to reach the vertical position from an initial supine position. Each motor response controls one degree of freedom of one actuator independently: the former moves the cart on the horizontal plane to minimize the error in the pole inclination. The latter is mapped into the torque of the top joint and it is activated by the output of the first motor response in order to compensate for the postural perturbation introduced by its actions. We propose an experimental procedure to test the evolution of the acquired responses in terms of error minimization, response timings and energy cost.

Results suggest that a chained control architecture can efficiently account for action synergy. Interestingly a progressive reduction of delays between consecutive motor responses correlates with a progressive minimization of the error and the energy necessary to accomplish the task. Coherently with previous behavioral studies [9,2,3], these results suggest that synchronization of multiple actions could be an important strategy to achieve optimal motor control. Finally we discuss the results obtained in the light of neuro-scientific evidences supporting the role of the cerebellum in adaptive motor response acquisition and its versatility in fine-grained time dynamics. Implications for biomimetic robot control are also taken into account.

2   Methods

Setup. In order to study the possible role of cerebellum in the acquisition of synergistic motor responses in a postural task we devise a simulated physics based setup[1,2] implementing the double-joint cart-pole dynamics (Fig. 1, *left*). A simulated agent has to acquire a motor pattern synchronizing two motor responses. The goal is to dynamically modify the body configuration and stand vertically in a position of equilibrium ($\Theta_{tgt1}=0.0$, $\Theta_{tgt2}=0.0$) starting from a supine position ($\Theta_{init1}=1.3$, $\Theta_{init2}=1.5$) . The agent is equipped with two actuators where, according to a simplified human posture model [2], the bottom joint (ankle joint) mimics the ankle, while the top joint (hip joint) mimics the hip. The agent is equipped with sensors computing the angle of the two joints with respect to the target. The agent can control the ankle joint performing antagonist side-wards movements modifying the relative inclination of the above pendulum with respect to the body position. It can also control the angular force applied to the hip joint adjusting the pendulum head inclination with respect to the bottom segment. The overall control is performed via the actuation of two connected joints (one degree of freedom each) in order to set and maintain the pendulum to a vertical equilibrium, accounting for gravity and self-induced perturbing

---

[1] http://www.processing.org/

[2] http://www.ricardmarxer.com/fisica/

forces. Differently from similar studies[10,11], the present setup does not imply a formal description of the model of the double inverted pendulum problem. The dynamics underlying the control of the system therefore are not explicitly accessible to the controller which takes the current position of the joints and the target angle, used to generate the teaching signal driving the learning process, as sole inputs (see section *Computational Architecture*).

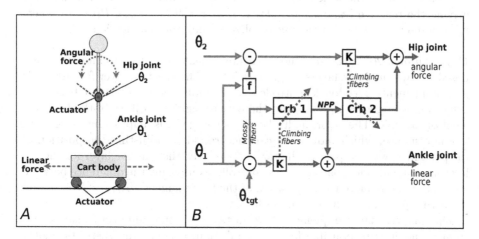

Fig. 1. *Left.* Double-joint cart-pole setup. The agent represented in equilibrium position (Θ_{tgt1}=0.0, Θ_{tgt2}=0.0). Passive joints are represented in green, actuators and forces in red, degrees of freedom in blue. *Right.* Computational Architecture. Sensory inputs are represented in green. Motor outputs are represented in red. Adaptive component is represented by the link of $CRB1$ and $CRB2$ via NPP. Mossy fibers represent the microcircuit input stage, while Climbing fibers provide the teaching signal (dashed line). K represents the gain of the proportional controller that converts the sensory input into motor output. f represents the function that sets the relative target angle for the hip joint, such as $\Theta_{tgt2} = arcsin(m_2l_2sin(\Theta_1))/m_1l_1 + m_2l_2$, where m_1 and m_2 represent the mass of the first and the second segment respectively, and l_1 and l_2 represent the length of the first and the second segment respectively.

Learning Algorithm. The bio-mimetic learning algorithm at the core of the behavior of the agent is based on an analysis-synthesis adaptive filter implementation mimicking the learning strategy of the cerebellar microcircuit [12,13]. The cue signal, representing the input conveyed by mossy fibers, is decomposed into several signals mimicking the expansion of information into cortical basis, which occurs within the cerebellar granular layer. The signal of the cortical basis is generated producing a fast excitatory component and a slow inhibitory one. Each component consists of a double exponential convolution with time constants randomly drawn from two flat probability distributions in a range of appropriate values. The value obtained after the two convolutions is then thresholded and scaled for each basis.

The output of the cerebellar controller (CR) is given by: $CR(t) = [\mathbf{p}(t)^T\mathbf{w}(t)]$ where $\mathbf{w}(t)$ is the vector of weights and $\mathbf{p}(t)$, the vector of basis, both in column

G. Maffei et al.

form. The weights are updated using the de-correlation learning rule: $\Delta w_j(t) = \beta\, E(t)\, p_j(t - \delta)$, where β is the learning rate and $E(t)$ is the error signal, computed by the inferior olive output. δ provides the latency of the nucleo-olivary inhibition. The value of δ determines how much the adaptive action anticipates the reactive one, and how much it has to exceed the feedback delay [14]. Finally, the error signal (E) for the cerebellar system is computed as the difference between the scaled cerebellar output (CR) and the sensory signal (US) driving the reactive response as follows: $E(t) = US(t) - k_{noi}CR(t - \delta)$, where k_{noi} represents the gain of the nucleo-olivary inhibitory connection [13].

Computational Architecture. The agent implements a control architecture composed by two modular layers (Fig. 1, *Right*). The reactive layer for the ankle joint implements a feedback controller which computes the difference between a given target angle (Θ_{tgt1}=0.0) and the actual angular position of the first segment of the pole. The error, multiplied by a gain (K), is mapped into a reactive motor response, which moves the cart accordingly readjusting the position of the pendulum with respect to the target. Similarly, the reactive layer for the hip joint implements a feedback reactive controller computing the difference between a target angle, relative to the position of the first segment ($\Theta_{tgt2} = f(\Theta_1)$), and the actual position of the pendulum head.

The adaptive layer implements two instances of the same cerebellar microcircuit, and it is responsible for the acquisition of adaptive synergistic motor responses. The first module $(CRB1)$ controls the adaptive motor response at the level of the ankle joint. The cue signal is given by the ankle angle absolute error ($|\,\Theta_1 - \Theta_{tgt1}\,|$). The signal to be learned (teaching signal) is given by the output of the reactive controller, encoding the action necessary to compensate for the error of the first segment of the pendulum. The output of the controller is an acquired compensatory motor response acting in a feed-forward manner and summing to the output of the reactive controller. The second instance of the same cerebellar microcircuit $(CRB2)$ is responsible for issuing adaptive motor responses at the level of the hip joint. Importantly, the cue signal is given by an efferent copy of the output signal of the first cerebellar module $(CRB1)$. This connection mimics the nucleo-pontine projection (NNP) linking the output stage of one microcircuit to the input stage of a second microcircuit, allowing the trigger of sequential synchronized responses. Similarly to $CRB1$, $CRB2$ is fed with a teaching signal given by the output of the hip reactive controller, encoding the action necessary to compensate for the error of the second segment of the pendulum. The output of the controller is an acquired feed-forward motor response summing to the output of the hip reactive controller.

Experimental Design. The goal of the system is to acquire appropriate motor responses in order to vertically stand from an initial supine position. The experimental session proceeds on a trial by trial base, having the agent set at a given supine position at the beginning of every trial (Θ_{init1}=1.3, Θ_{init2}=1.5). During each trial (6 seconds duration), the agent has to perform a motor pattern in order to modify the body configuration and stand vertically in a position of equilibrium ($\Theta_{tgt1}=0.0$, $\Theta_{tgt2}=0.0$). We run a session composed by a total of 300 trials. We divide the session into two parts: trials 0-50 are considered as the *acquisition period*, in which adaptive responses are learned. Trials 51-300 are considered as the *optimization period*, in which the previously learned adaptive responses are recursively fine-tuned and optimized. We therefore analyze the evolution of the learned responses comparing the amplitude and the timings with an increase in accuracy and reduction of error.

3   Results

The goal of the experiment is to test the learning capabilities of the proposed architecture in reaching a vertical position of equilibrium from a supine initial position. We train the agent to acquire two synergistic adaptive responses controlling the position of two connected joints with the goal of minimizing the deviation of the pendulum from a given target angle. The training session lasts 300 trials, 6 seconds per trial, where the initially untrained agent has to progressively acquire an appropriate motor pattern to achieve a desired position. We divide the whole session into two parts: *acquisition period* (trial 0-50) and *optimization period* (trial 51-300). During the acquisition period the adaptive responses are progressively acquired from the activity of the reactive controllers. During this phase the adaptive motor outputs are initially absent and progressively grow over time showing, however, a high variability ($std=2.23$). For this reason we focus mainly on the results obtained during the optimization phase, where the acquired responses can be considered already at and advanced stage. We are particularly interested in measuring the evolution, along the whole session, of errors, total energy cost and the time relationship between the ankle joint and the hip joint motor responses.

Acquisition. During early trials of the acquisition period the motor responses are driven by the sole reactive controllers. Such motor responses are not sufficient to lift the pendulum toward the target position causing the pendulum itself to fall side-wards (Fig. 2, *right*). Toward the end of the training session we notice that the acquired adaptive responses for both the ankle joint and the hip joint (Fig. 2, *left*) are sufficiently strong to lift the pendulum. However, they exhibit reciprocal motor patterns that induce a prominent oscillatory behavior not adequate to minimize the deviation from the target position (Fig. 2, *right*)(Fig. 3, *left*). This kind of pattern is the result of the interaction between the antagonist responses switching the direction of the force vector applied to the ankle joint. This could mean that the provoked oscillation is not accurately counterbalanced by the position of the hip joint controller, which despite outputting an opposite motor response, is still acting in a compensatory fashion rather than an anticipatory one.

Optimization. During early trials of the optimization period we notice a slight decrease of error characterized by high variance ($std=0.12$) (Fig. 3, *left*). Such

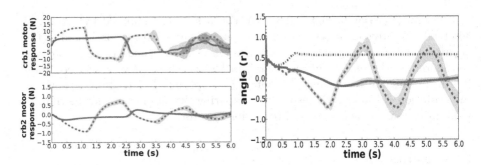

Fig. 2. *Left.* Top. Output of $CRB1$ mapped to the ankle joint. Bottom. Output of $CRB2$ mapped to the hip joint. Dashed red line represents the mean output for the *acquisition period*: trials 40-50. Solid green line represents the mean output for the *optimization period*: trials 290-300. Transparencies (same color code) represent standard deviation. *Right.* Pendulum angle. Dash-dotted blue line represents the values for trial 1. Dashed red line represents the mean values at the end of the *acquisition period*: trials 40-50. Solid green line represents the mean output at the end of the *optimization period*: trials 290-300.

values could be explained by the high variance also encountered in the total motor response energy ($std=1.94$). Generally we notice a trend towards minimization of delay between the peaks in the derivative of the responses (Fig. 3, *right*).

At the end of the optimization period, we notice a considerable reduction of the error, which stabilizes around a value of approximately 0.05. From this result we can assess that the system ultimately learned to stably stand in a vertical position (Fig. 3, *left*). Interestingly, a considerable decrease of total energy is also achieved (Fig. 3, *right*). The sum of the total energy for the two controllers amounts, at the end of the session, to approximately half of the values encountered at the beginning of the optimization period. Variance for both error ($std=0.008$) and total energies ($std=0.85$) are also greatly minimized. Finally the time difference between response peaks is notably reduced (Fig. 3, *right*). From a mean distance of approximately 190 ms at the beginning of the optimization period, they show at the end of the same session a time lag of approximately 110 ms. This result might confirm a trend exhibit by the system towards the synchronization of the responses.

Moreover, we report a high statistical correlation between error decrease and total energy decrease (*Spearman's* $\rho = 0.91$, *p value*<0.01), error decrease and response time differences (*Spearman's* $\rho = 0.73$, *p value* <0.01), and finally between total energy decrease and response time differences (*Spearman's* $\rho = 0.84$, *p value*<0.01). Globally considered, the experimental session shows a significant, positive trend in the evolution of accuracy, time precision and reduction of the total energy required to accomplish the task. This could mean that the interaction of two motor responses trained to reach a target configuration of two joints can benefit form a direct feed-forward link, such as the one proposed by our architecture.

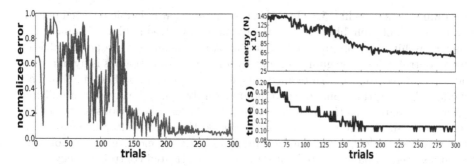

Fig. 3. *Left.* Normalized Error. Red line: the *acquisition period*. Green line: the *optimization period*. Error per trial as the normalized mean of the absolute difference between the absolute sum of Θ_1 and Θ_2 and their respective target positions (time window: 3-6 seconds). *Right. Top.* Total Energy. Sum of the cumulative sums of the absolute value of each individual motor response ($CRB1+CRB2$) per time step. *Right. Bottom.* Time difference among responses: to calculate the time relationship between motor responses we calculate the derivative of the output per each controller individually. We therefore compute the absolute time difference between the first peak of the derivative of the $CRB1$ output and the first peak of the derivative of the $CRB2$ output.

In the proposed architecture, indeed, a motor response triggers a following motor response therefore facilitating the acquisition of a sequence of actions. However, the internal dynamics according to which the two responses are evolving are relatively independent, and adjust over time according to the interaction with the environment and the specific performed task. We consider therefore the fine synchronization and the increase in efficiency an emergent property of the architecture. More specifically, we interpret such result according to the ability of the second cerebellar microcircuit ($CRB2$) to initially learn to compensate and finally to anticipate the consequences of the actions triggered by the first microcircuit ($CRB1$). This would underlie a progressive tendency of $CRB2$ to rely less on the sensory feedback, which encodes the effects of the actions of $CRB1$ and more on the prediction of such effects, in order to firstly compensate and secondly to anticipate. Such process ultimately results in an almost perfectly synchronized motor action which helps to reduce the overall energy costs for the control of the agent and improve the accuracy in the task performance.

4 Discussion and Conclusion

Within this study we are interested in the computational mechanisms underlying the acquisition of action sequences and coordinated motor responses. We propose that the cerebellum could be the neural substrate responsible for the acquisition of well-timed sequential motor responses. In order to test such prediction, we implemented a control architecture based on two connected instances of a cerebellar microcircuit with the goal of balancing a simulated double inverted pendulum, as shown in [15,3]. In [15] it is proposed that a double inverted pendulum model could reliably represent the dynamics involved in bipedal stance. The control architecture proposed by the authors implements pure proportional feedback control, where the gain mapping sensory input to actuators is dynamically adjusted in accordance with biomechanical constraints and body dynamics. This model might capture some of the fundamental dynamics involved in a multiple joint coordination, as the ones depending on sole reactive responses and underlying processes driven by slow sensory feedback. However, feedback control alone seems to be too slow to account for a rapid and efficient motor coordination, and it might need to be integrated with a more sophisticated control strategy.

Feedforward dynamics in motor control are widely established [16,17] and there is also evidence that the cerebellum could be in charge for the implementation of adaptive control strategies in the brain [14,12,13]. Multiple studies on physiology and modeling have shown that the learning mechanisms found in the cerebellum can be trained, in a trial by trial manner, in order to output predictive motor responses in a variety of motor tasks including postural control. Coherently with this view, we suggest that a feed-forward control strategy needs to be taken into account in order to explain coordination of multiple motor responses. Extending the hypothesis formulated in [8,7,6], here we propose that the cerebellar predictive properties can be exploited by sequentially linking two microcircuits together to perform the acquisition of a motor sequence, which could explain multiple joints coordination in a postural task. As in [6,7], we also hypothesized that the Nuclo-Pontine projection connecting the output stage of the cerebellar nuclei to its input stage (via Pons) could represent the neural substrate for such link. Our results support the functional role of this connection suggesting that a recursive feed-forward signal can optimize motor control and reduce motor performance error. Optimization in this case could be driven by the acquired ability of the two microcircuits to reciprocally learn the effects of their own actions, where an initial reactive behavior is progressively replaced by an adaptive one in order to efficiently fulfill a common goal. Such predictive motor responses could ultimately explain some of the dynamics encountered in multiple joint coordination for the maintenance of balance and equilibrium. Moreover, we show that this strategy can reduce the energy cost necessary to perform a sequence of movements.

Interestingly, the reduction of energy has been found to highly correlate with the synchronization pattern among joints responses. Synchronization seems to be an important aspect of postural control. In-phase or anti-phase patterns are found in the temporal oscillatory dynamics of coordinated movement in the hip and in the ankle of human healthy subjects quietly standing. In [18] it has been proposed that such pattern, necessary for equilibrium maintenance, can be an emergent property of the interaction of multiple joints. According to this view, it seems that multiple joints oscillation would tend to converge to either a phase or an anti-phase pattern according to an effect of mutual synchronization. Sole peripheral feedback control would therefore be in charge for coordination of movements as result of the mechanical interaction of joints and muscles. Similarly, a tendency to synchronization is also encountered in the behavior of our system, where a progressive shift of the peaks of the two responses could represent such transition toward an anti-phase stable state.

Despite exhibiting similar temporal dynamics, our proposal differs form the one stated by [18]. We indeed support that a form of control responsible for the temporal dynamics in action sequences can also be found at the level of the central nervous system. We suggest that the cerebellum could be in charge for this task given its adaptive properties in the temporal domain. Physiological studies on internal cerebellar dynamics support the role of the cerebellar microcircuit in the acquisition of finely grained temporal responses [19]. This is explained at the neural level by the recurrent excitatory-inhibitory connection found between Golgi cells and Granule cells, which modulates intensity and timings of the input provided via mossy fibers to the cerebellar cortex. Temporal dynamics in action coordination could therefore been explained by an adaptive learning mechanism such as the cerebellum, where synchronicity would be a special case of a more general mechanism for accurate reaction time modulation, useful to enhance motor performance and optimize energy costs.

The advantages described above could finally benefit robotic architectures. The proposed bio-mimetic approach might allow a more efficient adaptive control of posture in humanoid robots and, in general, a minimization of errors during navigation and manipulation tasks. The importance of learning to efficiently coordinate multiple actions, as found in humans and animals, can therefore be directly applied to agents able to learn useful sensory-motor contingencies from the interaction with the environment. The present architecture indeed differs from the ones proposed in similar studies (such as [10,11]), as the control strategy does not imply a formal description of the double inverted pendulum problem but still performs competitively in terms of control time (below 3 seconds). An advantage of such strategy can be found in the ability of the system to learn to accomplish a motor task without any built-in knowledge of the body scheme or the environment, which is progressively acquired by experience. A limitation of the system however is that the acquired motor responses efficiently apply only to the conditions in which the controller is trained, therefore not allowing a high degree of flexibility. A possible solution to be tested in future works could be the implementation of a set of cerebellar microcircuits trained in different conditions and appropriately triggered depending on the system state.

Acknowledgments. This work was supported by eSMC FP7-ICT- 270212.

References

1. Smith, A.M.: Does the cerebellum learn strategies for the optimal time-varying control of joint stiffness? Behavioural Brain Research 19, 399–410 (1996)
2. Sasagawa, S., Ushiyama, J., Kouzaki, M., Kanehisa, H.: Effect of the hip motion on the body kinematics in the sagittal plane during human quiet standing. Neuroscience Letters 450(1), 27–31 (2009)

3. Aramaki, Y., Nozaki, D., Masani, K., Sato, T., Nakazawa, K., Yano, H.: Reciprocal angular acceleration of the ankle and hip joints during quiet standing in humans. Experimental Brain Research 136(4), 463–473 (2001)
4. Timmann, D., Horak, F.B.: Perturbed step initiation in cerebellar subjects. 1. Modifications of postural responses. Experimental Brain Research 119(1), 73–84 (1998)
5. Penhune, V.B., Zatorre, R.J., Evans, A.C.: Cerebellar Contributions to Motor Timing: A PET Study of Auditory and Visual Rhythm Reproduction. Journal of Cognitive Neuroscience 10(6), 752–765 (1998)
6. Brandi, S., Herreros, I., Sánchez-Fibla, M., Verschure, P.F.M.J.: Learning of Motor Sequences Based on a Computational Model of the Cerebellum. In: Lepora, N.F., Mura, A., Krapp, H.G., Verschure, P.F.M.J., Prescott, T.J. (eds.) Living Machines 2013. LNCS, vol. 8064, pp. 356–358. Springer, Heidelberg (2013)
7. Herreros, I., Brandi, S., Verschure, P.F.: Acquisition and execution of motor sequences by a computational model of the cerebellum. BMC Neuroscience 14(suppl. 1), P409 (2013)
8. Maffei, G., Sanchez-fibla, M., Herreros, I., Paul, F.M.J.: The role of a cerebellum-driven perceptual prediction within a robotic postural task. In: SAB - Simulated Adaptive Behavior Conference (accepted, 2014)
9. Creath, R., Kiemel, T., Horak, F., Peterka, R., Jeka, J.: A unified view of quiet and perturbed stance: simultaneous co-existing excitable modes. Neuroscience Letters 377(2), 75–80 (2005)
10. Formal'skii, A.M.: On stabilization of an inverted double pendulum with one control torque. International Journal of Computer and Systems Sciences 45(3), 337–344 (2006)
11. Zhang, S., An, R., Shao, S.: A New Type of Adaptive Neural Network Fuzzy Controller in the Double Inverted Pendulum System. In: Deng, H., Miao, D., Lei, J., Wang, F.L. (eds.) AICI 2011, Part II. LNCS, vol. 7003, pp. 149–157. Springer, Heidelberg (2011)
12. Dean, P., Porrill, J.: Evaluating the adaptive-filter model of the cerebellum. The Journal of Physiology 589(Pt. 14), 3459–3470 (2011)
13. Herreros, I., Verschure, P.F.: Nucleo-olivary inhibition balances the interaction between the reactive and adaptive layers in motor control. Neural Networks 47, 64–71 (2013)
14. Miall, R.C., Weir, D.J., Wolpert, D.M., Stein, J.F.: Is the cerebellum a smith predictor? Journal of Motor Behavior 25(3), 203–216 (1993)
15. Park, S., Horak, F.B., Kuo, A.D.: Postural feedback responses scale with biomechanical constraints in human standing. Experimental Brain Research 154(4), 417–427 (2004)
16. Kawato, M.: Internal models for motor control and trajectory planning. Current Opinion in Neurobiology 9, 718–727 (1999)
17. Wolpert, D.M., Miall, R.C., Kawato, M.: Internal models in the cerebellum. Trends in Cognitive Sciences 2(9), 338–347 (1998)
18. Bardy, B., Stoffregen, T., Bootsma, R.: Postrual Coordination Modes Considered as Emergent Phenomena. Journal of Experimental Psychology 25(5), 1284–1301 (1999)
19. Crowley, J.J., Fioravante, D., Regehr, W.G.: Dynamics of fast and slow inhibition from cerebellar golgi cells allow flexible control of synaptic integration. Neuron 63(6), 843–853 (2009)

I-CLIPS Brain:
A Hybrid Cognitive System for Social Robots

Daniele Mazzei, Lorenzo Cominelli, Nicole Lazzeri,
Abolfazl Zaraki, and Danilo De Rossi

Research Center "E. Piaggio"
via Diotisalvi 2, 56126, Univ. of Pisa, Italy
mazzei@di.unipi.it
http://www.faceteam.it

Abstract. Sensing and interpreting the interlocutor's social behaviours is a core challenge in the development of social robots. Social robots require both an innovative sensory apparatus able to perceive the "social and emotional world" in which they act and a cognitive system able to manage this incoming sensory information and plan an organized and pondered response. In order to allow scientists to design cognitive models for this new generation of social machines, it is necessary to develop control architectures that can be easily used also by researchers without technical skills of programming such as psychologists and neuroscientists. In this work an innovative hybrid deliberative/reactive cognitive architecture for controlling a social humanoid robot is presented. Design and implementation of the overall architecture take inspiration from the human nervous system. In particular, the cognitive system is based on the Damasio's thesis. The architecture has been preliminary tested with the FACE robot. A social behaviour has been modeled to make FACE able to properly follow a human subject during a basic social interaction task and perform facial expressions as a reaction to the social context.

Keywords: Social robots, humanoids, cognitive systems, artificial intelligence, hybrid control architectures, expert systems.

1 Introduction

Since the ancient times, humans have always been curious about understanding and simulating the human nature. Nowadays, thank to the rapid advances in robotics, engineering and computer science, this curiosity has become reality. A new generation of robots with an anthropomorphic body and human-inspired senses is enjoying increasing popularity as research tool.

The efforts beyond these human-like robots aim at creating robots that communicate with humans in the same way that humans communicate with each other and work in close cooperation with humans sharing the same environment. Therefore it is not surprising that these robots are equipped with a human body to take advantage of the human-centered design of the environment and a set of

A. Duff et al. (eds.): Living Machines 2014, LNAI 8608, pp. 213–224, 2014.

sensors that mimics the human senses making the robot capable of intuitively communicate with humans [11, 12].

Some requirements are mandatory for developing a social and emotional intelligence of a humanoid robot: a sensory apparatus able to perceive the social and emotional world, an actuation and animation system able to properly control the robot's movements and gestures, but also a "smart brain" able to manipulate the flow of information generating fast and suitable responses. All these features make these robots powerful research tools suitable for studying human intelligence and behavioural models by investigating the social dynamic of human-robot interaction [3–5].

The presented architecture was designed on the basis of the Damasio's thesis [7, 8] and its formalization proposed by Bosse [2]. These theories explain and formalize perception, action and planning capabilities of the human nervous system that conceptually inspired our work. The presented architecture is based on a hybrid deliberative/reactive paradigm in which a reactive subsystem ensures that the robot can handle real-time low-level control challenges while a deliberative subsystem performs high-level planning tasks that require reasoning process [18]. The deliberative subsystem is based on a rule-based engine through which is possible to control the behaviour of the robot using an intuitive natural-like programming language. Thus, this architecture provides neuroscientists, psychologists and human behaviour researchers with an innovative control system for social robots that can be used as a tool for designing and testing behavioural models without requiring deep knowledge of programming or engineering.

The implemented architecture was preliminary tested controlling a humanoid robot called FACE (Facial Automaton for Conveying Emotions) in simulated social scenarios [13, 15, 14, 21, 20].

2   System's Architecture

"My view then is that having a mind means that an organism forms neural representations which can become images, be manipulated in a process called thought, and eventually influence behavior by helping predict the future, plan accordingly, and choose the next action."

In this seminal sentence of his book, *"Descartes' Error"*, Antonio Damasio describes the mind as a process in which **inputs** from sensors are converted into **knowledge structures** that allow **reasoning**. The result of the reasoning process are internal or external **actions** that together with the *new generated knowledge* drive feelings, emotions and behaviours of human beings.

Humans perceive the world and their internal state through multiple sensory modalities that in parallel acquire an enormous amount of information creating internal representations of the perceived world. Moreover, behaviours and skills are not innate knowledge but are assimilated by means of a knowledge acquisition process [6] and by emotional influences. This is also supported by the evidence that pure rational reasoning is not sufficient for decision making as demonstrated by studies conducted on subjects with affective and emotional deficits due to functional and behavioural disorders [7].

Social robots are nowadays equipped with a rich set of sensors aimed at acquiring information similarly to the human sensory apparatus [1, 9]. Raw data are processed and organized creating "meta-maps", i.e. structured descriptions, of the robot's body state *(proprioception)*, of the perceived environment *(exteroception)*, and of the perceived scene social meaning *(social perception)*. This knowledge representation as structured objects allows the system to manipulate the information in a flexible way with a high level of abstraction through a rule-based simple programming language.

As in human nervous system, planning is the slower part of the control architecture. Rule based expert systems can deal with a huge amount of rules but require time to compute all the possible solutions. In the meanwhile sensors and actuators must deal with quick reactive actions requiring fast communication channels and analysis algorithms [18]. For this reason a hybrid deliberative/reactive architecture which integrate a rule-based deliberative system with a procedural reactive system is a good solution for integrating a real-time reactive control with a more versatile and easy-to-use deliberative planning system.

The proposed architecture, shown in Fig. 1, is composed of three main functional blocks: SENSE, PLAN and ACT.

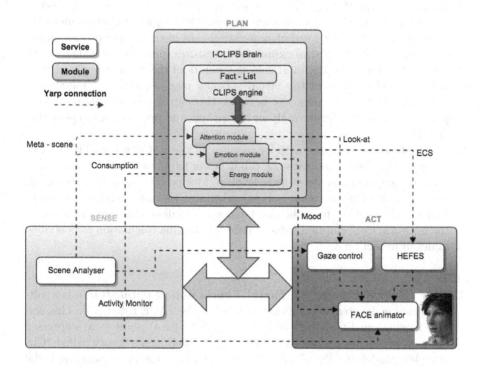

Fig. 1. Schema of the proposed hybrid deliberative/reactive architecture

The architecture includes a set of *Services*, standalone applications interconnected through the network. Each service collects and processes data gathered from sensors or directly from the network and sends new data over the network. Information flowing through the network is defined by XML packets that represent a serialised form of structured data objects. This information management through structured data packets makes possible to create a modular and scalable architecture by developing services that can receive and send data through the network using different programming languages and hardware devices.

The network infrastructure is based on YARP, an open-source middleware designed for the development of distributed robot control systems [16]. YARP manages the connections by using special *Port* objects. A port is an active object managing multiple asynchronous input and output connections for a given unit of data. Each service can open many different YARP ports for sending and receiving data through the network. Each structured data object is serialised as XML packet and sent over the network through a dedicated YARP port. Vice versa, each structured object received from the network through a YARP port is deserialised in the corresponding structured object.

The current stage of the architecture includes the following services (Fig. 1):

SENSE

Scene Analyzer: it is the core of the SENSE block. It processes the information acquired through the Microsoft Kinect Camera[1] to extract a set of features used to create a *meta-scene* object. The extracted features include a wide range of high-level verbal/non-verbal cues of the people present in the environment, such as facial expressions, gestures, position, speaking identification, and a set of the most relevant points of the scene calculated from the low-level analysis of the visual saliency map. Finally, the meta-scene is serialised and sent over the network through its corresponding YARP port. Details of the Scene Analyzer algorithms and processes are reported in [21, 20].

Activity Monitor: it is the activity monitor of the FACE robot. This service manages the connection with the robot power supply and monitors the current consumption and the voltage levels of four power channels of the robot. The activity monitor supply service calculates the robot energy consumption in Watt with a frequency of 1 Hz and serialises this information over the network. Consumption value is used for stamina calculation as described in detail in section 3.

ACT

HEFES (Hybrid Engine for Facial Expressions Synthesis): it is a software engine devoted to emotional control of the FACE robot [15]. This service receives an ECS (Emotional Circumplex Space) point (v,a) expressed in terms of *valence* and *arousal* according to the Russel's theory called "Circumplex Model of Affects" [19, 17] and calculates the corresponding facial

---

[1] www.microsoft.com

expression. i.e. a configuration of servo motors, that is sent over the network to the FACE animator.

Gaze Control: it is the control system of the robot's neck and eyes [21, 20]. This module receives a meta-scene object which contains a list of the persons in the field of view of the robot, each of them identified by a unique ID and associated with spatial coordinates (x,y,z). The Gaze control service is also listening to the *"look at"* YARP port used by the deliberative subsystem to send the ID of the subject towards which the robot must focus its attention (Attention process is described in detail in Sec. 3).

FACE Animator: it is the low-level control system of the FACE robot. This service receives multiple requests coming from the other services such as *facial expressions* and *neck movements*. Since the behaviour of the robot is inherently concurrent, parallel requests could generate conflicts. Thus, the animation engine is responsible for blending multiple actions taking account of the time and priority of each incoming request.

PLAN

I-Clips Brain: it is the core of the PLAN block. This service embeds the CLIPS (C Language Integrated Production System) rule-based expert system and works as a gateway between the procedural and the deliberative subsystems [10].

In the proposed architecture ACT, SENSE and PLAN blocks are only descriptive constructs. The virtual link created by the connections between ACT and SENSE services represents the reactive subsystem. Conversely, the deliberative subsystem is represented by the connections between the I-CLIPS Brain (PLAN) service and all the other services. Indeed this connections are not used for fast stimulus-response control loops but only for reasoning and decision making.

3 I-CLIPS Brain: A New Way of Thinking

I-CLIPS Brain is the most relevant service of the proposed architecture representing the contact point between the reactive and the deliberative subsystems. Similar to the other services, I-CLIPS Brain is a standalone application connected to the YARP network for sending and receiving data objects. The peculiarity of the I-CLIPS Brain service is represented by the interface with CLIPS, the rule-based expert system specifically intended to model human expertise or knowledge [10].

In CLIPS *facts* represent pieces of information and are the fundamental unit of data used by *rules*. Each fact is recorded in the *fact-list*. CLIPS supports the definition of *templates*, structured facts defined as lists of named fields called slots. Templates in a declarative language are structured data similar to objects in a procedural language therefore it is possible to convert objects in CLIPS templates and vice versa. The decision making process is based on the evaluation of rules. Each rule is composed of two parts: Left Hand Side (*LHS*) that contains all the conditions to make the rule trigger, and Right Hand Side (*RHS*) that contains all the actions that will be fired if the *LHS* conditions are satisfied. The *RHS* can contain both function calls or assertion of new facts. Assertion of new facts generates new knowledge that can be sent to the other services through the network or used as input for the other rules.

If the LHS of a rule is satisfied, that rule is not executed immediately but it is marked as *activated*. Activated rules are arranged in the *agenda*, a list of rules ranked in descending order of firing preference. Rules order in the agenda drives the execution order. It is possible to prioritize the rules disposition in the agenda by defining a *saliency* so that rules with higher saliency will be placed at the top of the execution list. In case of rules with the same saliency they will be ordered depending on the selected *conflict resolution strategy*. CLIPS makes available the selection of various conflict resolution strategies among which the *depth strategy* has been selected for its similarity to the typical human reasoning strategy. In the depth strategy the last rule activated by the facts is the first to be executed generating a behaviour that is reactive and influenced by the most recent events.

The saliency parameter was used to create *"agenda layers"*, subsets of the agenda containing concurrent rules. Rules with the same saliency belong to the same layer. In each layer conflicts between rules are solved according to the selected resolution strategy, in this case the depth strategy.

In this preliminary implementation of the rule-set, three layers have been defined using the following saliency values:

Analysing Rule Set (**Saliency=1000**): it includes rules devoted to the analysis of data received by the services (*primary facts*). Rules belonging to this layer generally entail the assertion of new *secondary facts* creating a higher level of knowledge. For example, the rule *find-subjects* triggers the assertion of the secondary fact *subject-in-scene* that makes the robot aware of the presence of people in the scene.

Planning Rule Set (**Saliency=100**): this layer contains rules involved in attentive and emotional decision making process. Planning rules have a LHS containing both primary and secondary facts. These rules assert new facts used as input for the Executing Rule Set. For example *subject-rising-hand* is activated if *subject-in-scene* and *subject-rising-hand* are asserted and its execution will update the fact *winner* with the ID of the subject who has been identified.

Executing Rule Set (**Saliency=10**): This layer includes rules devoted to send planned actions and data to the other services. For example, the rule *look-at-fun* sends the *Look-at* data packet containing the ID of the selected subject over the network.

The I-CLIPS brain structure is also modular and it manages the loading and unloading of various I-CLIPS behavioural modules. Each I-CLIPS behavioural module is composed of a procedural set of instructions and a CLIPS file containing rules and data structures. Modules loading event creates a link between the procedural and the deliberative worlds allowing the information available on YARP ports to be asserted in the CLIPS knowledge base through the corresponding data structures. Once new information is asserted in the I-CLIPS Brain, the activation of rules can generate new knowledge used by the rule engine itself to activate additional rules or to be sent over the network.

At the current stage the architecture includes the following I-CLIPS Brain modules (Fig. 1):

Attention Module: it is devoted to the robot's attention control. It analyses the meta-scene from a social and affective point of view by selecting and streaming over the network the ID of the subject defined as the most socially important person. The Attention module is based on a set of rules defined on the basis of a previous work in which social attentive cues were identified and ordered [20, 21].

Energy Module: it takes account of the energy consumption detected by the FACE robot energy monitor and asserts the secondary fact *Stamina*. *Stamina* is not used in this preliminary implementation but is conceived as the robot's energy level indicator that will influence the execution of behavioural and social rules.

Emotion Module: it is responsible for the emotional control of the FACE robot. This module drives the robot's mood and facial expression according to the perceived social and affective inputs. The Emotion module analyses the meta-scene defining the *face-emotion* template. *Face-emotion* is composed of two slots: *FacialECS* and *Mood*. *FacialECS* represents valence and arousal coordinates of the planned robot's facial expression. This slot is received by the HEFES service which generates the robot facial expressions. *Mood* is used for driving the robot's behaviour. At the current stage the slot *Mood* is included but not yet used by any services. *Mood* will be used by the FACE animator service to control autonomic activities such as blinking, micro facial movements and other behavioural actions.

An example of the I-CLIPS decision making process is shown in Figure 2. The Attention Module receives the meta-scene streamed by the Scene Analyser service asserting the corresponding templates. Rules concurrent fire comport the assertion of the fact *look-at* containing the ID of the subject towards which the robot must direct its attention. Finally, the corresponding *Look-at* data packet is send over the network through a dedicated YARP port. The Gaze control service is listening on the *Look-at* port for receiving the subject's ID. This information is used to navigate through the meta-scene received directly by the Scene Analyser service and extract the subject's coordinates to calculate the positions of the robot's eye and neck motors.

At the same time, the Emotion module analyses the meta-scene using a set of rules devoted to the identification of the subjects' facial expressions. If a subject shows a particular facial expression or performs an important social cue like "hand rising" or "speaking" the slot *FacialECS* is updated with the pleasure and arousal values selected by the corresponding behaviour planning rules. The

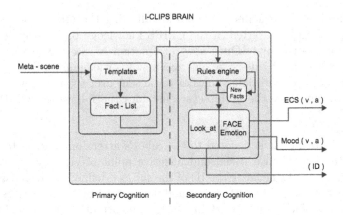

Fig. 2. Detail of the I-Clips Brain knowledge analysis and planning command generation

HEFES service is listening to the *FACE-emotion* port for receiving the pleasure and arousal coordinates and selecting the new facial expression.

4 Preliminary Tests and Results

Preliminary tests of the architecture were conducted by a student without programming skills who created the rule set of the modules described above. The rule set implemented a social behavioural model to make the FACE robot able to follow the subjects in the scene and react to real-time social cues with proper facial expressions.

The robot's behavioural model is explained in the following points:

- If no subjects are present in the robot's field of view, the robot is annoyed and looks at the most salient point (in term of colours and shapes) of the perceived scene;
- If someone is present in the scene, the robot's facial expression becomes neutral and the robot starts to follow the most important subject identified according to a ranking of the following social cues [20]: recognized hand gesture, distance, speaking probability, recognized facial expressions;
- If someone invades the robot's intimate space, it changes its facial expression to dislike keeping the attention on the subject. ...

In Fig. 3, the graphs show the subject head position tracked by the Scene Analyzer together with the robot's head orientation and the ECS values used for the generation of facial expressions, acquired during a 40 second long preliminary test. In A the robot is alone in the room and the scene analyser subject's position is set at default (0, 0) Fig. 4-top. The robot has an annoyed expression (v=0.27;a=-0.84) focusing the attention on the most salient point of the scene

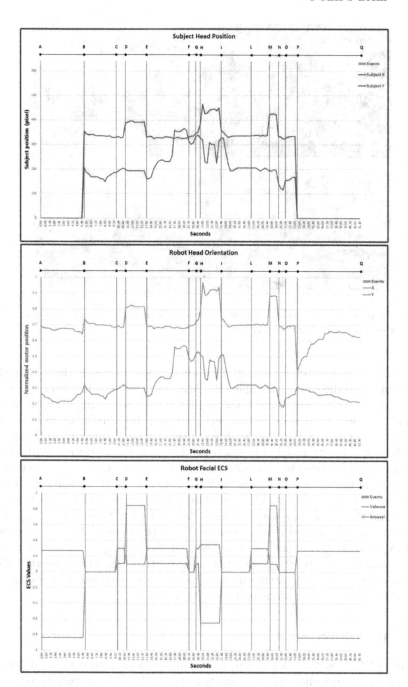

Fig. 3. Data acquired during the system preliminary test. (top) Subject's head position acquired by the Scene Analyzer (in pixel on an image of 640X480); (center) robot's head orientation expressed as Yaw (motor X) and Pitch (motor Y) normalized coordinates; (bottom) ECS values sent to the HEFES service for the generation of FACE expressions.

Fig. 4. Snapshot of the preliminary social interaction test acquired with the Scene Analyzer describe din detail in [21, 20]

identified by the scene analyzer. In B someone enters in the room, the robot shifts the attention to the subject changing its facial expression to neutral (v=0;a=0) (snapshot reported in Fig. 4). In C the subject speaks presenting itself, the robot still focuses the attention on the subject changing the facial expression to "interest" (v=0.11;a=0.30). In D the subject raises the hand speaking to the robot, the robot changes its facial expression to "Happiness" (v=0.85;a=0.10). In E hand gesture is removed but the subject is still speaking. In F the subject stops to speak, and the robot's face goes back to "neutral". In H the subject invades the robot's intimate space moving too close; the robot attention is kept on the subject and the facial expression becomes "fear/disgust" (v=-0.65;a=0.35). In I the subject leaves the robot's personal space and the robot shifts to a "neutral" facial expression. At the end of the session, in P, the subject leaves the room. The scene analyser subject's position is reset to (0, 0) Fig. 4-top) and the robot is alone again; the robot attention shifts back to the most salient point of the scene and its facial expressions becomes "annoyed".

5   Conclusions

Nowadays FACE and similar social robots are considered potential and powerful tools for neuroscience and psychology research. In order to become real test beds for human brain models and theories these systems have to be usable not only by engineers and computer scientists with technical programming skills but especially by psychologists and neuroscientists. For this reason this work was focused on the development of a robot control architecture that allows researchers without programming skills to design their own behavioural models to be tested on a real emotional humanoid robot.

The hybrid design of the architecture makes possible a fast control of the reactive processes required for the control of the robot's low-level actions but also the definition of complex behavioural control strategies that can be defined using a very intuitive programming language.

In the reported preliminary tests a student without technical programming skills developed a simplified behavioural model providing the robot with a set of basic social interaction capabilities. The system allowed the student to test various models using only the rule-based deliberative part of the control system without requiring any knowledge of the procedural language.

The system makes possible to interact in real-time with the rule-set providing experimenters with the possibility to interactively add or remove rules during the execution of the behavioural models and investigate how the robot's behaviour is influenced.

The system demonstrated to be a very powerful tool for modelling and implementing robot's behavioural and affective models. However future works aimed at extending the implemented rule-set testing additional control models and paradigms are required. New modules and services will be developed for including the use of the mood and of the energy consumption parameters in the management of the robot's actions and behaviours. The system will be also integrated with the new version of the Microsoft Kinect camera that tracks up to six subjects' skeleton simultaneously. Improved tests with more than one subject involved in the social scenarios will be performed in order to analyze the various user identification strategies implemented in the Scene Analyzer and in the Attention Module.

Acknowledgment. This work was partially based on the thesis "iClips: an interaction system for controlling a robotic FACE" by Nadia Vetrano (2012) supervised by Dr. Antonio Cisternino. The authors would like to thank Roberto Garofalo for his strong contribution to the development of the system.

This work was partially founded by the European Commission within the Project EASEL (Expressive Agents for Symbiotic Education and Learning) (FP7-ICT-611971).

References

1. Bar-Cohen, Y., Breazeal, C.L.: Biologically inspired intelligent robots, vol. 122. Spie Press (2003)
2. Bosse, T., Jonker, C.M., Treur, J.: Formalisation of damasios theory of emotion, feeling and core consciousness. Consciousness and Cognition 17(1), 94–113 (2008)
3. Breazeal, C.: Designing Sociable Robots. MIT Press, Cambridge (2002)
4. Breazeal, C.: Emotion and sociable humanoid robots. International Journal of Human-Computer Studies 59(1), 119–155 (2003)
5. Breazeal, C.: Socially intelligent robots. Interactions 12(2), 19–22 (2005)

6. Brooks, R.A., Breazeal, C., Marjanovic, M., Scassellati, B., Williamson, M.M.: The cog project: Building a humanoid robot. In: Nehaniv, C.L. (ed.) Computation for Metaphors, Analogy, and Agents. LNCS, vol. 1562, pp. 52–87. Springer, Heidelberg (1999)
7. Damasio, A.: Descartes' Error: Emotion, Reason, and the Human Brain. Grosset/Putnam, Random House, New York (1994, 2008)
8. Damasio, A.: The feeling of what happens (1999)
9. Fong, T., Nourbakhsh, I., Dautenhahn, K.: A survey of socially interactive robots. Robotics and Autonomous Systems 42(3-4), 143–166 (2003)
10. Giarratano, J.C., Riley, G.: Expert systems. PWS Publishing Co. (1998)
11. Gibson, J.: The concept of affordances. Perceiving, acting, and knowing, pp. 67–82 (1977)
12. Horton, T.E., Chakraborty, A., St. Amant, R.: Affordances for robots: a brief survey. AVANT. Pismo Awangardy Filozoficzno-Naukowej (2), 70–84 (2012)
13. Lazzeri, N., Mazzei, D., Zaraki, A., De Rossi, D.: Towards a believable social robot. In: Lepora, N.F., Mura, A., Krapp, H.G., Verschure, P.F.M.J., Prescott, T.J. (eds.) Living Machines 2013. LNCS (LNAI), vol. 8064, pp. 393–395. Springer, Heidelberg (2013)
14. Mazzei, D., Billeci, L., Armato, A., Lazzeri, N., Cisternino, A., Pioggia, G., Igliozzi, R., Muratori, F., Ahluwalia, A., De Rossi, D.: The face of autism. In: The 19th IEEE International Symposium on Robot and Human Interactive Communication, RO-MAN 2010, pp. 791–796. IEEE Computer Society Publisher (2010)
15. Mazzei, D., Lazzeri, N., Hanson, D., De Rossi, D.: Hefes: an hybrid engine for facial expressions synthesis to control human-like androids and avatars. In: 4th IEEE RAS & EMBS International Conference on Biomedical Robotics and Biomechatronics (BioRob 2012), pp. 195–200. IEEE Computer Society Publisher (2012)
16. Metta, G., Fitzpatrick, P., Natale, L.: Yarp: Yet another robot platform. International Journal of Advanced Robotic Systems 3(1) (2006)
17. Posner, J., Russell, J.A., Peterson, B.S.: The circumplex model of affect: An integrative approach to affective neuroscience, cognitive development, and psychopathology. Development and Psychopathology Null, 715–734 (September 2005)
18. Qureshi, F., Terzopoulos, D., Gillett, R.: The cognitive controller: a hybrid, deliberative/reactive control architecture for autonomous robots. In: Orchard, B., Yang, C., Ali, M. (eds.) IEA/AIE 2004. LNCS (LNAI), vol. 3029, pp. 1102–1111. Springer, Heidelberg (2004)
19. Russell, J.A.: The circumplex model of affect. Journal of Personality and Social Psychology 39, 1161–1178 (1980)
20. Zaraki, A., Mazzei, D., Giuliani, M., De Rossi, D.: Designing and evaluating a social gaze-control system for a humanoid robot. IEEE Transactions on Human-Machine Systems PP(99), 1–12 (2014)
21. Zaraki, A., Mazzei, D., Lazzeri, N., Pieroni, M., De Rossi, D.: Preliminary implementation of context-aware attention system for humanoid robots. In: Lepora, N.F., Mura, A., Krapp, H.G., Verschure, P.F.M.J., Prescott, T.J. (eds.) Living Machines 2013. LNCS (LNAI), vol. 8064, pp. 457–459. Springer, Heidelberg (2013)

Change of Network Dynamics in a Neuro-robotic System

Irene Nava[*], Jacopo Tessadori[*], and Michela Chiappalone

Department of Neuroscience and Brain Technologies, Istituto Italiano di Tecnologia,
Genova, Italy
{jacopo.tessadori,michela.chiappalone}@iit.it,
navairen@gmail.com

Abstract. In the past, tetanic stimulation has been used in several different instances to induce changes in the firing patterns of neural networks *in vitro*. In this paper, we ran a new experimental campaign to verify if this protocol induced lasting changes and if those changes were predictable. We found out that our stimulation protocol led to different results in cortical and hippocampal preparations: in the first case, stronger connections were weakened, resulting in a reduction of bursting activity and late evoked response; in the case of hippocampal preparations, single strong connections underwent strong changes but, on average, remained unchanged. In both preparations, the geometry of induced changes remains largely uncorrelated with the actual site of stimulation delivery.

Keywords: micro-electrode array, in vitro networks, tetanic stimulation, closed-loop stimulation, effective connectivity.

1 Introduction

Primary neuronal cultures coupled to Micro-Electrode Arrays (MEAs) represent a simple but interesting model system of the neuronal functions [1-6].These networks are spontaneously active, and their firing rates change with the age of the culture [7, 8]. Starting from the second week of in vitro development, spikes tend to cluster into bursts, a pattern that persists throughout the time in culture, representing the mature state of the network [9]. These bursts can be found in both hippocampal [10] and cortical [11] cultures. This mode of activity can be also modulated in vitro by appropriate electrical [2, 12], optical [13] and/or chemical manipulation [14-16]. Interestingly, Hebbian plasticity, in the form of long-term potentiation (LTP) and depression (LTD) has been reported in neural preparations coupled to MEAs [17-21], thus demonstrating learning capabilities of dissociated cultures [22, 23].

In a previous work, we took advantage of a 'tetanic stimulation' (i.e. short periods of high-frequency stimulation) paradigm [24-26] in order to develop a learning mechanism for a hybrid experimental setup. In these experiments, a biological neuronal network acted as a controller for a small mobile robot, by generating motor commands from the received sensory information. We observed that, indeed, immediately

---

[*] Equal contribution.

A. Duff et al. (eds.): Living Machines 2014, LNAI 8608, pp. 225–237, 2014.

after the delivery of tetanic stimulation, each subsequent stimulation elicited a stronger response than in the absence of tetanic stimulation. However, no systematic analysis in terms of network dynamics has been performed. In this work, we describe a new set of experiments aimed at unraveling possible changes in terms of electrophysiological behavior induced by the tetanic stimulation protocol. It is worth noting that the delivery of tetanic stimulation does not occur at regular intervals, but, rather, it is triggered by specific events (namely, the robot touching an obstacle). As pointed out in literature [27, 28], the distribution of the incoming stimuli is likely to have a rather profound impact in the network response. In particular, the questions we wanted to ask concerned the spatial relationship between site of delivery of stimulation and occurred changes; the ties between pre- and post- stimulation connectivity strength; the variations of relative firing times within the network, expressed in the forms of bursting patterns during spontaneous activity and late-responses to probing stimulations. Finally, in previous works [25, 26] we noted how cortical and hippocampal preparations tended to present different firing patterns to similar stimulation inputs, therefore in this work we analyzed cultures from both regions as distinct groups.

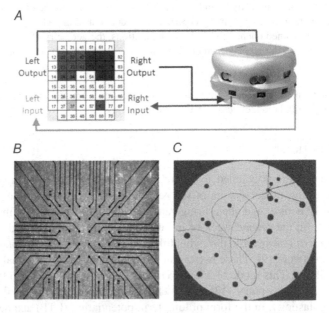

Fig. 1. A. Schematical illustration of the input/output electrode configuration in the MEA device: the light and dark green electrodes are selected as inputs, i.e. they are used to deliver stimuli as a function of the sensors readings from the robot. The light and dark red sets of electrodes are instead used as control regions for the wheels: the speed of the left wheel will be a function of the average activity detected on the light red electrodes, while the right wheel will be controlled by the dark ones. B. A hippocampal culture grown on a standard MEA device. C. The virtual realization of the Khepera robot (pink circle) is moving within the virtual arena, while readings from its sensors (black lines departing from the robot) collect information on distances from obstacles. Each green circle represents an obstacle of different diameter. The red line highlights the trajectory followed by the robot.

2   Materials and Methods

2.1   Neuronal Preparation and MEA Electrophysiology

Dissociated neuronal cultures were prepared from hippocampi and cortices of 18-day old embryonic rats (Charles River Laboratories), as previously described [29]. All experimental procedures and animal care were conducted in conformity with institutional guidelines, in accordance with the European legislation (European Communities Directive of 24 November 1986, 86/609/EEC) and with the NIH Guide for the Care and Use of Laboratory Animals. Cells were afterwards plated onto standard MEAs (Multichannel Systems, Reutlingen, Germany) previously coated with poly-D-lysine and laminin to promote cell adhesion (final density around 1200 cells/mm2) and maintained with 1 ml of nutrient medium (Fig. 1A). They were then placed in a humidified incubator having an atmosphere of 5% CO_2 and 95% air at 37°C. Recordings were performed on cultures between 25 and 46 Days In Vitro (DIV).

The electrophysiological activity of a culture was recorded through the MEA, which consists of 60TiN/SiN planar round electrodes (30 µm diameter; 200 µm center-to-center inter-electrode distance) arranged in an 8×8 square grid excluding corners. Each electrode provides information on the activity of the neural network in its immediate area. The amplified 60-channel data is conveyed to the data acquisition card which samples them at 10 kHz per channel and converts them into 12 bit data.

2.2   Experimental Protocol

The experimental protocol used for the experiments presented in this report, consists of 7 consecutive phases, for a total duration of around 3 hours. Specifically, we recorded: (1) Spontaneous activity (30min); (2) Test stimulus from 10 sites; (3) Robot navigation (30 min); (4) Robot navigation, with tetanic stimulation (30 min); (5) Robot navigation (30 min); (6) Test stimulus from 10 sites (same as phase 2); (7) Spontaneous activity (30 min).

In phases 1 and 7, spontaneous activity of the network is subject to observation, in order to determine, empirically, which electrodes are the most likely candidate as 'input' sites (i.e. sites from which stimulation must be delivered). Typical features to look for in this phase are a sustained mean firing rate (>0.1 spikes/s) and patterns of activity not synchronous with other regions. The best candidates (a set of 10 sites) are then selected for the second step of the experiment, to test the evoked response of the network. From each of the candidate 'input' channel, in turn, a 500 µs, 1.5V peak-to-peak, bipolar square wave is delivered every 5 seconds, until a total of 40 stimuli per channels have been delivered, while spiking activity is detected from other electrodes. In phases 2 and 6, for every stimulation electrode involved, 59 Post Stimulus Time Histograms (PSTH) are generated [30]: these graphs report the average number of spikes detected from each electrode in the 300 ms following each stimulation and therefore provide information on the strength of the connections in the culture. The generated PSTHs are then compared in order to look for areas that present a significant degree of specificity. In this way, it is possible to define an output (recording or motor) area that will respond mostly to stimulation from the corresponding input (stimulation or sensory) area, while remaining silent during stimulation from the opposite input area. Phases 4, 5 and 6 form the core of the experiment, during which the robot is left free to run in its arena (for details, [24, 26]). Phases 5 and 7 are effectively identical, while phase 6 implements the tetanic stimulation protocol: following each robot collision, a 2 second-long, 20 Hz stimulation may be delivered to the same-side input area.

2.3   Input and Output Electrodes Selection

The main disadvantage in dealing with dissociated cultures instead of experimental models with a preserved neural structure is the lack of predefined architecture. For this reason, before starting an experiment, a procedure has been performed to define the stimulation and recording areas of the network. During this procedure (i.e. phase 2 of the experimental protocol), we stimulated the cultures by delivering trains of 40 electrical stimuli from 10 sites in a serial way. Then, the PSTH area (i.e. the number of spikes in the 300 ms following each stimulation) between each pair of stimulation-recording electrodes is computed. The ideal case is the one where two stimulation electrodes and two groups of recording sites present very high ipsilateral (shortened to ipsi in the rest of the paper) connectivity and a very low contralateral (contra) one, i.e. stimulation from one input area will elicit a strong response from the output area it is associated to (high ipsi connectivity), but very little or no response from the other sensory area (low contra connectivity) and vice versa.

2.4   Burstiness Index

The most prominent feature of the electrical activity of dissociated cultures is their propensity for synchronized bursting (cfr. Introduction). In literature it is assumed that the persistence of global bursts in dissociated cortical cultures is a result of deafferentation [31]. It is not necessary to identify individual bursts to quantify the level of burstiness of a recording. We used a method reported in literature [2]: divide the recording into 100 ms long time bins and count the number of spikes in each bin, then compute the fraction of the total number of spikes accounted for by the 15% of bins with the largest counts. If f_{15} is the number of spikes accounted for by the 15% of bins with the largest count, the burstiness index (BI) is defined as:

$$BI = \frac{f_{15} - 0.15}{0.85} \qquad (1)$$

It therefore ranges from 0 in the case of perfectly tonic firing to 1 in the case of all spikes belonging to a burst.

2.5   Coding and Decoding of Sensory Information

In the employed neuro-robotic interface, sensory information is coded as an approximately linear function of the sensor readings of the robot. In particular, the sensors of the robot will provide an output of 1 when close to an obstacle, 0 when as far away from an obstacle as possible. The robot is equipped with three distance sensors per side, at different angle, and the readings of each side are averaged to provide only two inputs, one for the left and one for the right sides. More in detail:

$$s_{i,t} = \left(s_i^{MAX} - s_i^{min}\right)r_{i,t} + s_i^{min} \tag{2}$$

where subscript i denotes sensor side, $s_{i,t}$ is the stimulation rate of the ith input area at time t, and $r_{i,t}$ the sensor reading (r can range between 0 and 1) averaged over the sensors on the ith side of the robot, at time sample t, whereas s_i^{MAX} and s_i^{min} are user-set parameters fixing the maximum and minimum stimulation rate. For more details, see [24]. The decoding algorithm works in a slightly different way: each detected spike in one of the motor areas provides a contribution to the wheel of the corresponding speed, while this contribution decays exponentially in time when no spikes are detected:

$$\omega_{i,t} = \lambda_s \omega_{i,t-1} + k_s s_{i,t} \tag{3}$$

where λ_s indicates the decay constant, k the weights of each instantaneous contribution. $s_{i,t}$ is the number of spikes in the bin t of motor area i. In particular, the decay constant was set at a value that would cause $\omega_{i,t}$ to reduce by 10% for each second without detected spikes.

2.6   Evaluation of Robot Navigation

The employed decoding scheme supposed a more or less regular response to stimulation from the neural network in exam, with roughly the same number of action potentials evoked by the delivery of a stimulus. Therefore, a 'natural' way of benchmarking the robot navigation performance is that of comparing the actual recorded speed during the experiment with an ideal wheel speed computed by applying the decoding algorithm directly to the stimulation train: as mentioned before, if the network response to stimulation is reliable, the recorded and ideal wheel speeds should differ only by a multiplicative constant, since the average number of evoked spikes by a stimulation pulse will very likely be different from one.

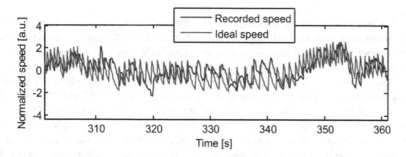

Fig. 2. Comparison of one minute of recorded (blue) and ideal (red) wheel speed traces. While the final temporal details do not exactly match, slower variations are preserved.

Obviously, the recorded speed will differ from the ideal one, due to delays in response and non-linearities in the behavior of neural networks (Figure 2). The speed traces are normalized in order to have zero mean and a standard deviation equal to one. For each experiment, we computed cross-correlations between the two stimulation trains and the two wheel speed traces, for a total of four stimulation-speed couples (two ipsi and two contra cross-correlations). Ideally, we would like ipsi correlations to be as high as possible (good response to same-side stimulation), while contra correlations should be zero (no response to opposite-side stimulation). For obvious reasons, it is impossible for the recorded speed to react instantly to stimulation. Therefore, for each correlation, we considered the peak within the 0-5 s time-lag interval and the performance P of each experiment has been defined as the difference between the average of the two ipsi stimulation-speed couples and that of the two contra couples.

3   Results

The first thing we considered is the performance of the robot in the various navigation experiments. Figure 3A shows the results of the performance index P obtained in the

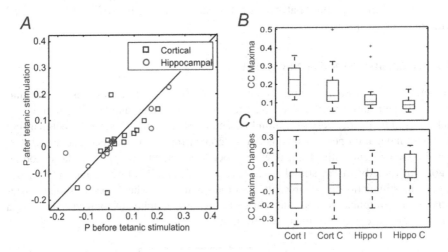

Fig. 3. A. Comparison of performance evolution following tetanic stimulation in two different cellular preparations (cortical and hippocampal cultures). On the X- and Y-axis, respectively, the performance index before and after the tetanic stimulation protocol are reported. B. Distributions of ipsi (I) and contra (C) cross-correlation (CC) maxima in the 0-5 s lag region (cfr. 2.6) for studied cortical (Cort, N = 14) and hippocampal (Hippo, N = 12) cultures. The central, red mark indicates the median value, the edges of the blue boxes the 25th and 75th percentiles, while the whiskers extend to the most extreme data points that are not outliers. Outliers are defined as points larger than $Q3 + 1.5 \times (Q3 - Q1)$ or smaller than $Q1 - 1.5 \times (Q3 - Q1)$. Outliers are represented as red crosses. C. Distributions of ipsi (I) and contra (C) cross-correlation (CC) maxima changes following tetanic stimulation for studied cortical (Cort, N = 14) and hippocampal (Hippo, N = 12) cultures. Meaning of graphs is the same as B.

different experiments before and after the tetanic stimulation. Several cultures proved to be unable to control the robot, with performances equal to 0 or less. In order to investigate the reason for these failures, we analyzed the single components of the P index, namely the ipsi and contra cross-correlation maxima. Their distributions are reported in Figure 3B. It is possible to note how cortical cultures tend to present responses that match, even if loosely, the stimulation train (median CC peak = 0.23). The reason of failure seems to be the impossibility to have well separated responses, as the ipsi and contro CC maxima are not significantly different (p-values of 0.079 and 0.151 for cortical and hippocampal cultures, respectively, Wilcoxon signed rank test).

We then moved on to analyze the changes brought about by the tetanic stimulation protocol (Figure 3A and C). None of the performance distributions significantly differs from 0 (p-values ranging from 0.151 to 0.470). We therefore decided to investigate whether the tetanic stimulation had a different effect compared to what originally supposed (proportional strengthening of already existing connections, starting from the electrodes of delivery and decreasing with distance) or simply if it had no lasting effect. Two different PSTHs were generated from each cultures, as described in 2.2: during phase 2 and during phase 6. Each culture was tested with stimuli from 10 different electrodes, while responses were collected from the remaining 49 non-ground electrodes. This leads to 490 effective connections investigated in each culture, for a grand total of 6860 effective connections investigated in cortical neurons and 5880 in hippocampal ones. Effective connectivity changes [32]were binned and analyzed according to two different characteristics: time occurred since delivery of stimulus and intensity before stimulus.

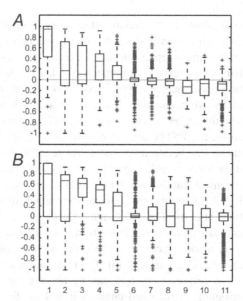

Fig. 4. Relationship between pre-tetanic stimulation intensities and observed changes in cortical (A) and hippocampal (B) cultures. The central, red mark indicates the median value, the edges of the blue boxes the 25th and 75th percentiles, while the whiskers extend to the most extreme data points that are not outliers. Outliers are defined as points larger than $Q3 + 1.5 \times (Q3 - Q1)$ or smaller than $Q1 - 1.5 \times (Q3 - Q1)$. Outliers are represented as red crosses.

Table 1. Intensity ranges of evoked responses (i.e. PSTH area) are listed in column two, with the matching labels from Figure 4 in column 1. Intensity of responses here is indicated as the average number of spikes recorded in the 300 ms following each probe stimulus (sp/st). The bin edges are logarithmically spaced and were chosen to have roughly the same number of points in each bin, for both hippocampal and cortical cultures. The third and fourth columns present the amount of effective connectivity considered in each class.

	sp/st	N Cortex	N Hippo
1	<0.1	1516	483
2	0.1-14	165	79
3	0.15-0.24	183	127
4	0.24-0.37	240	126
5	0.37-0.57	304	167
6	0.57-0.89	1292	828
7	0.89-1.36	951	369
8	1.36-2.10	1080	273
9	2.10-3.24	379	256
10	3.24-5.0	266	287
11	>5	365	2350

Table 1 and Figure 4 illustrate the relationship that is observed between the strength of each effective connection and the change it undergoes following tetanic stimulation. The data in the graph represent the difference between the PSTH areas after and before the delivery of tetanic stimulation divided by their sum, so that a value of 1 indicates the appearance of a previously non-existing connection and a value of -1 denotes an effective connection that ceases to exist. Table 1 lists the pre-tetanic connections strength ranges (expressed as average number of spikes in response to a probe stimulus, i.e. PSTH area) that were used to bin the data. The bin edges are logarithmically spaced between 0.1 and 5. Those values were chosen so that the number of points in each bin would at least be roughly in the same order of magnitude. The last two columns of Table 1 present the number of points within each bin.

The two examined cellular preparations results in rather different connection changes: weak connections in cortical cultures seem to undergo a mild increase, while the strongest ones suffer a modest decrease. In contrast, weak connection in hippocampal cultures undergo a very marked increase, while stronger connections undergo large variations but their mean value changes very little from the pre-tetanus condition.

The temporal pattern of response changes can be observed in Figure 5B. Each graph displays the distribution of recorded changes in a 4 ms time bin at different delays from the evoking stimulus. Normalization has been performed as in the previous case. In particular, the two 'sensory' graphs report data obtained following a probing stimulus delivered from one of the two electrodes used for tetanic stimulation

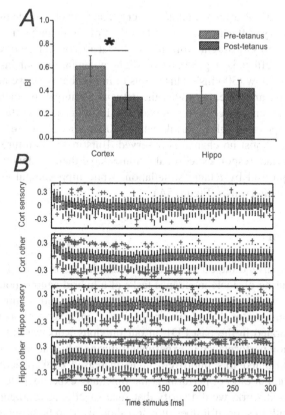

Fig. 5. A. Burstiness index before and after closed-loop experiments in different cellular prepa-rations. Error bars represent standard deviation, asterisk denotes p-value < 0.05, paired t-test. B. Average PSTH change following closed-loop experiments. From top to bottom, the four graphs represent data from: cortical cultures, stimulated from sensory areas; cortical cultures, stimu-lated from other areas; hippocampal cultures, stimulated from sensory areas; hippocampal cultures, stimulated from other areas (see text for further details). The PSTH is computed in 4 ms bins, each time point contains the distribution of the data recorded from all the recording channels, in all cortical (or hippocampal) experiments, following stimulation from sensory (or other) areas (for a grand total of, respectively, $49 \times 14 \times 2$, $49 \times 14 \times 8$, $49 \times 12 \times 2$, $49 \times 12 \times 8$ points per time bin). The central, red mark indicates the median value, the edges of the blue boxes the 25th and 75th percentiles, while the whiskers extend to the most extreme data points that are not outliers. Outliers are defined as points larger than $Q3 + 1.5 \times (Q3 - Q1)$ or smaller than $Q1 - 1.5 \times (Q3 - Q1)$. Outliers are represented as red crosses.

(cfr. 2.2), while the 'other' graphs contain data from PSTHs obtained with the probing stimulus delivered from the 8 remaining test electrodes. This data suggests that the cell type is much more relevant than the relative position of probing and tetanic sti-mulation electrodes. The two graphs on cortical data follow roughly the same shape, as do the graphs showing data from hippocampal cultures. In particular, it is possible to note how the only change consistent in all four cases is the rather large increase in early responses (4-20 ms after stimulus delivery) and a brief "no-change" window at ~20 ms after delivery. After this interval, the effect of the tetanic stimulation on connectivity is remarkably different from hippocampal to cortical cultures: in the first case, the resulting effect is a generalized, slight potentiation that lasts beyond the considered time window (300 ms). This seems to imply that what we are recording is not evoked response anymore, but rather the network resuming its (increased) spontaneous activity between consecutive stimulations. In the case of cortical cultures, on the other hand, responses slightly shrink until ~180 ms after response, while after this mark, on average, almost no change is observed. Burstiness in cultures follows very closely the fate of late responses: cortical cultures have their tendency to have bursts significantly suppressed by tetanic stimulation, while hippocampal cultures tend to burst slightly more afterwards (Figure 5A).

4   Discussion

In this work, we analyzed the effects of tetanic stimulation on cortical and hippocampal cultures in during obstacle-avoidance tasks of a hybrid neurorobotic system. The results we observed were quite different from what expected: no lasting benefits to the navigational capabilities of the robot were observed, in spite of short-term improvements as demonstrated in the past [33]: in our previous work we did observe an increase of navigation performance in experiments during which tetanic stimulation was delivered. We attributed this improvement to increased evoked activity following tetanic stimulation, but did not perform, at the time, any experiment to test this hypothesis. In the current work, we did verify that such specific strengthening effect does not occur in the cultures, or, if it does, it is too short lived to be observed by comparing connection maps obtained minutes apart from the tetanic stimulation itself.

Effective connectivity within the analyzed neural networks proved to change significantly and with some measure of regularity: weaker connections in both hippocampal and cortical cultures resulted strengthened, in different degrees, by the tetanic stimulation protocol, while stronger connections were, on average, weakened in cortical cultures. In hippocampal cultures, on the other hand, stronger connections proved to undergo very large swings, but without a clear change in average strength.

A similar rearrangement seems to take place also in the temporal structures of responses, with an increase in the immediate post stimulus window (up to ~20 ms) observed in both cortical and hippocampal cultures. The effects on longer time frames differed from hippocampal to cortical cultures: in the first case, a generalized, weak, increase of responses is observed, whereas cortical cultures display a reduction in their response to stimulus in the 20-200 ms interval after stimulus and no apparent change afterwards. Confronting evoked and spontaneous activity in both hippocampal and cortical cultures seems to point out that late-responses in evoked stimulations are strongly tied to bursting activity during spontaneous recordings, as also described in recent literature [34]: burstiness indexes and late-responses amplitudes change in unison in the observed experiments, slightly increasing as consequence of tetanic stimulation in hippocampal cultures, while decreasing in cortical ones.

In our opinion, the observed diversity of behavior in hippocampal and cortical cultures are not linked to differences in single neurons, but rather to their spatial organization: a recurring dissimilarity between the two preparations seems to be the fact that hippocampal cultures tend to form very dense clusters of cells with sparse interconnectivity, while cortical preparations are usually more homogeneous (data not shown). Similarly, we find reasonable to believe that the observed weakening in strong connections within cortical cultures might be the cause of the subsequent significant drop in bursting rate and, consequently, late responses to stimulation.

In conclusion, we observed quite unexpected changes in the functional connectivity maps of neural networks following a tetanic stimulation protocol. Those changes, especially in their time distribution following a probing stimulus, seem to be remarkably common in the observed cultures and, to our knowledge, they have not been described before in literature. While the proposed protocol demonstrated able to induce lasting changes in dynamic activity of networks, it is unsuitable, as it is, to provide a learning mechanism for hybrid architectures, because of our inability (so far) to selectively strengthen a given connection. A possible way to achieve this goal might be that of delivering a tetanic stimulation paired to low-frequency stimulation as proposed, for instance, in [32].

Acknowledgments. The authors wish to thanks PhD student Marta Bisio for culturing and maintaining hippocampal and cortical networks over Micro-Electrode Arrays and Dr Marina Nanni and Dr Claudia Chiabrera from NBT-IIT for the technical assistance for the dissection and dissociation procedures.

References

[1] Beggs, J.M., Plenz, D.: Neuronal avalanches in neocortical circuits. The Journal of Neuroscience 23, 11167–11177 (2003)

[2] Wagenaar, D.A., Madhavan, R., Pine, J., Potter, S.M.: Controlling bursting in cortical cultures with closed-loop multi-electrode stimulation. The Journal of Neuroscience 25, 680–688 (2005)

[3] Eytan, D., Marom, S.: Dynamics and effective topology underlying synchronization in networks of cortical neurons. The Journal of Neuroscience 26, 8465–8476 (2006)

[4] Wagenaar, D.A., Pine, J., Potter, S.M.: An extremely rich repertoire of bursting patterns during the development of cortical cultures. BMC Neuroscience 7, 11 (2006)

[5] Pasquale, V., Massobrio, P., Bologna, L., Chiappalone, M., Martinoia, S.: Self-organization and neuronal avalanches in networks of dissociated cortical neurons. Neuroscience 153, 1354–1369 (2008)

[6] Raichman, N., Ben-Jacob, E.: Identifying repeating motifs in the activation of synchronized bursts in cultured neuronal networks. Journal of Neuroscience Methods 170, 96–110 (2008)

[7] Van Pelt, J., Corner, M.A., Wolters, P.S., Rutten, W.L.C., Ramakers, G.J.A.: Longterm stability and developmental changes in spontaneous network burst firing patterns in dissociated rat cerebral cortex cell cultures on multi-electrode arrays. Neurosci. Letters 361, 86–89 (2004)

[8] Bologna, L.L., Nieus, T., Tedesco, M., Chiappalone, M., Benfenati, F., Martinoia, S.: Low-frequency stimulation enhances burst activity in cortical cultures during development. Neuroscience (2010)

[9] Marom, S., Shahaf, G.: Development, learning and memory in large random networks of cortical neurons: lessons beyond anatomy. Quarterly Reviews of Biophysics 35, 63–87 (2002)

[10] Bonifazi, P., Ruaro, M.E., Torre, V.: Statistical properties of information processing in neuronal networks. Eur. J. Neurosci. 22, 2953–2964 (2005)

[11] Chiappalone, M., Vato, A., Berdondini, L., Koudelka, M., Martinoia, S.: Network dynamics and synchronous activity in cultured cortical neurons. Int. J. Neural. Syst. 17, 87–103 (2007)

[12] Maeda, E., Robinson, H.P.C., Kawana, A.: The mechanism of generation and propagation of synchronized bursting in developing networks of cortical neurons. The Journal of Neuroscience: the Official Journal of the Society for Neuroscience 15, 6834–6845 (1995)

[13] Dal Maschio, M., Difato, F., Beltramo, R., Blau, A., Benfenati, F., Fellin, T.: Simultaneous two-photon imaging and photo-stimulation with structured light illumination. Optic Express 18, 18720–18731 (2010)

[14] Frega, M., Tedesco, M., Massobrio, P., Pesce, M., Martinoia, S.: 3D neuronal networks coupled to micro-electrode arrays: an innovative in vitro experimental model to study network dynamics. In: 8th International Meeting on Substrate-Integrated Microelectrode Arrays, pp. NMI

[15] Gramowski, A., Jügelt, K., Stüwe, S., Schulze, R., McGregor, G.P., Wartenberg-Demand, A., Loock, J., Schröder, O., Weiss, D.G.: Functional screening of traditional antidepressants with primary cortical neuronal networks grown on multielectrode neurochips. European Journal of Neuroscience 24, 455–465 (2006)

[16] Keefer, E.W., Gramowski, A., Gross, G.W.: NMDA receptor-dependent periodic oscillations in cultured spinal cord networks. Journal of Neurophysiology 86, 3030–3042 (2001)

[17] Jimbo, Y., Robinson, H.P.C., Kawana, A.: Strengthening of synchronized activity by tetanic stimulation in cortical cultures: application of planar electrode arrays. IEEE Transactions on Biomedical Engineering 45, 1297–1304 (1998)

[18] Jimbo, Y., Tateno, Y., Robinson, H.P.C.: Simultaneous induction of pathway-specific potentiation and depression in networks of cortical neurons. Biophysical Journal 76, 670–678 (1999)

[19] Stegenga, J., le Feber, J., Marani, E., Rutten, W.L.: Phase-dependent effects of stimuli locked to oscillatory activity in cultured cortical networks. Biophys. J. 98, 2452–2458 (2010)

[20] Tateno, T., Jimbo, Y.: Activity-dependent enhancement in the reliability of correlated spike timings in cultured cortical neurons. Biological Cybernetics 80, 45–55 (1999)

[21] Madhavan, R., Chao, Z.C., Potter, S.M.: Plasticity of recurring spatiotemporal activity patterns in cortical networks. Physical Biology 4, 181–193 (2007)

[22] Shahaf, G., Marom, S.: Learning in networks of cortical neurons. The Journal of Neuroscience: the Official Journal of the Society for Neuroscience 21, 8782–8788 (2001)

[23] le Feber, J., Stegenga, J., Rutten, W.L.: The effect of slow electrical stimuli to achieve learning in cultured networks of rat cortical neurons. PloS One 5, e8871 (2010)

[24] Tessadori, J., Mulas, M., Martinoia, S., Chiappalone, M.: A neuro-robotic system to investigate the computational properties of neuronal assemblies. In: 2012 4th IEEE RAS & EMBS International Conference on Biomedical Robotics and Biomechatronics (BioRob), pp. 332–337. IEEE (2012)

[25] Tessadori, J., Venuta, D., Kumar, S.S., Bisio, M., Pasquale, V., Chiappalone, M.: Embodied neuronal assemblies: A closed-loop environment for coding and decoding studies. In: 2013 6th International IEEE/EMBS Conference on Neural Engineering (NER), pp. 899–902. IEEE (2013)

[26] Tessadori, J., Venuta, D., Pasquale, V., Kumar, S.S., Chiappalone, M.: Encoding of stimuli in embodied neuronal networks. In: Lepora, N.F., Mura, A., Krapp, H.G., Verschure, P.F.M.J., Prescott, T.J. (eds.) Living Machines 2013. LNCS, vol. 8064, pp. 274–286. Springer, Heidelberg (2013)

[27] Gal, A., Marom, S.: Entrainment of the intrinsic dynamics of single isolated neurons by natural-like input. The Journal of neuroscience 33, 7912–7918 (2013)

[28] Mainen, Z.F., Sejnowski, T.J.: Reliability of spike timing in neocortical neurons. Science 268, 1503–1506 (1995)

[29] Frega, M., Pasquale, V., Tedesco, M., Marcoli, M., Contestabile, A., Nanni, M., Bonzano, L., Maura, G., Chiappalone, M.: Cortical cultures coupled to micro-electrode arrays: a novel approach to perform in vitro excitotoxicity testing. Neurotoxicology and Teratology 34, 116–127 (2012)

[30] Chiappalone, M., Vato, A., Berdondini, L., Koudelka-Hep, M., Martinoia, S.: Network dynamics and synchronous activity in cultured cortical neurons. International Journal of Neural Systems 17, 87–103 (2007)

[31] Madhavan, R., Chao, Z.C., Potter, S.M.: Plasticity of recurring spatiotemporal activity patterns in cortical networks. Physical Biology 4, 181 (2007)

[32] Chiappalone, M., Massobrio, P., Martinoia, S.: Network plasticity in cortical assemblies. European Journal of Neuroscience 28, 221–237 (2008)

[33] Tessadori, J., Bisio, M., Martinoia, S., Chiappalone, M.: Modular neuronal assemblies embodied in a closed-loop environment: toward future integration of brains and machines. Frontiers in Neural Circuits 6 (2012)

[34] Weihberger, O., Okujeni, S., Mikkonen, J.E., Egert, U.: Quantitative examination of stimulus-response relations in cortical networks in vitro. Journal of Neurophysiology 109, 1764–1774 (2013)

Enhanced Locomotion of a Spherical Robot Based on the Sea-urchin Characteristics

Jorge Ocampo-Jiménez, Angelica Muñoz-Meléndez,
and Gustavo Rodríguez-Gómez

Instituto Nacional de Astrofísica, Óptica y Electrónica, Mexico
soytuc@ccc.inaoep.mx, {munoz,grodrig}@inaoep.mx

Abstract. The objective of this research is to present the design of a robot able to act in highly unstructured environments. To do so, a spherical robot loosely based on the sea-urchin characteristics is introduced; to synthesize such characteristics a set of actuators were incorporated into a pendulum-based spherical robot in order to handle obstacles and hollows successfully, as demonstrated in the simulation results that are shown and discussed.

1 Introduction

This research focuses on the design of locomotion and reconfiguration mechanisms that enable mobile robots and autonomous vehicles to act in highly unstructured environments, such as those involved in natural and intentional disasters, where the use of wheeled vehicles is unfeasible. In these situations, the deployment of robust technologies for inspection and recognition might be crucial for planning a prompt response and saving lives.

The robots object of this research have a spherical shape. Spherical robots are known to be robust platforms with omnidirectional motion, that are able to deal with irregular surfaces up to $\frac{1}{10}$th of their diameter. These robots are suitable for dealing with highly unstructured environments and there are still many challenges and opportunities to improve their design.

Nature is rich in locomotion and reconfiguration mechanisms that can certainly be used as metaphors to inspire or improve the design of robust mobile robots and autonomous vehicles. In this research we are interested in some skills of sea urchins for a number of reasons summarized below:

- The inner structure of these organisms is hidden and protected by a hard shell. This feature is important when designing robots and vehicles that are intended to operate in extreme conditions.
- The supportive systems of echinoderms is economical, condition that contributes to a low energy consumption compared with other animals [3]. This feature is relevant when designing robots with an acceptable speed/power tradeoff. In effect, the energy required for moving light spines or propelling a sphere is less considerable than the energy that is needed for moving a set of legs or wings, common solutions for field robots.

A. Duff et al. (eds.): Living Machines 2014, LNAI 8608, pp. 238–248, 2014.

This investigation makes two main contributions to the design of mobile robots. First, the synthesis and combination of two features of sea urchins: a spherical morphology and a coverage of flexible spines that are used as both, a deterrent and an aid for motion. And second, the incorporation of the synthesized features in the design of a mobile robot to enhance its locomotion for soil and terrestrial unstructured environments.

The approach of this research is to study of a remarkable functional natural system in order to improve the design of our robots and vehicles. To the best of our knowledge, there is no similar design reported in the literature. Simulation is considered in this research as a powerful tool to improve current and future capabilities of complex robots [10]. For that, our model has been simulated using the simulation environment Webots©.

It is worth to remark that the simulation of behavior or modeling of aquatic echinoderms is beyond the scope of our research.

The rest of this article is organized as follows. Section 2 gives on overview of related work. Section 3 introduces our proposal of extended spherical robot. Section 4 presents results. Finally Section 5 discusses conclusions and perspectives of this research.

2   Related Work

In this section, first some works inspired by sea urchins' skills are reviewed. Then, significant works concerned with the design of spherical vehicles are reviewed and compared.

2.1   Bioinspiration from Sea Urchin Skills

Sadeghi et al. [6] have proposed an adhesion mechanism inspired from the tube feet of sea urchins. This mechanism is very useful for underwater robots that have to climb surfaces, grasp and manipulate objects; as well as for swimming robots that need to get hold of a fixed position.

To the best of our knowledge, this in the only work about robotic systems inspired from sea urchins' skills.

2.2   Basic Models of Spherical Robots

According to the literature, the most popular methods to propel spherical robots are those which change the center of mass that enable the robot to roll in a desired direction. Spherical robots are usually propelled by a pendulum, a hamster ball, and the change of mass as explained below.

The pendulum based propulsion system is based on an axle that goes through the spherical skeleton across the diameter of the sphere, with the rest of mechanisms attached to that axle. With the center of gravity underneath the sphere, torque can be applied to the sphere by moving the pendulum forward, which moves the center of mass in the same direction. Previous action results in a move of the center of mass in front of the contact point between the sphere and the support surface, which causes the robot to roll forward. When the pendulum moves towards a specific angle continuously, it provides the needed impulse that enables the sphere to keep in movement. In this model, illustrated in Fig. 1(a), there are only two points of contact of the sphere with the support surface. That means that both, the external and internal materials of the sphere are not considered relevant in this model.

The hamster ball propulsion system comprises a wheeled vehicle lying on the internal base of the sphere. The principle is similar to the pendulum based model, the vehicle located inside of the sphere moves, changing the center of mass of the sphere. This happens because most of the sphere's mass is concentrated in the internal wheeled vehicle. An interesting property of the hamster ball based model, illustrated in Fig. 1(b), is that it can be omnidirectional if the internal wheeled vehicle has differential locomotion. However, an internal wheeled vehicle with enough grip to the internal wall of the sphere, as well as a uniform internal wall are required to ensure constant motion of the sphere.

Finally, the change of masses propulsion system has a set of weights distributed on radial axes inside the sphere. These weights move to fixed positions along their respective axes, which results in a change of the center of mass, and consequently in the movement of the sphere. The propulsion using this system is far from being trivial, because of the complex configuration of weights required to bring the center of mass of the sphere towards specific positions. However, vehicles propelled by change of masses are completely omnidirectional. Fig. 1(c) illustrates this model.

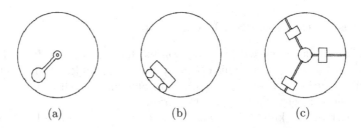

(a)   (b)   (c)

Fig. 1. Cross-section of spherical robots propelled by a pendulum (a), a "hamster" ball (b), the shifting of masses (c)

2.3   Extended Models of Spherical Robots

In spite of its numerous advantages, a spherical robot is constrained by its low torque capacity required for changing its center of mass. Robots controlled by the models described in previous section are capable of going upward on slopes up to $30°$, and dealing with obstacles and hollows of up to $\frac{1}{10}$th of their diameter [8]. In this section we review some extended models of spherical robots oriented to surpass these limits. These extensions can be divided into two classes, deformable and non deformable models.

Deformable robots are characterized by their flexible exoskeleton. The change of shape of the skeleton provides the energy for jumping and leaving hollows, as well as for overtaking some horizontal obstacles. These models are however limited by their material of construction and their weight. Figs. 2(a) and 2(b) illustrate two different models of deformable spherical robots proposed in [1] and [9].

Non deformable robots are characterized by their solid exoskeleton. Below we review four interesting models reported in the literature.

Thistle [2] is a spherical robot equipped with a transversal axle inside which is capable of propelling the sphere vertically when it is stuck. Thistle is able to deal with obstacles greater than $\frac{1}{10}$th of its diameter on flat surfaces. Nevertheless, this solution is only feasible when the sphere weights less than 10 Kg, which limits the number of sensors and actuators carried by the sphere. See Fig. 2(c).

Hex-a-ball [5] is a hexapod robot equipped with a spherical exoskeleton, each leg of the robot is arranged in such a way that the robot can become a sphere when required. Hex-a-ball combines a lot of engineering and computing skills to achieve a hybrid locomotion. The motors have dual roles that control the legs of the robot or propel the sphere alternatively. The main constraint of Hex-a-ball is its speed: the robot is only capable of doubling its speed when it has the spherical form. See Fig. 2(d).

Sun [11] is a spherical robot equipped with two manipulators, one at each pole of the sphere, which can be used to propel the sphere when its is stuck or to go up stairs (see Fig. 2(e)). And Schroll [8] is a spherical robot that uses the angular momentum for increasing its torque to surpass steps and large obstacles. The accumulation of momentum is achieved by a rotating disc inside of the sphere

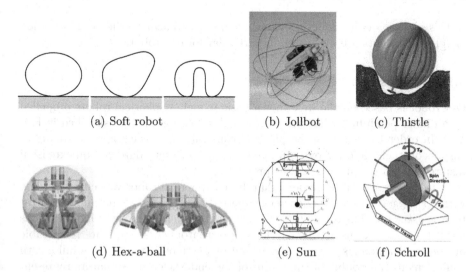

(a) Soft robot   (b) Jollbot   (c) Thistle

(d) Hex-a-ball   (e) Sun   (f) Schroll

Fig. 2. Extended models of deformable (a-b) and non deformable (c-f) spherical vehicles

Table 1. Comparison of models of spherical robots

Model	slopes > 30⁰	obstacles up to $\frac{1}{10}$ of its diameter	extension	advantages	disadvantages
Pendulum	no	no	none	few & simple mechanisms	low torque
Hamster ball	no	no	none	holonomic	restricted materials
Shift of masses	no	no	none	holonomic	complex control, low torque.
Deformable models	partially	yes	deformable skeleton	able to leave hollows	limited materials and weight of the robot, fixed initial unstucking state
Hex-a-ball	partially	yes	manipulators	able to leave hollows	slow, fixed initial unstucking state
Thistle	partially	yes	manipulators	simple extension, able to leave hollows	limited weight of the robot, fixed initial unstucking state
SUN	no	yes	manipulators	able to climb stairs, object manipulation	short slopes, fixed initial unstucking state
Schroll	partially	partially	internal disc	able to leave hollows and climb slopes	short slopes
Our robot	not considered	yes	retractable manipulators	able to leave hollows	number of manipulators

that steers the robot towards a specific direction. Schroll is unable to accumulate enough momentum for dealing hollows with lengthy slopes (see Fig. 2(f)).

Table 1 summarizes the features of robots revised in this section.

3 Design Considerations

In this section, design issues previously identified to achieve the desired motion of spherical robots are discussed, in particular for our robot.

3.1 Composition

Our robot is based on the well-known basic model of spherical robots propelled by an inner pendulum, that was introduced in Section 2.2 and sketched in Fig. 1(a). In order to overcome the physical constraint of basic spherical robots to surpass obstacles and hollows of more than $\frac{1}{10}$th of their diameter, an extension inspired by sea urchins' shells was proposed.

In this extension, a basic pendulum-based spherical robot was circumscribed within a bigger sphere. The former is referred to as *endoskeleton* and the latter to as *exoskeleton*. The role of the exoskeleton is to shelter the set of modifications needed for this extension, namely for new actuators. Actuators are retractable or telescopic antennas, each one activated by one motor, that have a full length equivalent to the double of the radius of the endoskeleton. Actuators represent the spines of the sea urchin and can be used for a better leverage. These actuators are uniformly distributed over the surface of the exoesqueleton, using the algorithm suggested in [7]. Fig. 3 shows the main components of the extended robot.

The previous nested structure was chosen to keep balanced the weight of the robot and, in spite of the extension, still be able to propel the robot by moving its center of mass by controlling a pendulum. The number of actuators needed to provide enough torque for dealing with obstacles is directly related to the height of hollows and obstacles. For instance, for hollows and obstacles with a height of up to 25% or 50% the radius of the endoskeleton at least 24 and 30 actuators are required, respectively. This relation was determined by simulation and is described in detail in [4].

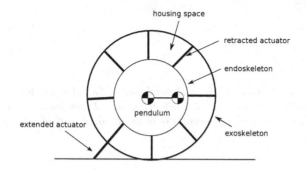

Fig. 3. Cross-section of the extended robot

3.2 Motion

The motion of the robot is controlled through four variables: the pendulum's speed, the main axle's position, and both the pendulum's position and expansion of actuators, as explained below.

- **Control of the pendulum's speed.** By moving repeatedly its pendulum the robot is capable of self-propelling longitudinally. The pendulum is activated by one motor located in the center of the endoskeleton that has only 1 DOF. The faster the frequency of the pendulum's movement, the faster the robot. In this case, the maximum speed reached by the robot is 4 rad/seg. See Fig. 4(a).
- **Control of the main axle's position.** By tuning the position of the pendulum along the central axle, combined with the control of the pendulum's speed, the robot is capable of propelling laterally. This movement is obtained by a motor that can slightly slide the position of the pendulum and has the effect of changing the orientation of the central axle. See Fig. 4(b).
- **Control of the pendulum's position and expansion of actuators.** By keeping the pendulum in a fixed position the robot is able to move its center of mass towards that position, and as a result, it will roll in the corresponding direction. This control is applied when an obstacle or hollow is detected and

the robot tries to surpass it. For a better leverage, the actuators that fulfill two conditions are expanded: (1) they are closer to the support surface, and (2) they are opposed to the pendulum's direction. See Fig. 4(c).

The details of the implementation of these mechanisms, as well as technical details of the model can be consulted in [4].

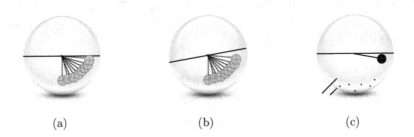

(a)   (b)   (c)

Fig. 4. Motion. (a) Control of the pendulum's speed for achieving longitudinal movement. (b) Control of the main axle for achieving lateral movement. (c) Control of the pendulum's position to change the center of mass and dilatation of actuators for better leverage.

Table 2. Simulation parameters

Meaning	Value
Gravity constant	$9.81\frac{m}{s^2}$
Coulomb Friction	1
Coefficient of restitution	0.5
Bounce Velocity	0.01
Force dependent slip	0
Linear actuator force	15 N
Pendulum moment of inertia	5.145×10^{-4} kg m^2
Sphere moment of inertia	4.23×10^{-2} kg m^2
Pendulum mass	0.512 kg
Sphere mass (including 30 linear actuators)	2.008 kg
Distance from the center of mass to the bob	0.0317 m
Radius of the sphere	0.1778 m
Simple linear actuator length	0.0889 m

4   Simulation Results

These experiments were performed using a simulated extended spherical robot in the Webots© environment version 7.4.0. Experiments were performed on a computer equipped with an Intel Core i7-4700MQ processor, 8GB RAM, graphic card NVIDIA GeForce 765M, under Windows 8.1. Table 2 summarizes the parameters used in all the simulations described in this section.

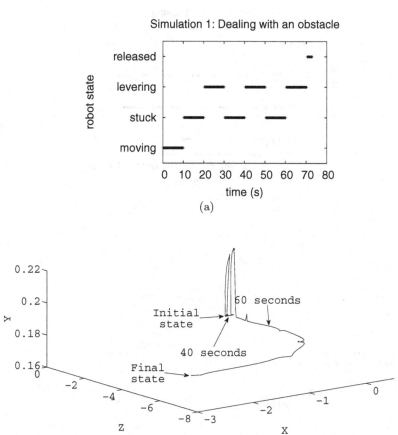

(a)

(b)

Fig. 5. States (a) and trajectory (b) followed by the robot during a simulation consisting in surpassing an obstacle with height equal to $\frac{1}{4}$ of the radius of the robot

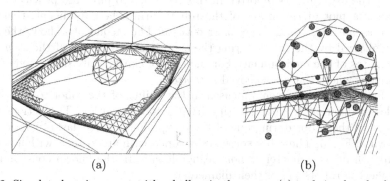

(a)   (b)

Fig. 6. Simulated environment with a hollow in the center (a) and simulated robot (b)

Simulation 2: Leaving a hollow

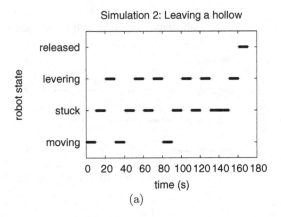

(a)

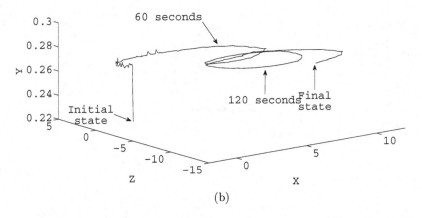

(b)

Fig. 7. States (a) and trajectory (b) followed by the robot during a simulation consisting in leaving a hollow with height equal to $\frac{1}{8}$th of its diameter

For all the experiments reported in this section, two plots are provided. The first one is a plot of the states of the robot during the experiment, that are classified as *moving, stuck, levering,* or *released.* The second plot shows the trajectory followed by the robot during the experiment, in terms of the x, y and z coordinates of the environment. For the sake of clarity, only the region of the environment where the robot moved is plotted.

The first set of experiments focused in the ability of the robot to deal with obstacles of different heights, starting from a static position. The actions and trajectory of the robot during one of these experiments that lasted 80 seconds are shown in Fig. 5. The robot successfully coped with an obstacle with a height equal to $\frac{1}{8}$th of its diameter, a non negligible gain to the known constraint of basic spherical robots ($\frac{1}{10}$ of their diameter).

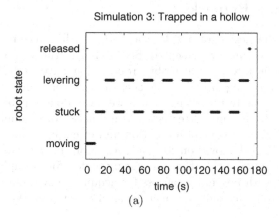

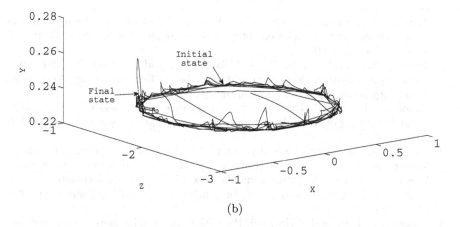

(b)

Fig. 8. States (a) and trajectory (b) followed by the robot during a simulation where the robot is able to leave a hollow where it was trapped. The height of the hollow is equal to $\frac{1}{8}$th of its diameter. Time is not indicated in graph (b) for emphasizing the persistent attempts of the robot to leave the hollow.

The second set of experiments were more complicated than the former and focused on the ability of the extended spherical robot to deal with hollows of a height equal to $\frac{1}{8}$th of its diameter. The environment used in these experiments as well as a snapshot of the robot are shown in Fig. 6.

The results of two experiments are depicted in Figs. 7 and 8. In both cases, the robot succeeded to leave a hollow of a height equal to $\frac{1}{8}$th of its diameter. However, in the second case the robot performed several attempts before getting untrapped. Note that the robot persistently tries to solve the situation as long as it is in the hollow, until the situation is solved.

5   Concluding Remarks

In this article, the design of a spherical robot with a reconfigurable system inspired from sea urchins is presented. Sea urchin features were taken here as a metaphor to gain flexibility in mobile robots that have to deal with difficult problems such as moving in highly unstructured environments.

As results are decidly promising so far, we have already plan to extend this work in the near future, in matters such as a scheme for dynamic activation of the robot's actuators to control its motion and orientation not only when it is stuck but also when it moves on flat surfaces, more in the sense that real sea urchins use their spines. Or further interesting topics; for example the extension of robot's spines with proximity sensors for acquiring environmental data. And last (but not least), the building of a physical robot prototype based on this model.

References

1. Armour, R., Paskins, K., Bowyer, A., Vincent, J., Megill, W.: Jumping robots: a biomimetic solution to locomotion across rough terrain. Bioinspiration & Biomimetics 2(3) (2007)
2. Jakubik, P., Suomela, J., Vainio, M., Ylikorpi, T.: Ariadna AO4532-03/6201 biologically inspired solutions for robotic surface mobility. Technical report, Automation Technology Laboratory. Helsinki University of Technology (2004)
3. Motokawa, T., Sato, E., Umeyama, K.: Energy expenditure associated with softening and stiffening of echinoderm connective tissue. The Biological Bulletin 222(2), 150–157 (2012)
4. Ocampo, J.: Model of spherical robot for the locomotion on irregular surfaces. Master's thesis, Inst. Nal. de Astrofísica, Óptica y Electrónica (2010) (in Spanish)
5. Phipps, C., Minor, M.: Introducing the Hex-a-ball, a hybrid locomotion terrain adaptive walking and rolling robot. In: Climbing and Walking Robots, pp. 525–532 (2006)
6. Sadeghi, A., Beccain, L., Mazzolai, B.: Design and development of innovative adhesive suckers inspired by the tube feet of sea urchins. In: Proceedings of the 4th IEEE RAS/EMBS International Conference on Biomecanical Robotics and Biomechatronics, pp. 617–622 (2012)
7. Saff, E., Kuijlaars, A.: Distributing many points on a sphere. The Mathematical Intelligencer 19, 5–11 (1997)
8. Schroll, G.: Design of a spherical vehicle with flywheel momentum storage for high torque capabilities. Master's thesis, Department of Mechanical Engineering, Massachusetts Institute of Technology (2008)
9. Sugiyama, Y., Shiotsu, A., Yamanaka, M., Hirai, S.: Circular/spherical robots for crawling and jumping. In: Robotics and Automation, ICRA 2005, pp. 3595–3600 (2005)
10. Žlajpah, L.: Simulation in robotics. Mathematics and Computers in Simulation 79, 879–897 (2008)
11. Zhuang, W., Liu, X., Fang, C., Sun, H.: Dynamic modeling of a spherical robot with arms by using kane's method. In: Fourth International Conference on Natural Computation 2008, vol. 4, pp. 373–377 (2008)

Design of a Control Architecture for Habit Learning in Robots

Erwan Renaudo[1,2], Benoît Girard[1,2], Raja Chatila[1,2], and Mehdi Khamassi[1,2]

[1] Sorbonne Universités, UPMC Univ Paris 06, UMR 7222, Institut des Systèmes
Intelligents et de Robotique, F-75005, Paris, France
[2] CNRS, UMR 7222, Institut des Systèmes Intelligents et de Robotique, F-75005,
Paris, France

Abstract. Researches in psychology and neuroscience have identified multiple decision systems in mammals, enabling control of behavior to shift with training and familiarity of the environment from a goal-directed system to a habitual system. The former relies on the explicit estimation of future consequences of actions through planning towards a particular goal, which makes decision time longer but produces rapid adaptation to changes in the environment. The latter learns to associate values to particular stimulus-response associations, leading to quick reactive decision-making but slow relearning in response to environmental changes. Computational neuroscience models have formalized this as a coordination of model-based and model-free reinforcement learning. From this inspiration we hypothesize that it could enable robots to learn habits, detect when these habits are appropriate and thus avoid long and costly computations of the planning system. We illustrate this in a simple repetitive cube-pushing task on a conveyor belt, where a speed-accuracy trade-off is required. We show that the two systems have complementary advantages in these tasks, which can be combined for performance improvement.

Keywords: Adaptive Behaviour, Habit Learning, Reinforcement Learning, Robotic Architecture.

1 Introduction

Researches in the field of instrumental conditioning in psychology have shown that rodents learning to press a lever in order to get food progressively shift from a goal-directed decision system to a habitual system [7,8]. After moderate training, devaluation of the outcome (e.g. pairing it with illness) leads the animal to quickly stop pressing the lever. In contrast, after extensive training the animal perseveres with pressing the lever even after outcome devaluation - hence "habit" [1,21]. This has been hypothesized to enable the animal to avoid slow and costly decision-making through planning by shifting to reactive decision-making when the stability of the environment makes habits reliable, a capacity which is shared with humans and other mammals [2].

A. Duff et al. (eds.): Living Machines 2014, LNAI 8608, pp. 249–260, 2014.
© Springer International Publishing Switzerland 2014

In contrast, current robots are still rarely equipped with efficient online learning abilities and mostly rely on a single planning decision-making system, thus not providing alternative solutions to motion planning in situations where such strategy is limited [17]. Indeed the planning strategy can be approximative when coping with uncertainties, *e.g.* when there is perceptual aliasing [4], and can also require high computational costs and long time to propagate possible trajectories through internal representations [10]. We have previously shown that taking inspiration from the way rodents shift between different navigation strategies – a capacity which has been shown to be analogous to the shifts between goal-directed and habitual decision systems [14] – can be applied to a robotic platform to enable to automatically exploit the advantages of each strategy [4,3]. However, these experiments only involved navigation behaviors from one location to another. To our knowledge, no application has yet been made of the coordination of goal-directed and habitual systems to robotic tasks.

In this work, we illustrate the application of a decision architecture combining a goal-directed expert with a habitual one to a simple task where a simulated robot have to learn to repeat the less costly sequence of actions to push a series of cubes arriving in front of him on a conveyor belt. We build our algorithm on computational neuroscience models which have shown that combining model-based and model-free reinforcement learning can accurately reproduce properties of the competition between goal-directed and habitual systems [5,13,9]. In these models, the goal-directed system is modelled with model-based reinforcement learning in the sense that the system plans sequences of actions towards a particular goal by using the transition and reward functions. In parallel, the model-free reinforcement learning progressively learns by trial-and-error the Q-values associated to different state-action couples. The criterion for switching from one system to the other is based on the measure of uncertainty in the model-free system: the less variance there is in the Q-values, the more reliable the model-free habitual system is considered and the more likely it will control the behavior of the simulated agents.

In contrast to these previous computational neuroscience models, we do not a priori give the transition and reward functions (i.e. the considered model of the task) to the algorithm but rather make it learn it automatically by observing experienced transitions and rewards. Moreover, we arbitrate without bias between systems, as the selection of each one is random and equiprobable. The task that we simulate requiring a certain balance between speed and accuracy so as not to skip some cubes coming on the conveyor belt, our simulations show that the two systems have complementary advantages that can be combined for a highest performance. In a first series of simulations where the systems are controlling individually the agent, we characterize their performances in a constant belt velocity and constant distance between cubes setup and when the belt velocity is changed during the simulation. We show that each system is performing differently to these conditions as the model-free is more efficient than the model-based to exploit the stability in the environment, but the model-based adapts quickly to condition changes in the environment. We then show how combining the two systems and switching control among them, even with a basic rule, can improve the robot policy and gives it the ability to perform well both in a stable environment and during transitional phases to another stable setup, with the same architecture.

2   Materials and Methods

2.1   Global Architecture

Our Decision Layer [10] consists of two Experts that learn a policy and a Meta-Controller that supervises the Experts' performance (Fig. 1a). The Flexible Expert is a Model-Based Reinforcement Learning agent and the Habitual Expert is a Model Free reinforcement learning agent [19].

These modules receive the current State $S \in \mathcal{S}$ from a Perception Module and choose their actions in a set $\mathcal{A}$. Each Expert decides, from the current State and their knowledge, which action to take. In parallel, the Meta-Controller decides which Expert is the most efficient in the current State and allows it to send its action choice to be executed.

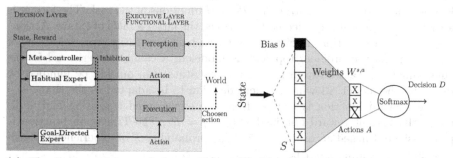

(a) The inner structure of the Decision Layer and its connection with the Executive and Functional Layers.

(b) The Habitual Expert neural network. Black X size is proportional to activity in the network.

Fig. 1. Robotic organisation of modules and Habitual Expert structure

2.2   Habitual Expert

The Habitual Expert (MF) is implemented as a 1-layer neural network. It learns directly the relevant state-action policy without an internal representation of transitions between states of the world (hence the term Model Free). Propagating the values from input S (and bias b) to action output A is computationally cheap, but learning the whole policy is long: only the experienced state-action value is updated. Learning a new policy to adapt to a new environment configuration is longer than just learning the first policy so this expert is reluctant to changes.

$$A_t(i) = W_t \cdot S_t + b_t(i) \ . \tag{1}$$

The connection weights W_t that learn the State-Action association are updated according to a Qlearning rule [20]: the connection between input neurons encoding the previous State S_{t-1} and the output neuron of the action done is modified with an amount depending on the reward R obtained.

$$\delta = R_t(s_{t-1}, a_{t-1}) + \gamma_{MF} \cdot max_a\left(W_{t-1}(a) \cdot S_t\right) - \left(W_{t-1}(a_{t-1}) \cdot S_{t-1}\right) \ . \quad (2)$$

$$W_t(a_{t-1}) = W_{t-1}(a_{t-1}) + \alpha_{MF} \cdot \delta \ . \quad (3)$$

α_{MF} : learning rate, γ_{MF} : decay factor.

Each action activity is interpreted as the probability $P(A(i))$ of taking action $A(i)$, using a Softmax rule (4). The decision is taken stochastically in the resulting distribution (τ_{MF} : temperature.).

$$P_t(A_t(i)) = \frac{\exp\left(\frac{A_t(i)}{\tau_{MF}}\right)}{\sum_j \exp\left(\frac{A_t(j)}{\tau_{MF}}\right)} \ . \quad (4)$$

2.3   Flexible Expert

The Flexible Expert (MB) is a Model Based Reinforcement Learning agent. It learns a model of the *Transition* and *Reward* functions of the task. The former is a *cyclic graph* of States connected by Actions, the latter a *table* of (State, Action) and Reward association. Decisions are taken based on these representations of the world. As the problem topology is modeled, a change experienced in the environment (ie. a transition leads to a new state) can quickly be handled by updating the model, allowing the next decision to be adapted to the changes.

The Reward function is learned from the experienced transition and is directly the instant reward obtained $R_t(S, A) = R_t$. The Transition function is progressively learned according to (5). The probability T of experienced transition $S \xrightarrow{A} S'$ is updated at learning rate α_{MB}.

$$T_t(S, A, S') = T_{t-1}(S, A, S') + \alpha_{MB} \cdot \left(1 - T_{t-1}(S, A, S')\right) \ . \quad (5)$$

Planning with the models consists in computing the Quality Q(S,A) of performing action A in the given state S. It is done iteratively by propagating the known rewards and refining the estimated Quality value according to the Transition function until convergence (γ_{MB} : decay factor):

$$Q_t(s, a) = max\left(R_t(s, a), (\gamma_{MB} \cdot \sum_{s'} T_{t-1}(s, a, s') \cdot \max_{a'} Q_t(s', a'))\right) \ . \quad (6)$$

The Decision is also taken with the *softmax rule* (cf. Eq. (4)).

The drawback of such a method comes from the increasing size of the transition model. Planning becomes more and more time consuming and the Expert is less and less reactive. As the environment evolves, even in a predictable way, the action decided from the perceived State at S_{t-1} may be irrelevant when acting in State S_t. To improve the Flexible Expert performance and keep a manageable model while dimensionality increases, the following features are implemented :

1. Planning in the graph is bounded in time : if planning is longer than a certain time chosen in agreement with the task dynamics, the computation of Quality is stopped and the approximated values are used for decision, as it is more important to be reactive enough than having accurate values in this task.
2. As the best policy is learnt, fewer and fewer states are visited, producing a peaked distribution V_S of states visits. The allocated computation time being limited, planning should only consider the most visited states. These states are hypothetized to be the most interesting for the Expert, as the policy focuses on a subset of all experienced states. A subgraph of the N most visited states is extracted to have their Q-values computed as a priority. To have a relevant value for N, we compute the entropy of the V_S distribution, getting a measure of the model organisation :

$$H(V_S) = -\sum_{i \in S} P(i) \cdot log_2\left(P(i)\right) \text{ with } P(i) = \frac{Card(V_{S_i})}{Card(V_S)} \ . \tag{7}$$

We compare this measure to the maximal entropy of the model, deducing a ratio R_c of the compressibility of the State distribution representation :

$$R_c = \frac{H(V_S)}{H_{max}(V_S)} \text{ with } H_{max}(V_S) = log_2|\mathcal{S}| \ . \tag{8}$$

This method guides the planification to the most visited states. The drawback is that it may erroneously limit the use of the model during early states of learning where the number of states is small but the distribution already presents a contrasted shape. In this case, planning in the full graph is still possible at reasonable cost. To avoid this behaviour, R_c is transformed into a Ratio R_n - depending on the known number of nodes - given the following function (9).

$$R_n = (1 - \omega) + \omega \cdot R_c \text{ with weight } \omega = \frac{1}{1 + e^{-\sigma|\mathcal{S}|}} \ . \tag{9}$$

The final number of states to plan on is a proportion of the number of known states $|\mathcal{S}|$:

$$N = R_n \cdot |\mathcal{S}| \ . \tag{10}$$

2.4   Meta-Controller

The Meta-Controller gives the control to one of the Experts given a criterion. It allows only one of the Experts to send its decision to the Execution Layer. It also sends back the decision to both Experts, such that they can update their knowledge about its relevance in the current state according to the feedback, and cooperate in learning the best policy. The criterion considered in this work is an equiprobable random selection of each Expert, as a proof of concept of the interest of combining the two.

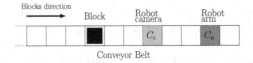

Fig. 2. The experimental setup : a discrete conveyor belt is carrying blocks in front of the robot. The robot's camera points at space C_c and its arm can reach the space in C_a. Blocks are going from left to right such that a block can be first seen and then touched.

3   Results

3.1   Experiment Description

We evaluated our Architecture performance in the simulation of a simple task of block pushing. The system has been implemented using the ROS middleware [18]. Our simulated robot is facing a conveyor belt on which are placed blocks. These blocks are characterized by their velocity (BS) and the distance between two blocks (inter-block distance, or IBD). These simulation parameters may be constant or evolve during the experiment, leading to four different cases. In this work, we focused on :

1. Regular case : inter-block distance is constant, speed of blocks is constant.
2. Speed Shift case : IBD is constant, BS changes during experiment.

In our setup, acting is required to update the perception (see Sect. 3.2 for Perception Module description). The robot has three available actions :

1. **Do nothing (DN)** : this action doesn't modify the environment nor bring perceptual information. It is a waiting action with no cost ($R_t = 0$) when executed.
2. **Look Cam (LC)** : this action doesn't modify the environment but updates the view modality about the presence of a block in C_c. It has a cost of $R_t = -0.03$.
3. **Push Arm (PA)** : this action can modify the environment : if a block is in C_a and PA is done, it is removed from the belt. The contact modality is updated about the perceived block. The action costs $R_t = -0.03$ but brings a positive reward when a block is pushed for a final reward of $R_t = 0.97$).

3.2   Perception Module

The Perception Module (Fig. 3) transforms Perceptions into States. Our simulated robot is equipped with a visual block detector simulated camera (signal p^{bs}) and a tactile binary sensor on its arm (signal p^{bt}). When the corresponding action is selected (LookCam for p^{bs} and PushArm for p^{bt}), these informations update memories where each element is one step further in the past. Each modality has its own memory where older block perceptions are recorded. Memories

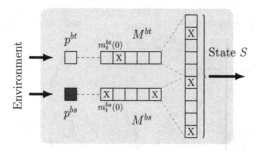

Fig. 3. Perception module for our task. The X corresponds to a memorized block. The system visually perceives a block and updates the corresponding memory. *bs* and *bt* stand for *block seen* and *block touched*.

(M^{bs}, M^{bt}) have a finite length (8 elements) such that the system only considers closest perceptions. Each configuration of both memories defines a unique State, used by the Experts.

The perceptive input are binary and their perceptions are determined according to (11).

$$p^{bt} = C_a \cdot \text{PushArm}, \quad p^{bs} = C_c \cdot \text{LookCam} . \tag{11}$$

Memories can evolve in two ways : if enough time has elapsed (State Max Duration = 0.1s here) or when a new perception brings information. In both cases, all elements from the memory are shifted to the next timestep. Information that exceeds memory length is forgotten. The first memory element is populated with the relevant perceptive data.

$$\begin{cases} m_t^{bs,bt}(i) = m_{t-1}^{bs,bt}(i-1) \ \forall i \in |M| . \\ \\ m_t^{bs,bt}(0) = p^{bs,bt} . \end{cases} \tag{12}$$

3.3 Parameters Search

In the following, we consider : $BS = 8$ spaces/s (optimal policy : DN-DN-DN-DN-PA), $IBD = 4$ spaces/block.

We first searched for the best parametrization for both Experts controlling individually the robot, in the Regular case. An Expert is performant if it maximises the obtained Cumulative Reward (CR) and minimizes the standard deviation of CR over runs. We tested for each Expert the combination of 3 to 5 values (for MB and MF : $\alpha_{MB,MF} \in \{0.01, 0.05, 0.1, 0.5, 0.9\}, \gamma_{MB,MF} \in \{0.5, 0.98, 0.9999\}, \tau_{MB,MF} \in \{0.05, 0.1, 0.5, 0.9\}$ plus $\tau_{MB} = 0.01$) :

For the best solutions, we favor the most rewarding in mean and then the less varying. We choose $\alpha_{MB} = 0.5, \gamma_{MB} = 0.5, \tau_{MB} = 0.1$ and $\alpha_{MF} = 0.1, \gamma_{MF} = 0.9999, \tau_{MF} = 0.05$.

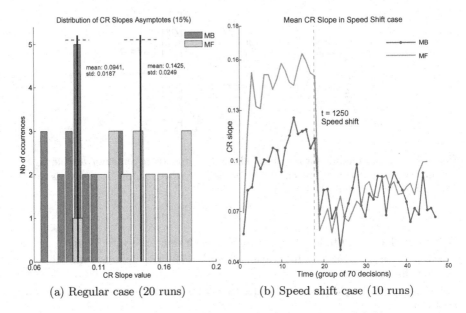

(a) Regular case (20 runs)   (b) Speed shift case (10 runs)

Fig. 4. Policy evaluations. (4a) histogram of approximated slopes. Solid lines are means of CR slope for each Expert, dashed lines the standard deviation (4b) Mean CR slope over time. The slope is approximated every 70 decisions.

3.4   Individual Experts Performances

Each Expert is tested individually in the Regular and Speed Shift cases (simulation parameters : see Sect. 3.3 ; in Speed Shift case, we change BS to 13.2 spaces/s at 1250 decisions, which correspond to an optimal DN-DN-PA policy). The Cumulative Reward is linearly approximated from its 15% last values to evaluate the discovered policy. For the Speed shift case, we also approximate CR before the speed shift (on the same duration) and compare it to the first approximation to evaluate the sensitivity of the policy to speed shift. The slope of these approximations measures the quality of the policy as it depends on the obtained rewards. From figure 4a, we observe that the MF discovers and follows better policies in mean but tends to be more exploring than the MB with a larger deviation in slopes values. In the Regular case, the MF is more relevant than the MB to obtain the best performance as possible. This is due to the low cost and high precision of the Q-values acquired by the MF. In contrast, the MB learns a model of the task whose number of states rapidly grows, making the planning process slow, costly and relying on approximations of action values.

In the Speed Shift case, we observe a break in Cumulative reward for the MF. Figure 4b shows that in mean, the MB performance is less sensitive to the change than the MF : the environmental change induces a loss that is more than twice higher in the MF than in the MB. After the shift and until the end of simulation, both MB and MF are performing similarly. This behavior can be explained by the long time required by the MF to relearn the Q-values of a new efficient action sequence, what doesn't happen in the given time. In contrast, a single exposure of the MB to the new sequence of events imposed by the speed shift enables it to change its model of the task and thus to plan a new sequence of actions, but it still suffer from the approximation of action values to find a better policy.

Both Experts exhibit a complementary role : while the MF is best suited to optimize the policy in stable conditions, the MB can better handle transient phases following environmental changes.

3.5 Combination of Experts

The whole architecture (MB+MF, supervised by Meta-Controller) is then tested on both cases, with the same setup. In the Regular case, figure 5a shows that the strategy of selecting stochastically each Expert improves the mean performance of the robot compared to using only the MB. On the other hand, as the Experts are chosen randomly, the robot is not relying only on the MF, which is the most efficient strategy in the Regular case. This explains that the MB+MF performance is still worse than the MF only. In the Speed Shift case, figure 5b shows that the change in the environment doesn't affect significantly the robot's performance, as it benefits from the MB ability to quickly replan an adapted policy. The Combination of Experts robustness to changes compensates for the advantage gained by the MF before the shift. At the end of the simulation, the

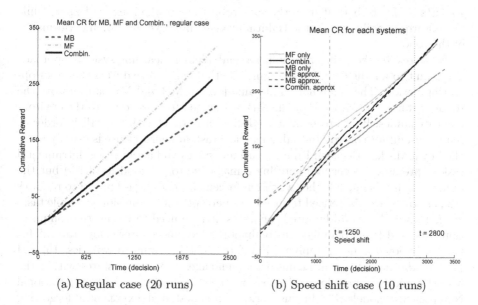

(a) Regular case (20 runs)   (b) Speed shift case (10 runs)

Fig. 5. Mean Cumulative Reward obtained from individual Experts and their Combination (solid line)

MF hasn't found a policy that is at least as good as before the shift (though the task allows a higher rate of reward, as there are more blocks to push on).

4 Discussion

This work presented the decision layer of a robotic control architecture able to learn habits, taking inspiration from computational neuroscience models [13] and multiple reinforcement learning systems applied to navigation [9,4]. The model has two different Experts, or strategies, one habitual – that learns State-Action association, and make quick decisions, but slowly adapts its policy when the environment changes – and one flexible – that maintains a representation of the environment and the task, can adapt quickly but is slow in deciding as it evaluates action outcomes. These strategies are selected depending on an arbitration criterion by a Meta-Controller. The criterion used in this work is an random equiprobable selection of Experts, as a proof-of-concept of the interest of combining the two.

We first highlighted each Expert properties in a Regular and a Speed Shift cases. We showed that, as expected, the Habitual Expert learns better policies than the Flexible Expert when the environment is stable, but a transient phase like a shift in belt speed, making the policy less appropriate, will result in a long lower performance period. The learnt Flexible Expert policies are less performant in mean as the computation time constraint and the focused planning lead to less precise Q-values when the number of states grows, and a sub-optimal policy. On the other hand, updating its model allows the Flexible Expert to be less affected by the speed shift. We then showed that a random selection of each Experts is able to benefit from the shift robustness of the Flexible Expert while the rewarding policies from the Habitual Expert improve the global performance of the robot.

These results show that the multiple reinforcement learning systems approach is relevant to handle complex environments that can evolve during the robot operating period. The combination at same level of MB and MF can improve the robot autonomy provided that the MB is designed to be reactive to the environment dynamics. It has to be able to decide in parallel with the MF in order to remain useful for control. Indeed, a classical task in neuroscience is usually modelled by a Markov Decision Process with few states and actions (e.g. intrumental task of pressing a lever and entering a magazine to get food [6,13,16]) but the dimensionality is much higher when reinforcement learning is applied to robotics [12], as states are discretized from robot's perceptions. In our not so simple task, we end up with several hundreds of states and we need to bound the computation time and focus planning on the hypothesized most interesting states. This is a well-known issue of applying Reinforcement Learning to robotics [15] and justifies the need for a mechanism that manages the known states within the MB system, a proposition which has recently been applied to Computational Neuroscience models [11]. In this work, we first tested an exponential forgetting mechanism on the transition model to remove unvisited paths. As this mechanism doesn't strongly affect performance and planning time is still increasing with the growth of states, we switched to the time constraint and focused planning mechanism (described in Sect. 2.3). The increase in performance suggests that planning with a complex model requires a budgeted approach that only considers the relevant sub-model, instead of pruning parts of the model related to irrelevant old experiences. Our Experts' parametrization has been optimized for the regular case. In a first approach, we choose the parameters to be constant but in this task, the MB should be exploring enough such that the agent gets out of the initial attractor state and then exploit its model to quickly adapt to environmental changes. Here, both the Experts have an exploiting temperature, and the lower adaptability of the MF comes from the way it learns (updating only the experienced actions).

This work also generalizes the concepts from [4] for the control of robots. The multiple reinforcement learning systems approach can be applied not only for navigation but also on a wider variety of tasks, provided that the robot is able to perform the relevant actions. As our system can rely on the Habitual Expert, our architecture can benefit from its properties of being quick to decide the next action when the task is stationary. On the other hand, our Flexible Expert can be compared to the robotic decision-making systems, that are based mostly on planning algorithms that use a representation of the world [17]. The latter usually rely on a provided representation whereas our Flexible Expert learns its model and updates it according to changes in the task. This enhances the behavioral adaptability of the robot in non-stationary environments. We need to further investigate the arbitration criterion between Experts to get the optimal alternations and benefit from the whole architecture.

Acknowledgements. This work has been funded by a DGA (French National Defence Agency) scholarship (ER), by the Project HABOT from Ville de Paris and by French Agence Nationale de la Recherche ROBOERGOSUM project under reference ANR-12-CORD-0030.

References

1. Balleine, B.W., Dickinson, A.: Goal-directed instrumental action: contingency and incentive learning and their cortical substrates. Neuropharmacology 37, 407–419 (1998)
2. Balleine, B.W., O'Doherty, J.P.: Human and rodent homologies in action control: corticostriatal determinants of goal-directed and habitual action. Neuropsychopharmacology 35, 48–69 (2010)
3. Caluwaerts, K., Favre-Félix, A., Staffa, M., N'Guyen, S., Grand, C., Girard, B., Khamassi, M.: Neuro-inspired navigation strategies shifting for robots: Integration of a multiple landmark taxon strategy. In: Prescott, T.J., Lepora, N.F., Mura, A., Verschure, P.F.M.J. (eds.) Living Machines 2012. LNCS, vol. 7375, pp. 62–73. Springer, Heidelberg (2012)
4. Caluwaerts, K., Staffa, M., N'Guyen, S., Grand, C., Dollé, L., Favre-Félix, A., Girard, B., Khamassi, M.: A biologically inspired meta-control navigation system for the psikharpax rat robot. Bioinspiration and Biomimetics (2012)

5. Daw, N.D., Niv, Y., Dayan, P.: Uncertainty-based competition between prefrontal and dorsolateral striatal systems for behavioral control. Nature Neuroscience 8(12), 1704–1711 (2005)
6. Dezfouli, A., Balleine, B.W.: Habits, action sequences and reinforcement learning. European Journal of Neuroscience 35(7), 1036–1051 (2012)
7. Dickinson, A.: Contemporary animal learning theory. Cambridge University Press, Cambridge (1980)
8. Dickinson, A.: Actions and habits: The development of behavioural autonomy. Phil Trans Roy Soc B: Biol Sci 308, 67–78 (1985)
9. Dollé, L., Sheynikhovich, D., Girard, B., Chavarriaga, R., Guillot, A.: Path planning versus cue responding: a bioinspired model of switching between navigation strategies. Biological Cybernetics 103(4), 299–317 (2010)
10. Gat, E.: On three-layer architectures. In: Artificial Intelligence and Mobile Robots. MIT Press (1998)
11. Huys, Q.J., Eshel, N., O'Nions, E., Sheridan, L., Dayan, P., Roiser, J.P.: Bonsai trees in your head: how the pavlovian system sculpts goal-directed choices by pruning decision trees. PLoS Computational Biology 8(3) (2012)
12. Kaelbling, L.P., Littman, M.L., Moore, A.W.: Reinforcement learning: a survey. Journal of Artificial Intelligence Research 4, 237–285 (1996)
13. Keramati, M., Dezfouli, A., Piray, P.: Speed/accuracy trade-off between the habitual and goal-directed processes. PLoS Computational Biology 7(5), 1–25 (2011)
14. Khamassi, M., Humphries, M.D.: Integrating cortico-limbic-basal ganglia architectures for learning model-based and model-free navigation strategies. Frontiers in Behavioral Neuroscience 6, 79 (2012)
15. Kober, J., Bagnell, D., Peters, J.: Reinforcement learning in robotics: A survey. International Journal of Robotics Research (11), 1238–1274 (2013)
16. Lesaint, F., Sigaud, O., Flagel, S.B., Robinson, T.E., Khamassi, M.: Modelling Individual Differences in the Form of Pavlovian Conditioned Approach Responses: A Dual Learning Systems Approach with Factored Representations. PLoS Comput Biol 10(2) (February 2014)
17. Minguez, J., Lamiraux, F., Laumond, J.P.: Motion planning and obstacle avoidance. In: Siciliano, B., Khatib, O. (eds.) Handbook of Robotics, pp. 827–852. Springer, Heidelberg (2008)
18. Quigley, M., Conley, K., Gerkey, B.P., Faust, J., Foote, T., Leibs, J., Wheeler, R., Ng, A.Y.: Ros: an open-source robot operating system. In: ICRA Workshop on Open Source Software (2009)
19. Sutton, R.S., Barto, A.G.: Introduction to Reinforcement Learning, 1st edn. MIT Press, Cambridge (1998)
20. Watkins, C.: Learning from Delayed Rewards. PhD thesis, King's College, Cambridge, UK (1989)
21. Yin, H.H., Ostlund, S.B., Balleine, B.W.: Reward-guided learning beyond dopamine in the nucleus accumbens: the integrative functions of cortico-basal ganglia networks. Eur. J. Neurosci. 28, 1437–1448 (2008)

Dynamic Model of a Jet-Propelled Soft Robot Inspired by the Octopus Mantle

Federico Renda[1], Frederic Boyer[2], and Cecilia Laschi[1]

[1] BioRobotics Institute, Scuola Superiore Sant'Anna, Pisa, Italy
[2] Institut de Recherche en Communication et Cybernetique de Nantes,
Ecole des Mines de Nantes, Nantes, France
f.renda@sssup.it
http://sssa.bieroboticsinstitute.it/

Abstract. This article addresses the study of cephalopods locomotion. As a first step, we here propose an analytical model of the cephalopod mantle. The approach is based on the geometrically exact shell theory used in nonlinear structural dynamics and exploits the symmetric shape of the mantle. Once all the mathematical background is reminded we propose a first use of this model by using the constitutive laws of the shell as control laws able to reproduce the axisymmetric contractions observed in cephalopods. Further numerical exploitation of these theoretical results are to date in progress along with works in fluid mechanics.

Keywords: Dynamics, Soft Robots, Continuum Robots, Octopus.

1 Introduction

To self propel forward, cephalopod as squids and octopus have invented an original solution based on the cyclic contraction of a soft cavity named mantle. During each cycle, the mantle dilates while water runs into it and then brutally contracts in order to expel water and to generate by reaction the forward thrust [1]. The contraction of the mantle is ensured by a network of circular muscles symmetrically positioned all around the body [2]. Remarkably, these animals have developed strategies exploiting the softness of their body in order to re-capture energy from ambient flow that might otherwise be lost as this is the case of rigid rockets which also propel by jet reaction [3]. Recently these animals have represented an important source of inspiration for the development of new underwater thruster composed of soft material, capable of working in cramped environment with fast acceleration and great manoeuvrability. To reach this goal some crucial points have to be addressed, new design solutions have to be thought that take into account the use of soft materials and new models and control strategies [4] [5] have to be developed for such soft structures. In particular a pulsed jet instead of a steady state propulsion have to be considered. A first example of pulsed jet soft robot is presented in [6] where a cable driven underwater thruster has been described, inspired by the octopus mantle (fig. 1).

A. Duff et al. (eds.): Living Machines 2014, LNAI 8608, pp. 261–272, 2014.
© Springer International Publishing Switzerland 2014

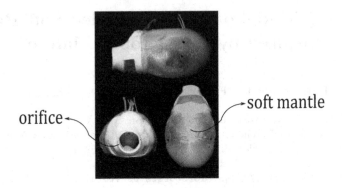

orifice

soft mantle

Fig. 1. A jet-propelled soft robot inspired from cephalopods mantle [6]

In order to study the animal swimming mechanisms in detail and reproduce them on robotics artefacts, we need to state locomotion models of these jet-propelled soft animals. Ideally, these models would be analytical models allowing to idealize the principles of locomotion. However, such a modelling effort is highly challenging, and can be divided in two parts depending if we consider the fluid (outside and inside the mantle) or the body. As a first step, we here propose to address the analytical modelling of the mantle body alone, i.e. in vacuum. This model will be next used to address the difficult problem of the analytical modelling of the flow outside but also inside, a problem rarely addressed so far. To derive a model of the mantle, we use the framework of the geometrically exact theory of shells in finite transformations [7], [8].

In this kind of approach the shell is considered as a Cosserat medium i.e. a continuous assembly of rigid micro-solids whose rigid overall motions are considered from the beginning without any approximations. This aspect is crucial to tackle locomotion problems where we not only need to model the internal strains occurring in the body but also the net rigid overall motions in space. While in Cosserat beams, the micro-solids are the cross sections of the beam to each of which is attached a rigid frame (3 unit vectors), in the case of shells, the micro-solids are rigid fibers transversally attached to the mid-surface of the shell to which, one can only attach a single vector named "director" [9]. Based on this Cosserat model, several internal kinematics can be adopted depending whether the shell is thin or not. In the first case, the directors remain perpendicular to the mid-surface while in the second case, they can rotate freely with respect to the mid-surface with two additional degrees of freedom which induce two further strain fields named "transverse shearing". The first kinematics correspond to the so called Kirchhoff model of shells while the second correspond to the Reissner model [10]. The model of the mantle here proposed is based on this second theory which leads to partial differential equations of minimum order. Furthermore, taking inspiration from cephalopods, the shell will be considered as axisymmetric. To derive such a model we follow the uses of nonlinear structural dynamics by first defining the shell kinematics. From these kinematics, we will build a set of strain measures and will derive the dynamic balance equations in their Cauchy form, i.e. in terms of internal stresses. Finally, this picture will be completed of the constitutive laws in section 6. All these developments will be achieved in the case of an axisymmetric shell. Based on this closed formulation, we will propose a first way of implementing it in the locomotion problem of the squid. The implementation uses the constitutive law as a control law where the radius of the axisymmetric strips of mantle are specified by a time law aimed at reproducing the cyclic muscular contractions of the animal. Finally, the article will end with a few concluding remarks and future perspectives.

2   Mantle Model

The mantle is a piece of living tissues forming a cavity which opens into water through an orifice (see Figure 2). In the rest of the article, it is modelled by a shell of thickness $2h$ made of a soft hyperelastic isotropic material of density ρ.

Adopting the Reissner model, the configuration space of the shell can be first defined by:

$$\mathcal{C} := \{(X^1, X^2, t) \in B \times \mathbb{R} \mapsto (\varphi, \mathbf{b}) \in \mathbb{R}^3 \times S^2\} \qquad (1)$$

where $B \subset \mathbb{R}^2$ defines the material domain of the mid-surface, φ represents the field of position of the points of B and $\mathbf{b}$ stands for the field of unit vectors attached to the shell fibers, i.e. the directors living in the two dimensional unit sphere S^2.

In the following, we consider axisymmetric shells. As a result, the configuration space can be reduced further. In the next section, this reduction is detailed step by step.

2.1   Kinematics of Axisymmetric Shells

Mathematically, an axisymmetric surface or "surface of revolution", is obtained by rotating a planar curve or "profile" around a fixed axis named "symmetry axis". This rotation changes the "profile curve" into any of the meridian curves that constitute the shell. The ambient Euclidean space is endowed with a fixed base of orthogonal unit vectors $(\mathbf{e}_1, \mathbf{e}_2, \mathbf{e}_3)$, where $\mathbf{e}_3$ supports the symmetry axis. Denoting by ϕ the angle of revolution, the orthogonal basis $(\mathbf{e}_r, \mathbf{e}_\phi, \mathbf{e}_3)$ fixed to the meridian of angle ϕ is defined by:

$$g1 \in SE(3):$$

$$\phi \in [0, 2\pi[\mapsto g1(\phi) = \begin{pmatrix} R_{\mathbf{e}_3}(\phi) & 0 \\ 0 & 1 \end{pmatrix}$$

where $R_{\mathbf{e}_3}(\phi)$ is a rotation ϕ around the axis $\mathbf{e}_3$. For the sake of convenience, we introduce another reference frame $(\mathbf{e}_r, \mathbf{e}_3, -\mathbf{e}_\phi)$ adding the transformation

$$g2 \in SE(3): g2 = \begin{pmatrix} R_{\mathbf{e}_r}(\pi/2) & 0 \\ 0 & 1 \end{pmatrix}$$

Now, denoting by X the material coordinate along the meridian curves, the profile of the shell is defined by the curve $\mathbf{r} : (X, t) \in [0, L] \times [0, \infty) \mapsto \mathbf{r}(X, t) = (r(X, t), z(X, t), 0)^T$ that lies in the plane $(\mathbf{e}_r, \mathbf{e}_3)$ and for which, $r(.)$ and $z(.)$ are two smooth functions which define the radius and the altitude of the point X on the profile (see Figure 2). Let us call $\theta(X, t)$ the angle between $\mathbf{e}_3$ and the shell fiber located at any X along the ϕ-meridian, then the so called *director orthogonal frame* $(\mathbf{a}, \mathbf{b}, -\mathbf{e}_\phi)$ with centre $\mathbf{r}$ is defined by:

$$g3 \in SE(3) : (X, t) \in [0, L] \times [0, \infty) \mapsto$$

$$g3(X, t) = \begin{pmatrix} R_{-\mathbf{e}_\phi}(\theta) & \mathbf{r} \\ 0 & 1 \end{pmatrix}$$

Finally, putting them all together, the shell configuration space is

$$g \in SE(3) : g(X, \phi, t) = g1g2g3 = \begin{pmatrix} R & \varphi \\ 0 & 1 \end{pmatrix}$$

As a result, we can now introduce the following definition of the configuration space of an axisymmetric shell:

$$\overline{C} := \{ (X, \phi, t) \in \overline{B} \times \mathbb{R} \mapsto g \in SE(3) \} \tag{2}$$

where, referring to the more general context of (1), $X^1 = X$ and X^2 has been replaced by ϕ ($\overline{B} \subset \mathbb{R} \times S^1$) with $X^2 = r^o(X)\phi$, and $r^o(X)$ the value of $r(X)$ in the reference configuration of the shell (before any deformation).

The tangent plane on the surface $g(X, \phi, t)$ is represented by two vector field: $\widehat{\xi^1}(X, t) = g^{-1}g_{,X} = g^{-1}g\prime = g3^{-1}g3\prime$ and $\widehat{\xi^2}(X, t) = g^{-1}g_{,r^o\phi}$. The hat represents the isomorphism between the twist vector space $\mathbb{R}^6$ and the Lie algebra $se(3)$. Below their components are specified:

$$\widehat{\xi^1} = \begin{pmatrix} \widetilde{\mathbf{k}}^1 & \mathbf{q}^1 \\ 0 & 0 \end{pmatrix} \in se(3)$$

$$\widehat{\xi^2} = \begin{pmatrix} \widetilde{\mathbf{k}}^2 & \mathbf{q}^2 \\ 0 & 0 \end{pmatrix} \in se(3)$$

$$\xi^1 = (\mathbf{k}^{1T}, \mathbf{q}^{1T})^T = (0, 0, \mu, \lambda, \eta, 0)^T \in \mathbb{R}^6$$

$$\xi^2 = (\mathbf{k}^{2T}, \mathbf{q}^{2T})^T = (\frac{\sin(\theta)}{r^o}, \frac{\cos(\theta)}{r^o}, 0, 0, 0, \frac{-r}{r^o})^T \in \mathbb{R}^6$$

where the tilde is the usual isomorphism between a vector of $\mathbb{R}^3$ and the corresponding skew-symmetric matrix.

The time evolution of the configuration curve g is represented by the twist vector field $\varsigma(X, t) \in \mathbb{R}^6$ defined by $\widehat{\varsigma} = g^{-1}\delta g/\delta t = g^{-1}\dot{g} = g3^{-1}\dot{g3}$, where, thanks to the axisymmetry, only $g3$ depend on time. Let us specify the component of

$$\widehat{\varsigma} = \begin{pmatrix} \widetilde{\mathbf{w}} & \mathbf{v} \\ 0 & 0 \end{pmatrix} \in se(3)$$

$$\varsigma = (\mathbf{w}^T, \mathbf{v}^T)^T = (0, 0, w, v_a, v_b, 0)^T \in \mathbb{R}^6$$

Finally, from the last equivalence, the kinematic equation is

$$\dot{g3} = g3\hat{\varsigma} \tag{3}$$

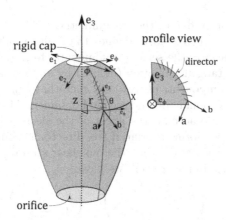

Fig. 2. Axisymmetric shell kinematics (left). Profile view, beam-like parametrization (right)

2.2 Compatibility Equations

We have seen above that $g3\prime = g3\widehat{\xi^1}$. By taking the derivative of this equation with respect to time and recalling that $\dot{g3} = g3\hat{\varsigma}$, we obtain the following compatibility equation between velocity and deformation variables: $\dot{\widehat{\xi^1}} = \widehat{\varsigma\prime} + \widehat{\xi^1}\hat{\varsigma} - \hat{\varsigma}\widehat{\xi^1}$. In terms of twist vectors it can be written as:

$$\dot{\xi^1} = \varsigma\prime + ad_{\xi^1}(\varsigma) \tag{4}$$

where

$$ad_{\xi^1} = \begin{pmatrix} \tilde{\mathbf{k}}^1 & 0 \\ \tilde{\mathbf{q}}^1 & \tilde{\mathbf{k}}^1 \end{pmatrix}$$

is the adjoint map, that represents the action of the Lie algebra on itself.

2.3 Strain Measures

Before all, let us introduce two quadratic forms named: $h(X)$ and $k(X)$, that represent the first and the second fundamental forms of a Reissner shell respectively.

The first fundamental form of a surface is a quadratic form that determines how the Euclidean metric of $\mathbb{R}^3$ is induced on the surface in any of its points. In our case, it is practically defined in each point (X, ϕ) of the surface by a tensor whose components are deduced from the scalar products of all the vectors of the field of basis (said natural basis) $(\mathbf{q}^1, \mathbf{q}^2)(X, \phi)$. For our revolution surface it can be simply detailed as:

$$h = \begin{pmatrix} h_{11}(X) & 0 \\ 0 & h_{22}(X) \end{pmatrix} = \begin{pmatrix} \lambda^2 + \eta^2 & 0 \\ 0 & \frac{r^2}{r^{o2}} \end{pmatrix}$$

which depends only on X due to the axisymmetry.

While the first fundamental form defines the scalar product of any tangent vectors to the mid surface, the second fundamental form defines the curvature of the surface in any tangent direction (defined by an unit tangent vector). For a Reissner shell the components of $k(X)$ are such that $k_{\alpha\beta} = \mathbf{q}^\alpha \cdot \mathbf{b}_{,X^\beta}$.

Contrary to the traditional first and second fundamental form of a surface, this "special" forms take into account the effect of the shear between two material elements, where the two forms overlap if the director $\mathbf{b}$ points in the direction normal to the mid-surface of the shell (i.e. no shear strain).

For an axisymmetric Reissner shell, we have:

$$k = \begin{pmatrix} k_{11}(X) & 0 \\ 0 & k_{22}(X) \end{pmatrix} = \begin{pmatrix} \mu\lambda & 0 \\ 0 & \frac{r\sin(\theta)}{r^{o2}} \end{pmatrix}$$

Following [7], the following strain tensor field has been adopted:

$$e = \frac{1}{2}(h - h^o) = \begin{pmatrix} 1/2(\lambda^2 + \eta^2 - 1) & 0 \\ 0 & 1/2(\frac{r^2}{r^{o2}} - 1) \end{pmatrix}$$

to describe the membrane strain state in the mid-surface. As regards the shear strain state, we use the following strain vector:

$$s = \begin{pmatrix} \mathbf{q}^1 \cdot \mathbf{b} - \mathbf{q}^{1o} \cdot \mathbf{b}^o \\ \mathbf{q}^2 \cdot \mathbf{b} - \mathbf{q}^{2o} \cdot \mathbf{b}^o \end{pmatrix} = \begin{pmatrix} \eta(X) \\ 0 \end{pmatrix}$$

while the flexural strain state is parameterized by using the following tensor field:

$$d = k - k^o = \begin{pmatrix} \mu^o - \mu\lambda & 0 \\ 0 & \frac{\sin(\theta^o)}{r^o} - \frac{r\sin(\theta)}{r^{o2}} \end{pmatrix}$$

In all the above definitions, the upper index o represents a field when it is evaluated in the reference relaxed configuration. Furthermore, by taking as material coordinate X, the Euclidean curvilinear abscissa along the corresponding meridian when it is in the reference configuration, we have $h_{11}^o = 1$, while it is natural to consider that there is no transverse shearing in the reference resting con figuration, i.e. $\eta^o = 0$.

2.4   Dynamics

The dynamic model of the shell is given by the balance of kinetic momenta, i.e. by Newton's laws or a variational principle. In any case, this model takes the form of a set of partial differential equations (p.d.e.'s) which govern the time evolution of the system (here the shell) on its configuration space. With the definition (1) of the configuration space of a shell (not forcedly axisymmetric), these p.d.e.'s have been derived in [7] as follows:

$$1/j(j\mathbf{n}^\alpha)_{,X^\alpha} + \overline{\mathbf{n}} = 2\rho h\ddot{\varphi}$$
$$1/j(j\mathbf{m}^\alpha)_{,X^\alpha} + \varphi_{,\alpha} \times \mathbf{n}^\alpha + \overline{\mathbf{m}} = 2\rho h\mathbf{b} \times \ddot{\mathbf{b}} \tag{5}$$

where $j = \sqrt{\det(h)} = r/r^o\sqrt{\lambda^2 + \eta^2}$, $X^1 = X$, $X^2 = r^o(X)\phi$, while the vectors $\mathbf{n}^\alpha$ and $\mathbf{m}^\alpha$ are respectively the resultant of internal stress forces and couples transmitted applied by the left part ($x < X^\alpha$) into the right part ($x \geq X^\alpha$) of the shell, across the section X^α, normalized with the surface Jacobian j, $\overline{\mathbf{n}}$ and $\overline{\mathbf{m}}$ are the external force and couple per unit of mid-surface area. For the repeated α the Einstein convention has to be used as in the rest of the paper. As expected, these equations give the time-evolution of all the pairs $(\varphi, \mathbf{b})$ over the shell as a function of the external load and the internal stress.

With respect to the local reference frame equation (5) can be written, in a geometric notation, as:

$$\frac{1}{j}\frac{\partial(j\zeta_i^\alpha)}{\partial X^\alpha} + \zeta_e = \frac{\partial}{\partial t}(\Gamma\varsigma) \tag{6}$$

Here ζ_i^α is the wrench of internal forces, ζ_e is the external wrench of distributed applied forces and Γ is the screw inertia matrix.

Due to the axisymmetry, the internal and external wrench fields take the particular form [8]:

$$\zeta_i^1(X,t) = (0,0,M_X,N_X,H,0)^T \in \mathbb{R}^6$$

$$\zeta_i^2(X,t) = (M_\phi,0,0,0,0,-N_\phi)^T \in \mathbb{R}^6$$

$$\zeta_e(X,t) = (0,0,l,f_a,f_b,0)^T \in \mathbb{R}^6$$

and the screw inertia matrix is equal to: $\mathbb{R}^6 \otimes \mathbb{R}^6 \ni \Gamma = diag(0,0,\rho J,2\rho h,2\rho h,0)$.

Inserting this particular form of force wrenches and expanding the derivative in (6), gives the dynamic equations of an axisymmetric shell in the form:

$$\Gamma\dot{\varsigma} = 1/j(j\zeta_i^1)_{,X^1} - ad^*_{\xi\alpha}(\zeta_i^\alpha) + \zeta_e + ad^*_\varsigma(\Gamma\varsigma) \tag{7}$$

where

$$ad^*_{\xi\alpha} = \begin{pmatrix} \widetilde{\mathbf{k}}^{\alpha T} & \widetilde{\mathbf{q}}^{\alpha T} \\ 0 & \widetilde{\mathbf{k}}^{\alpha T} \end{pmatrix} \qquad ad^*_\varsigma = \begin{pmatrix} \widetilde{\mathbf{w}}^T & \widetilde{\mathbf{v}}^T \\ 0 & \widetilde{\mathbf{w}}^T \end{pmatrix}$$

are the co-adjoint map.

In the equations above ρ is the body density, h is the half of the shell thickness and J is the second moment of the cross sectional line equal to $J = h^2/3$.

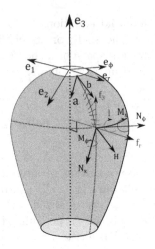

Fig. 3. Internal (dark) and external (red) loads exerted in (X, ϕ)

2.5   Constitutive Equations

According to [7], for a hyper-elastic isotropic material, the general constitutive equations of a shell are defined in terms of the strain measures of as follows:

$$\tilde{n}^{\beta\alpha} = \frac{2Eh}{1-\nu^2} H^{\beta\alpha\gamma\delta} e_{\gamma\delta} \ , \ \tilde{m}^{\beta\alpha} = \frac{2EhJ}{1-\nu^2} H^{\beta\alpha\gamma\delta} d_{\gamma\delta}$$

$$\tilde{q}^\alpha = 2Ghh^{o\alpha\beta} s_\beta, \tag{8}$$

where E is the Young modulus, G the shear modulus and ν is the Poisson modulus; $h^{o\alpha\beta}$ are the elements of h^{o-1} and $H^{\beta\alpha\gamma\delta}$ define a four order (Hook-like) tensor given by:

$$H^{\beta\alpha\gamma\delta} = [\nu(h^{o\beta\alpha}h^{o\gamma\delta}) + 1/2(1-\nu)(h^{o\beta\gamma}h^{o\alpha\delta} + h^{o\beta\delta}h^{o\alpha\gamma})]. \ \bullet$$

The $\tilde{n}^{\alpha\beta}$, $\tilde{m}^{\alpha\beta}$ and $\tilde{q}^\alpha$ stand respectively for the components of the *effective resultant traction, couple and shearing stresses*[1], which are related to the stresses of (7) through the relations:

$$M_\phi = -\frac{r}{r^o}\tilde{m}^{22}, \qquad M_X = -\lambda\tilde{m}^{11},$$

$$N_X = \lambda\tilde{n}^{11} - \mu\tilde{m}^{11}, \qquad N_\phi = r\tilde{n}^{22} - \frac{\sin(\theta)}{r^o}\tilde{m}^{22},$$

$$H = \tilde{q}^1 + \eta\tilde{m}^{11}.$$

---

[1] The word "effective" here stands to underline that these functions can be directly related to the constitutive equations of the three dimensional theory [7].

With these relations and the constitutive equations (8), the constitutive equations for our internal stresses (i.e. $N_X(X)$, $N_\phi(X)$, $H(X)$, $M_X(X)$ and $M_\phi(X)$) are given by:

$$
\begin{aligned}
N_X &= \tfrac{2Eh}{1-\nu^2}\left[\lambda\left(e_{11}+\nu e_{22}\right)-J\mu\left(d_{11}+\nu d_{22}\right)\right], \\
N_\phi &= \tfrac{2Eh}{1-\nu^2}\left[\tfrac{r}{r^o}\left(e_{22}+\nu e_{11}\right)-J\tfrac{\sin(\theta)}{r^o}\left(d_{22}+\nu d_{11}\right)\right] \\
H &= 2h\eta\left[G+\tfrac{E}{1-\nu^2}\left(e_{11}+\nu e_{22}\right)\right], \\
M_X &= -\tfrac{2EhJ}{1-\nu^2}\lambda\left(d_{11}+\nu d_{22}\right) \\
M_\phi &= -\tfrac{2EhJ}{1-\nu^2}\tfrac{r}{r^o}\left(d_{22}+\nu d_{11}\right),.
\end{aligned}
\tag{9}
$$

From a mechanical point of view, we can now state a closed form of the axisymmetric shell dynamics by gathering the p.d.e's (7), the definition of the strains of section IV and the constitutive laws (9). In order to avoid the polar singularity ($r^o = 0$) in the constitutive law, the mantle is assumed to be connected to a small rigid spherical cap axisymmetric with respect to the mantle axis which crosses the cap in the cap's pole (see Figure 2). The inertia of the cap is assumed to be negligible and the former formulation has to be completed with the following geometric boundary conditions:

$$
g3(0) = g3_-,
\tag{10}
$$

where $g3_-$ is fixed by the boundaries of the cap in which the mantle is clamped. At the other tip of the mantle, we have the following natural boundary conditions:

$$
N_X(L) = f_{a,+}, \qquad H(L) = f_{b,+}, \qquad M_X(L) = -l_+,
\tag{11}
$$

where $f_{a,+}$, $f_{b,+}$ and l_+ denote the eventual external forces and torque along **a**, **b** and $-\mathbf{e}_\phi$ respectively and applied onto the sharp boundaries of the mantle orifice. Finally, if the resulting closed formulation is adapted to a passive shell, the case of the squid requires further modifications of this closed formulation.

3 Application to the Squid Locomotion

The previous formulation can be used to address the study of cephalopods locomotion. To that end, we consider that the squid is jet-propelled along the z-axis while the axisymmetric external loads cannot generate a net displacement in another dimensions of the Lie group $SE(3)$. Based on the use of a Galilean reference frame [11], the geometrically exact approach does not require to use a reference frame attached to the mantle. For instance, in our case, the net motions are directly given by the integration of the p.d.e.'s (7), in which $z(X)$, denotes the absolute displacement field of the X-circular strip of mantle along the $\mathbf{e}_3$-axis. Furthermore, let us remind that the squid is modelled by an axisymmetric shell internally actuated by a network of circular muscles organised in rings around the shell axis. As a result, we have replaced the reference strains of the shell 2.3

with a desired strains field in time obtained by the following constrain inputs: $\eta^d(X,t) = 0$, $\lambda^d(X,t) = 1$ and $r^d(X,t) = f(X,t)$ where $r^d(X,t)$ is the desired value of the radius of the axisymmetric strip of squid which passes through the point of abscissa X along the shell profile. Since the muscles can only contract, the rhythmic muscular activity can be modelled by taking $f(X,t)$ as a T-periodic function with two phases. In the first phase, or contraction phase, it decreases over a short part of $[0,T]$. In the second phase, or relaxation phase, $r_d(.,t) = r^o$, and the mantle recover its resting shape passively thanks to the internal restoring stresses (fig. 4). In these conditions, the constitutive law (9) is partly used as a kind of linear control law where the elastic coefficients stand for proportional control gains. Going further, we could emancipate ourselves from isotropic materials and consider more generally constitutive laws of the type (9) where the constant passive elastic coefficients are replaced by active time varying ones modelling the nonlinear muscular activity. In spite of all these possibilities, we restrict our investigations to the case of (9) with constant coefficients.

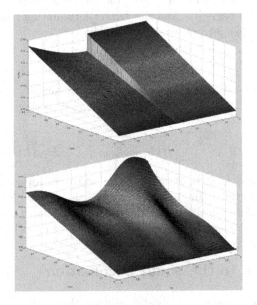

Fig. 4. Desired radius (up) input and resulted shell radius (down) for a conic shell with $r^d(L,t) = 0.15 - 0.03\sin(2\pi t/8)(t <= T/2) + r^o(L,t)(t > T/2)$

In this case, the dynamics of the shell can be put into the state-space form $\dot{x} = f(x,x',t)$ (with $x' = \partial x/\partial X$, and $\dot{x} = \partial x/\partial t$) as follows:

$$\dot{g}3 = g3\hat{\varsigma}$$
$$\dot{\xi}^1 = \varsigma\prime + ad_{\xi^1}(\varsigma)$$
$$\dot{\varsigma} = \Gamma^{-1}[1/j(j\varsigma_i^1),_{X^1} - ad_{\xi^\alpha}^*(\varsigma_i^\alpha) + \varsigma_e + ad_\varsigma^*(\Gamma\varsigma)]$$

(12)

In these forward dynamics, the state vector x is infinite dimensional since all its components (along with those of f) are some functions of the profile abscissa X. As a result, the above state equation has to be first space-discretised on a grid of nodes along $[0, L]$ before to be time integrated using explicit or implicit time integrators starting from the initial state $x(0)$. In this grid, all the space derivatives appearing in the f vector can be approximated by finite difference schemes, while the internal stress forces and couples are given by the constitutive law (9). As regards the boundary conditions in $X = L$, they are simply accounted for by directly taking (11) as the external loads applied on the corresponding node. Finally, removing f_a, f_b, l, f_{a+}, f_{b+} and l_+ from this state equations gives the model of the squid in vacuum, while the hydrodynamic forces will enter through these terms that we will seek as some functions of the current state of the body mantle. Going further into the details, f_a, f_b, l will model the effect of pressure forces applied inside and outside the mantle on any axisymmetric circular strip while f_{a+}, f_{b+} and l_+ can model eventual local effects as those produced by the vorticity shed at the edge of the mantle. Actually, the system (12) have been solved in matlab©through an *ad hoc* explicit second order finite difference scheme. In Figure 5 few snapshot of the result simulation given by the inputs of fig. 4 are shown for a conic shell of $E = 110Pa$, $\nu = 0.5$, $L = 80cm$ and $h = 2mm$.

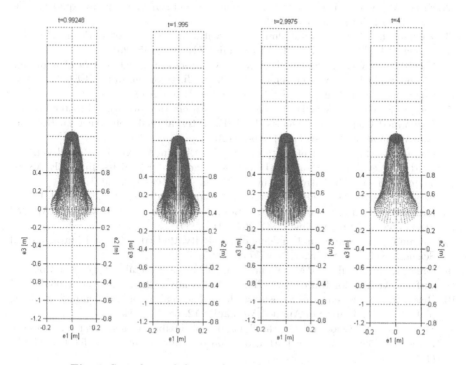

Fig. 5. Snapshots of the result simulation at $t = 1, 2, 3, 4secs$

4   Conclusion

In this work, we did a first step toward the analytical modelling of the jet-propelled locomotion inspired from squids and octopuses. To that end, we used the Reissner model of shells that we adapted to the axisymmetric shape of cephalopods mantle. Based on the Cosserat approach of continuous mechanics the solution allows to easily tackle the rigid overall motions which are crucial in locomotion while maintaining the exact modelling of finite deformations. At the end of the developments, we stated a closed formulation to address the mantle dynamics and we performed a numerical simulation by using the constitutive equations as a kind of linear control law.

On the long term, coupling the body and fluid models will help to understand the strategies that cephalopods discovered along their long history to improve their locomotion and in particular, those related to the exploitation of body softness, energy restoring and so on.

References

1. Gosline, J.M., DeMont, M.E.: Jet-propelled swimming in squids. Sci. Am. 252(1), 96–103 (1985)
2. Krieg, M., Mohseni, K.: New perspectives on collagen fibers in the squid mantle. J Morphol 273(6), 586–595 (2012)
3. Weymouth, G.D., Triantafyllou, M.S.: Ultra-fast escape of a deformable jet-propelled body. Journal of Fluid Mechanics 721, 367–385 (2013)
4. Giorelli, M., Giorgio Serchi, F., Laschi, C.: Forward speed control of a pulsed-jet soft-bodied underwater vehicle. In: Proceedings of the MTS/IEEE OCEANS Conference, San Diego, CA, USA, September 21-27 (2013)
5. Giorgio Serchi, F., Arienti, A., Laschi, C.: An elastic pulsed-jet thruster for Soft Unmanned Underwater Vehicles. In: IEEE International Conference on Robotics and Automation, Karlsruhe, May 6-10 (2013)
6. Giorgio Serchi, F., Arienti, A., Laschi, C.: Biomimetic Vortex Propulsion: Toward the New Paradigm of Soft Unmanned Underwater Vehicles. In: IEEE/ASME Transactions on Mechatronics, vol. 18(2), pp. 484–493 (April 2013)
7. Simo, J.C., Fox, D.D.: On stress resultant geometrically exact shell model. Part I: formulation and optimal parametrization. Journal Computer Methods in Applied Mechanics and Engineering 72(3), 267–304 (1989)
8. Antman, S.S.: Nonlinear Problems of Elasticity, 2nd edn. Applied Mathematical Sciences, vol. 107. Springer, New York (2005)
9. Green, A.E., Naghdi, P.M., Wainwright, W.L.: A general theory of a Cosserat surface. Archive for Rational Mechanics and Analysis 20(4), 287–308 (1965)
10. Reissner, E.: The effect of transverse shear deformation on the bending of elastic plates. ASME Journal of Applied Mechanics 12, 68–77 (1945)
11. Simo, J.C., Vu-Quoc, L.: On the dynamics of flexible beams under large overall motions - The plane case: Part I. Journal of Applied Mechanics 53(4), 849–854 (1986)

Hippocampal Based Model Reveals the Distinct Roles of Dentate Gyrus and CA3 during Robotic Spatial Navigation

Diogo Santos Pata[1], Alex Escuredo[1],
Stéphane Lallée[1], and Paul F.M.J. Verschure[1,2]

[1] Universitat Pompeu Fabra (UPF), Synthetic, Perceptive, Emotive and Cognitive
Systems group (SPECS)
Roc Boronat 138, 08018 Barcelona, Spain
http://specs.upf.edu
[2] Institució Catalana de Recerca i Estudis Avançats (ICREA)
Passeig Llus Companys 23, 08010 Barcelona, Spain
{diogo.pata,alex.escuredo,stephane.lallee,paul.verschure}@upf.edu
http://www.icrea.cat

Abstract. Animals are exemplary explorers and achieve great navigational performances in dynamic environments. Their robotic counterparts still have difficulties in self-localization and environment mapping
tasks. Place cells, a type of cell firing at specific positions in the environment, are found in multiple areas of the hippocampal formation.
Although, the functional role of these areas with a similar type of cell
behavior is still not clearly distinguished. Biomimetic models of navigation have been tested in the context of computer simulations or small
and controlled arenas. In this paper, we present a computational model
of the hippocampal formation for robotic spatial representation within
large environments. Necessary components for the formation of a cognitive map [1], such as grid and place cells, were obtained through attractor dynamics. Prediction of future hippocampal inputs was performed
through self-organization. Obtained data suggests that the integration of
the described components is sufficient for robotic space representation. In
addition, our results suggest that dentate gyrus (DG), the hippocampal
input area, integrates signals from different dorsal-ventral scales of grid
cells and that spatial and sensory input are not necessarily associated in
this region. Moreover, we present a mechanism for prediction of future
hippocampal events based on associative learning.

Keywords: spatial navigation, cognitive map, hippocampus, place cells,
CA3, dentate gyrus, robotics.

1 Introduction

Animals have the talent to alter their navigational method and plan optimized
novel trajectories after environmental changes. It has been argued that animals
form an internal cognitive map [1], which may potentiate rapid adaptation to

A. Duff et al. (eds.): Living Machines 2014, LNAI 8608, pp. 273–283, 2014.

their surroundings. The hippocampal formation in the medial temporal lobe of the mammalian brain has been associated with learning, memory and spatial navigation. Since the discovery of place cells [2], a type of cell showing firing activity at unique positions (place fields) of the explored environment, these type of cell have been considered to be the building blocks of the animals cognitive map. Place cells are found in the dentate gyrus (DG), CA3 and CA1 regions of the hippocampus. The formation of place cells is thought to be the result of summation and association of earlier inputs. Medial (MEC) and lateral (LEC) entorhinal cortex (EC) areas, one synapse upstream from hippocampus, feed the hippocampal formation with spatial (motion) and processed sensory information, respectively. Grid cells, a type of cell found in MEC layer II and III, spike at multiple positions of the explored environment and present a specific triangular tessellation pattern when plotted over space. Grid cells are considered the metric system needed for building a spatial representation of space. As described in [3], grid cells scale up progressively from the dorsal to ventral part of MEC. That is, the spacing between cells firing fields increase along the dorsal-ventral axis. In combination, MEC and LEC cells provide the hippocampus with specific information about the animal position. The DG is the input region of the hippocampus, receiving excitatory signals from EC layer II through the so-called perforant pathway. This pathway is divided into medial and lateral paths, each of them carrying signals from MEC and LEC respectively. The medial perforant path projects onto the middle one-third of the DG molecular layer, while the lateral perforant path terminates in the outer one-third of the DG molecular layer [4]. Consequently, the DG projects signals to the CA3 area. The CA3 region receives input from granule cells in the DG and from EC layer II cells. The extensive collateral ramification of CA3 pyramidal cells [5], make it a suitable network for associative memory and learning processes. It has been shown that both DG and CA3 areas contain cells which sensitivity is specific to unique positions of the environment. However, the roles for DG and CA3 areas in spatial navigation are still poorly distinguished. In this paper, we hypothesize two different roles for DG and CA3 during navigational tasks. On one hand, DG integrates multiple information arriving from different levels of the dorsal-ventral axis of MEC, but does not necessarily mix outer and middle one-third layer signals. However, DG is still capable of presenting position specific activity. On the other hand, CA3 region acts as an associative network, integrating motion and sensory signals, and is capable of performing spatial and sensory predictions. With that in mind, we have developed a computational model of the EC-DG and EC-CA3 connections and tested it on a mobile robot.

2   Methods

We construct a set of neural populations resembling the activity of each of the components of our system, namely: MEC, DG and CA3. Grid cells (MEC) were built through attractor mechanisms on a toroidal network as described in [6]. Specific evidence of continuous attractor dynamics in grid cells has been shown physiologically [7] which make attractor models a plausible mechanism for generating the behavior of such cell. Four sub-populations of grid cells with different gains (0.09; 0.07; 0.05 and 0.03), thus progressively bigger grid scales, were categorized as: Dorsal, Medial-1, Medial-2 and Ventral, respectively (see figure 2 and 4). As in [6], synaptic weights were updated accordingly with the robot movement, bumping the activity in a direction based on the robot's movement direction DG population was defined as a matrix of N neurons with equal size as MEC networks. The neurons of the network are initialized by

$$A_j = \frac{\sum_{c=1}^{M} A_c}{M} \tag{1}$$

where the activity of each neuron j is determined by the average sum of activity in the correspondent network position for every grid cell network, being M the amount of grid cell networks projecting to DG (see figure 1). A low-dimensional continuous attractor mechanism was implemented in the DG network, and it is initialized after MEC signals arrive to DG (figure 1). Given that DG network has an all-to-all type of connectivity, the activity of each cell at time t + 1, i.e. $A_i(t+1)$, in between MEC arrivals, is defined by

$$A_i(t = 1) = \sum_{i=1}^{N} A_j(t) w_{ij} \tag{2}$$

where N is the number of neurons in the network and w_{ij} is the synaptic weight connecting cell j to cell i, with $i, j \in \{ 1, 2, ..., N\}$. The synaptic strength is defined by the Gaussian weight function

$$W_{ij} = I_{exp} \left(\frac{||c_i - c_j||^2}{\sigma^2} \right) - T \tag{3}$$

where $c_i = (c_{i_x}, c_{i_y})$ is the Cartesian position of the cell i in the network (with $i_x \in \{ 1, 2, ..., N_x \}$ and $i_y \in \{ 1, 2, ..., N_y\}$), and where N_x and N_y are the number of columns and rows in the network. As in [6], the strength of the synapses is defined by the parameter I, whereas σ is the size of the Gaussian and T is the shift parameter defining excitatory and inhibitory zones. Contrary to [6], the norm $||.||^2$ is the Euclidean distance between cell i and cell j in non-periodic boundary conditions (see table 1).

Because MEC and DG show theta- and gamma-frequencies [8] [9], we scaled the update time for each population accordingly with their oscillatory frequency, 4-12Hz and 30-90Hz, respectively (see figure 1 bottom-right). That is, the DG attractor cycle was set to update fourteen times faster than MEC signal arrivals. This time delay is sufficient for the activity in DG to be attracted towards a stable point. Every time MEC networks project onto DG, the later is reset and updated conformably equation 1. Attractor models have been used to model place cells in the DG (see [10], for example). Their type of attraction is based on the animal movement, such as the previously described mechanism of attraction

Table 1. Values of the parameters used in DG model

Parameter	Value	Unit
N	100	[cell]
N_x	10	[cell]
N_y	10	[cell]
I	0.3	[no unit]
σ	0.36	[meter]
T	0.05	[no unit]

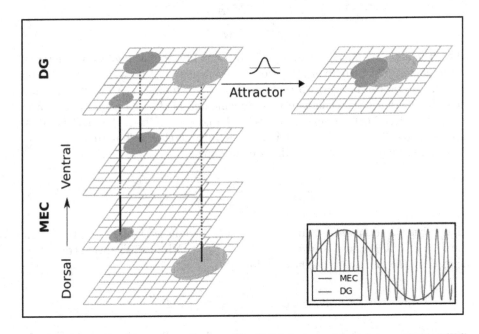

Fig. 1. Place cells attractor mechanism. The DG integrates input from multiple MEC grid cell networks at different scales of the dorsal-ventral axis. The attractor bumps network activity to a single point. Bottom-right: Every time the oscillation reaches its maximum a time step is performed in its network. The higher frequency in the DG relative to MEC ensures that the DG attractor updates in between MEC arrivals.

for grid cell generation in MEC layer II (see [6]). However, attractor models do not describe the characteristics of DG cells. Contrary to those models, place cells do not present a topographical organization. Also, such models do not take into account MEC inputs. Instead, they reproduce a grid cell activity with a grid scale sufficiently large to fire at a single position of the environment. Thus, the gain factor, a parameter modulating the spacing between firing fields, needs to be defined in accordance with the size of the arena. To validate the hypothesis that CA3 accounts for a hippocampal input prediction mechanism, we used the Convergence Divergence Zone Framework as described in [11] to learn associations of the input space. This framework, based on the self-organizing map algorithm, is trained through unsupervised learning, generating low-dimensional representations of multiple inputs. Since CA3 region receives input from both lateral and medial areas of the entorhinal cortex, and is thought to be a highly associative type of network, we projected grid cell network and visual (sensory) signals onto our CA3 network. Thus, error prediction can be quantified by the difference between expected and real inputs.

In order to validate the hypothesis that place cells in the DG emerge through the integration of multiple MEC inputs, and that associative networks in the CA3 area account for hippocampal input prediction, we implemented the described system on a mobile robot setup (Figure 2) and tested it in a 3x4 meters open arena, during 45 minutes. The robot was programmed to randomly explore the environment, using its proximity sensors to avoid collision with walls. Visual input was obtained through a fixed wireless camera placed in the front of the robot. Given that sensory information is highly processed when it arrives to LEC areas, we simulated such processing by applying an edge-detection algorithm to the visual input before it is projected to CA3. LEC network activity was simulated from a one-to-one mapping of the pixel activity in the filtered visual input matrix in normalized gray-scale with dimensions of 100x100 pixels. The AnTs video tracking system [12] tracked the robots position during the whole session. The robots orientation was obtained from the tracking system and used to feed grid cell networks. Two 32bits computers (Linux Ubuntu 13.10 and Windows XP) were connected to the system network and used to run the system components. The system was developed using C++ and Python programming languages, and system connections were established through the YARP middleware [13]. During robot spatial exploration, grid cell networks (MEC) projected their activity to both DG and CA3 populations, while filtered visual inputs only projected to CA3.

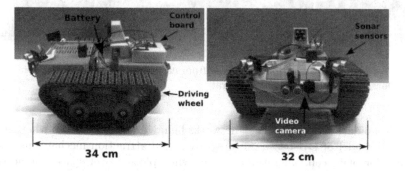

Fig. 2. Robotic platform. Left: side view; right: front view. Two bidirectional continuous track wheels sustain the platform. Three sonar sensors placed in the front of the robot allow it to perform collision-avoidance. A wireless video camera records the robot visual field. A control board orchestrates sensory and motor signals.

3   Results

Grid cell activity from the multiple dorsal-ventral networks reveal the triangular tessellation pattern of their firing fields when plotted over the explored environment (figure 3 and 4). Grid scale and firing field size was in accordance with electrophysiological studies [3]. Grid cells belonging to different dorsal-ventral MEC axis presented a progressive increase in both grid scale and firing fields size (figure 4). These results show that the mechanism of grid cell generation based on attractor dynamics is sufficiently robust to generate the triangular tessellation pattern within large-scale environments.

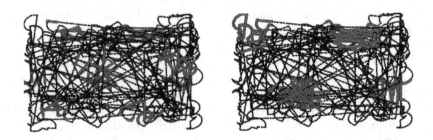

Fig. 3. Triangular tessellation pattern of two grid cells. Black dots represent the robots trajectory. Red dots represent positions where the cell activity was higher than 75 percent of its maximum activity. Left: Medial-1 grid cell. Right: Medial-2 grid cell.

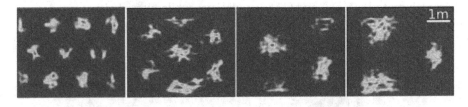

Fig. 4. Autocorrelogram of four grid cells from each dorsal-ventral level. From left to right: Dorsal, Medial-1, Medial-2, Ventral.

After validation of the grid cell networks behavior, we questioned whether the proposed model of DG based on integration and attraction mechanisms would account for place cell generation. Indeed, when plotted over space, the activity of DG network cells reveal sensitivity to specific positions of the environment (figure 5). Contrary to grid cells, no scale or firing field size pattern was observed. We must notice that no homogeneous shape of the firing field was observed. Indeed, different cells can have completely different firing field characteristics. For instance, cell 87 in figure 5 has a much broader place field when compared with cell 15 of the same figure. Another interesting feature of our models DG place cells is that despite being sensitive to certain positions of the explored environment, the number of positions to which they are sensitive to, vary from cell to cell. In order to quantify the amount of place fields to which each cell reacts, an iterative hierarchical clustering algorithm analyzed all data points, in x-y coordinates, that presented activity superior to 75 percent of the cells higher activity. The clustering threshold rule was determined to the maximum of 30 percent (150cm) of the environment maximum length (500cm). Clustering results (figure 7) indicate that 52 percent of the cells in this network are sensitive to one single position of the environment, while 28 percent encode sensitivity in two regions of the environment. Silent cells, a type of cell that do not spike significantly at any position of the environment, constitute 12 percent of the neural population. A few amount of cells, 8 percent, present sensitivity at 3 or 4 different regions.

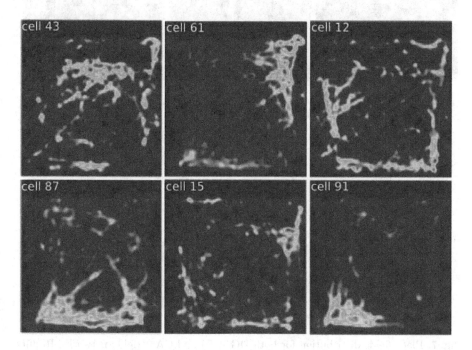

Fig. 5. Dentate gyrus place cells activity autocorrelogram.Place fields of six cells.

To further validate the role of CA3 during navigational tasks, we tested whether our CA3 model accounts for spatial encoding as it happens in the same region of the mammalian brain. As for DG, CA3 cells are also revealed to be spatially sensitive. Cells become active accordingly to the robots position (figure 6). Indeed, when comparing with the distribution of DG cells per amount of place fields, CA3 cells shown to have more confined place fields. Results from the hierarchical clustering analysis shows that 71 percent of the cells are active

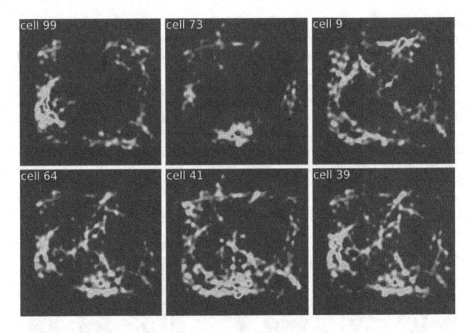

Fig. 6. CA3 place cells activity autocorrelogram. Place fields of six cells.

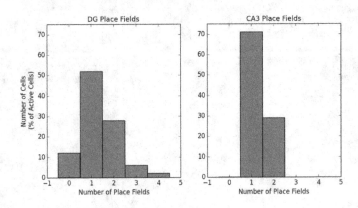

Fig. 7. Place fields distribution for both DG (left) and CA3 (right) networks. Results from hierarchical clustering.

at unique positions of the environment, while 29 percent are sensitive to two different positions (e.g. cell 39 in figure 6). Such results suggest that learned associations between MEC and LEC inputs are sufficient and accurate enough for place sensitivity.

Despite its role in place sensitivity and memory formation, we hypothesized that CA3 would account as a mechanism for hippocampal input prediction. To test this hypothesis, we projected MEC and LEC signals to the CA3 region during robot exploration, aiming that the Convergence Divergence Zone Framework would learn associations from the input state and predict future inputs. As suggested by [14] for the DG, we used a substantially larger amount of MEC inputs into the CA3 than those of LEC. Specifically, 70 percent of signals associated in CA3 were provided by grid cell networks, while only 30 percent were projected from LEC. Quantification of the prediction error was performed by the difference, in percentage, between the real and the expected input for each modality (figure 7). Naturally, the error tends to decrease for every modality as the robot explores the arena. As we would expect, prediction error progressively decreases from dorsal to ventral grid cells input. The reason might be that more dorsal networks bump the activity around the cyclic network faster when compared with more ventral networks. With slower bumping movements of the activity, ventral networks tend to be more stable. Interestingly, the prediction error for the visual input stabilized faster than any other modality, but never reached the minimum of Medial-2 or Ventral errors. The fact the visual inputs are relatively stable and easily predicted when the robot moves straight but not when it turns, might explain why it stabilizes faster but does not improve after stabilization.

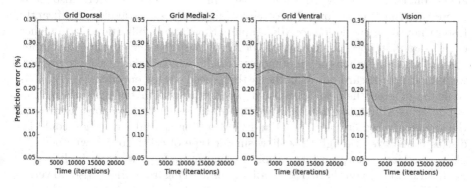

Fig. 8. Prediction error for each modality

4   Discussion

In this article, we have presented an integrated hippocampal model that accounts for spatial representation and input prediction, embedded on a real mobile agent. We have shown that a previously described model of grid cells based on attractor dynamics is sufficiently robust for generating the same cell behavior in a living machine navigating in a large environment. As previously shown in rats, the grid scale increased progressively from the dorsal to ventral areas of the MEC. We have tested the hypothesis that place cells in the DG are a result from the integration of multiple grid cell networks with different scales. To do that, we have presented a model of the DG place cells based on both integration of earlier projections and attractor mechanisms, in which the gating of inputs and network dynamics work at different frequencies. A network of non-periodic boundaries conditions is reset with sparse activity every time that MEC networks project their inputs. Because the DG recurrently processes internal signals at a higher frequency than the frequency at which external signals arrive, the mechanisms of attraction are fast enough to stabilize activity at any given point in the network. We have shown that this model is capable of tuning the receptive field of each cell to specific positions of the environment. Specifically, 52 out of 100 cells were shown to be active at unique positions of the environment, while 28 were sensitive for two firing fields, and only 12 were considered to be silent cells. We have also hypothesized that even though both DG and CA3 hippocampal areas contain apparently similar types of cells (place cells), the role of CA3 during navigational tasks would not only be relevant for associative learning, but may also be crucial for prediction of future inputs. With that in mind, we tested a previously described associative learning framework within the context of CA3 input associations. The same grid cell networks used for the DG model were used in combination with processed visual input to feed the CA3 network. Results suggest that input error prediction tends to be smaller for grid cell inputs at the ventral areas when compared with dorsal areas. The fact that activity movements within the network are faster for dorsal than for ventral areas could explain why inputs are easier predicted for the later. Also, it is interesting that prediction error of visual input tends to stabilize significantly faster than those predictions of MEC inputs. However, stabilization is at a higher error value than posterior predictions of MEC. Despite the fact that visual input changes smoothly when the robot moves straight, which make it easier to predict, we should not discard the hypothesis that speed vector encoded by grid cells could be of great help for the visual prediction, but that visual information is poorer as a hint of grid cells space state. In conclusion, we have addressed the question whether roles of DG and CA3 during spatial navigation could be distinguished even though they show similar type of cells. To test whether DG place cells are part of an integration mechanism or not, we have built a new model of DG based on the attraction of signals arriving from MEC. Also, to verify that MEC and LEC associations not only show place sensitivity, but could also account for a prediction mechanism, we have used a previously described model of associative learning using inputs resembling those of hippocampal formation. Both models were tested on a mobile robot freely navigating in an open arena. Results suggest that integration of MEC signals in the DG are sufficient to generate place cells like type of activity but with lower precision than CA3 place cells. Also, when fed with hippocampal type inputs a map of self-organization was able to predict future states. To our knowledge, this was the first time that the question of role distinction in hippocampal areas with similar cell behavior has been addressed and explained with concrete neural mechanisms. Also, a novel attractor model for place cells in the DG built through MEC inputs and non-predefined gain values was presented. In future work, we plan to combine both DG and CA3 models as a mechanism for path integration correction. Closing the entorhinal-hippocampal loop has great advantages for studying phenomena such as global- and rate-remapping in both grid and place cells.

References

1. Tolman, E.C.: Cognitive maps in animals and man. Psychologival Review 55, 189–208 (1948)
2. O'Keefe, J., Dostrovsky, J.: The hippocampus as a spatial map. preliminary evidence from unit activity in the freely-moving rat. Brain Research 34(1), 171–175 (1971)
3. Hafting, T., Fyhn, M., Molden, S., Moser, M.B., Moser, E.I.: Microstructure of a spatial map in the entorhinal cortex. Nature, 801–806 (2005)
4. Hjorth-Simonsen, A., Jeune, B.: Origin and termination of the hippocampal perforant path in the rat studied by silver impregnation. J. Comp. Neurol, 215–231 (1972)
5. Engel, J.: Epilepsy: A Comprehensive Textbook in Three Volumes. Lippincott Williams & Wilkins, Philadelphia (2008)
6. Guanella, A., Verschure, P.F.J.: A model of grid cells based on a path integration mechanism. In: Kollias, S.D., Stafylopatis, A., Duch, W., Oja, E. (eds.) ICANN 2006. LNCS, vol. 4131, pp. 740–749. Springer, Heidelberg (2006)
7. Yoon, K., Buice, M.A., Barry, C., Hayman, R., Burgess, N., Fiete, I.R.: Specific evidence of low-dimensional continuous attractor dynamics in grid cells. Nat Neurosci (2013)
8. Alonso, A., Garcia-Austt, E.: Neuronal sources of theta rhythm in the entorhinal cortex of the rat. Exp Brain Res (1987)
9. Akam, T., Oren, I., Mantoan, L., Ferenczi, E., Kullmann, D.: Oscillatory dynamics in the hippocampus support dentate gyrus-ca3 coupling. Nature Neuroscience (2012)
10. Samsonovich, A., McNaughton, B.L.: Attractor map model of the hippocampus. J. Neurosci. 17(15) (1997)
11. Lallee, S., Dominey, P.F.: Multi-modal convergence maps: From body schema and self-representation to mental imagery. Adaptive Behavior (2013)
12. Bermudez, S.: http://sergibermudez.blogspot.com.es/p/ants-downloads.html
13. Metta, G., Fitzpatrick, L., Natale, P., Natale, L.: Yarp: yet another robot platform. Journal on Advanced Robotics Systems (Special Issue on Software Development and Integration in Robotics) (2006)
14. Rennó-Costa, C., Lisman, J.E., Verschure, P.F.M.J.: The mechanism of rate remapping in the dentate gyrus. Neuron 68(6), 1051–1058 (2010)

Trajectory Control Strategy
for Anthropomorphic Robotic Finger

Shouhei Shirafuji, Shuhei Ikemoto, and Koh Hosoda

Department of Multimedia Engineering, Graduate School of Information Science
and Technology, Osaka University, 2-1, Yamadaoka, Suita, Osaka, 565-0871, Japan
{shirafuji.shouhei,ikemoto,koh.hosoda}@ist.osaka-u.ac.jp

Abstract. This paper proposes a trajectory control strategy for a
tendon-driven robotic finger based on the musculoskeletal system of the
human finger. First, we analyzed the relationship between the stereotypical trajectory of the human finger and joint torques generated by the
muscles, and hypothesized that the motion of the human finger can be
divided into two categories: one following a predetermined trajectory and
the other changing the trajectory, which is mainly caused by the action of
intrinsic muscles. We applied this control method to an anthropomorphic
tendon-driven robotic finger and observed the change in motion caused
by adjustments in the actuator's pattern, which corresponds to human
intrinsic muscles.

1 Introduction

Many humanoid robotic hands and fingers have been developed over the decades,
most of which use the tendon-driven mechanism to drive their joints. This mechanism makes it possible to place actuators in the distance, instead of on the
corresponding joint, and reduce the size and weight of the end-effector. This
mechanism has been widely studied, and tendon-driven manipulators can be
controlled in the same way as standard manipulators, whose motors are placed
on the corresponding joint to drive them [1,2]. However, tendon-driven manipulators can be controlled on the basis of their characteristic that the motion of an
actuator affects several joints through the connected tendon. For example, Zollo
et al. [3] developed a robotic hand where three joints of each finger with springs
were driven by a cable connected to an actuator. This resulted in human-like
motion using only one motor. As this example shows, the tendon-driven mechanism is useful for creating a lightweight robotic hand and efficiently controlling
several joints.

The musculoskeletal system of the human hand provides a good example for
the design of a tendon-driven robotic hand and a control method using the
coordinated motion of the joints caused by tendons, because the human hand,
especially the human finger, can be regarded as a tendon-driven manipulator. In
this context, we have developed a tendon-driven robotic finger based on the bio-
mechanical musculoskeletal model of the human finger [4].The control method
that utilizes some advantages of the human musculoskeletal structure will have

A. Duff et al. (eds.): Living Machines 2014, LNAI 8608, pp. 284–295, 2014.

an efficient way to control the robotic finger, which will provide good insight into the control system of the human hand. Therefore, this paper analyzes the relation between the stereotypical trajectory of the human finger during reaching motion reported by Kamper et al. [5] and joint torques exerted by the human finger muscles. We propose a method to control the developed robotic finger based on this analysis.

In Section 2, we present the human finger structure and the developed robotic finger based on its model. Section 3 analyzes the properties of human muscles during typical human motion and proposes a strategy to control the trajectory of the anthropomorphic robotic finger. We control the motion of the developed robotic finger with the proposed method in section 4 and conclude the study in section 5.

2 Anthropomorphic Robotic Finger

We have developed an anthropomorphic robotic finger based on the bio-mechanical model of a human finger proposed by Leijnse and Kalker [6].In this section, we briefly introduce the bio-mechanical model of the human finger. Then, we describe our developed robotic finger and the concept of virtual degrees of freedom (DoF), which we previously proposed to apply the common analysis method for tendon-driven manipulators to the developed robotic finger [4].

Fig. 1 shows the planar model of the human middle finger [6]; five muscles: the flexor digitorum profundus (FDP), flexor digitorum superficialis (FDS), extensor digitorum communis (EC), interosseous (IN), and lumbrical (LU) muscle, and three joints: the metacarpophalangeal (MP), proximal interphalangeal (PIP), and distal interphalangeal (DIP) joints. The FDP, FDS, and EC muscles are extrinsic muscles, and the IN and LU muscles are intrinsic muscles. The LU muscle is inserted parallel to the tendon of the FDP muscle in the model. The tendon for each of the EC, IN, and LU muscles branches after passing the MP joint, and after branching, each tendon is called the lateral and medial bands. The EC, IN, and LU muscles insert on the common lateral and medial band. The moment arm between each tendon and each joint is basically assumed as a constant, but the moment arm between the lateral band and PIP joint changes according to the angle of the PIP joint.

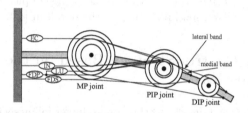

Fig. 1. Bio-mechanical planar model of human finger proposed by Leijnse and Kalker [6]

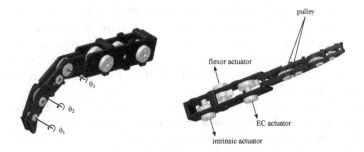

Fig. 2. Cad model of the robotic finger we developed in our previous work [4]

In our previous study, we proposed the design of a tendon-driven robotic finger with features of the human finger model, and developed the prototype based on this design with three actuators replicating flexor muscles (FDP and FDS muscles), EC muscle, and intrinsic muscles (IN and LU muscles), as shown in Fig. 2, to show that we can control the robotic finger with at least three actuators [4].

n a common tendon-driven manipulator, the kinematics are usually described using a Jacobian matrix, called a tendon Jacobian (or coupling) matrix $\boldsymbol{J}$ [1, 7], as follows:

$$\dot{l} = \boldsymbol{J}\dot{\boldsymbol{\theta}}, \tag{1}$$

where l is the vector of tendon displacement, and $\boldsymbol{\theta}$ is the vector of the joint angle. The relation between each joint torque and tensile force of each tendon can also be represented using the tendon Jacobian matrix, and the fact that at least $N + 1$ tendons are needed to exert any set of joint torques in the N-DoF tendon-driven manipulator is derived from this relation. A robotic finger with 3-DoF can be controlled by three actuators because of the constraint caused by the branching tendons. This constraint generates the coupled motion in the PIP and DIP joints, and these joints can be virtually regarded as one joint. Leijnse and Kalker used this fact to analyze the function of muscles in the human finger [6]. We generalized the kinematics of the coupled motion caused by branching tendons as the relation between the tensile force of each tendon and each virtual joint angle using a Jacobian virtual tendon in the same manner as equation (1), and derived the equation to keep the coupled motion to control this type of tendon-driven manipulator (see the detail in [4]).

3   Trajectory Control Strategy Based on the Structure of Human Muscles

It is well known that the fingertip of the human finger follows a certain trajectory with respect to the base frame of the finger during a reaching task, and the trajectory can be described by a logarithmic spiral [5]. Given the trajectory

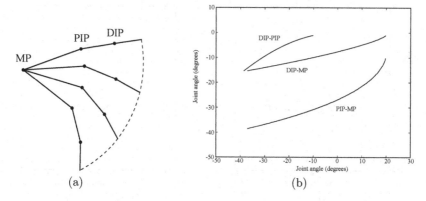

Fig. 3. (a)Typical trajectory of the human fingertip (dotted line) and the postures of our robotic finger when it follows this trajectory. (b)Trajectory of three pairs of joints of the robotic finger: MP-PIP, MP-DIP and PIP-DIP.

of the fingertip, the posture trajectory of the robotic finger is also specified because the DoF of the robotic finger is virtually two when the end motion of two joints is coupled by the branching tendon, as described in the previous section. Fig. 3(a) shows the typical human fingertip trajectory and the transition of the robotic finger posture when its end-point follows this trajectory. Fig. 3(b) shows the relations between each joint of the robotic finger during this motion. It can be seen that the relation is nonlinear in the early part of the motion, but becomes linear in the latter part. This linear relation has been reported in several neuroscientific papers [8–10].

The relation among three joints, shown in Fig. 3(b), can be transformed to that between two variables as virtual joint angles using the method introduced in the previous section and in our previous paper [4]. Therefore, the change in these variables can be represented in a plane, and the dotted line in Fig. 4(a) represents the trajectory of these variables during a typical motion. This representation is useful to see the relation between the motion and joint torques exerted by each human muscle (or each actuator of the manipulator). The vectors, illustrated as arrows in Fig. 4(a), represent the joint torques exerted by each muscle at a point on the trajectory of the motion when each muscle exerted a unit force (1 N), where we refer to the values of each moment arm between each joint and tendon in the bio-mechanical literature [6, 11] as our previous work [4]. These torque vectors are the same as those used by Leijnse [12]to analyze the redundancy of the muscles and function of the LU muscle. It can be seen that the trajectory with the motion slopes upward from left to right, and the torque vector of the IN muscle points to the upper left. From the relation between velocity and force, in case the vector of instantaneous velocity at a point on the trajectory and the torque vector are orthogonal to each other, the corresponding muscle does no work to the instantaneous motion. This fact indicates that this muscle does not move with the motion as long as these vectors are orthogonal.

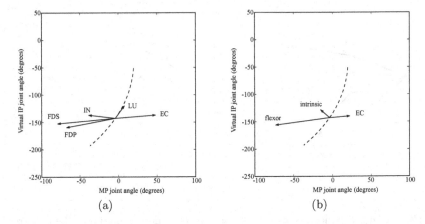

Fig. 4. (a) Trajectory of the virtual joint angles and joint torque vectors of muscles. (b) Trajectory of the virtual joint angles and joint torque vectors of the developed robotic finger where two pairs of muscles are replaced as two actuators by summing their moment arms.

Therefore, the movements of the IN muscle are smaller than those of other muscles with typical finger motion. In addition, we can generate torque that is completely orthogonal to the velocity at any point in the trajectory using the linear combination of the IN and LU muscles, and at least one of the IN and LU muscle is needed to generate a complete orthogonal torque vector. For example, these two muscles are designed as actuators connected to a tendon whose moment arms are the sum of the moment arms in the developed robotic finger as the "intrinsic" actuator, and the torque vector of this actuator can be seen in Fig. 4(b) (additionally, the "flexor" actuator is the superposition of the FDP and FDS muscles). It can be seen that this vector is approximately perpendicular to the latter part of the trajectory, which we mentioned above as a linear relation. Therefore, this actuator moves little during a typical motion, especially in the latter part. Conversely, the fact that some muscles do no work to a certain trajectory indicates that these muscles can change this trajectory efficiently. Note that there is no meaning in other relations between the direction of the torque vector and velocity vector, except their orthogonality (e.g., parallelism), because these relations change depending on the definition of the virtual angles.

Notable property of human muscles that can be inferred from Figs. 4(a) and (b) is not only that the work done by some muscles is quite small during stereotypical motions, but also that the directions of the torque vector for some muscles are symmetrical to each other. For instance, the direction of the torque vector of the FDS muscle is opposite to that of the EC muscle, as shown in Fig. 4(a). The pair of the FDP and EC muscles has almost the same relation in their torque vector direction, but the direction of the FDP is slightly shifted from that of the FDS. This implies that the rows of Jacobian virtual tendons related to these muscles are close to being linearly dependent. In controlling the trajectory of the finger, this has an important meaning: the displacement of these muscles is mostly not affected by changes in the IN and LU muscles when the intrinsic muscles move to keep the relationship in which the two muscles can be seen as one muscle. It is because of this that the posture of a tendon-driven manipulator with 2-DoF is uniquely determined if it has at least two tendons, and the motions of these tendons are independent of each other only in this case. The assumption that the IN and LU muscles are regarded as one muscle may be seen as a strong assumption, but it is very beneficial for controlling the developed robotic finger and design of its actuation system. In addition, original human finger has 4-DoF in three-dimensional space including abduction and adduction of the MP joint. It means that the motions of IN, LU and other muscles are independent of each other in this case without the assumption described above.

If the motion of the intrinsic muscles (the IN and LU muscles) does not affect the motion of the extrinsic muscles (the FDP, FDS, and EC muscles), the finger's motion can be separated into two categories. One is a stereotypical open-close motion, mainly controlled by the extrinsic muscles, and the other motion, which is controlled by the intrinsic muscles, adjusts this motion. This property makes it possible to provide the robotic finger with appropriate actuators for each motion. For instance, high-powered actuators are appropriate for controlling the open-close motion because they are frequently used in various tasks, whereas the actuators for adjusting this motion should have higher reactivity and fine position resolution. Furthermore, control of the joint stiffness using the nonlinear elasticity of the actuators is easier with the property of the two motions described above. The co-activation of the muscles with nonlinear elasticity to control joint stiffness is well known as a remarkable human ability [13], and some tendon-driven or humanoid robots use the same method to control their joint stiffness [14, 15]. In this stiffness-control method, the desired displacement of each actuator is derived from the desired posture and desired stiffness of the robot, and typically, there is a need to recalculate them when the desired trajectory is changed. By contrast, if two motions are completely independent, i.e., if the intrinsic muscles can be regarded as one muscle and the torque vectors of each extrinsic muscle are rigidly parallel to each other, the recalculation for joint stiffness is not needed once the trajectory of displacement for each actuator is determined for the open-close motion. This is because the motion of the group, including the intrinsic muscles, does not affect the displacement of each muscle in another group, and the stiffness change is caused by the co-activation of extrinsic muscles. This latter fact is because the direction of the torque vector of the EC muscle is opposite to that of the FDP and FDS muscles. In fact, these directions are not strictly parallel, as mentioned above, and the trajectory is not significantly affected by the stiffness change. While it could be concluded that their directions should be strictly opposite, in the structure of human muscles, a slight deviation in their directions is needed to apply force to the joints in the opposite direction of the force exerted by the intrinsic muscles.

The fact that the motion of the intrinsic muscles has almost no effect on the extrinsic muscles can be confirmed by the following simple example. Consider

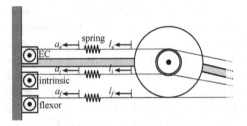

Fig. 5. Model of the robotic finger with three tendons representing the EC, flexor, and intrinsic muscle

the quasi-static motion of a robotic finger with three tendons representing the EC, flexor (the superposition of FDP and FDS), and intrinsic (the superposition of LU and IN) muscle, and each tendon connects to a motor through a spring, as shown in Fig. 5. For simplicity of calculation, we assume that the elastic property of the spring is linear. The vector of torque applied to each joint by these tendons corresponds to those illustrated in Fig. 4(b). Let a be the displacement of each motor; l be the displacement of each tendon; and the subscripts f, e, and i represent the corresponding muscle: the flexor, EC, and intrinsic, respectively. Each l is determined by the equilibrium between the elastic force exerted by the corresponding springs depending on a and the external force. When the motors are controlled to follow the stereotypical trajectory while keeping a constant tensile force of the flexor in the quasi-static condition without any external force, l changes, as shown by solid lines in Fig. 6, and we represent the trajectory of the motor corresponding to the intrinsic muscle in this motion as $a_i^*(t)$. We do not care about the velocity of the motors because that is not the point to be discussed here. When two motors for the flexor and EC are driven by the same amount, but the motor for the intrinsic muscle is driven by a constant amount greater than that mentioned above, l changes, which is shown by the dashed lines in Fig. 6. The changes in l when the motor for the intrinsic muscle is driven less are illustrated as dash-dotted lines in Fig. 6. It can be seen that the change in the tendon displacement corresponding to the intrinsic muscle is less than that for other tendons. This is because the direction of the torque vector of this motor is nearly vertical to the trajectory of the joint angles during stereotypical motion, as mentioned above. Furthermore, the trajectories of the tendon displacements corresponding to the flexor and EC muscles are almost unchanged when the driving amount of the motor corresponding to the intrinsic muscle is changed. This is because the directions of the torque vectors of the two motors are almost parallel, and the intrinsic muscles are replaced by a motor connected to a tendon. Fig. 7 shows the change in the trajectory of the robotic finger based on the motion of the motor corresponding to the intrinsic muscle.

On the basis of these discussions, we propose a control strategy for an anthropomorphic robotic finger that we normally control with predefined motor motions following the stereotypical fingertip trajectory and adjust it as needed

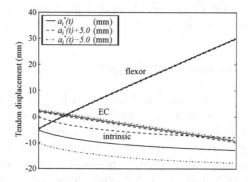

Fig. 6. Trajectory of each tendon displacement when the motors are driven to follow the stereotypical trajectory of human fingertip (solid), the motor corresponding to the intrinsic muscle is driven more (dotted), and it is driven less (dash-dotted)

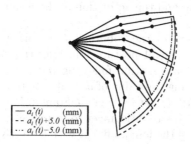

Fig. 7. Change in the trajectory of the robotic finger according with the motion of the motor corresponding to the intrinsic muscle

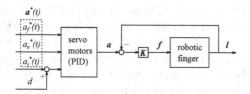

Fig. 8. Block diagram of the proposed strategy for control of trajectory

by changing the motion of the motor corresponding to the intrinsic muscle. It can be represented as a block diagram in Fig. 8, where $a^*(t)$ is the predetermined trajectory of motors, and a, f, l, and K are the vectors of consequential motor displacement, tensile force, tendon displacement, and the matrix composed of the spring constants, respectively. We control d to accomplish the desired motion. This is not only the control method of the robotic finger, but also the hypothesis of the human strategy used for neuro-muscular control of the hand.

4   Control of the Robotic Finger

In this section, we apply the proposed scheme of trajectory control to the developed robotic finger to validate it. A typical performance of the human finger in which the trajectory is actively controlled is typing. We adjusted the trajectory of the fingertip to contact the desired key of the keyboard while typing. This motion is a superior human ability representing the dexterity of the human hand, and has been studied in the field of neurophysiology [16,17]. Therefore, we controlled the robotic finger based on the electromyographic pattern measured by Kuo et al. [17]. As reported by them, the finger is first lifted up by the EC muscle during typing, and the downward motion begins with decreasing the activity of the EC muscle. The intrinsic muscles begin to contract after the start of the downward motion, and the activation of the flexor muscles follows after a short interval.

Based on this, we control the robotic finger with the following steps: (1) upward motion from a neutral posture following the stereotypical trajectory, (2) beginning the downward motion following the stereotypical trajectory, and (3) increasing or decreasing the value of d gradually to change the stereotypical trajectory. In this case, $a^*(t)$ in the block diagram in Fig. 8 is the predetermined motor trajectory driving the robotic finger to be raised and lowered following the stereotypical trajectory of the human finger with a positive tensile force, and d is a linearly increasing or decreasing function of time during the downward motion and zero during the upward motion.

The actual motion of the robotic finger when we control it using the proposed method for five patterns of change in d is shown in Fig. 9. In the experiments, the robotic finger was placed horizontally on the ground, for simplicity, to avoid the effect of gravity. In addition, we controlled it slowly enough so that its motion can be regarded as quasi-static. Figs. 9(a) and (b) show the common upward motion in the various motions according to the change in d. Fig. 9(e) shows the downward motion when d is always zero, and it follows the stereotypical finger trajectory. Figs. 9(c) and (d) show the downward motions when d increases linearly. The motion in which the MP joint flexes and the PIP and DIP joints extend can be seen, which is quite similar to the measured motion of a human finger during typing in [17]. On the other hand, Figs. 9(f) and (g) show the downward motion when d decreases linearly. The increasing rate of d in Fig. 9(c) is larger than (d), and the decreasing rate in Fig. 9(g) is larger than that in (f). We cannot discuss the similarity between this motion and the actual human motion for the case when d decreases gradually or its increasing or decreasing rate is different because Kuo et al. [17] did not deal with these situations. However, from this result, it was confirmed that the trajectory of a robotic finger can be controlled by adjusting the parameter d.

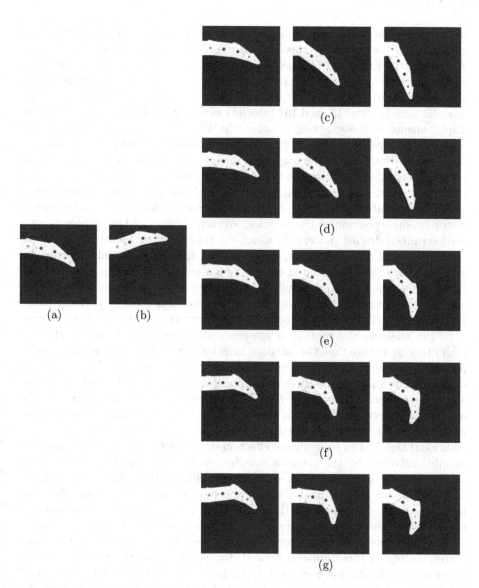

Fig. 9. Motions of the robotic finger when it is controlled by the proposed method in five patterns of the change of d. (a) Initial position of the robotic finger. (b) Posture of the robotic finger after the upward motion. (c)(d) Successive pictures during the downward motion when the value of d increases linearly. (e) Successive pictures during the downward motion when the value of d is zero. This is the same with the stereotypical trajectory of human fingertip. (f)(g) Successive pictures during the downward motion when the value of d decreases linearly.

5   Conclusion

In this paper, the relationship between the stereotypical trajectory of the human fingertip and the torque vectors of each muscle was analyzed. In the analysis, we showed the following: (1) the torque vector of the IN muscle is nearly perpendicular to the stereotypical trajectory, and it can exert in a direction exactly perpendicular to the trajectory by combining with the LU muscle; (2) if the intrinsic muscles (the IN and LU muscles) can be regarded as one muscle, the displacements of the extrinsic muscles (the FDP, FDS, and EC muscles) are not significantly influenced by the change in the displacements of these muscles because the directions of the torque vectors of the extrinsic muscles are nearly parallel to each other. Based on these facts, we proposed a strategy to control the trajectory of a tendon-driven robotic finger developed in our previous study, and in this strategy, the IN and LU muscles are replaced as a tendon-driven motor by superposition of the muscles. We usually control the motors using a predetermined driving pattern, in which the robotic finger follows a stereotypical trajectory of a human finger, and we change the consequential trajectory by adjusting the driving pattern of one of the motors corresponding to the human intrinsic muscles. Finally, we applied this strategy to the developed robotic finger and confirmed that the trajectory can be controlled according to the required tasks based on the observation of the resultant trajectory of the robotic finger.

In the analysis, the motion of muscles can be regarded as independent of each other assuming that the intrinsic muscles can be regarded as one muscle. As noted before, this assumption is unnecessary when we consider the motion of the human finger in the three-dimensional space because the DoF related to abduction-adduction motion of the MP joint exists in the human finger. The moment arm between the axis of the abduction-adduction motion and extrinsic muscles is quite small compared to the intrinsic muscles. It means that this additional DoF has an insignificant effect on the stereotypical open-close motion, mainly controlled by the extrinsic muscles. As a result, the proposed method can be directly applicable to the anthropomorphic robotic finger with 4-DoF. In addition, we incorporated the intrinsic muscles into our robotic finger as their simple superposition, but these muscles can exert torques with a direction exactly perpendicular to the stereotypical trajectory by adjusting the proportion of activity between them. Although more complex control between the actuators corresponding to these muscles is needed after designing them in the robotic finger separately, efficient control of the trajectory would be possible. This would help answer questions about neural control of the human hand. Finally, from the perspective of robotics, we can design an ideal robotic manipulator based on the discussion in this paper with two types of strictly independent motions, a primal motion and an additional motion adjusting it. This would be a useful concept for a dexterous robotic hand.

Acknowledgments. This work was supported by JSPS KAKENHI Grant Numbers 24-3541.

References

1. Murray, R.M., Li, Z., Sastry, S.S.: A mathematical introduction to robotic manipulation. CRC Press (1994)
2. Tsai, L.W.: Robot analysis: the mechanics of serial and parallel manipulators. Wiley, New York (1999)
3. Zollo, L., Roccella, S., Guglielmelli, E., Carrozza, M., Dario, P.: Biomechatronic design and control of an anthropomorphic artificial hand for prosthetic and robotic applications. IEEE/ASME Transactions on Mechatronics 12(4), 418–429 (2007)
4. Shirafuji, S., Ikemoto, S., Hosoda, K.: Development of a tendon-driven robotic finger for an anthropomorphic robotic hand (in press). The International Journal of Robotics Research (in press 2014)
5. Kamper, D., Cruz, E., Siegel, M.: Stereotypical fingertip trajectories during grasp. Journal of Neurophysiology 90(6), 3702–3710 (2003)
6. Leijnse, J., Kalker, J.J.: A two-dimensional kinematic model of the lumbrical in the human finger. Journal of Biomechanics 28(3), 237–249 (1995)
7. Kobayashi, H., Hyodo, K., Ogane, D.: On tendon-driven robotic mechanisms with redundant tendons. The International Journal of Robotics Research 17(5), 561–571 (1998)
8. Soechting, J.F., Lacquaniti, F.: Invariant characteristics of a pointing movement in man. The Journal of Neuroscience 1(7), 710–720 (1981)
9. Dejmal, I., Zacksenhouse, M.: Coordinative structure of manipulative hand-movements facilitates their recognition. IEEE Transactions on Biomedical Engineering 53(12), 2455–2463 (2006)
10. Friedman, J., Flash, T.: Trajectory of the index finger during grasping. Experimental Brain Research 196(4), 497–509 (2009)
11. Spoor, C.W.: Balancing a force on the fingertip of a two-dimensional finger model without intrinsic muscles. J. Biomech. 16(7), 497–504 (1983)
12. Leijnse, J.: Why the lumbrical muscle should not be bigger–a force model of the lumbrical in the unloaded human finger. Journal of Biomechanics 30(11-12), 1107–1114 (1997)
13. Hogan, N.: Adaptive control of mechanical impedance by coactivation of antagonist muscles. IEEE Transactions on Automatic Control 29(8), 681–690 (1984)
14. Kobayashi, H., Ozawa, R.: Adaptive neural network control of tendon-driven mechanisms with elastic tendons. Automatica 39(9), 1509–1519 (2003)
15. Wimbock, T., Ott, C., Albu-Schaffer, A., Kugi, A., Hirzinger, G.: Impedance control for variable stiffness mechanisms with nonlinear joint coupling. In: The IEEE/RSJ International Conference on Intelligent Robots and Systems (IROS), pp. 3796–3803. IEEE (2008)
16. Jindrich, D.L., Balakrishnan, A.D., Dennerlein, J.T.: Effects of keyswitch design and finger posture on finger joint kinematics and dynamics during tapping on computer keyswitches. Clinical Biomechanics 19(6), 600–608 (2004)
17. Kuo, P., Lee, D.L., Jindrich, D.L., Dennerlein, J.T.: Finger joint coordination during tapping. Journal of Biomechanics 39(16), 2934–2942 (2006)

Neuromechanical Mantis Model Replicates Animal Postures via Biological Neural Models*

Nicholas S. Szczecinski, Joshua P. Martin,
Roy E. Ritzmann, and Roger D. Quinn

Case Western Reserve University,
11111 Euclid Avenue, Cleveland, Ohio, USA 44106

Abstract. A neuromechanical model of a mantis was developed to explore the neural basis of some elements of hunting behavior, which is very flexible and context-dependent, for robotic control. In order to capture the complexity and flexibility of insect behavior, we have leveraged our previous work [1] and constructed a dynamical model of a mantis with a control system built from dynamical neuron models, which simulate the flow of ions through cell membranes. We believe that this level of detail will provide more insight into what makes the animal successful than a finite state machine (FSM) or a recurrent neural network (RNN). Each of the model's walking legs has six degrees of freedom. Each joint is actuated by an antagonistic pair of muscles, controlled by a custom designed variable-stiffness joint controller based on insect neurobiology. The resulting low-level control system serves as the groundwork for a more complete behavioral model of the animal.

Keywords: Neuromorphic control, neuromechanical modeling.

1 Introduction

Insects have served as models for robot locomotion for some time [2,3]. They exhibit rich, context-dependent behavior that is desirable for robots in many applications. We have begun studying the praying mantis, an insect that uses a repertoire of movements depending on the context in which it is pursuing prey. Here we focus on lunging and pivoting movements that orient the animal toward prey before striking [4]. We have produced a neuromechanical model to explore these motions.

Several simulation models have been successfully used to investigate insect locomotion. The presented work follows from that of Szczecinski et al. [1], which constructed a neural locomotion controller for the cockroach *Blaberus discoidalis*. A similar model is that of Daun-Gruhn et al., which examines rhythm generation

---

* This work was supported by NASA Space Technology Research Fellowship NNX12AN24H. Further support was provided by AFOSR grant FA9550-10-1-0054, as well as NSF Grant IOS-1120305.

and inter-leg communication in the stick insect [5,6]. Both of these explore how neural signals generate muscle activity and ultimately motion. However, both are concerned with generating rhythms for walking, which is not the scope of this paper. To simulate the precise movements that mantises make while hunting, the presented model pays very close attention to what is known about neuromuscular control in still-standing insects, and is not presently concerned with rhythm generation for walking.

The most complete insect posture and locomotion neural control model is Walknet [3]. Walknet is an artificial neural network (ANN) model based on the stick insect. It has been used to control kinematic simulations and robots. Walknet can produce many insect-like motions, and switch between different behaviors based on environmental conditions. Despite its capabilities, Walknet does not necessarily describe what the insect is doing; it is composed of unrealistic neuron models, which are trained to match data rather than neuroanatomy. We believe that our model will eventually produce many of the same behaviors, and provide a basis for emulating them through more realistic neural structures and models.

Focusing on the hunting behaviors of the mantis lets us capitalize on a rich array of behaviors beyond locomotion. Mantises visually stalk and strike at prey with hunger-dependent aggression [7]. We justify extending our previous work with cockroaches to praying mantises because they are closely related, descending from a cockroach-like common ancestor [8]. Thus many components of cockroach locomotion and neural and muscular anatomy are similar to those in mantises. The main difference is that while cockroaches walk with all six legs most of the time, mantises hold their forelegs against the body and use their mid- and hindlegs to walk, pivot and lunge [4,9]. The kinematics of the lunging maneuver are similar to leg movements seen as cockroaches escape from a predator [10], and can be similarly classified as rotational or translational stationary turns with distinct patterns of underlying leg movements [4]. Our model replicates these motions and serves as the basis for future research on the descending control of these precise movements.

We present a neuromechanical simulation of the praying mantis capable of making the same pre-strike postural adjustments as the animal. It contains approximately 500 conductance-based neurons connected by about 1000 synapses, used to control 48 muscles that actuate a 25 link rigid body model of a mantis. Unlike our previous cockroach model, this system pays very close attention to the control mechanisms at each joint. All of the parameters for each neuron, synapse and muscle have been specifically set for the function of the related leg joint, producing stable, compliant position control for every joint. The model has reproduced various mantis poses, and has maintained stability under perturbation. The resulting model serves as the groundwork for future behavior-related work.

2   Methods

2.1   Modeling Components

All simulation was performed with AnimatLab 2, a neuromechanical modeling program [11]. All optimization was performed in MATLAB (Mathworks, MA, USA) using custom optimizers. Triangulated meshes were generated in Blender (Stichting Blender Foundation, Amsterdam, Netherlands) for each leg segment based on measurements made on a male mantis (*Tenodera sinensis*). Each segment is assumed to have the same density, the total mass of the specimen divided by its total volume. They were then assembled in AnimatLab and used to simulate rigid body dynamics.

Muscle attachment points and muscles themselves were added to the model in plausible locations. Cockroach muscular anatomy is better known than that of the mantis [12], so this information was used because of their taxonomical similarity. Each joint is actuated by one antagonistic pair of muscles, modeled as Linear Hill muscles. This model treats a muscle as an ideal linear actuator in parallel with a spring of stiffness k_p and linear damper with coefficient c. This assembly is in series with another spring, stiffness k_s. The tension on the muscle, T, develops according to:

$$\frac{dT}{dt} = \frac{k_s}{c}\left(k_p x + c\dot{x} - \left(1 + \frac{k_p}{k_s}\right) \cdot T + Act \right) \tag{1}$$

where $x(t) = l_{muscle}(t) - l_{rest}$, and Act is the muscle activation, which is a sigmoidal function of the motor neuron (MN) voltage:

$$Act(V_{MN}) = A \cdot \left(1 + exp\big(B \cdot (C - V_{MN})\big)\right)^{-1} \tag{2}$$

where A, B, and C are constants that describe the maximum active tension, the steepness of the sigmoid, and the input voltage offset, respectively. The full details of the model can be found in [1].

Neurons are implemented as leaky conductance models with no additional ion channels. This means they are first-order integrators, like resistive-capacitive circuits, that have a persistent reference voltage E_{rest}. Their membrane voltage changes according to:

$$C_{mem}\frac{dV}{dt} = G_{mem} \cdot (E_{rest} - V) + \sum_{i=1}^{n} g_{syn,i} \cdot (E_{syn,i} - V) + I_{app} \tag{3}$$

where C_{mem} and G_{mem} are the capacitance and conductance (inverse of the resistance) of the neuron membrane, respectively. All values labeled E are constant voltage values, and are properties of the network. Each of a neuron's n incoming synapses has a reference voltage, E_{syn}, which is applied to the neuron with a gain equal to the conductance of the synapse:

$$g_{syn} = g_{max} \cdot min\left(max\left(\frac{V_{pre} - E_{lo}}{E_{hi} - E_{lo}}, 0\right), 1\right) = min\big(max(aV_{pre} + b, 0), g_{max}\big) \tag{4}$$

where V_{pre} is the voltage of the presynaptic neuron, and E_{lo} and E_{hi} are properties of the synapse. These can be combined into variables a and b, which will simplify analysis in section 2.2. It should be clear that $0 \leq g_{syn} \leq g_{max}$. Again, more details about this model and its behavior are described in [1].

2.2   Joint Controller Design

Although each joint in each leg has a similar construction, every joint serves a different purpose and requires different strength. Therefore we designed a stereotypical joint controller from insect neurobiology and solved for the unique numerical parameters required for each. The result is a process that uses leg kinematics and dynamics to reliably and repeatably design legs of any morphology, provided that muscle attachments will allow the desired joint rotations.

Insect leg muscles may have three types of motor neurons: slow, fast, and inhibitory [13]. However, slow motor neurons are responsible for most locomotion tasks [14]. For our purposes, modeling only a slow motor neuron for each muscle is an acceptable simplification.

In order to make precise movements like those observed in mantises, each joint must possess a position feedback controller. Insects are known to control posture via negative position feedback at their joints [15]. It is also known that increasing the load on a leg will increase the muscle activity [16]. Therefore we designed a neuromuscular position feedback controller whose gain is modulated by perceived load.

Part of this is a subsystem in which the change in the motor neuron's voltage is proportional to the difference between two other neurons' voltages, which encode the desired joint rotation and the actual joint rotation. In addition, the proportionality is a function of another neuron's voltage, which codes for feedback gain. The network shown in Fig. 1 achieves this, and this can be shown by examining the steady state voltage of some of its neurons. Let us first examine the steady state voltage of V_{Add}, V_{Add}^*:

$$V_{Add}^* = \frac{G_{mem} \cdot E_{rest} + g_{Des} \cdot E_{Des} + g_{Act} \cdot E_{Act}}{G_{mem} + g_{Des} + g_{Act}} \tag{5}$$

Lower case gs are synaptic conductances, while capital Gs are constant membrane conductances. Now let us make the following substitutions: let $G_{mem} = 1$, $e = V_{Des} - V_{Act}$ (the error signal that we are interested in), $V_1 = 1/2 \cdot (V_{Des} + V_{Act})$, $E_1 = 1/2 \cdot (E_{Des} + E_{Act})$, and $\Delta E_1 = 1/2 \cdot (E_{Des} - E_{Act})$. Let us also use the second form in Equation (4) to describe synapse conductance. This yields:

$$V_{Add}^* = \frac{E_{rest} + 2(aV_1 + b)E_1 + a\Delta E_1 e}{1 + 2(aV_1 + b)} \tag{6}$$

If we set $E_1 = E_{rest}$, rename $2(aV_1 + b) = g_1$, and make the substitution that $U_{Add}^* = V_{Add}^* - E_{rest}$, we are left with:

$$U_{Add}^* = \frac{a\Delta E_1}{1 + g_1} \cdot e = k_{pos} \cdot e \tag{7}$$

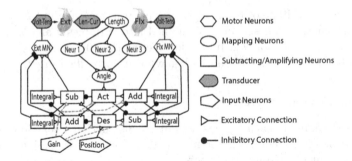

Fig. 1. Schematic of a joint controller. The only structural difference between joints is the number of mapping neurons; 1 to 10 neurons are tested, and the best performing network is selected.

This means that the change in a neuron's voltage, which is then turned into muscle force, is linearly proportional to the displacement in the angle encoded in e, as long as the above substitutions are enforced in the model. One important note is that g_1, and therefore k_{pos} is not constant, but varies with the average of V_{Des} and V_{Act}. Thus the "Gain" neuron, which excites both "Act" and "Des," can change k_{pos}. This is desirable, since we want our system to change feedback gain when forces act on the leg. When using biologically plausible values, k_{pos} can be made to vary by a factor of about 2. We therefore repeat this structure with V_{Add} and V_{Sub} as inputs to the motor neurons (Fig. 1) to increase k_{pos}'s range to 4.

This position control network is more complicated than those in related work [1]. This extra complexity makes the controller much more linear, and thus more predictable than simpler designs. Such predictability makes the whole system more reliable, and makes the pivots and lunges replicated in this paper possible.

A proportional controller's steady state error can be driven to 0 when the integral of the error is also fed back into the system [17]. Therefore we also add a pathway (Fig. 1, "Integral") whose neurons have time constants 200 times longer than the others in the network. Our experiments show that this reduces error, but is not perfect since a neuron is not an ideal integrator. This effect is evident in Fig. 2A.

How does a neural system measure the rotation of a joint for control? Insects detect the stretch in chordotonal organs (COs), which are often in series with the extensor muscle, to detect joint rotation. Dozens of neurons receive input from a single CO, and each has different ranges of sensitivity [18]. This idea was abstracted to design a unique network for each joint that converts the length of the extensor muscle into joint rotation. In order to accelerate tuning, the steady-state network activity, rather than simulation data, is optimized. As in Equation (5), the network output is computed and compared to the known muscle length at 10 expected joint rotations. A genetic algorithm is first used to estimate E_{syn}, E_{hi}, and E_{lo} for each synapse in the network, and then a BFGS (Broyden

Fletcher Goldfarb Shanno) minimization is used to refine the parameters. An example of such a network's performance is shown in Fig. 2B.

Once the feedback system is tuned, muscle parameters must be set. For each leg, the manipulator Jacobian is computed. A genetic algorithm then explores the configuration space and the orientation of a unit force applied to the foot to find the maximum possible torque on each joint. Kinematics are used to compute the resulting muscle tension. This maximum force is used to set the muscle tension amplitude A in Equation (2). k_s represents the stiffness of the tendon-like apodeme, which is very stiff [19], so the k_s required to hold the maximum force with a joint deflection of only 10 degrees is computed and applied to the model. Passive stiffness and damping are also known to be large in insects [19], so k_p is set to 20% of k_s, and c is set to 20% of k_p.

Finally, the muscle activation steepness B and voltage offset C (Equation (2)) are computed. For a set of desired joint positions and stiffnesses, the steady state of the feedback system and muscles is solved numerically. For each case, the equilibrium position and the torque required for a small deflection of the limb are computed. The difference between expected and desired joint angles and stiffnesses is minimized first by a genetic algorithm, and then refined with a BFGS least squares routine.

2.3   Animal Experiments

Mantises were observed as they hunted prey (2nd to 3rd instar cockroach nymphs) in a $40cm$ by $40cm$ acrylic arena. A high-speed camera (A602f, Basler AG, Germany) was positioned above the arena and collected video data at 100fps. Videos were analyzed using custom scripts in MATLAB. The two-dimensional overhead image of the insect was overlaid with a ball-and-stick model, aligning the segments of the mid-(T2) and hindlegs (T3). Angles were measured at the femur-tibia joint (FT, the most distal joint) and at the intersection of a line extending from the visible portion of the femur and a line down the middle of the body (FB). These angles are labeled in Fig. 4.

Mantises possess four to five active joints on the proximal segment of the leg invisible to an overhead camera, so these two measurements do not give the animal's exact motion the same way that three-dimensional underside video with joint tracking would [20]. Our two measurements, however, let us quantify animal and model behavior in a simple way.

2.4   Generating Model Poses

Since 3-dimensional kinematic data of all of a mantis' joint motions is not yet available, making the model perform the pivots observed in the animal required that we solve an inverse kinematics problem. From a known starting position, translating or rotating (or both) the body while the feet remain planted yields the necessary foot-to-body vector for each leg. Videos of the underside of mantises informed which joints were most obviously contributing to pivots. Then, the configuration of those joints was optimized with a BFGS minimization to solve for *a* (not *the*) set of joint angles that would produce the desired body motion. This method has been used to solve similar issues in other bio-inspired robots [21]. We must stress that these solutions may not be what the animal is doing, but our model requires that we approximate their activity. Feeding joint rotations needed for these desired positions into our joint controllers did produce the observed body and limb motion in most cases, suggesting that some or most may be correct, but further biological research is necessary to determine exactly how the animal executes such motions.

Three main experiments were run with the model. First, joint controllers were tested for position error for different feedback gains. Once the controllers were verified, a resting posture was approximated from videos of mantises. This position was tested for stability, that is, to make sure the model's body returned to its original orientation after perturbation torques about all 3 axes. Finally, the model was made to mimic the postures observed in animals. By focusing on the low-level performance and behavior, we mimicked the observed animal motions.

3   Results

3.1   Position Controller Validation

Each joint was given an eight-step input signal, each step being 200ms long and commanding an extension 12.5% of its range of motion. This experiment was run with different controller stiffnesses. The pathway connecting load and the gain control neuron was removed, so the stiffness neuron could be directly stimulated. Fig. 2A shows the performance of the most distal joint on the left hind leg. Increasing the feedback gain clearly decreases the time constant of the motion, meaning that the system is stiffer. The integral pathway is not an ideal integrator, so the system still has a nonzero DC gain, resulting in reasonably large steady state errors at low gains.

Figure 2B shows how the joint rotation mapping network performs. This is qualitatively similar to sensory neurons that interface with COs; each neuron has a different range of sensitivity, and amplifies the signal differently. These sum such that V_{Angle} equals the actual joint angle.

3.2   Postural Stability

Once joint controllers were tested, the entire model was assembled, and a resting posture was estimated from underside videos of mantises. This configuration was tested for stability first by ensuring the model could support its own weight, and then by applying impulsive torques about each axis. This nonlinear system is likely not asymptotically stable, and our system's complexity does not lend itself well to analytical methods. Therefore we numerically showed the stability of this pose by plotting the quantity $0.5 \cdot \omega^T \omega$ for each perturbation experiment, as seen in Fig. 3. The velocity of the model clearly decays to 0 rapidly, showing that

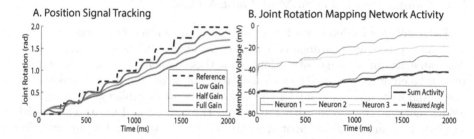

Fig. 2. A. Controller performance for the most distal joint on the hind leg. Increasing the feedback gain accelerates the dynamics. B. The three mapping neurons in Fig. 1 have different ranges and amplitudes of activity, but they sum up to yield an accurate approximation of the joint angle.

the posture selected is stable. Perturbation experiments with all of the postures presented in the next section yielded similar results (data not shown).

The observed stability is due to the feedback controllers, whose gains are modulated by the force acting on the leg. The model effectively has height position control through the sum of joint position controllers. Applying a vertical force to any leg will not only move it from its reference height, but will also increase the gain of the joint controllers. In this way each leg has negative height feedback, but positive force feedback. The model increases muscle activity when loaded, as seen in insects [16]. Some tests were run to compare this implementation to other options. The load pathways were disabled, which meant that force could not change the joint controller stiffnesses. This meant that applying loads caused larger deflections, and reduced stiffness throughout caused the model to oscillate with large, destabilizing amplitudes. Using force feedback to stiffen joints is important for rejecting unwanted motions, and we believe this will make our model more adaptive as further work is pursued.

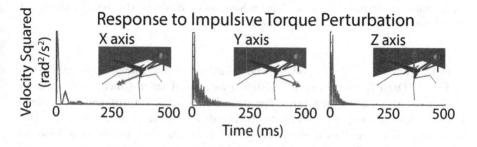

Fig. 3. Velocity squared of the body when subjected to impulsive torque perturbations. The decaying velocity numerically suggests local stability.

3.3   Comparison between Model and Animal

The scope of this modeling work is to produce a model that can make the same stationary pre-strike postures as a mantis. Animal motions were analyzed to determine maximum ranges of motion during these maneuvers ($\pm 0.5rad$, $10mm$). Inverse kinematics were used to find plausible joint angles required to make these poses, and were used as input for the individual joint controllers. The resulting model motions are shown with those from the animal in Fig. 4. The model supported its weight throughout the entirety of each trial, and maintained its stability when perturbed in the same way as in Fig. 3. Top projections of the angles shown in Fig. 4 were measured as both the model and the animal made pivot or translation maneuvers.

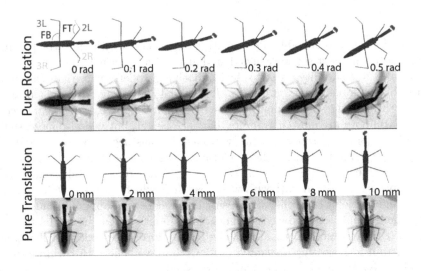

Fig. 4. Top view of model (top in each row) and animal (bottom in each row) as it performs a pivot (top) and translation (bottom). Body angle with the horizontal or displacement from rest is labeled. Images have been aligned so the feet (which do not move) are collinear from image to image.

Data from many mantis pivots and translations are shown as scatter plots in Fig. 5. Data from the model is shown as a solid trace (linear interpolation of discrete rotations/translations) because the model is deterministic. The data for body rotation appear to capture the trends seen in the animal, for all angles measured. The simulation even accounts for points that might otherwise be labeled outliers (FT angle versus rotation, top left). The translation data from the animal appear to be less predictable (bottom), and as a result the model is less accurate in predicting them. There may be two reasons for this discrepancy. The first is that the animal is unlikely to make a straight translation. The data

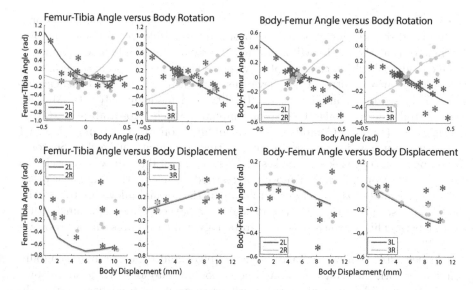

Fig. 5. Comparison of top-projected angles labeled in 4. Model data, which is not stochastic, is plotted as lines. Data points are from animal trials ($n = 20$ for rotations, $n = 11$ for translations). To help distinguish data sets, left leg data are plotted as asterisks, and the right leg data as circles.

presented here were from motions with less than 10 degrees of body rotation. The second is that the animal's starting leg positions may not match the model's, in which case the angles will not change in the same way.

4 Discussion

We present a neuromechanical simulation of the praying mantis, capable of mimicking animal postures by actuating limbs with muscles controlled by biological neural models. The design process we present is integral to this system's success, because we do not simply optimize parameters, but design them based on the model's kinematics and dynamics. The actuators and controllers are broken into smaller subsystems, whose parameters are solved for based on desired steady state behavior. Since most of the methods used are not stochastic, the results are repeatable, and do not require testing many parameter combinations. In fact, the parameters for each leg were computed in less than forty minutes, an exceptionally fast alternative to using a stochastic method to optimize simulation output. In addition, the results are more meaningful, since they are designed to achieve static and dynamic capabilities, not to mimic desired system output in a few specific cases. We believe this method simultaneously helps us design effective robotic controllers and learn more about insect nervous systems.

The similarities between animal and model kinematics (Fig. 5) are a good example of this. The muscles and joint controllers were not designed for the specific behaviors of pivots or translations; they were designed for adaptive-force position control, which can be used to produce animal-like posture and stability. We believe that the model will perform similarly as we include new behaviors.

One might ask why such a complex system is necessary to obtain something as simple as position control. We believe this biological controller is beneficial for three main reasons: compliance, adaptability, and modeling integrity. Muscles and controller design make the system compliant, which helps it absorb external disturbances (Fig. 3). Compliance also simplifies control by handling internal disruptions, such as joint angle commands that are slightly infeasible. Sensory-driven position feedback gain makes the system adaptable, allowing muscles to use low activation unless that motion is resisted, either by the ground or other influences. Finally, paying closer attention to the animal allows us to use the model as a scientific tool, testing biological hypotheses and informing future experiments. This advances insect neurobiology, which inspires further robotics work [2].

Our next step will be to make the model pivot and translate more generally using a network that converts desired body posture into low-level position commands. This is ideal for interfacing with proprioception from head and body joints, which would allow the model to move in a context-dependent way [22]. In addition, we want to examine quadruped walking, which would involve constructing an inter-joint coordination system like that in [1] on top of the joint controllers presented here. Stepping rules could be informed through 3D kinematics [20].

5   Conclusion

As computational power becomes more ubiquitous, neural modeling is an increasingly relevant tool for animal modeling. Over three decades of research have been devoted to understanding the neural basis of insect locomotion, and enough is known to make a plausible model. From a robotics standpoint, this means better understanding how and why insects are such effective locomotors, and also produces more detailed copies of how the animal controls its motion. We think that the system presented here is the first step toward a more complete insect locomotion model, and ultimately a robot controller.

References

1. Szczecinski, N.S., Brown, A.E., Bender, J.A., Quinn, R.D., Ritzmann, R.E.: A Neuromechanical Simulation of Insect Walking and Transition to Turning of the Cockroach Blaberus discoidalis. Biological Cybernetics (2013)
2. Ritzmann, R.E., Quinn, R.D., Watson, J.T., Zill, S.N.: Insect walking and biorobotics: A relationship with mutual benefits. Bioscience 50(1), 23–33 (2000)
3. Schilling, M., Hoinville, T., Schmitz, J., Cruse, H.: Walknet, a bio-inspired controller for hexapod walking. Biological Cybernetics 107(4), 397–419 (2013)

4. Cleal, K.S., Prete, F.R.: The Predatory Strike of Free Ranging Praying Mantises, Sphodromantis lineola (Burmeister). II: Strikes in the Horizontal Plane. Brain Behavior and Evolution (48), 191–204 (1996)
5. Knops, S.A., Tóth, T.I., Guschlbauer, C., Gruhn, M., Daun-Gruhn, S.: A neuromechanical model for the neural basis of curve walking in the stick insect. Journal of Neurophysiology, 679–691 (November 2012)
6. Daun-Gruhn, S., Tóth, T.I.: An inter-segmental network model and its use in elucidating gait-switches in the stick insect. Journal of Computational Neuroscience (December 2010)
7. Prete, F.R., Hurd, L.E., Branstrator, D., Johnson, A.: Responses to computer-generated visual stimuli by the male praying mantis, Sphodromantis lineola (Burmeister). Animal Behaviour 63(3), 503–510 (2002)
8. Grimaldi, D., Engel, M.S.: Evolution of the Insects. Cambridge University Press, Cambridge (2005)
9. Rossel, S.: Foveal Fixation and Tracking in the Praying Mantis. Journal of Comparative Physiology A Neuroethology Sensory Neural And Behavioral Physiology 139, 307–331 (1980)
10. Nye, S.W., Ritzmann, R.E.: Motion analysis of leg joints associated with escape turns of the cockroach, Periplaneta americana. Journal of Comparative Physiology. A, Sensory, Neural, and Behavioral Physiology 171(2), 183–194 (1992)
11. Cofer, D., Cymbalyuk, G., Reid, J., Zhu, Y., Heitler, W.J., Edwards, D.H.: AnimatLab: a 3D graphics environment for neuromechanical simulations. Journal of Neuroscience Methods 187(2), 280–288 (2010)
12. Carbonell, C.S.: The Thoracic Muscles of the Cockroach Periplaneta americana (L.), vol. 107. Smithsonian Institution, Washington, D.C (1947)
13. Pearson, K.G., Iles, J.F.: Innervation of coxal depressor muscles in the cockroach, Periplaneta americana. The Journal of Experimental Biology 54(1), 215–232 (1971)
14. Watson, J.T., Ritzmann, R.E.: Leg kinematics and muscle activity during treadmill running in the cockroach, Blaberus discoidalis: II. Fast running. Journal of comparative physiology. A, Sensory, Neural, and Behavioral Physiology 182(1), 23–33 (1998)
15. Büschges, A., Gruhn, M.: Mechanosensory Feedback in Walking: From Joint Control to Locomotor Patterns. Advances In Insect Physiology 34(07), 193–230 (2007)
16. Zill, S.N., Schmitz, J., Büschges, A.: Load sensing and control of posture and locomotion. Arthropod Structure & Development 33(3), 273–286 (2004)
17. Dorf, R.C., Bishop, R.H.: Modern Control Systems. Pearson Education, Inc., Upper Saddle River (2008)
18. Field, L.H., Matheson, T.: Chordotonal Organs of Insects. Advances In Insect Physiology 27, 1–230 (1998)
19. Zakotnik, J., Matheson, T., Dürr, V.: Co-contraction and passive forces facilitate load compensation of aimed limb movements. The Journal of Neuroscience: the Official Journal of the Society for Neuroscience 26(19), 4995–5007 (2006)
20. Bender, J.A., Simpson, E.M., Ritzmann, R.E.: Computer-assisted 3D kinematic analysis of all leg joints in walking insects. PloS one 5(10) (2010)
21. Nelson, G.M., Quinn, R.D.: Posture control of a cockroach-like robot. IEEE Control Systems 19(2), 9–14 (1999)
22. Yamawaki, Y., Uno, K., Ikeda, R., Toh, Y.: Coordinated movements of the head and body during orienting behaviour in the praying mantis Tenodera aridifolia. Journal of Insect Physiology 57(7), 1010–1016 (2011)

A Natural Movement Database for Management, Documentation, Visualization, Mining and Modeling of Locomotion Experiments

Leslie M. Theunissen[1,4], Michael Hertrich[1], Cord Wiljes[2,4], Eduard Zell[3,4],
Christian Behler[3], André F. Krause[1,4], Holger H. Bekemeier[1], Philipp Cimiano[2,4],
Mario Botsch[3,4], and Volker Dürr[1,4]

[1] Biological Cybernetics,
[2] Semantic Computing,
[3] Computer Graphics,
[4] Cognitive Interaction Technology,
Center of Excellence (CITEC), Bielefeld University, Germany
volker.duerr@uni-bielefeld.de

Abstract. In recent years, experimental data on natural, un-restrained locomotion of animals has strongly increased in complexity and quantity. This is due to novel motion-capture techniques, but also to the combination of several methods such as electromyography or force measurements. Since much of these data are of great value for the development, modeling and benchmarking of technical locomotion systems, suitable data management, documentation and visualization are essential. Here, we use an example of comparative kinematics of climbing insects to propose a data format that is equally suitable for scientific analysis and sharing through web repositories. Two data models are used: a relational model (SQL) for efficient data management and mining, and the Resource Description Framework (RDF), releasing data according to the Linked Data principles and connecting it to other datasets on the web. Finally, two visualization options are presented, using either a photo-realistic rendering or a plain but versatile cylinder-based 3D-model.

1 Introduction

Transferring insights on animal locomotion to technical systems is one of the oldest themes in biomimetics. Today, the control of legged locomotion is a prolific field of research, both in the life sciences and in engineering. Nevertheless, the remaining obvious performance deficits of walking machines – with regard to their biological paragons – give rise to the question why insights on animal locomotion could not be transferred more efficiently to walking machines. While there are several likely reasons for this, one of them is the problem of integrating information from different model organisms (e.g., human, cat, cockroach, stick insect), experimental methodologies (e.g., neurophysiology, biomechanics, motion capture, modeling), and levels of approach (e.g., neural networks, reduced preparations, unrestrained intact animals). In the face of the immense growth of knowledge on natural legged locomotion, it will be

A. Duff et al. (eds.): Living Machines 2014, LNAI 8608, pp. 308–319, 2014.

essential to prepare experimental results for public access, integrate results from different labs, and improve their amenability through appropriate visualization and documentation, e.g., through Linked Data. The main objective of this paper is to present a case example of such an integrated data publication with immediate relevance for the control of spatial coordination in climbing.

The case example that we choose comprises motion capture datasets on unrestrained walking and climbing trials of three species of stick insects: *Carausius morosus, Aretaon asperrimus* and *Medauroidea extradentata = Cuniculina impigra*. All three of these species are commonly used in physiological studies on invertebrate locomotion [1, 2] and have served as paragons for artificial neural network controllers of six-legged locomotion (e.g., [3]) and walking machines [4, 5]. A demonstrator sample of our data is registered for open access publication (doi:10.4119/unibi/citec.2013.3) and has been made available in May 2014. Our case example focuses on kinematic data (marker trajectories, joint angle time courses, along with a generic description of the kinematic chains), but the database structure is already laid out for inclusion of any other experimental time-course measurements, such as recordings of ground reaction forces, muscle or nerve activity, and many others.

Concerning the potential for shared use of datasets, public-access databases are the current state of the art. As yet, though public-access databases of motion capture data have been available for some time, all of them appear to concentrate on human movement sequences. Furthermore, the available datasets are particularly appropriate for studies on computer graphics and animation but, typically, are inappropriate for addressing neuroscience-related questions. This inappropriateness may be grounded in the problem that the movement of markers cannot be related to the variables controlled by the nervous system, e.g., specific degrees of freedom of real joints.

Two prominent motion capture databases are the *CMU motion capture database* of the Carnegie Mellon University (Pittsburgh, USA; http://mocap.cs.cmu.edu/) and the *HDM05 database* of the "Hochschule der Medien" in Stuttgart, Germany (http://www.mpi-inf.mpg.de/resources/HDM05/). Both of these databases have been used for various research problems in computer science, such as generating naturalistic human movement sequences from low-dimensional, behavior-specific data spaces [6], segmenting motion sequences into distinct behaviors [7], analyzing the structure of the stored motion capture data by similarity metrics [8], content-based retrieval [9, 10] and annotation [11]. Despite these achievements in computer science, existing databases have very little to offer to the neuroscience of locomotion and, therefore, to the transfer of knowledge to walking machines. In contrast, our own database comprises morphological description of the animal studied, time courses of all relevant joint angles, along with metadata about the experimental setup, etc.. As a result, the data can be compared among species, which, to the best of our knowledge, has not been done before in scientific motion capture databases. Data can be analyzed with any measure of spatial and temporal coordination within and among legs, and is suitable for benchmarking of technical walking systems.

To illustrate this, we will first present a showcase of animal locomotion data, acquired in a collaborative research project on biomimetics of autonomous locomotion (EU FP7 project EMICAB; www.emicab.eu). Based on this example, we will introduce the data structure used in the scientific analysis and the corresponding entity relationship model for a relational database (section 3). Finally, we will present two versions of visualization (section 4), and introduce the Linked Data approach (section 5).

2   Case Example: Comparative Kinematics of Insect Climbing

Insects can efficiently and reliably climb through a canopy, a substratum much more complex than any substratum that can be mastered by modern walking machines. Aiming for biomimetic control strategies for efficient and reliable climbing, we picked the whole-body kinematics of unrestrained walking and climbing insects as our case example. In the experiments, three species of stick insects were left to walk/climb across a set of stairs as described in detail by [12]. The species belong to different families of the order Phasmatodea and differ in both size and body-to-limb length ratio. Four setups with different stair height were used, thus altering the requirements for climbing in four experimental conditions.

Insects were labeled with 18 to 20 reflective markers of 1.5 mm diameter, and motion-captured using a Vicon MX10 system with eight T10 cameras (Vicon, Oxford, UK). As a result, the raw data in our database are Cartesian marker trajectories in a world-fixed coordinate system, with 200 Hz temporal and approximately 0.1 mm spatial resolution. Marker trajectories were further processed in Matlab (The Mathworks, Natick, USA), where morphological information about the animal and measurements of marker locations were included to calculate segment-fixed coordinate systems and joint angle time courses, along with secondary information such as sequences of ground contact phases (Fig. 1), step types (Fig. 2), and more. Further details about the method and calculation of secondary data are given in [12].

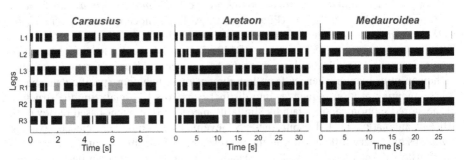

Fig. 1. Gait patterns of three species of stick insects during a climbing sequence. The colored stance phases indicate the first contacts on one of the two stairs. Same trials as in Fig. 2A.

The first three figures illustrate the kinematic complexity of unrestrained climbing behavior, reflecting the variety and detail of different kinds of kinematic analysis that are possible with our database. Fig. 1 shows a typical illustration of temporal coordination among legs. The so-called podograms show the alternating sequences of stance and swing movements of the six legs. During stance, a foot is in ground contact and the leg contributes to propulsion and stability. During swing, a foot is lifted off the ground and moved towards a new touch-down location. The colored bars mark the first stance movement in contact with a stair, thus indicating the start of a climbing sequence. Species comparison reveals that gaits are strongly irregular in all three species: stance durations and number of legs on the ground change continuously.

This irregularity is reflected also by different step types during the climbing sequences. To illustrate this, Fig. 2A shows the foot trajectories of a neighboring pair of left legs as they stepped across the setup. Here, we distinguish step types according to their lift-off and touch-down surfaces on the setup, allowing us to tell regular walking steps (type 1) from climbing steps (types 2 to 4). Note how the hind leg (HL, green) not only touched down almost exactly where its leading middle leg did (ML, red), but also how similar their trajectories were. Exceptions include two additional HL steps on stair 2 (blue arrows, *Carausius* and *Aretaon*), and two lower HL swing movements in *Medauroidea*. At present our complete database contains several thousand of walking and climbing steps, and Fig. 2B shows the relative frequencies of the different step types. Relative frequency of climbing steps (type 2-4) is lowest in long-legged *Medauroidea*. In contrast, it is highest for steps to the side walls (type 5).

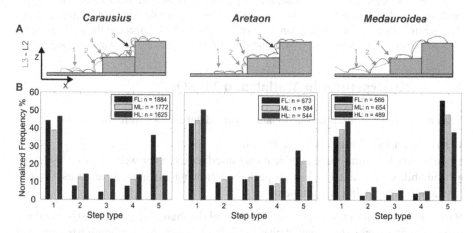

Fig. 2. Foot trajectories and step types of three insect species. A) Side views of foot trajectories of left hind leg (L3, green) and left middle leg (L2, red). Numbers label four step types, as determined by their lift-off and touch-down surfaces. Scale bars are 100 mm. B) Histograms of five step types for front, middle and hind legs (FL, ML, HL) and three insect species. Type 5 includes steps that either touched down on or lifted off from the side wall of the setup.

An important aspect of efficient climbing is the spatial coordination among legs. For stick insects, it is known that trailing legs follow the leading leg of the same body side (e.g., [13]), such that touch-down locations are very similar. It has been argued that the underlying coordinate transfer among neighboring legs is efficient because trailing legs will immediately find foothold successfully. With Fig. 3, we illustrate how our database allows comparison of spatial coordination among three species and two behavioral episodes. Based on the step type assignment as shown in Fig. 2, we can compare the accuracy of hind leg targeting for walking and climbing episodes. Accuracy is similar among species, but there is a characteristic medial offset in *Aretaon*.

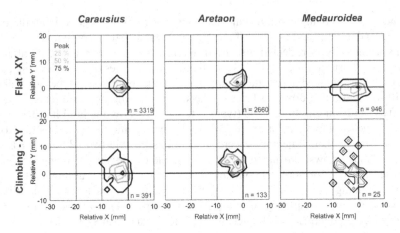

Fig. 3. Inter-species comparison of accuracy of spatial coordination. Data suitable for bench-marking of walking machines. Graphs relate pairs of leading middle (ML) and trailing hind legs (HL). Distributions of HL touch-down locations are shown with respect to the position of the ML (origin) during walking (step type 1) and climbing (step types 2/3). n: number of steps.

3   Data Structures in Matlab and MySQL

With regard to usability and functionality of the database, we were looking for

- a comprehensive, relational data model,
- suitability for multi-client graphical user interfaces, e.g. for web access,
- availability of a suitable Matlab package, allowing for immediate access from a scientific computing environment.

Based on these criteria, we opted for a relational database, using the *MySQL* database management system (http://www.mysql.com/), accessed by the administration tool *phpMyAdmin* (http://www.phpmyadmin.net) under the use of an *XAMPP* Apache server (http://www.apachefriends.org). For direct access of the database from Matlab, we chose the *MySQL database connector* (http://www.mathworks.com/matlabcentral/fileexchange/8663-mysql-database-connector).

As the main objective was to work with experimental data - in this case on animal locomotion - the data structure of the database needed to mirror the experimental procedures and to capture its conditions and results. Since each experiment may be carried out as a set of sessions, each of which comprises a number of trials, each ex-perimental result is related to a unique trial. This trial is further specified by a given setup and subject. Different experimental conditions are described by distinct setup descriptions (e.g., to capture the height of the two stairs) or by distinct subject de-scriptions (e.g., if certain ablations were made). Experimental data are stored either as time courses or as spatial trajectories. Here, time courses may be any kind of real-valued variable that is measured with a given, fixed sampling rate. Trajectories, on the other hand, are time-varying three-component vectors, e.g., for describing Cartesian coordinates of marker locations in space.

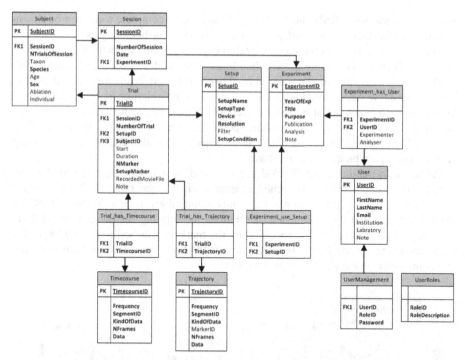

Fig. 4. Entity Relationship Model (ERM) of the database. Tables about metadata (e.g., Experiment, User, Setup) and measurements (e.g., Session, Trial, Timecourse, Trajectory). The visualization can directly access, e.g., the table Timecourse. For the complete ERM, refer to the complementary material (http://www.cit-ec.de/movement). PK: Primary Key; FK: Foreign Key.

Within the scientific computing environment Matlab, each trial was stored in a single data file. The data structure of this file was devised such that it could be applied to most, if not all motion capture experiments on insects, and could be transferred to other animal morphologies with minor adaptation (e.g., by changing segment names). The core of this data structure is the branched kinematic chain, consisting of the main body chain CS0–T3–T2–T1–Hd and up to eight side chains, namely T#.CS–R#–cox–fem–tib–tar for right legs, T#.CS–L#–cox–fem–tib–tar for left legs, and Hd.CS–ant#–scp–ped for antennae, where CS stands for 'coordinate system', T stands for 'thorax segment', ant stands for 'antenna', and # is a placeholder for the thorax segment (T1 to T3 in leg chains), or for the left and right body side (antR and antL in antenna chains). The acronyms Hd, cox, fem, tib, tar, scp and ped denote the segments head, coxa, femur, tibia, tarsus, scape and pedicel, respectively. The detailed specification of this data structure is supplied in the complementary material (http://www.cit-ec.de/movement). Most importantly, the data structure does not only store original raw data, but also computed secondary data (e.g., joint angles), as well as metadata about the experiment, user, session and setup.

Once an experiment has been analyzed in Matlab, all corresponding trial files can be uploaded into the *MySQL* database (Fig. 5), using a custom-written script that stores all variables in the corresponding database tables, according to the Entity

Relationship Model (ERM) shown in Fig. 4. Similarly, complete trials can be down-loaded from the database to Matlab, e.g., for revision or further analysis.

The core of the ERM is the metadata table *Experiment* that links to the tables *User*, *Setup* and *Session*. The table *Session* then links to the tables *Subject* and *Trial*. Finally, the table *Trial* links to the data tables *Timecourse* and *Trajectory*, but also links back to the table *Setup*, thus accounting for important relation to setup conditions. Most entity attributes are stored as numbers or strings, except for the large data vectors and matrices that contain floating-point numbers, as in trajectories or time courses. The latter are stored as binary large objects (so-called BLOBs), compressing them into compact binary data structures.

Further information about the *Subject* is stored as a kinematic body model (not shown). To account for the branched kinematic chains, the table *Body* links to the table *Chain*, which, in turn, has pairwise links to the tables *Node* and *Segment*. Nodes are further specified by a sequence of rotations, accounting for the degrees of freedom; Segments may be further specified by the markers attached to them. One of the main advantages of the integrated definition of the body model is the immediate use of this information for visualization purposes.

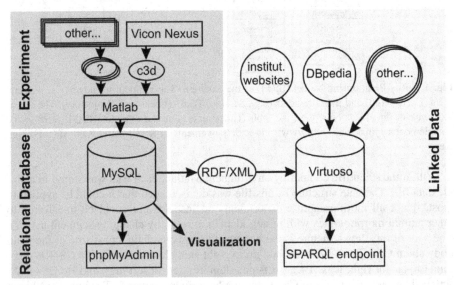

Fig. 5. Overview of our integrated approach to scientific data management, visualization and documentation. Experimental data from various sources are analyzed in *Matlab* and uploaded into a relational database (*MySQL*), from which it can be queried (*phpMyAdmin*) and visualized. Through the Linked Data approach, the database contents are transferred into a *Virtuoso* database, linking it to various web repositories. Queries are sent from a SPARQL endpoint.

4   Visualization

In order to visualize the recorded locomotion data, two different rendering frameworks have been developed: A photo-realistic offline visualization based on a geometrically accurate reconstruction of an *Aretaon asperrimus*, and a real-time visualization in a web browser, using geometric primitives (Fig. 6).

The high-quality geometric model was generated by first scanning an animal using a microCT scanner, yielding a high-resolution regular 3D array of density values. In this volumetric dataset, the insect was segmented and separated from background, and topological noise, such as small holes or tunnels, was removed by morphological operations. A triangle mesh of the outer surface of the insect was then extracted from the volumetric dataset using the Marching Cubes algorithm [14]. Since any physical measurement process inevitably introduces a certain amount of noise, the extracted triangle mesh contains high frequency geometric oscillations. This geometric noise has been removed by a few steps of Laplacian smoothing, which is a generalization of 2D diffusion flow to two-manifold triangle meshes [15]. An adaptive remeshing step [15] optimizes the triangulation to get rid of the low-quality skinny triangles produced by the Marching Cubes algorithm. The reconstructed and post-processed model is shown in Fig. 6, center. To increase the visual realism, photographs of the real insect were mapped as textures onto the surface, using the 3D modeling tool *Autodesk Maya*. This yields the final textured model shown in Fig 6, left, which consists of about 180k vertices and 360k triangles.

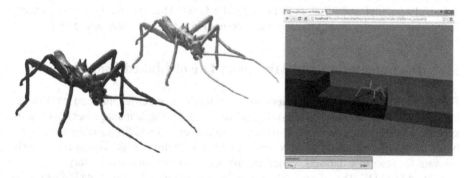

Fig. 6. Two types of visualization of insect locomotion. Left: Textured surface triangle mesh of the insect species *Aretaon asperrimus*, based on microCT scan data. Center: non-textured visualization. Right: Versatile cylinder model for real-time visualization of database content in a web browser. With its direct access to the database, WebGl visualization served as a utility test.

Also using Maya, this model was equipped with an interior control skeleton, which enables its easy animation by simply manipulating the joint angles. The skeleton was then articulated based on standard forward kinematics, and a smooth deformation of the surface mesh was computed from the updated skeleton using either linear blend skinning or dual quaternion skinning [16]. Using the recorded locomotion data to drive the skeleton articulation, and rendering the animated model using *Maya*'s global illumination technique finally results in a high-quality photo-realistic visualization, as shown in the accompanying video on the website http://www.cit-ec.de/movement.

The high-resolution model and the tool-chain described above are mainly intended for producing high-quality videos of pre-selected locomotion data, which typically requires several minutes up to an hour for rendering. As a consequence, this visualization is not suitable if a particular trial in the motion database was to be displayed.

However, modern web technology enables the real-time preview of locomotion data even in a web browser, for instance when trying to find a desired motion dataset through the web-interface to our MySQL database (Fig. 5).

The enabling key technology for visualization within a browser is WebGL (www.khronos.org/webgl/), which provides interactive real-time 3D graphics to be rendered into an HTML5 canvas element. Since WebGL is a subset of OpenGL, the de-facto standard API for interactive computer graphics, it requires only moderate effort to port a desktop-based graphics application to an interactive website enhanced by 3D graphics components.

The control flow of the 3D viewer widget was implemented in JavaScript, accessing the locomotion data from the database through PHP, and visualizing the requested insect motion using a simple, flexible and efficient cylinder model. To enhance the comprehensiveness of the animation, the spatial setup was also included (Fig. 6 right). Finally, simultaneous joint angle time courses are displayed on demand (not shown).

Since WebGL can directly access the graphics hardware (GPU) through the OpenGL driver, the rendering performance of a browser-based viewer is similar to a desktop application. However, JavaScript is still too slow for animation computations such as forward kinematics and cylinder transformations. Therefore, we implement these performance-critical components as shader programs, which are executed in a massively parallel and efficient manner on the GPU. This eventually allows interactive previewing of different motion datasets in an intuitive web interface (Fig. 6).

5   Documentation and Mining by Use of Linked Data

For data to be useful to other scientists, they have to be retrievable and well documented. While a relational database is a powerful tool for data management, its highly specific schema poses an obstacle to these requirements. In addition, integrating databases from different sources is a complex and labor-intensive task. This is particularly relevant for scientific datasets, which are diverse and often interdisciplinary.

Linked Data [17, 18] offers a solution to these challenges. Linked Data builds on the existing WWW and extends it with a semantic layer based on community-generated vocabularies. Linked Data can be processed by machines over large amounts of data thereby improving retrieval by search engines, and enabling queries which combine datasets from different sources. In addition, the grounding in commonly accepted vocabularies also serves human understanding of the data without the need for separate documentation. Linked Data standards (RDF, OWL, SPARQL) have been defined by the W3C consortium (http://www.w3.org/standards/semanticweb/data). Vocabularies are being developed for the description of domain-specific content. Numerous datasets from different domains are available in Linked Data.

Our goal was to translate the metadata about the stick insect locomotion experiments contained in the natural movement database into Linked Data and to test the usefulness of the approach by applying competency questions. In building the Linked Data representation we closely followed the database's ERM (Fig. 4), which had already identified the entities and possible relationships within the modeled domain. The main task in creating Linked Data is to locate suitable, established vocabularies and to identify relevant existing datasets that can be linked [19]. For convenient access of background information, our current Linked Data example integrates three sources of data:

- Metadata about the stick insect locomotion experiments
- Institutional data about researchers, organizations and publications at Bielefeld University
- DBpedia (http://dbpedia.org), a Linked Data representation of Wikipedia (e.g., supplying pictures of the species used in the experiments)

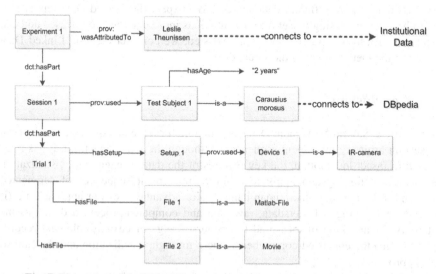

Fig. 7. Illustration of the Linked Data concept with regard to the case data (excerpt)

For data about the researchers involved, organizations and publications we used the *VIVO ontology* (http://sourceforge.net/apps/mediawiki/vivo), which combines several well-established vocabularies describing entities relevant for academic research communities. *DBpedia* already comes with its own ontology. In addition, we used the *W3C's Provenance Ontology* (http://www.w3.org/TR/2013/REC-prov-o-20130430/). Based on these vocabularies we generated the Linked Data directly from the database by means of a Perl script that exports the metadata about the experiments into RDF/XML. The export result was imported into the graph-based database *Virtuoso* (http://virtuoso.openlinksw.com/dataspace/doc/dav/wiki/Main/), which provides a SPARQL endpoint to query the data (Fig. 5). By combining the three data sources, one can answer queries that would have been impossible to answer before, such as:

1. "Give me all datasets about insects!"
2. "Which experiments were conducted by researchers from the Biological Cybernetics group?"
3. "Which publications were spawned by these motion tracking experiments?"

These queries combine information (1.) about an experiment's test species (e.g., *Carausius morosus*) with DBpedia entries such as the species' taxonomy (e.g., belongs to class Insecta), or (2./3.) about the experimenter with entries from institutional web-sites, e.g., CVs or publication lists.

Fig. 7 depicts an excerpt of the Linked Data created in this project. The vocabu-lary, the Linked Data itself, an endpoint to query it, and some sample queries are available on the project website http://www.cit-ec.de/movement. Because the RDF generation is fully automated, any new entries of future experiments into the Natural Movement Database will be automatically exported into Linked Data without addi-tional effort. With its current content we have successfully answered competency questions that require different data sources. We expect the Linked Data representa-tion to become increasingly useful in the future, as more datasets are released and can be integrated into this framework. In a next step, we consider creating Linked Data not just of the metadata but of the actual data.

6   Conclusions

We propose a Natural Movement Database suitable for large and diverse, multivariate datasets on arbitrary movement sequences. Although, in its present form, it has been devoted to insect locomotion, the key features of the data management, visualization and documentation system apply to arbitrary movement sequences of segmented limbs and/or bodies, e.g., the human hand. Major advantages of our approach are (i) the integrated storage of metadata, raw data and computed/processed data; (ii) the suitability for any kind of measured time courses, e.g., electrophysiological record-ings; and (iii) the enhanced comprehensiveness through visualization and the Linked Data approach.

The dataset used for our case example has been successfully used to evaluate a novel time series data mining method [20]. Apart from such research topics in com-puter science, we expect great benefit for various research issues in the neuroscience of motor control. For example, the potential for pooling data across labs   and experi-mental approaches may allow an integrated view on datasets, each of which may be limited to a small number of trials and/or relatively short recording periods (e.g., for methodological reasons). As one of the next steps, we invite colleagues to contact us about sharing their data for integration in our database. Thus, by increasing available sample sizes and numbers of parameter combinations, we hope to boost the power of statistical analyses in the face of natural variability of behavior.

Finally, the database is a valuable tool for benchmarking of modeling studies in software and hardware – revealing the discrepancies between natural and technologi-cal performance.

Acknowledgements. This work was supported by the EU project EMICAB (FP7-ICT, No. 270182) to VD, and by the Cluster of Excellence 277 CITEC funded in the framework of the German Excellence Initiative. The authors thank M. Nagel (Sie-mens AG) and R. Lapp (CT Imaging GmbH) for the microCT scans, and to A. Jagel, M. Koch, T. Vollbrecht, and G. Zentgraf for generating the high-quality visualization.

References

1. Cruse, H., Dürr, V., Schilling, M., Schmitz, J.: Principles of insect locomotion. In: Arena, P., Patanè, L. (eds.) Spatial Temporal Patterns for Action-Oriented Perception in Roving Robots, pp. 43–96. Springer, Berlin (2009)
2. Büschges, A.: Lessons for circuit function from large insects: towards understanding the neural basis of motor flexibility. Curr. Opin. Neurobiol. 22, 602–608 (2012)
3. Dürr, V., Schmitz, J., Cruse, H.: Behaviour-based modelling of hexapod locomotion: linking biology and technical application. Arthr. Struct. Dev. 33, 237–250 (2004)
4. Schmitz, J., Schneider, A., Schilling, M., Cruse, H.: No need for a body model: Positive velocity feedback for the control of an 18-DOF robot walker. Applied Bionics and Biomechanics 5, 135–147 (2008)
5. Schneider, A., Paskarbeit, J., Schäffersmann, M., Schmitz, J.: Biomechatronics for embodied intelligence of an insectoid robot. In: Jeschke, S., Liu, H., Schilberg, D. (eds.) ICIRA 2011, Part II. LNCS, vol. 7102, pp. 1–11. Springer, Heidelberg (2011)
6. Safonova, A., Hodgins, J.K., Pollard, N.S.: Synthesizing physically realistic human motion in low-dimensional, behavior-specific spaces. ACM Transactions on Graphics 23, 514–521 (2004)
7. Barbic, J., Safonova, A., Pan, J.-Y., Faloutsos, C., Hodgins, J.K., Pollard, N.S.: Segmenting motion capture data into distinct behaviors. In: Proc. Graphics Interface, GI 2004, pp. 185–194 (2004)
8. Parks, D.H.: Analyzing the structure of a motion capture database under a similarity metric. Technical Report CPSC533C: University of British Columbia. Vancouver, Canada (2008)
9. Müller, M., Röder, T., Clausen, M.: Efficient content-based retrieval of motion capture data. ACM Transactions on Graphics 24(3), 677–685 (2005)
10. Demuth, B., Röder, T., Müller, M., Eberhardt, B.: An information retrieval system for motion capture data. In: Lalmas, M., MacFarlane, A., Rüger, S.M., Tombros, A., Tsikrika, T., Yavlinsky, A. (eds.) ECIR 2006. LNCS, vol. 3936, pp. 373–384. Springer, Heidelberg (2006)
11. Müller, M., Baak, A., Seidel, H.-P.: Efficient and robust annotation of motion capture data. In: Proc. ACM SIGGRAPH/Eurographics Symposium on Computer Animation. (2009)
12. Theunissen, L.M., Dürr, V.: Insects use two distinct classes of steps during unrestrained locomotion. PLOS one 8, e85321 (2013)
13. Cruse, H.: The control of the anterior extreme position of the hindleg of a walking insect. Carausius morosus. Physiol. Entomol. 4, 121–124 (1979)
14. Lorensen, W.E., Cline, H.E.: Marching Cubes: A high resolution 3D surface construction algorithm. In: Proc. ACM SIGGRAPH, pp. 163–170 (1987)
15. Botsch, M., Kobbelt, L., Pauly, M., Alliez, P., Levy, B.: Polygon Mesh Processing. AK Peters (2010)
16. Kavan, L., Collins, S., Zara, J., O'Sullican, C.: Geometric skinning with approximate dual quaternion blending. ACM Transactions on Graphics 27(4) (2008)
17. Berners-Lee, T., Hendler, J., Lassila, O.: The Semantic Web: a new form of Web content that is meaningful to computers will unleash a revolution of new possibilities. Sci. Am. 284, 34–43 (2001)
18. Heath, T., Bizer, C.: Linked Data: Evolving the web into a global data space. Synthesis Lectures on the Semantic Web: Theory and Technology Morgan & Claypool (2011)
19. Wiljes, C., Cimiano, P.: Linked Data for the natural sciences: Two use cases in chemistry and biology. In: Proc. Workshop on the Semantic Publishing, SePublica 2012, pp. 48–59 (2012)
20. Schleif, F.-M., Mokbel, B., Gisbrecht, A., Theunissen, L., Dürr, V., Hammer, B.: Learning relevant time points for time-series data in the life sciences. In: Villa, A.E., Duch, W., Érdi, P., Masulli, F., Palm, G. (eds.) ICANN 2012, Part II. LNCS, vol. 7553, pp. 531–539. Springer, Heidelberg (2012)

Benchmarking Human-Like Posture and Locomotion of Humanoid Robots: A Preliminary Scheme

Diego Torricelli[1], Rahman S.M. Mizanoor [2], Jose Gonzalez[3], Vittorio Lippi[4],
Georg Hettich[4], Lorenz Asslaender[4], Maarten Weckx[3], Bram Vanderborght[3],
Strahinja Dosen[5], Massimo Sartori[5], Jie Zhao[6], Steffen Schütz[6],
Qi Liu[6], Thomas Mergner[4], Dirk Lefeber[3], Dario Farina[5], Karsten Berns[6],
and Jose Louis Pons[1]

[1] Bioengineering Group, Spanish National Research Center (CSIC), Madrid, Spain
diego.torricelli@csic.es
[2] Department of Mechanical Engineering, Vrije Universiteit Brussel, Belgium
mizanoor.rahman@vub.ac.be
[3] Technaid S.L., Madrid, Spain
jose.gonzalez@technaid.com
[4] Neurology, University of Freiburg, Neurozentrum, Freiburg, Germany
vittorio.lippi@uniklinik-freiburg.de
[5] University Medical Center Göttingen, Germany
strahinja.dosen@bccn.uni-goettingen.de
[6] Robotics Research Lab, University of Kaiserslautern, Germany
zhao@cs.uni-kl.de

Abstract. The difficulty in defining standard benchmarks for human likeness is a well-known problem in bipedal robotics. This paper proposes the conceptual design of a novel benchmarking scheme for bipedal robots based on existing criteria and benchmarks related to the sensorimotor mechanisms involved in human walking and posture. The proposed scheme aims to be sufficiently generic to permit its application to a wide range of bipedal platforms, and sufficiently specific to rigorously test the sensorimotor skills found in humans.

The achievement of global consensus on the definition of human likeness has a crucial importance not only in the field of humanoid robotics, but also in neuroscience and clinical settings. The EU project H2R is specifically dealing with this problem. A preliminary solution is here given to encourage the international discussion on this topic within the scientific community.

Keywords: Benchmarking, human likeness, bipedal robots, walking, standing, posture.

1   Introduction

The issue of benchmarking robotic locomotion has been receiving increasing attention during the last decade [1, 2]. The use of benchmarks may help standardize the robot designs and fabrication, ease the evaluation and comparison of the systems, and exploit the biological solutions for many unsolved problems such as the mechanical compliance, energy consumption and enhanced human-robot interactions.

A. Duff et al. (eds.): Living Machines 2014, LNAI 8608, pp. 320–331, 2014.

The main obstacle in identifying common benchmarks is that different methods and metrics are typically employed and reported for specific robotic systems and functional scenarios [1]. In the field of humanoid robots, the benchmarks previously proposed either focus only on the result of a specific task (e.g. open a door, stair climbing, [3]) or on specific problems related to intelligence (e.g. social and multi-agent interactions, [4]).

Stable locomotion represents one of the main challenges in humanoid robotics for which no well-defined standards and benchmarking schemes exist. Despite their potential for high mobility, most bipeds have never been tested outside the laboratory. The problem is mainly due to the fact that the control and mechatronics of a legged machine is intrinsically a complex issue and its evaluation and comparison is very difficult. Human likeness is considered a significant criterion for design excellence, under the perspective of smoothness, versatility and energy efficiency [5]. However, a generalized, well-accepted and complete benchmarking scheme of human-like locomotion is not yet available.

Within the EU project H2R [6], we are promoting the international discussion on the features to be included in the ideal benchmarking scheme. As a first step towards this goal, we are presenting a conceptual design of a benchmarking scheme for human-like walking and standing. The proposed scheme has a twofold goal. On the one hand it aims to be sufficiently generic and versatile to permit its application to a wide range of bipedal platforms and scenarios. On the other hand, it should be sufficiently specific to rigorously test the sensorimotor skills found in humans. We specifically omitted arm, articulated spine and head movements, although they can implicitly influence gait and postural behaviors.

2 The Proposed Benchmarking Scheme

2.1 General Concepts

Our goal is to identify a subset of the possible human-like features related to walking and standing, which are applicable to different robotic bipedal platforms, regardless of their weight, size, number of degrees of freedom, or control architecture. This proposal is limited to those features related to sensorimotor control. High-level cognitive abilities, such as path planning, prediction, external world recognition, and learning, are not included in this scheme.

To standardize a method for the evaluation of the different human-like features in the different domains, we propose a benchmarking scheme that is schematically resumed in Table 1 and Table 2. Each scheme is structured in three main sections:

A. *Sensorimotor tasks.* In this part, the benchmarking scheme reports the internal and external constraints of the problem, meaning the desired motor task (e.g. standing, walking), and the interaction with the external world (e.g. pushes, inclined surface, etc.). We refer to this as to sensorimotor tasks because they underlie feedback propagation and processing of somatosensory information including joint internal forces and whole-body position.

B. *Benchmarking criteria*. It includes the specific sensorimotor abilities to be measured (e.g. robustness, energy consumption) and the different benchmarks that are used to quantify them (e.g. speed, falling rate, etc.).

C. *Perturbation devices*. This third part includes possible external perturbation devices that can be used in a laboratory setting to further measure the benchmarks proposed in the previous part (item B).

2.2 Benchmarking Scheme for Standing

Posture control is the ability to maintain body center of mass (CoM) above the base of support, compensating external disturbances such as gravity, pushes or support surface movements. Posture control deficits severely affect walking performance. For this reason, testing postural skills has high relevance for the assessment of locomotion abilities, both in humans and robotic scenarios [7, 8]. Postural skills in humans are normally assessed by analyzing the excursion of center of pressure (CoP) and center of mass (CoM). Also kinematics and intersegmental forces can be measured by applying motion analysis techniques and mechanical modeling [9].

In this section, we present a number of methods and criteria that have been used either in human studies or in robotic scenarios, to measure the postural control ability of a bipedal structure. Among all the methods identified in the literature, we selected those that in our opinion are particularly suitable to reveal human-like features (Table 1).

Sensorimotor Tasks

We identified three relevant sensorimotor scenarios (see Table 1A): i) unperturbed standing, ii) perturbed standing, and iii) voluntary CoM displacements.

Unperturbed standing is normally assessed on static horizontal or inclined support surface. The inclination of the surface can be in the anterio-posterior direction (sagittal plane) or in the medio-lateral direction (frontal plane).

In the case of perturbed standing, the main proposed perturbations will be (a) pushes, (b) support surface tilts, and (c) support surface translations. Within the tilting perturbations, the body sway referenced platform (BSRP) mode is a useful technique to eliminate ankle proprioceptive feedback and selectively test the vestibular system in the absence of vision [10]. The directions of disturbances are distinguished in sagittal plane and frontal plane, except for BSRP mode, which is limited to sagittal plane due to ankle morphology.

Concerning voluntary movements, a typical test that we consider useful is the rhythmic weight shift (RWS). It is already used in clinical scenarios, and consists of performing sinusoidal rhythmic sways with different amplitudes and frequencies in sagittal and lateral directions [11]. Another test that should be considered is the maximum static leaning. This test is also used in clinical settings, under the name of Limits of Stability (LOS) test [12].

Table 1. The proposed benchmarking scheme for standing

A. Sensorimotor tasks

Motor task	Condition	Direction	
Unperturbed	Horizontal surface	-	
	Inclined surface	Sagittal	
		Lateral	
Perturbed	Pushes	Sagittal	
		Lateral	
	Surface tilt	Sagittal	
		Lateral	
		Body sway reference platform (BSRP)	
	Surface translation	Sagittal	
		Lateral	
Voluntary movements	Rhythmic Weight Shift (RWS) test	Sagittal	
		Lateral	
	Limits of Stability (LOS) test	Sagittal	
		Lateral	

B. Benchmarking criteria

Feature	Criterion	Measurements tools		
Stability	Max support amplitude/ frequency	X		X
	Max external force	X	X	
	Max voluntary CoM displacement/frequency	X		X
	Energy Stability Margin (EMS)		X	X
	Time standing on BSRP	X		
Motion	Joint kinematics			X
	Frequency Response Function (FRF)		X	
	Compliance to external forces		X	X
	Ankle-hip coordination		X	X
	Natural looking motion	X		
Energy efficiency	Energy per DoF*		X	X
	Energy per weight-height*		X	X

◁ = Visual inspection, ⊥ = Kinetics, ⌐ = Kinematics

* To be further defined

C. Perturbation devices

Function	Typology		
Support	Fixed	Sloped board	
	Passive	Rubber foam	
		Wheel board	
		Rocker board	
	Actuated	Tilt	
		4-bar mechanism	
		Stuart platform	
Pushing	Manual		
	Falling ball		
	Rope and winch		

Benchmarking Criteria

In the proposed scheme, we identified the following features to evaluate human-like robotic behavior (Table 1B): i) stability, ii) motion, and iii) energy efficiency.

Stability can be defined as the ability to maintain balance without falling. Stability can be defined as the ability to maintain balance without falling. The maximum magnitudes of the disturbance/condition that the robot can handle before falling can be considered as a quantitative metrics of stability [13]. In particular, the following metrics are proposed: for perturbed scenario, maximum amplitude and frequency of disturbance (tilt or translation), and maximum external force (for push condition) can be used. For voluntary dynamic movements (RWS test), the maximum amplitude and frequency of CoM displacements is proposed. An accepted approach to identify the limits of stability is the Energy Stability Margin (ESM) proposed by Messuri & Klein [14] and its extensions [15, 16], which describe the minimal potential energy necessary to tumble the robot over one of the boundary edges of the support polygon. In the case of BSRP mode, a possible metric of stability is the time the robot is able to stand without falling.

Human-like (or natural looking) motion can be assessed directly – by comparing human and robot trajectories in the space of joint positions – or globally – by measuring whole body motion or the relative contribution of different body parts. Whole body motion can be accurately quantified using the frequency response function (FRF), which allows to characterize the gain and phase of CoM displacement in response to the, typically pseudorandom tilt disturbance [7, 17]. Visual inspection from human observers, using a dot motion display, has been proposed as an effective alternative way to characterize human-like motion from a global perspective [18].

In human postural tasks, ankle and hip move coordinately to keep the CoM over the base of support. The relative contribution of hip and ankle strategy changes with the magnitudes and typology of the disturbance [13], as well as with age and sensory impairments. An appropriate quantification of this inter-joint coordination can be obtained by correlating sways during sine wave stimulus, as done in [19]. Humans are also characterized by compliant behavior, emerging in response to interactions with the environment and to external disturbances. We propose to estimate compliance during standing by measuring the CoM displacements in consequence of external pushes.

Balancing of upright stance in humans is a continuously active process, which therefore requires energy [20]. However, to our best knowledge, assessing energy costs during standing is not explicitly considered in bipedal robotics. We believe that efficiency should be taken into account in the ideal benchmarking scheme for standing, in particular during perturbed conditions. In this direction, a reasonable solution may be to adapt some of the methods commonly used for walking, such as the energy-per-DoF or the energy-per-height/weight (see Section 2.3).

Perturbation Devices

Perturbation devices may introduce important constraints in the application of benchmarking procedures, thus biasing the comparison between robots. We propose a number of solutions that can be implemented in most laboratory settings (Table 1C).

Support surface devices are classified in fixed and moving. Moving devices are distinguished in actuated (by motors or manually) and passive. Among the passive solutions, a rubber foam can be used to induce effective tilting disturbances, as it is currently done in clinical setting. A wheel board can be used to induce translational disturbances. The rocker board is an interesting solution that combines tilt perturbations and translations. Even if very simple, this condition is very challenging for robotic control systems. An effective device for translations is a parallel swing mechanism attached to the roof. Complex conditions like 3D or pseudorandom surface movements should be induced by a motion platform, often in the form of a Stuart platform. Finally, pushing disturbances could be performed manually by a swinging ball of known weight and release height, or by a rope and winch.

Safety mechanisms should be used in all scenarios to avoid robot breaking after falling. A simple rope attached to the roof, or similar solution, should be sufficient.

2.3   Benchmarking Scheme for Walking

Achieving stable and efficient walking is a crucial goal in humanoid design. In humans, walking emerges from the combination of several mechanisms, which include neural, biomechanical and morphological aspects [21]. As a result humans show very robust, versatile and energy efficient functional abilities in a vast range of conditions. It is generally believed that translating such human-like principles into robotic platforms may improve their functional performance. In human studies, huge quantitative information on limb kinematics and kinetics has been collected over the last century, and several metrics have been derived from it. Among these, we selected appropriate candidates that can describe typical human behavior and that can be easily applied to the robotic scenario. Results are depicted in Table 2.

Sensorimotor Tasks

Walking tasks are organized in three classes (Table 2A): unperturbed walking, perturbed walking, and more complex voluntary motion. In the unperturbed domain, horizontal flat walking with body weight support (BWS) is convenient to test rhythmic leg movements separately from balance. If BWS is removed, free over-ground walking is obtained, where all the basic functions of locomotion can be tested. Presence of slopes has been also included within the unperturbed scenario, because no changes in the environment occur. Three kinds of slopes have been included: upward, downward, and lateral.

Four perturbed scenarios are considered: pushes, rough terrain, changes in slope and weight bearing, in sagittal and lateral direction. In the weight-bearing scenario, the condition of transporting a backpack of unknown weight is considered.

The last class of benchmarking scenario is related to voluntary changes of locomotion parameters. Three cases will be considered. The first one is changing from standing to walking, and vice versa. This condition involves complex coordination, such as shifting body weight to one foot, inclining the upper body, and taking the first step. The second task is the online change of walking speed. The third voluntary task is to change direction during locomotion.

Table 2. The proposed benchmarking scheme for walking

A. Sensorimotor tasks

Motor task	External condition	Direction	
Unperturbed walking	Flat ground, with BWS	-	
	Flat ground	-	
	Continuous Slopes	Up	
		Down	
		Lateral	
Perturbed walking	Pushes	Sagittal	
		Lateral	
	Rough terrain	Sagittal	
		Lateral	
	Change slope	Sagittal	
	Weight bearing		
Voluntary transitions	Start-stop walking	-	
	Change speed	-	
	Curved path	-	

B. Benchmarking criteria

Feature	Criterion	Measures ◁	⏚	⌐
Stability	Max and min speed	X		X
	Max number of steps	X		
	Max slope	X		X
	Max ext. force	X	X	
	Max obstacles dimension	X		
	Max ground softness	X		
	Max load	X		
	Max centrifugal force	X		X
	Success rate of transitions	X		
Motion	Kinematic profiles			X
	Gait harmony		X	X
	Dynamic Time Warping			X
	Heel, ankle, forefoot rocker			X
	Natural looking motion	X		
	Compliance to ext. forces	X	X	X
Kinetics (DYNAMIC SIMILARITY)	Equal relative foot phase			X
	Equal duty factor			X
	Equal Froude number			X
	Equal force on foot		X	X
	Power prop. m*v		X	X
	Joint moments profiles		X	X
	Passive Gait Measure (PGM)		X	X
	Dynamic Gait Measure (DGM)		X	X
Energy efficiency	Spec energy cost transp. C_{et}		X	X
	Spec mech. cost transp. C_{mt}		X	X
	Energy per DoF		X	X

◁ = Visual inspection, ⏚ = Kinetics, ⌐ = Kinematics

C. Perturbation devices

Function	Typology		
Terrain	Clear	Static slope	
		Treadmill	
	Irregular	Hard & continuous	
		Soft & continuous	
		Sparse obstacles	
Pushing	Manual		
	Falling ball		
	Rope and winch		

Benchmarking Criteria

Stability. The most general criterion for robot (and human) stability is the robustness to falling. The condition of "falling" is easily detectable by visual inspection. As done in standing, to convert this discrete criterion into quantitative metrics, the maximum magnitudes of the disturbance/condition that the robot can handle before falling will be considered. For all walking conditions, the maximum and minimum speed, as well as the maximum number of steps before falling can be used. In more specific scenarios, the following metrics can be defined: for walking on slopes, the maximum ground inclination allowed; for pushes, the maximum external force tolerated by the biped during single- and double-stance phases; for irregular terrain, the maximum obstacle dimension, and in the case of soft terrain, the maximum ground softness; in weight bearing condition, the maximum permitted loading; in the case of curved path, an indicator of the maximum centrifugal force, e.g. the ratio between speed and radius of curvature. For voluntary transitions, stability can be hardly quantified on a continuous scale, because these can be either performed or not during a single trial. For this reason, we propose to measure the success rate of achievement across a fixed amount of trials.

Kinematic Analysis. At the kinematic level, the basic procedure to compare robots with human is direct correlation of joint kinematic profiles. In particular, knee bending, pelvis movements and foot placement seem to be strong indicators of human likeness. On a more global level, the gait harmony has been proposed as a metric to measure the synchrony and symmetry of whole-body gait movements [22], obtained by either spatiotemporal parameters or acceleration harmonics of the CoM. A further good candidate for comparing kinematics across robots and humans is the Dynamic Time Warping (DTW) method [23], which measures the similarity between temporal sequences that vary in time or speed. It is typically applied in speech recognition, and no application to walking has been found in literature. Foot motion is also a crucial aspect in walking. In the ideal benchmarking scheme, the assessment of basic wheel-like mechanisms of the foot - namely heel, ankle, and forefoot rockers – should be included [24, 25]. An alternative way to evaluate human-like motion is by visual inspection using the dot motion display, as discussed in the section on standing. Alexander [26] postulated five criteria to assess dynamic similarity across bipeds of any size. Three of these criteria concern motion: I) equal contralateral relative foot phase, which measures the symmetry of movement, II) equal duty factor, which indicates the percentage of stance phase within a gait cycle, and III) equal Froude number [27]. Previous described methods are normally applied during unperturbed conditions. Perturbed scenario may also be useful to test one typical feature of human motion: compliance to external disturbances. To our knowledge, no well-defined methods have been proposed in literature to quantify how the compliant characteristics of the human body affect the reaction to unexpected forces.

Kinetic Analysis. Human locomotion is characterized by passive and active dynamic behaviors interspersed throughout a gait cycle. A classic approach to analyze kinetics is to measure and then compare the joint moment profiles. Recently, the Passive Gait Measure (PGM) and Dynamic Gait Measure (DGM) have been proposed to quantify the passive and dynamic characteristics of human-like locomotion [28]. Internal and external forces in human walking are repeatable variables, which could be used as gold standards for comparison with humanoid counterparts. Alexander also postulated two criteria for dynamic similar gaits concerning forces, in addition to the previous three motion-based criteria: IV) forces on feet are equal multiples of body weight; and V) power outputs are proportional to body weight times speed.

Energy Efficiency. Energy consumption is one of the most important factors that mark the difference between humans and robots. A widely used benchmark of walking efficiency is the specific cost of transport (c_t) [5, 29], defined as the ratio of the energy consumed and the weight times the distance travelled. The specific energetic cost of transport comprises the total energy consumed, including positive and negative work of actuators and energy costs related to electronics. The specific *mechanical* cost of transport (c_{mt}) is needed when one wants to isolate the positive mechanical work, and more reliably compare the energy costs of different robotic platforms independently from the control aspects. As an application example, the specific costs of transport of three robots compared to human values were reported in [5]. Energy and power consumption can be also measured at joint level, by measuring torques and joint speeds.

Perturbation Devices

A successful benchmarking scheme should be easily applicable to different platforms and laboratory scenarios. To this aim, we propose some practical solutions, taking into account that perturbations should be applied in a wide range of magnitudes and timing, and that the device should be easy to use, as low cost as possible, and possibly available in the market.

In the case of inclined walking, the sagittal slope can be reproduced by a static support (e.g. wooden) or by an off-the-shelf inclined treadmill. For lateral slopes, the static support may be the preferred solution. The perturbation device should allow for inclinations up to at least 10° upwards and downwards.

Possible solutions for pushing disturbances are a pendulum made of a known mass released from a predefined height, or a rope connected to a sensorised winch, similarly to what proposed in standing condition. In both cases, the robot should be placed on a treadmill, since both ball and rope should be conveniently attached to static supports. Also manual pushes can be considered, which may allow for a very fast and easy to use test, yet more difficult to measure in a quantitative way.

Reproducing rough terrain should take into account the following variables: i) dimension of the obstacle with respect to the dimension of the biped, ii) shape of the obstacle (e.g. presence of inclined surfaces), iii) rigidity of the obstacle (hard or soft), iv) number of obstacles, and v) distance between obstacles. Among the infinite possible combinations of these variables, we propose three different scenarios: 1) continuous irregular terrain, to test the capability of the ankle to adapt to the terrain during the weight acceptance; 2) continuous soft terrain, to test the ability to maintain balance and generate propulsion, and 3) sparse obstacles located on the ground, to test the reaction to collisions and unexpected deviations of limb motion.

3   Discussion

With respect to current benchmarking in robotic competitions – e.g. DARPA robotic challenge and Robocup - this scheme represents a complementary tool specifically focused on human-like behavior rather than absolute performance. In this respect, it is worth mentioning that higher human likeness does not necessarily mean higher functional or stability performance. For instance, during standing, robots could easily outperform human performance in terms of stability. Conversely, during walking, achieving human like performance would translate into increased stability, versatility and efficiency. Another relevant advantage with respect to competition contexts is that testing specific sensorimotor functions instead of global behavior may constitute a useful approach to better identify specific cause-effect relationships.

The development of this benchmarking scheme is still at a very early stage. With this paper we want to provide a robust and generic benchmarking structure as well as stimulate the international discussion on the relevant features and methods to be included. As a practical effort in this direction, we are currently developing an open source software that will be disseminated in the near future within the robotics community. The software is made of two parts. In the first part, the user can select the motor tasks to test, as well as the limitations on measurement/perturbation technology available in the laboratory. With these inputs, the software will define the type of benchmarks that can be applied to the robotic platform. The second part of the software allows calculating and reporting the benchmark values obtained by the particular robot, and store the data used to obtain these results. The first release of this software will also serve as a beta-test for the evaluation of usability across different laboratory contexts. Through a web-based application, feedback from users will be collected on the following issues: i) relevance of the proposed benchmarks, ii) usability of the software, and iii) ideas for improvements, e.g. new benchmarks, protocols, or devices. The user will also have the opportunity to share the values obtained by the specific robot. These data will be useful to start building appropriate scales for the correct evaluation of human likeness, through the comparison with values obtained by other laboratories and by human experiments in the same conditions. A mailing list has been already created [6] to collect ideas and disseminate events specifically focused on this topic.

Further efforts should be devoted to define the actual reference values necessary to develop meaningful quantitative scales. The general rationale is to assume the human behavior as golden standard, namely 100% human-like. Appropriate database of human data in all conditions should thus be made available. The generation of human data not available yet may constitute an interesting goal for human experimental research. An additional open issue is the definition of disturbance magnitudes, which should be appropriately scaled – according to the existing scaling laws [30] – to allow its usage across bipeds with different sizes and weights.

In this paper, we did not include an extensive analysis of the state of the art of current benchmarking methods, due to space limitations. This will be possibly included in a further extended publication, and used to propose first guess estimations on quantitative reference ranges.

4   Conclusion

We believe that benchmarking human-like posture and locomotion is a major challenge and unresolved key issue for humanoid robotics. The major goal of our efforts is to produce a benchmarking solution that can be realistically used by most research groups around the world. For this reason, we propose a scheme based on a modular structure, which can permit to select the subgroups of conditions, benchmarks and devices that are more suitable to the specific robotic platform and laboratory setting.

Future efforts should be devoted to: i) the inclusion of new ideas for benchmarks and metrics arising from the community, ii) the definition of a minimum subset of benchmarks among all proposed, iii) the establishment of a reference database of human data for all conditions, iv) the definition of absolute ranges of all metrics proposed. To facilitate this process, the H2R project is currently developing an open source software that will be soon shared within the international community.

Acknowledgements. The research activity presented in this paper has been funded by the European Seventh Framework Programme FP7-ICT-2011-9, under the grant agreement no 60069 - H2R "Integrative Approach for the Emergence of Human-like Robot Locomotion".

References

1. Behnke, S.: Robot Competitions Ideal Benchmarks for Robotics Research. In: IROS 2006 - Workshop on Benchmarks in Robotics Research (2006)
2. del Pôbil, P.: A.: Why do We Need Benchmarks in Robotics Research? In: IROS2006 - Workshop on Benchmarks in Robotics Research (2006)
3. DARPA Robotics Challenge Trials (DRC), http://www.theroboticschallenge.org/
4. Rahman, S.M.M.: Evaluating and Benchmarking the Interactions between a Humanoid Robot and a Virtual Human for a Real-World Social Task. In: Papasratorn, B., Charoenkitkarn, N., Vanijja, V., Chongsuphajaisiddhi, V. (eds.) IAIT 2013. CCIS, vol. 409, pp. 184–197. Springer, Heidelberg (2013)
5. Collins, S., Ruina, A., Tedrake, R., Wisse, M.: Efficient bipedal robots based on passive-dynamic walkers. Science 307, 1082–1085 (2005)
6. H2R, Integrative approach for the emergence of human-like locomotion, FP7-ICT-2011-9 Agreement no60069, http://www.h2rproject.eu
7. Peterka, R.J.: Sensorimotor integration in human postural control. J. Neurophysiol. 88, 1097–1118 (2002)
8. Mergner, T., Schweigart, G., Fennell, L.: Vestibular humanoid postural control. J. Physiol. Paris 103, 178–194 (2009)
9. Winter, D.A.: Biomechanics and motor control of human gait: normal, elderly and pathological (1991)
10. Ishida, A., Imai, S., Fukuoka, Y.: Analysis of the posture control system under fixed and sway-referenced support conditions. IEEE Trans. Biomed. Eng. 44, 331–336 (1997)
11. Cheng, P.-T., Wang, C.-M., Chung, C.-Y., Chen, C.-L.: Effects of visual feedback rhythmic weight-shift training on hemiplegic stroke patients. Clin. Rehabil. 18, 747–753 (2004)

12. Wallmann, H.W.: Comparison of elderly nonfallers and fallers on performance measures of functional reach, sensory organization, and limits of stability. J. Gerontol. A. Biol. Sci. Med. Sci. 56, M580–M583 (2001)
13. Horak, F., Macpherson, J.: Postural orientation and equilibrium. Compr. Physiol. (1996)
14. Messuri, D., Klein, C.: Automatic body regulation for maintaining stability of a legged vehicle during rough-terrain locomotion. IEEE J. Robot. Autom. 1 (1985)
15. Lin, B.-S., Song, S.-M.: Dynamic modeling, stability and energy efficiency of a quadrupedal walking machine. In: Proc. IEEE Int. Conf. Robot. Autom. (1993)
16. Hirose, S., Tsukagoshi, H., Yoneda, K.: Normalized energy stability margin and its contour of walking vehicles on rough terrain. In: Proc. 2001 ICRA. IEEE Int. Conf. Robot. Autom (Cat. No.01CH37164), vol. 1 (2001)
17. Goodworth, A.D., Peterka, R.J.: Contribution of sensorimotor integration to spinal stabilization in humans. J. Neurophysiol. 102, 496–512 (2009)
18. Troje, N.F.: Decomposing biological motion: a framework for analysis and synthesis of human gait patterns. J. Vis. 2, 371–387 (2002)
19. Schweigart, G., Mergner, T.: Human stance control beyond steady state response and inverted pendulum simplification. Exp. Brain Res. 185, 635–653 (2008)
20. Miles-Chan, J.L., Sarafian, D., Montani, J.-P., Schutz, Y., Dulloo, A.: Heterogeneity in the energy cost of posture maintenance during standing relative to sitting: phenotyping according to magnitude and time-course. PLoS One 8, e65827 (2013)
21. Bernstein, N.A.: Dexterity and its development (1996)
22. Iosa, M., Fusco, A., Marchetti, F., Morone, G., Caltagirone, C., Paolucci, S., Peppe, A.: The golden ratio of gait harmony: repetitive proportions of repetitive gait phases. Biomed Res. Int. 2013, 918642 (2013)
23. Lemire, D.: Faster retrieval with a two-pass dynamic-time-warping lower bound. Pattern Recognit. 42, 2169–2180 (2009)
24. Perry, J.: Gait analysis: normal and pathological function. SLACK Inc. (1992)
25. Hansen, A.H., Childress, D.S., Knox, E.H.: Roll-over shapes of human locomotor systems: effects of walking speed. Clin. Biomech. 19, 407–414 (2004)
26. Alexander, R.M.: The Gaits of Bipedal and Quadrupedal Animals. Int. J. Rob. Res. 3, 49–59 (1984)
27. Duncan, W.J.: Physical similarity and dimensional analysis, London (1957)
28. Mummolo, C., Kim, J.H.: Passive and dynamic gait measures for biped mechanism: formulation and simulation analysis. Robotica 31, 555–572 (2012)
29. Gabrielli, G., von Kármán, T.: What price speed? Specific power required for propulsion of vehicles. Mech. Eng. ASME 72, 775–781 (1950)
30. Dermitzakis, K., Carbajal, J.P., Marden, J.H.: Scaling laws in robotics. Procedia Computer Science, 250–252 (2011)

The Influence of Behavioral Complexity on Robot Perception

Vasiliki Vouloutsi, Klaudia Grechuta,
Stéphane Lallée, and Paul F.M.J. Verschure

1 Universitat Pompeu Fabra (UPF), Synthetic, Perceptive,
Emotive and Cognitive Systems group (SPECS)
http://specs.upf.edu
2 Institució Catalana de Recerca i Estudis Avançats (ICREA)
Passeig Llus Companys 23, 08010 Barcelona, Spain
http://www.icrea.cat

Abstract. Since robots' capabilities increase, they will soon be present in our daily lives and will be required to interact with humans in a natural way. Furthermore, robots will need to be removed from controlled environments and tested in public places where untrained people will be able to freely interact with them. Such needs raise a number of issues: what kind of behaviors are considered important in promoting interaction and how these behaviors affect people's perception regarding the robot in terms of anthropomorphism, likeability, animacy and perceived intelligence. In this paper, we propose a motivational and emotional system that drives the robot's behavior and test it against six interaction scenarios of varying complexity. In addition, we evaluate our system in two different environments: a controlled (laboratory) environment and a public space. Results suggest that the perception of the robot significantly changes depending on the complexity of the interaction but does not change depending on the environment.

Keywords: human-robot interaction, behavioral modulation, allostatic control, social robots.

1 Introduction

As technology advances, developing robots that will be part of our daily lives became a topic of great interest. With time, robots become more skilled and competent and soon they will not only serve as scientific research platforms, or industrial tools, but also as personalized assistants for people. Indeed, there seems to be a steady shift from industrial robots to more social ones, designed for business, education, health, or entertainment. Examples of such robots include robotic cleaning appliances [1], toys [2][3], tour guides [4][5], educational robots [6] or social partners [7].

Depending on the task and the purpose of the robot, different behaviors and levels of interaction with humans are required. We believe that it is important for

A. Duff et al. (eds.): Living Machines 2014, LNAI 8608, pp. 332–343, 2014.
© Springer International Publishing Switzerland 2014

robots which interact with humans to display complex behaviors and social characteristics. Our hypothesis is that the user's perception of robots that interact with humans is positively correlated with the complexity of their behavior.

Until now, robots that interact with humans have been confined in highly controlled environments where they mostly interact with trained users. Although the need to assess a robotic system in a public place, with untrained users is gradually satisfied [8][9][10][11], further studies regarding the behavioral components of a robot necessary for a natural human-robot interaction are required. Since the purpose of such robots is to interact with humans outside of laboratory conditions, we believe that the interaction environment (laboratory, museum, exhibition etc.) does not affect the human perception of these robots.

To test our hypothesis, we designed six different interaction scenarios of increasing behavioral complexity and evaluated them using the Godspeed Human-Robot Interaction Questionnaire [12] that measures the user's perception regarding anthropomorphism, likeability, animacy and perceived intelligence. Anthropomorphism refers to the attribution of human characteristics, behaviors or figures to non human things. Likeability expresses the positive impression one can attribute to another person, or an animal. Animacy reflects to life-like movement and intentional behavior while perceived intelligence depends on the robot's competence and behavioral coherence.

We expected that the behavioral complexity of the robot has a positive effect on how users perceive it. Additionally, we assessed the most complex behavioral scenario in a public space to investigate the influence, if any, the environment has on the perception of the robot. We also evaluated whether the perception of the robot differs between conditions where a user directly interacts with the robot or simply observes an interaction.

2 Setup

In the following section we describe the experimental setup used to assess the human perception of the robot. The setup consists of the humanoid robot iCub, a wheeled platform (iKart), a tangible table (Reactable) and a RGB/depth sensor (Kinect). The Kinect is used to provide information regarding the location of the human within the view of the robot. An example of the setup scenario is shown in Figure 1.

2.1 iCub and iKart

The iCub is a humanoid robot with dimensions similar to a 3.5 years old child. It has 53 actuated degrees of freedom distributed over the hands, arms, head and legs [13]. Its sensory inputs are provided by tactile, cameras and microphones. The robot is equipped with novel artificial skin covering the hands (fingertips and the palm) and the forearms. The facial expressions, which include the mouth and the eyebrows, are projected from behind the face panel using lines of red LEDs. The combination of the control architecture and the iCub we call H5Walpha.

The iCub is mounted on the iKart, a holonomic mobile platform designed to provide autonomous navigation capabilities in structured environments. Navigation capabilities are provided using odometry coupled with automatic recalibration in the critical section of the interaction (i.e before manipulating objects over the table).

The previously mentioned control architecture is based on Distributed Adaptive Control (DAC) [14], and tailored towards dyadic human robot interactions in a range of scenarios [15].

2.2 Reactable

The Reactable is a tabletop tangible interface [16] composed by a round table, with a translucent top and a back projector, which displays the properties of objects through the translucent top and a camera. Object recognition is performed using the reacTIVision tracking system. We took advantage of the Reactable by implementing several games that can be played by two players through object manipulation.

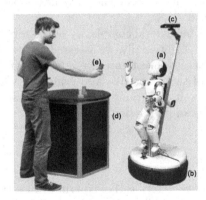

Fig. 1. Example of the proposed scenario: the humanoid iCub (a) is mounted on the iKart (b) to navigate within the environment. On top of the iCub, the Kinect sensor is placed (c) to provide information regarding the location of the human. The iCub can interact with the human in different interaction scenarios, including playing games and music using the Reactable (d) by manipulating objects (e).

2.3 Evaluation of the Different Scenarios

The six tested scenarios differ according to the complexity of the robot's behavior: "Still face", "Yoga", "Gaze", "Interpersonal Distance Regulation" (IDR), interaction without the Reactable ("Interaction NRT") and finally interaction with the Reactable ("Interaction RT"). Each scenario has a varying degree of complexity with respect to the previous one (see Table 1) varying from no behavior ("Still Face") to full interaction ("Interaction RT"). The scenarios are

Table 1. The table shows the behavioral parameters used for each of the six interaction scenarios. The complexity of each scenario is defined by the number of the parameters used.

	Interaction Scenarios				
	Still Face Yoga	Gaze	IDR	Interaction NRT	Interaction RT
Body Motion	Yes Yes	Yes	Yes	Yes	
Eye Contact	Yes	Yes	Yes	Yes	
Distance Regulation		Yes	Yes	Yes	
Speech			Yes	Yes	
Touch			Yes	Yes	
Proactive Behavior			Yes	Yes	
Playing Games				Yes	
Facial Expressions	Neutral Neutral	Neutral	Varying	Varying	Varying

defined by explicit parameters of the robot control in order to investigate the impact of the complexity of H5W's performance and the interpretation of the robot's capabilities by the human observer.

Still Face. In the "Still Face" (SF) scenario, the robot remains completely still, looking at a fixed point in the center of its view, and displays a neutral facial expression.

Yoga. The "Yoga" scenario consists of the robot performing a repeated sequence of pre-recorded body postures. The facial expression is set to neutral.

Gaze. Here, the iCub is reacting to the movement of the participant by directing its gaze (head movement) to the estimated location of the participant's head, therefore facilitating eye-contact. The robot does not engage in any other form of interaction or movement. In this scenario, the iCub's facial expression is set to neutral.

IDR. In the "Interpersonal Distance Regulation" (IDR) scenario, the iCub is reacting to the participant's movement by looking at him as well as maintaining a pre-defined interpersonal distance. The robot's facial expressions change depending on the perceived distance of the human: if the distance is too long, the robot is surprised and moves forward until it reaches the pre-defined distance; if the distance is too short, the robot moves backwards and displays the expression of fear, whereas if the distance between the robot and the participant is within the pre-defined range, the robot is happy. If no human is detected, the iCub's facial expression is set to neutral. The pre-defined distance is set to 0.9 meters with a deviation of $\pm$ 0.125 meters. The robot is always facing the partner by controlling the angular speed using a simple proportional controller. The robot is moving using the iKart with a linear speed of 0.075 m/s and an angular speed of 3.5 degrees per second.

Interaction NRT. In the "Interaction NRT" (INRT) scenario, the robot's behavior is based on a model of drives and emotions developed in [15] (see section 3): the robot's needs drive its behavior whereas external events affect the its emotional state. Hence, the robot is not passively acting or reacting to the human's actions but proactively engages in an interaction by commenting through speech about the status of the drives that need to be satisfied (out of homeostasis). Such a descriptive action creates a temporary satisfaction of the associated drive. The robot's abilities include maintaining interpersonal distance and tracking people using gaze, expressing emotions through facial expressions, gestures and prosody, discriminating between different types of touch and understanding generic spoken statements, questions and orders about a limited predefined vocabulary centered on the objects manipulated or displayed on the table and several people names.

Interaction RT. "Interacion RT" (IRT) scenario exceeds the previous one by adding the module of playing games. Here, the iCub's need to play games engages the human in two different activities. In each activity the robot is responding to the human's actions with speech, facial expressions and gestures. The first activity is a competitive game, namely Pong (see Fig. 2). Pong is a 2D simulated table tennis game where the ball is limited within a virtual rectangle. In order to win this game the player has to defeat the opponent by scoring more points. A point is scored after each ball passes the opponent's paddle line. In this scenario, the robot is commenting on various game's events, for example when the robot or the human has scored a point or has successfully hit the ball with the paddle.

The second activity is the musical DJ game (see Fig. 3). Here, both human and robot are collaborating to produce music. Each player has three different musical loops (bass, melody and drums) that they can activate by modifying each loop's volume. In this interaction scenario there is no fixed goal other than producing music. The robot's goal is to reach "musical symbiosis": a musical outcome that the human partner will like. This is achieved through a voting system (vote up, vote down), where the human partner can vote if the robot's musical loop choice was nice or not. The robot not only learns from the interaction but also comments on different events like *"i like that too"* if the human liked the robot's loop choice, or *"rise those beats up!"* if the human has been idle for sometime to motivate him to play more.

In both games the robot additionally uses modalities from the previously described scenarios, namely, body motion, eye contact, speech, touch, proactive behavior as well as facial expression.

Since the iCub is mounted on the iKart for navigation purposes, none of the above scenarios includes movement of the iCub's legs.

To evaluate the perception of the robot depending on the interaction scenario, we asked 82 students of the Pompeu Fabra University (UPF) to interact with the robot in one of the six scenarios. All participants, regardless of the type of interaction, received the same instructions: to enter the room where the iCub is located and interact with the robot in the most natural way. They were free to observe it, play with it, touch it or talk to it and could leave the experiment whenever they wanted. After the interaction, all participants were asked to fill out a questionnaire that measures the perception of the robot in the following domains: anthropomorphism, likeability, animacy and perceived intelligence (5-point Likert scale as defined in [12]). In the IRT scenario, participants were exposed to both games.

Usually, in the field of HRI, the interaction scenarios involve a robot and a human partner, and are conducted within a controlled (laboratory) environment [17]. Here we wanted to test whether our most complex system (IRT), originally designed for dyadic interactions with no direct implementation of handling more than one user, can also be used with multiple users interacting simultaneously with the robot, since a personalized interaction could lead to a different perception than the one where the participant only observed an interaction. Therefore we tested the H5Walpha with visitors to the "Barcelona Robotics Meeting 2014" held at the Mobile World Centre in Placa Catalunya in Barcelona for the.

Fig. 2. The pong game displayed on the Reactable

Fig. 3. The musicDJ game displayed on the Reactable

3   Behavioral Modulation System

As the behavior of the robot gradually becomes more complex (consisting of different behavioral modules), a system that coordinates the interaction between these modules is needed. To achieve such complex and coherent set of behaviors we propose a motivational system composed of drives and emotions. We argue that agents endowed with such a motivational system could be perceived as social agents, hence contributing to the human's perception of the robot regarding anthropomorphism, likeability, animacy and perceived intelligence.

The proposed system of drives and emotions (H5Walpha) is implemented in the INRT and IRT scenario and it is based on the well established cognitive architecture "Distributed Adaptive Control" (DAC) [18]. H5Walpha's behavior is autonomous and informed by its own drive system. Drives are part of a homeostatic mechanism [19] [20] used by an organism to maintain equilibrium. A closely related concept is allostasis, i.e. the process of achieving stability with the environment through change, which goes in the opposed direction of homeostasis, which achieves stability through constancy. Combining these two levels of control animals can perform real-world tasks like foraging, regulating their internal state and maintaining a dynamic stability with their surroundings [21]. Although there are various opinions about the nature of a drive, it is generally accepted that an organism is governed by multiple drives [22] as classically defined in Maslows hierarchy of needs [23]. Following the Ada control structure [24], and for the purpose of the interaction, our system implements four drives that control the robot's behavior: 1) the need for social interaction, 2) the need for physical interaction, 3) the need for spoken interaction and finally 4) the need for entertainment (playing games) with the human. The three former drives are implemented to the INRT whereas all of them are implemented in the IRT scenario.

Apart from the motivational mechanism, agents should be also endowed with emotions that not only define communicative signals by expressing one's internal state (external/utilitarian), but also organize behaviors (internal/epistemic) [25]. Here we propose that emotions, both epistemic and utilitarian, form a critically dependent drive reduction, i.e. emotions reflect the ability of the agent to achieve its goals. Furthermore, the utilitarian purposes of emotions are essential in human-robot interaction as they communicate internal states of the robot.

A more detailed explanation of the proposed system and architecture can be found in [15], and a video of the proposed interaction in [26].

4   Results

Here we investigate the impact of behavioral complexity in which H5W engages on human social perception. In particular we performed an exploratory data analysis was conducted to determine if the participants evaluation of the robot was normally distributed. Results for the Kolmogorov-Smirnov test for normality indicated that the distribution of anthropomorphism ($p = .0016$, SD $= .79$), animacy ($p = .001$, SD $= .85$), likeability ($p = .001$, SD $.86$) and intelligence ($p = .006$, SD $= .74$) deviate significantly from a normal distribution. We therefore used non-parametric tests to further analyze the data. In order to avoid a Type I error we applied a Bonferroni correction.

A Spearman Rank Order Correlation was run to determine the relationship between the participants perception of the robot regarding anthropomorphism, animacy, likeability as well as perceived intelligence and the type of interaction (SF, Yoga, Gaze, IDR, INRT, IRT). We found a positive statistically significant correlation (Fig. 4) between the interaction type and the four measurements: anthropomorphism ($\rho(121) = . 412$, $p < .001$), animacy ($\rho(121) = . 616$, $p < .001$), likeability ($\rho(121) = . 513$, $p < .001$) as well as perceived intelligence ($\rho(121) = . 552$, $p < .001$).

We ran a Kruskal-Wallis test to compare the perception of anthropomorphism, likeability, animacy and perceived intelligence between the least and most complex scenario. The results show significant differences in the perception of the robot between SF, Yoga, Gaze, IDR, INRT and IRT in anthropomorphism (H(5) $= 21.56$, $p = .001$), animacy (H(5) $= 50.31$, $p < .001$), likeability (H(5) $= 34.51$, $p < .001$) and perceived intelligence (H(5) $= 37.11$, $p < .001$).

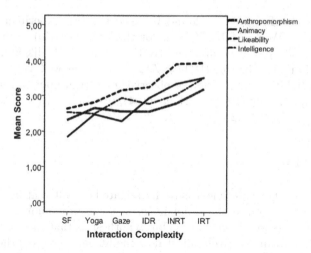

Fig. 4. The graph shows the mean scores by type of interaction for anthropomorphism, animacy, likeability and perceived intelligence

We conducted Mann-Whitney tests to follow up the analysis. The results show a significant difference in the perceived animacy between SF (p <.001), Yoga (p <.001), Gaze (p <.001), IDR (p = .005) and IRT but not between INRT and IRT. Similarly, we found a significant difference regarding the perceived intelligence between SF, Yoga (p <.001), Gaze (p = .001), IDR (p = .005) and IRT (Fig. 5a) but again, not between INRT and IRT. For the parameter of likeability, SF (p <.001), Yoga (p <.001) and Gaze (p = .002) scored significantly lower in comparison to IRT whereas, for anthropomorphism, SF (p = .003), Gaze (p = .005) and IDR (p = .007) scored significantly lower in comparison to IRT (Fig. 5b).

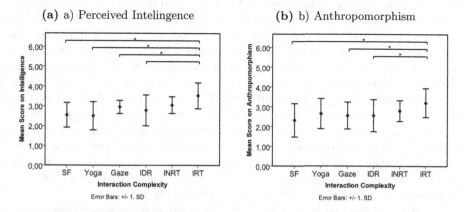

(a) a) Perceived Intelingence   **(b)** b) Anthropomorphism

Fig. 5. The graphs represent the mean score on the Linkert scale for each type of interaction for anthropomorphism, animacy, likeability and perceived intelligence. Stars (*) indicate significance level of (p<.01).

No statistical difference was found between the two experimental environments (SPECS - World Mobile Center for the IRT scenario) in anthropomorphism($H(1) = .42$, $p = .51$), animacy ($H(1) = 0$, $p = .99$), likeability ($H(1) = 1.49$, $p = .22$) and intelligence ($H(1) = .78$, $p = .37$) , as well as between people that participated in an interaction or observed one: anthropomorphism ($H(1) = 1.61$, $p = .29$), animacy ($H(1) = .21$, $p = .64$), likeability ($H(1) = .04$, $p = .82$) and intelligence ($H(1) = .26$, $p = .60$).

5 Discussion

The main goal of this experiment is to investigate how different behaviors affect the human's perception of the robot in terms of anthropomorphism, animacy, likeability and perceived intelligence. Our hypothesis is that the behavioral complexity of a robot can positively affect how the human perceives the robot during an interaction. We define as behavioral complexity the number of different modules that run simultaneously during an interaction. Such modules include body motion, eye contact (tracking a person's face), distance regulation, speech production and comprehension, tactile discrimination, displaying of emotions through facial expressions and proactive behavior. To test our hypothesis we devised six scenarios involving different levels of behavioral complexity starting from no interaction (the robot does not move/react) to full interaction (the robot's behavior is guided by its motivational system and is also reacting to the environment) and asked participants to evaluate the robot using the Godspeed Human-Robot Interaction Questionnaire. This questionnaire allows us to compare the results from each scenario and monitor how each measurement is affected.

The analysis of the data collected through the questionnaires show that there is a significant, though moderate, positive correlation between behavioral complexity and the perception of the robot in the four tested measurements. Therefore, more complex behaviors score higher in anthropomorphism, animacy, likeability and perceived intelligence. As robots will be meant to function in environments other than a laboratory, we tested the most complex scenario in two different environments: our laboratory (SPECS, Universitat Pompeu Fabra, Barcelona, Spain) and the "Mobile World Centre" in Placa Catalunya in Barcelona, Spain. Results show that there is no significant difference between the place of the interaction and the users' perception of the robot. No differences in the users' perception of the robot were also found when one is directly interacting with the robot (in the most complex scenario) or simply observing an interaction. These results are interesting as they allow us to assess an interaction in uncontrolled environments, without affecting the human's perception. Thus, it seems that one does not need to directly interact with a robot to form an opinion about it.

Further statistical analysis shows that anthropomorphism, which refers to the attribution of human characteristics, behaviors or forms to the robot, differed significantly between SF, Gaze, IDR and IRT but not between Yoga or INRT and IRT . Although Gaze, IDR and Yoga imply some sort of body motion, in the first two cases the robot only moves its head (and navigates in space with the iKart in the case of IDR), while its hands remain in the default position. For the iCub, the default hand position is not the same as in humans (straight, facing the ground) but form a 90-degree angle between the hand, shoulder and hip facing towards the front. In the Yoga scenario the robot is not tracking the human but shows a smooth hand-arm coordination and motion. These findings suggest that a coherent, human-like body motion can account for higher perception of anthropomorphism. To further investigate this assumption, we will run the Gaze and IDR experiment with a resting hand position that is closer to the one of humans. A further investigation of the influence of gaze on anthropomorphism is necessary, as so far it seems that it does not directly affect it.

In terms of perceived intelligence, the robot was evaluated significantly higher in the last scenario (IRT) in comparison to the first four (SF, Yoga, Gaze, IDR) but not compared to INRT. Since speech, touch and proactive behavior were added to INRT and IRT scenario simultaneously, we cannot report on the influence of the individual components, but only of their combination. The main difference between INRT and IRT is the addition of the playing games module, which is hard to dissociate into separate behavioral parameters. Overall, the most complex interaction scenario differed significantly from the first four scenarios in almost all measurements, but not the fifth one (INRT). We need to further investigate the exact components of the proactive behavior, touch and speech provided by the INRT that account for no difference between the four first scenarios as well as the IRT. Furthermore, as a single step, the ability to play games with a human is not enough to cause perceptual changes, but can contribute to higher perception of intelligence compared to the first four scenarios.

Our results show that the perception of the robot is indeed affected by its behavioral complexity. While self-reporting measurements provide an introspective view on the interaction, much more data can be collected through an external evaluation. In order to better understand which parameters, or which combinations cause modulations in human perception, the analysis of behavioral data is essential. Future work will aim at extracting an external point of view about the subject by video-coding the interaction. More specifically, the coding of the subject emotion could lead to investigation about the level of empathy generated by a given interaction or component.

Whether these results can be generalized to other types of robots is yet to be studied. We all well aware that further tests are need to understand which parameters affect the human's perception of the robot. From the knowledge acquired, we realized that smaller steps between each experimental setup are necessary to be able to clearly identify each parameter. Nonetheless, this preliminary study allows us to verify the fact that behavioral complexity is important and opens the road to identify which behavioral characteristics such as eye contact, body motion, speech, touch or proactive behavior, if manipulated, will contribute to a meaningful human-robot interaction.

Acknowledgments. This work is supported by the EU FP7 project EFAA (FP7-ICT- 270490), the EU FP7 project WYSIWYD (FP7-ICT-612139) and the Spanish Plan Nacional TIN2010-16745 (FAA-Arquitectura Cognitiva Biomimetica para un Funcional Ayudante de Androide Socialmente en Activo).

References

1. Sung, J., Christensen, H.I., Grinter, R.E.: Robots in the wild: understanding long-term use. In: 2009 4th ACM/IEEE International Conference on Human-Robot Interaction (HRI), pp. 45–52. IEEE (2009)
2. Fernaeus, Y., Håkansson, M., Jacobsson, M., Ljungblad, S.: How do you play with a robotic toy animal?: a long-term study of pleo. In: Proceedings of the 9th International Conference on Interaction Design and Children, pp. 39–48. ACM (2010)
3. François, D., Powell, S., Dautenhahn, K.: A long-term study of children with autism playing with a robotic pet: Taking inspirations from non-directive play therapy to encourage children's proactivity and initiative-taking. Interaction Studies 10(3), 324–373 (2009)
4. Thrun, S., Beetz, M., Bennewitz, M., Burgard, W., Cremers, A.B., Dellaert, F., Fox, D., Haehnel, D., Rosenberg, C., Roy, N., et al.: Probabilistic algorithms and the interactive museum tour-guide robot minerva. The International Journal of Robotics Research 19(11), 972–999 (2000)
5. Nourbakhsh, I., Kunz, C., Willeke, T.: The mobot museum robot installations: A five year experiment. In: Proceedings. 2003 IEEE/RSJ International Conference on Intelligent Robots and Systems, IROS 2003, vol. 4, pp. 3636–3641. IEEE (2003)
6. Kanda, T., Hirano, T., Eaton, D., Ishiguro, H.: Interactive robots as social partners and peer tutors for children: A field trial. Human-computer interaction 19(1), 61–84 (2004)
7. Breazeal, C.L.: Designing sociable robots. MIT Press (2004)
8. Kanda, T., Shiomi, M., Miyashita, Z., Ishiguro, H., Hagita, N.: A communication robot in a shopping mall. IEEE Transactions on Robotics 26(5), 897–913 (2010)
9. Satake, S., Kanda, T., Glas, D.F., Imai, M., Ishiguro, H., Hagita, N.: How to approach humans?-strategies for social robots to initiate interaction. In: 2009 4th ACM/IEEE International Conference on Human-Robot Interaction (HRI), pp. 109–116. IEEE (2009)
10. Wada, K., Shibata, T.: Living with seal robots in a care house-evaluations of social and physiological influences. In: 2006 IEEE/RSJ International Conference on Intelligent Robots and Systems, pp. 4940–4945. IEEE (2006)
11. Sabanovic, S., Michalowski, M.P., Simmons, R.: Robots in the wild: Observing human-robot social interaction outside the lab. In: 2006. 9th IEEE International Workshop on Advanced Motion Control, pp. 596–601. IEEE (2006)
12. Bartneck, C., Kulić, D., Croft, E., Zoghbi, S.: Measurement instruments for the anthropomorphism, animacy, likeability, perceived intelligence, and perceived safety of robots. International journal of social robotics 1(1), 71–81 (2009)
13. Metta, G., Sandini, G., Vernon, D., Natale, L., Nori, F.: The icub humanoid robot: an open platform for research in embodied cognition. In: Proceedings of the 8th Workshop on Performance Metrics for Intelligent Systems, pp. 50–56. ACM (2008)
14. Verschure, P.F.: Distributed adaptive control: A theory of the mind, brain, body nexus. Biologically Inspired Cognitive Architectures (2012)

15. Vouloutsi, V., Lallée, S., Verschure, P.F.M.J.: Modulating behaviors using allostatic control. In: Lepora, N.F., Mura, A., Krapp, H.G., Verschure, P.F.M.J., Prescott, T.J. (eds.) Living Machines 2013. LNCS, vol. 8064, pp. 287–298. Springer, Heidelberg (2013)

16. Geiger, G., Alber, N., Jordà, S., Alonso, M.: The reactable: A collaborative musical instrument for playing and understanding music. Her&Mus. Heritage & Museography (4), 36–43 (2010)

17. Breazeal, C.: Toward sociable robots. Robotics and Autonomous Systems 42(3), 167–175 (2003)

18. Verschure, P.F., Kröse, B.J., Pfeifer, R.: Distributed adaptive control: The self-organization of structured behavior. Robotics and Autonomous Systems 9(3), 181–196 (1992)

19. Cannon, W.B.: The wisdom of the body. The American Journal of the Medical Sciences 184(6), 864 (1932)

20. Seward, J.P.: Drive, incentive, and reinforcement. Psychological Review 63(3), 195 (1956)

21. Sanchez-Fibla, M., Bernardet, U., Wasserman, E., Pelc, T., Mintz, M., Jackson, J.C., Lansink, C., Pennartz, C., Verschure, P.F.: Allostatic control for robot behavior regulation: a comparative rodent-robot study. Advances in Complex Systems 13(03), 377–403 (2010)

22. McFarland, D.: Experimental investigation of motivational state. Motivational Control Systems Analysis, 251–282 (1974)

23. Maslow, A.H.: A theory of human motivation. Published in (1943)

24. Eng, K., Klein, D., Babler, A., Bernardet, U., Blanchard, M., Costa, M., Delbrück, T., Douglas, R.J., Hepp, K., Manzolli, J., et al.: Design for a brain revisited: the neuromorphic design and functionality of the interactive space 'Ada'. Reviews in the Neurosciences 14(1-2), 145–180 (2003)

25. Arbib, M.A., Fellous, J.M.: Emotions: from brain to robot. Trends in Cognitive Sciences 8(12), 554–561 (2004)

26. Lallée, S., Vouloutsi, V., Wierenga, S., Pattacini, U., Verschure, P.: EFAA: a companion emerges from integrating a layered cognitive architecture. In: Proceedings of the 2014 ACM/IEEE International Conference on Human-Robot Interaction, pp. 105–105. ACM (2014)

Design and Control of a Tunable Compliance Actuator*

Victoria Webster, Ronald Leibach, Alexander Hunt,
Richard Bachmann, and Roger D. Quinn

Case Western Reserve University,
Department of Mechanical and Aerospace Engineering
10900 Euclid Ave. Cleveland, OH, USA

Abstract. TCERA (Tunable Compliance Energy Return Actuator) is a
robotic actuator inspired by properties and behavior of the human knee
joint, in that it utilizes antagonistic contraction to vary torsional stiffness and joint angle. The actuator is an electrically activated artificial
muscle which uses two constant air-mass pneumatic springs configured
antagonistically about the knee joint. The positions of the actuator insertion points are controlled in order to change the effective torsional
stiffness about the knee axis. Additionally, the rigid member to which
the insertion points are attached is able to rotate in order to alter the
static equilibrium of the knee system. Furthermore, the system has been
simulated and an artificial neural network (ANN) has been trained to
determine the control bar angle and attachment location required based
on the desired angle, stiffness, and predicted torque state of the joint.
Using this controller we have achieved actuator position and stiffness
control both with and without load.

Keywords: biologically inspired, tunable compliance, pneumatic muscles, compliant transmission, air springs.

1 Introduction

Legged robots are notoriously energy inefficient. Big Dog has remarkable balance
control and functionality, but it has been reported to use two gallons of fuel per
mile [1]. Boston Dynamics' hydraulic bipeds also have remarkable strength and
promise great agility, but they use massive off-board power and cooling systems.
Honda's ASIMO is capable of performing numerous complicated tasks. However,
it has a max running speed of only 9km/h, and can only operate for 30 min on
a single battery [2]. Hubo, another electric robot, can walk for two hours using
on-board batteries, however it walked slowly on a treadmill in this test [3]. In
larger legged robots such as these, actuation is typically achieved by electric,

---

* This material is based upon work supported by the National Science Foundation
Graduate Research Fellowship under Grant No. DGE-0951783. Any opinion, findings, and conclusions or recommendations expressed in this material are those of the
authors and do not necessarily reflect the views of the National Science Foundation.

A. Duff et al. (eds.): Living Machines 2014, LNAI 8608, pp. 344–355, 2014.

hydraulic or pneumatic actuators. It is clear that with direct joint drive with motors or hydraulics, improved mobility requires higher energy consumption. The energy cost of locomotion for legged robots could be greatly reduced with inspiration from animal systems.

Braided pneumatics (air muscles) are attractive because of their gross muscle-like properties. When used in an antagonistic pair they can provide a joint with tunable passive compliance that is independent of joint torque [4]. The disadvantage of any pneumatic system is that compressing air is not energy efficient. Mechanical compressors are approximately 20% efficient when gas cooling is considered. This energy efficiency is very poor when compared to a gearmotor that typically has an efficiency greater than 70%, and it is even worse when considering that the compressor is an additional stage of energy conversion (and loss) after a gearmotor.

Compared to robots, many animals are much more energy efficient over a wide range of speeds. The energy efficiency of a running human is approximately 45% [5]. However, initial inquiry into muscle properties reveals they are approximately 25% efficient[5] which does not compare favorably with electric motor actuation. If it is not the actuators themselves, then how do large animals achieve such high gait efficiencies? Humans and other larger animals have muscle-tendon systems that allow them to absorb, store, and return energy on a step-by-step basis. Specifically, humans are remarkably capable of capturing, storing, and releasing energy at each step while running and thereby reducing the required energy input. In fact, muscle-tendon structures can return as much as 93% of the work required to stretch them during running [5].

In addition to energy storage and release, muscle-tendon systems act to reduce energy expenditure by reducing necessary control output [6, 7]. When a robot without passive compliant mechanisms in its joints is perturbed, the controller must run actuators to reject the disturbances. Animals have the further advantage that their nervous systems can tune the compliance of their joints to more efficiently reject disturbances, which is especially useful during locomotion over irregular terrain. Robots working in real world environments are constantly perturbed, and the ability of an actuator to passively absorb these disturbances will greatly increase robotic energy efficiency.

In systems using electric motors, compliance is often implemented through the use of series elastic actuators [8]. While the standard instantiation of series elastic systems implements constant joint stiffness, recent work has developed series elastics which allow tunable stiffness (For a review see [9]). Actuation systems have been developed which implement antagonistic elastic schemes. Such systems have either resulted in very complicated control laws [10] or require complicated mechanisms to produce the nonlinear effect needed [11–13]. By implementing braided pneumatic actuators (BPAs) as constant air-mass springs in a biomechanically inspired antagonistic-pair transmission, we have developed a tunable compliance actuator, which we have named TCERA (Tunable Compliance Energy Return Actuator). In contrast to passive energy return prosthetics [14], TCERA allows separate position and stiffness control of the associated joint in the unloaded state. Additionally, a controller is presented to allow nearly independent angle and stiffness control given a loaded configuration.

2   Mechanical Design

Taking inspiration from the hamstring/quadriceps system in mammals, the actuator design is based on the principle of antagonistic co-contraction of muscles (Fig. 1). Mammalian systems change joint angle by independently contracting both members of an antagonistic pair and change stiffness by co-contraction. TCERA changes the neutral position of the joint via rotation of a bar to which the muscles attach (control bar), and changes stiffness by varying the attachment location of each muscle simultaneously (attachment slider) (Fig. 2).

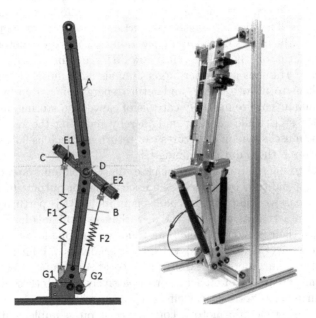

Fig. 1. TCERA uses two antagonistic constant air-mass pneumatic springs (F1,F2) attached via a pin connection (G1,G2) to the distal leg segment (B). The origin of each air muscle is attached to a slider mechanism (E1,E2) on the control bar (C) at the joint (D). The control bar is driven relative to the proximal leg segment (A) due to the action of a motor on the same segment. The distance of each slider from the joint is controlled by rotation of a threaded rod housed in the control bar. The angle of the control bar can be varied independently from the position of the sliders.

This design overcomes several critical issues found in antagonistic-pair, series elastic, and pneumatic actuator systems. TCERA eliminates the need for bulky onboard compression and air storage systems by using constant air-mass pneumatic muscles (FESTO air muscles), which only require a compressor once for

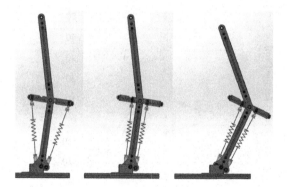

Fig. 2. By simultaneously changing the distance between each slider and the joint pivot, the stiffness of the joint can be changed independently of the joint angle without a pneumatic air exchange system (left to center). The neutral position of the joint can be changed with minimal transient effect on stiffness by rotating the control bar (center to right).

initial charging. Additionally, the dynamics of the system are not affected by the depressurizing process known to bottleneck active pneumatic systems [9]. In comparison to series elastic transmissions, the use of air springs rather than traditional metal springs reduces overall actuator weight and improves safety in case of failure.

One of the advantages of TCERA becomes apparent when observing the human knee joint during a walking gait. During stance the joint goes through loading and unloading at a much higher stiffness (260 N-m/rad) than during swing (20-60 N-m/rad)(Fig. 3) [15, 16]. Inspired by this, TCERA is designed to vary its torsional stiffness independently of joint position. In order to investigate the development of TCERA for humanoid robotic applications we have designed and constructed a test stand based on the configuration of the human knee. The limb dimensions have been chosen based on the biometrics of a 50th percentile adult female, with a tibia length of 0.375 m and a femur length of 0.468 m [17].

2.1   Air Spring Parameter Selection

Several factors were taken into account in the selection and implementation of the air springs that act as the spring elements in TCERA. The stiffness of the air springs was chosen to produce a range of torsional stiffnesses about the joint which would include the maximum seen during human walking. Due to the difference in mass of the constructed leg (6.1 kg) and that of the average female leg (10.92 kg [18]), the target stiffness range of the joint has been scaled by 0.55. This results in a desired stance stiffness of 143 N-m/rad. The pressure required to achieve this stiffness and to provide the necessary strain based on system geometry was determined experimentally to be 206.8 kPa. When initially charged at a pressure of 206.8 kPa, the air spring underwent a maximum strain of

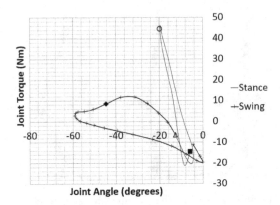

Fig. 3. Analysis of knee joint torque vs knee angle for a 75 kg human reveals a significantly higher stiffness during stance (from the solid black square to the hollow circle and back) than during swing (from the hollow triangle to the solid diamond and back). Positive values represent extension, and negative values represent flexion.

13.6 % with a measured stiffness of 8171 N/m. Based on the actuator geometry this results in a maximum joint stiffness of 167 N-m/rad.

3   Simulation and Control

3.1   Actuator Modeling

The equations governing the actuator torque and stiffness can be found as a function of joint angle and actuator attachment locations using Lagrange's equations for the static system:

$$T = \frac{\partial V}{\partial \alpha} \tag{1}$$

where T is joint torque, α is the small angle between the leg segments, and V is the potential energy of the system. Assuming the mass of the air spring elements to be small with respect to the attached body components and desired forces, the potential energy can be found as:

$$
\begin{aligned}
V = &\frac{K_1}{2}(l_c + r_1 - \sqrt{x^2 + y_1^2 - 2xy_1 \cos(\alpha + \beta - \phi_1)})^2 + \\
&\frac{K_2}{2}(l_c + r_2 - \sqrt{x^2 + y_2^2 + 2xy_2 \cos(\alpha + \beta + \phi_2)})^2
\end{aligned} \tag{2}
$$

where K_1 and K_2 are the spring stiffnesses, r_1 and r_2 are the spring resting lengths, l_c is the length of the air spring connector caps, x is the distance between each muscle attachment and the actuated joint, α is the joint angle and β is the angle of the control bar relative to the proximal leg segment. y_i is the distance from the distal muscle attachment to the actuated joint. ϕ_i can be calculated as:

$$\phi_i = \arcsin\left(\frac{d_i}{y_i}\right) \tag{3}$$

where d_i is the offset distance of the distal muscle attachment perpendicular to the distal leg segment.

Taking the derivative of Eq. 2 with respect to α results in the following expression for torque:

$$T = \frac{K_2 x y_2 \sin(\alpha + \beta + \phi_2)(l_c + r_2 - \sigma_2)}{\sigma_2} - \frac{K_1 x y_1 \sin(\alpha + \beta - \phi_1)(l_c + r_1 - \sigma_1)}{\sigma_1} \tag{4}$$

where:

$$\sigma_1 = \sqrt{x^2 + y_1^2 - 2x y_1 \cos(\alpha + \beta - \phi_1)} \tag{5}$$

$$\sigma_2 = \sqrt{x^2 + y_2^2 + 2x y_2 \cos(\alpha + \beta + \phi_2)} \tag{6}$$

The joint stiffness can then be found by:

$$K = \frac{\partial T}{\partial \alpha} \tag{7}$$

Resulting in:

$$K = \frac{K_1 x^2 y_1^2 \epsilon_1}{\epsilon_3} + \frac{K_2 x^2 y_2^2 \epsilon_2}{\epsilon_4} + \frac{K_2 x y_2 \epsilon_6 \left(l_c + r_2 - \sqrt{\epsilon_4}\right)}{\sqrt{\epsilon_4}} - \frac{K_1 x y_1 \epsilon_5 \left(l_c + r_1 - \sqrt{\epsilon_3}\right)}{\sqrt{\epsilon_3}} + \frac{K_2 x^2 y_2^2 \epsilon_2 \left(l_c + r_2 - \sqrt{\epsilon_4}\right)}{\epsilon_4^{\frac{3}{2}}} + \frac{K_1 x^2 y_1^2 \epsilon_1 \left(l_c + r_1 - \sqrt{\epsilon_3}\right)}{\epsilon_3^{\frac{3}{2}}} \tag{8}$$

where:

$$\begin{aligned}
\epsilon_1 &= \sin(\alpha + \beta - \phi_1)^2 \\
\epsilon_2 &= \sin(\alpha + \beta + \phi_2)^2 \\
\epsilon_3 &= x^2 + y_1^2 - 2\epsilon_5 x y_1 \\
\epsilon_4 &= x^2 + y_2^2 + 2\epsilon_6 x y_2 \\
\epsilon_5 &= \cos(\alpha + \beta - \phi_1) \\
\epsilon_6 &= \cos(\alpha + \beta + \phi_2)
\end{aligned} \tag{9}$$

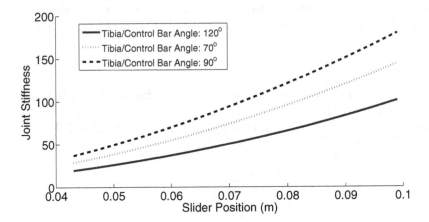

Fig. 4. The stiffness of the actuator is a function of both slider position and the angle between the control bar and the tibia. During operation with small loads any change in the control bar angle will result in a corresponding change in the tibia angle, thereby maintaining the tibia/control bar angle at 90°. It is only when loads are applied that the tibia/control bar angle deviates significantly from 90°.

The stiffness of the actuators is therefore a function of both the attachment slider position and the angle between the distal leg segment and the control bar (Fig. 4). The controller must determine the control bar angle and the slider position based on the desired angle (α_d), desired stiffness (K_d), and torque about the joint. An artificial neural network (ANN) has been developed and trained to compute the required parameters.

3.2 Artificial Neural Network Training and Control

In order to develop the training data necessary for using an ANN for TCERA position and stiffness control, a simulation of the actuator has been implemented in Python. The use of a simulation to develop training sets allows fast proto-typing of the system. This simulation develops a configuration space by taking in joint angle, control bar angle, and slider distance in order to calculate the corresponding torque and stiffness. The configuration space is output to Excel as two arrays: the input array $[\alpha_d, K_d, T]$, and the output array $[x, \beta]$. This data is then used to train an ANN using MATLAB's Neural Net Toolbox for data fitting. The dataset is randomly divided into training (70%), validation (15%), and testing (15%) subsets prior to training and utilizes a basic network architecture of one hidden layer. A convergence test has been performed to determine the number of hidden nodes needed. As a result, the ANN has one input layer, one hidden layer consisting of 100 nodes, and one output layer.

3.3   Simulation

A physical simulation was developed in MATLAB in order to test the effects of physical parameters on actuator performance. The equations of motion are developed using quasicoordinates [19]. The ankle joint is fixed to the base. The top of the femur segment (where the hip would be in a larger system) is constrained to vertical translation and all joints are revolute. The equations of motion can be determined in absolute quasicoordinates as:

$$M_G \begin{bmatrix} \ddot{R} \\ \dot{\omega} \end{bmatrix} + C_G \omega - V_G g = \begin{bmatrix} F \\ M \end{bmatrix} \tag{10}$$

where R is the translation vector of the base point, ω is the angular velocity vector in absolute quasicoordinates, g is gravitational acceleration, M is the vector of absolute moments applied to each joint, and:

$$M_G = \begin{bmatrix} I_V^T \bar{m} I_V & I_V^T \bar{m} \bar{\Phi} \\ \bar{\Phi}^T \bar{m} I_V & \bar{\Phi}^T \bar{m} \bar{\Phi} + I^* \end{bmatrix}$$

$$C_G = \begin{bmatrix} I_V^T \bar{m} \dot{\bar{\Phi}} \\ \bar{\Phi}^T \bar{m} \dot{\bar{\Phi}} + \tilde{\omega} I^* \end{bmatrix} \tag{11}$$

$$V_G = \begin{bmatrix} I_V^T \bar{m} I_V \\ \bar{\Phi}^T \bar{m} I_V \end{bmatrix}$$

Eq. 10 can be converted to relative true coordinates (motor coordinates) through the use of rotation and constraint matrices:

$$D_1^T D_0^T C_0^{RT} M_G C_0^R D_0 D_1 \ddot{\alpha}_G + D_1^T D_0^T C_0^{RT} C_G C_0^R D_0 D_1 \dot{\alpha}_G$$
$$- D_1^T D_0^T C_0^{RT} V_G g = \begin{bmatrix} F \\ T \end{bmatrix} \tag{12}$$

where α is the position and angle vector in relative coordinates, T is the relative torque at each joint, C_0^R is the global rotation matrix describing the transformation from absolute quasicoordinates to relative quasicoordinates, D_0 is the constraint matrix transforming relative quasicoordinates to relative true coordinates, and D_1 is a constraint matrix that constrains the degrees of freedom so that there is no translation of the ankle joint:

$$C_0^R = \begin{bmatrix} I & 0 & 0 \\ 0 & I & 0 \\ 0 & C_{21} & I \end{bmatrix} \quad D_0 = \begin{bmatrix} 1 & 0 & 0 & 0 \\ 0 & 1 & 0 & 0 \\ 0 & 0 & 0 & 0 \\ 0 & 0 & 0 & 0 \\ 0 & 0 & 0 & 0 \\ 0 & 0 & 1 & 0 \\ 0 & 0 & 0 & 0 \\ 0 & 0 & 0 & 0 \\ 0 & 0 & 0 & 1 \end{bmatrix} \quad D_1 = \begin{bmatrix} 0 & 0 \\ 0 & 0 \\ 1 & 0 \\ 0 & 1 \end{bmatrix} \tag{13}$$

Finally, the constraint of the hip is implemented as a high stiffness horizontal spring force acting at the end of the femur.

3.4 Controller Implementation

Control of the physical test stand was achieved via Arduino through Simulink. The trained ANN was ported into Simulink using the built-in Simulink diagram generator in the MATLAB Neural Network Toolbox. The slider position and control bar angle can be received as analog inputs from potentiometers using the available Simulink support for the Arduino Mega, which allows serial control of the microcontroller. This information is used to calculate the torque about the joint at the desired position based on the physical parameters of the system and a given applied force. The torque, along with the desired angle and stiffness, is then processed by the ANN in Simulink to determine the required slider position and control bar angle. The motors are then driven via PID control on the Arduino to reach these positions through a Sabertooth 2x25 V2 motor controller.

4 Results

4.1 Simulation

In simulation, the actuator is able to accurately respond to position and stiffness commands (Fig. 5). The steady state error for both position and stiffness in response to a commanded step function is less than 1% across the entire range of motion and stiffness. A change in commanded angle results in a slight change of stiffness during the angle transition period, but this quickly converges back to the commanded value.

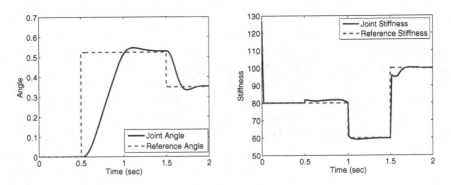

Fig. 5. Simulation - The ANN is able to control the angle (left) and stiffness (right) of the test stand

4.2 Physical System

Via the ANN, both position and stiffness of the system can be controlled (Fig. 6, Fig. 7). For prototyping and development purposes, control is achieved through a MATLAB/Arduino interface using serial communication. Unfortunately, this system has proven to result in a significant communication delay (2.7 s). This delay is not present for the controller alone, which runs in real time. As a result it is clear that this delay is a fundamental limitation of the MATLAB/Arduino interface. For future applications, the controller will be implemented on-board, eliminating the need for this interface.

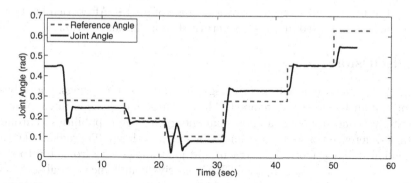

Fig. 6. Physical System - Joint angle reference tracking while the actuator is oriented in a horizontal position is shown. For clarity, the reference signal has been delayed a uniform 2.7 s to account for the serial communication delay.

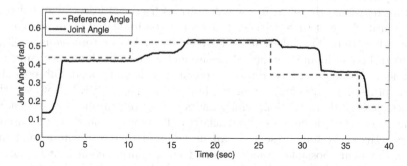

Fig. 7. Physical System - Joint angle reference tracking while the actuator is oriented in a vertical position is shown. For clarity, the reference signal has been delayed a uniform 2.7 seconds to account for the serial communication delay. The device stuck on the friction clutch briefly when commanded to 0.52 radians, resulting in slight plateaus before the integrator term of the PID allowed the motor force to overcome friction. Future iterations of this device will remove the clutching mechanism, eliminating this concern.

In low gear the device is capable of changing the slider position at a rate of 9.525 cm/s and in high gear this rate increases to 42.4 cm/s. This results in the ability of the device to change from the highest to lowest stiffness (6.3 cm) in

0.66 seconds in low gear and 0.15 seconds in high gear. Additionally, the maximum speed of the control bar angle is 10.7 rad/s, allowing the control bar to cover its entire range in a minimum of 0.098 seconds. The average steady state error during constant stiffness angle-reference tracking in the unloaded state is 0.017 radians. For angle-reference tracking under gravitational loading, the average steady state error is 0.014 radians.

In order to test the torsional stiffness of the joint, the leg was loaded and unloaded in five pound increments over a range of angles. After each load cycle the weight was unloaded and the leg returned to its initial position.

5   Discussion

A tunable compliance actuator, utilizing constant-mass air springs, and a basic control system have been developed. The actuator has demonstrated position and stiffness control as well as potential-energy storage capabilities. Throughout angle-reference tracking, position error was less than 0.054 radians. During mammalian walking, footfall placement and joint angles vary from step to step. Just as in mammalian systems, TCERA's passive compliance can absorb such small errors as well as external perturbations. During incremental loading experiments the average torsional stiffness was found to be 135.7 N-m/rad, which is comparable to the desired value of 143 N-m/rad. In order to improve stiffness predictability, further testing of air springs is underway.

For all incremental loading tests, the actuator demonstrated some positional return upon unloading. This indicates the potential for energy storage and return in future revisions. However, in some conditions frictional losses hindered complete return. Some of these frictional losses may be mitigated by improvements in the drive train. The inherent backdrivability of DC motors requires either current to be continuously supplied to the motor or a clutching mechanism to be implemented in order to maintain a constant angle under load. Subsequently, a constant friction clutching system was implemented on TCERA. However, this increases losses due to friction, which limits energy return and ease of control. Instead, an automatic clutch or nonbackdrivable transmission should be used.

The actuator system presented can be adapted to any revolute joint and as such has many possible applications in fields requiring compliant actuation, including mobile robotics and orthotics.

References

1. Raibert, M.: Adaptive Motion in Animals and Machines, Keynote Address (2008)
2. HondaWorldwide: ASIMO: The Honda Humanoid Robot ASIMO (2012)
3. Oh, J.H.: Development Outline of the Humanoid Robot: HUBO II. In: International Conference on Robotics and Automation, Keynote Address, St. Paul, MN, USA (2012)
4. Colbrunn, R.W., Nelson, G.M., Quinn, R.D.: Modeling of Braided Pneumatic Actuators for Robotic Control. Int. Conf. on Intelligent Robots and Systems. In: International Conference on Robots and Systems, IROS 2001, Maui, Hawaii, pp. 1964–1970 (2001)

5. Alexander, R.M.: Principles of animal locomotion. Princeton University Press, Princton (2003)
6. Loeb, G., Brown, I., Cheng, E.: A hierarchical foundation for models of sensorimotor control.. Experimental Brain Research 126, 1–18 (1999)
7. Jindrich, D.L., Full, R.J.: Dynamic stabilization of rapid hexapedal locomotion. J. Exp. Biol. 205, 2803–2823 (2002)
8. Pratt, G., Williamson, M.: Series elastic actuators. In: IEEE International Conference on Intelligent Robots and Systems, pp. 399–406 (1995)
9. Van Ham, R., Sugar, T.G., Vanderborght, B., Hollander, K.W., Lefeber, D.: Review of actuators with passive adjustable compliance/controllable stiffness for robotic rpplications. IEEE Robotics and Automation Magazine, 81–94 (September 2009)
10. Kolacinski, R.M., Quinn, R.D.: Design and mechanics of an antagonistic biomimetic actuator system. In: Proceedings. 1998 IEEE International Conference on Robotics and Automation (Cat. No.98CH36146), vol. 2, pp. 1629–1634. IEEE, Leuven (1998)
11. Hurst, J., Rizzi, A.: Series compliance for an efficient running gait. IEEE Robotics & Automation Magazine 15(3), 42–51 (2008)
12. Tonietti, G., Schiavi, R., Bicchi, A.: Design and control of a variable stiffness actuator for safe and fast physical human/robot interaction. In: International Conference on Robotics and Automation, Barcelona, Spain, pp. 1–6 (2005)
13. Polinkovsky, A., Bachmann, R.J.: An Ankle Foot Orthosis with Insertion Point Eccentricity Control. In: IEEE/RSJ International Conference on Intelligent Robots and Systems, pp. 1603–1608. IEEE, Vilamoura (2012)
14. Unal, R., Carloni, R., Behrens, S.M., Hekman, E.E.G., Stramigioli, S., Koopman, H.F.J.M.: Towards a fully passive transfemoral prosthesis for normal walking. In: IEEE RAS & EMBS International Conference on Biomedical Robotics and Biomechatronics (BioRob), pp. 1949–1954. IEEE, Roma (2012)
15. Perry, J.: Gait Analysis: Normal and Pathological Function. SLACK Inc., Thorofare (1992)
16. Winter, D.: The biomechanics and motor control of human gait: normal, elderly and pathological. Waterloo, 2nd edn. University of Waterloo Press (1991)
17. Japan, N.S.D.A.: ST-E-1321 Japanese Female Body Size. Technical report
18. Huston, R.: Principles of Biomechanics. Dekker Mechanical Engineering. Taylor & Francis (2008)
19. Nelson, G., Quinn, R.: A quasicoordinate formulation for dynamic simulations of complex multibody systems with constraints. In: Dynamics and Control of Structures in Space (1996)

An Experimental Eye-Tracking Study for the Design of a Context-Dependent Social Robot Blinking Model

Abolfazl Zaraki[1], Maryam Banitalebi Dehkordi[2]
Daniele Mazzei[1], and Danilo De Rossi[1]

[1] Research Center "E. Piaggio", Faculty of Engineering, University of Pisa, Italy
a.zaraki@centropiaggio.unipi.it
[2] Perceptual Robotics Laboratory, Scuola Superiore Sant'Anna, Pisa, Italy

Abstract. Human gaze and blinking behaviours have been recently considered, to empower humanlike robots to convey a realistic behaviour in a social human-robot interaction. This paper reports the findings of our investigation on human eye-blinking behaviour in relation to human gaze behaviour, in a human-human interaction. These findings then can be used to design a humanlike eye-blinking model for a social humanlike robot. In an experimental eye-tracking study, we showed to 11 participants, a 7-minute video of social interactions of two people, and collected their eye-blinking and gaze behaviours with an eye-tracker. Analysing the collected data, we measured information such as participants' blinking rate, maximum and minimum blinking duration, number of frequent (multiple) blinking, as well as the participants' gaze directions on environment. The results revealed that participants' blinking rate in a social interaction are qualitatively correlated to the gaze behaviour, as higher number of gaze shift increased the blinking rate. Based on the findings of this study, we can propose a context-dependent blinking model as an important component of the robot's gaze control system that can empower our robot to mimic human blinking behaviour in a multiparty social interaction.

Keywords: blinking model, eye-tracking study, gaze behaviour, humanlike robot, social human-robot interaction.

1 Introduction

Nowadays, social humanlike robots are being developed widely, for various tasks and human-centered scenarios, in which they should appropriately interact with humans. Due to their humanlike appearances, humanlike robots can play potentially the role of human in society, and become ultimately a human companion in future; however, their acceptability, that highly depends on their appearance and behaviour, is still challenging [1] [2] [3].

One of the important issues in robot's acceptability is the capability of the robot in expressing context-aware behaviours, when interacts with human. For

A. Duff et al. (eds.): Living Machines 2014, LNAI 8608, pp. 356–366, 2014.

Fig. 1. The humanlike robot engaged in a multiparty social interaction

example, an acceptable robot should be able to actively understand current social-context analysing human-relevant features and display a realistic gaze toward the most prominent target at the proper time, accordingly. For this reason, we have previously presented a context-aware gaze control system [4] [5] that enables a humanlike robot to analyse its surrounding environment and display a believable gaze toward multiple people, who are interacting with each other and with the robot (as shown in Fig. 1). The system collects visual-auditory streams, extracts several high-level features of human (e.g., body gesture, head pose, social events, etc.), and identifies target of environment through analysing high-level features. It then generates the robot's gaze parameters (e.g., head-eyes movement amplitudes, velocities, latencies, etc.), in order to move the gaze of the robot in a humanlike manner. Although the proposed system enables the robot to replicate human gaze behaviour in a believable manner, the *acceptability* of the robot can be further increased, if in addition to the gaze, the robot mimics the human blinking behaviour, during a multiparty social interaction.

Simulating blinking behaviour requires a human-level blinking model that should be derived from real data of human. Several investigations on eyelid dynamics have revealed that human blinking behaviour highly depends upon human cognitive processes and emotional state [6]. Bentivoglio et al. [7] analysed human blinking rate during different cognitive tasks such as reading, resting, and etc., however, they did not report the blinking duration and the relation between blinking and gaze behaviours. In an accurate study, Vanderwerf et al. [8] examined the kinematics and neurophysiological aspects of eyelid movements of healthy human subjects during different conditions (e.g., spontaneous, voluntary, air puff, etc.). In addition, they investigated the effect of eye position on the eyelid movement; however, they did not evaluate the blinking behaviour of human subjects that engage in a social interaction. The linkage between blink and gaze shift was investigated in several works (e.g., [9] [10]), which is also the focus of this work.

The above-mentioned works investigated different aspects of the problem; however, the relation between blinking and gaze behaviours in a social human-human interaction needs to be taken into consideration. This correlation between *eye-blinking* behaviour and a *task* that human is involved is important in designing a blinking model for a humanlike robot.

In this work, in order to model human eye-blinking behaviour for our humanlike robot (called FACE [11] [12] [13]), we consider the dynamic of eyelid movements and its relation to the attention/gaze behaviour in a social interaction. For this reason, in an accurate eye-tracking study, we showed to 11 participants, a 7-minute video of social interactions of two people and collected their eye-blinking and gaze behaviours with a DIKABLIS eye-tracker device [14]. Using an annotation software, we analysed the collected data and measured parameters such as participants' blinking rate, maximum and minimum blinking duration, number of frequent (multiple) blinking, as well as the gaze direction on environment, during the experiment. In addition, we investigated the relation between the blinking and gaze behaviour of participants. The preliminary results show that participants' blinking rate in a social interaction are qualitatively correlated with the gaze behaviour, as higher number of gaze shift increased the blinking rate.

2   Eye-Tracking Study

In order to measure the human blinking parameters (e.g., blinking rate, maximum and minimum blinking duration, number of frequent blinking, etc.) and their relationship to gaze behaviour, an eye-tracking study was performed. This experiment was carried out at the Technical University of Munich, in which a total number of 11 participants of both genders (with mean age of 27.3) took part.

2.1   Experiment Procedure

In this experiment a 7-min video was shown to the participants, while their pupil motion, blinking behaviour, and gaze behaviour were recorded using a DIKABLIS eye-tracking system. The video of the experiment was displayed using a 23-inch monitor and the participants sat 75 cm away from it (as shown in Fig. 2). The eye-tracker captured the participants' eye and also the surrounding environment, using an infrared camera and an RGB camera, respectively. The DIKABLIS eye-tracker was calibrated prior performing the experiment. The calibration time allowed the participants' eye to be adapted to the environment.

The video presented during the experiment showed a social interaction between two persons that enter a room separately, sit down on the chairs and talk to each other and to the camera (participants). The people in the video, executed diverse social signals (e.g., stretching while being seated, raising an arm,

Fig. 2. The experiment setup: the participant watched a video while an eye-tracking device recorded the eye-blinking and gaze behaviours

getting up from the chair to get a drink, and retrieving a smart phone from their pocket) in order to guide the participant attention/gaze to different directions on the environment. Finally, the two actors of the video left the room, separately. To close the interaction loop between a participant and the video, the actors of the video interacted with participant time to time during the experiment. In short, when the participant watched the video, he/she was assumed as the third person in the social interaction scenario.

2.2 Data Analysis

As described, the eye-tracker device generated two separated video files: (i) the participant's pupil motion as proxy of gaze and blinking behaviour, and (ii) the environment seen by the participant. Employing DIKABLIS analyser software, the gaze directions of participant in global coordinates, were estimated. The software then generated a single video file that showed the gaze behaviour of participant on the environment during the experiment. Using the generated video, we precisely tracked the participant's gaze on the video. Fig. 3 shows participant is wearing the eye-tracker and watching the social human-human interaction video.

In order to measure the blinking parameters, as well as the gaze behaviour, we annotated the recorded videos of 11 participants. Annotation was performed employing ELAN[1] [15] [16], an annotation software developed by the Max Planck

---

[1] http://tla.mpi.nl/tools/tla-tools/elan

Fig. 3. (a) A DIKABLIS eye-tracker with an infrared camera (to capture the participant's pupil) and an RGB camera (to capture the field video). (b) DIKABLIS analyser merges two recorded videos into a single video, and estimates the participant's gaze direction on the environment. The participant's gaze direction is indicated by a green-cross on the screen.

Institute for Psycholinguistics. It is a powerful tools especially for annotation of visual-auditory resources, data creation and management. As the results, we obtained the participants' gaze position, gaze shift frequency, and blinking parameters. The gaze is considered as the moments that participant looked at either the first person, the second person or the environment in the video (as shown in figure 3). The gaze shift is detected when the participant changed the current line of sight from one point to another (e.g., from first person to second person in the video).

Through the data analysis, we noticed that participants showed a variety of blinking and gaze behaviours, and in fact, there is not a unique model to precisely express the human eye behaviours. Particularly, in the case of blinking behaviour, some of the participants showed much higher blinking rates compare to others. Although, participants showed different behaviours and defining a unique model

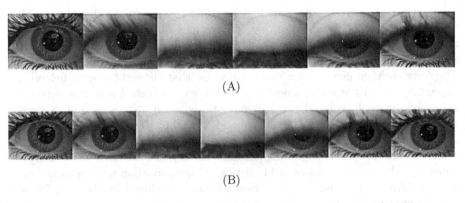

Fig. 4. Illustration of human single blink and frequent (multiple) blink

as *humanlike* is difficult, we can extract a general model from the human data. For this reason, we categorized the participants into two groups according to their total number of blinking, during the experiment: (i) people that their blinking rate is in the range of the blinking rate of normal subjects [17] [18], and (ii) people with higher blinking rate. In addition, some of the participants showed a number of frequent blinking (as shown in Fig. 4), which we considered in designing the blinking model for the humanlike robot.

3 Result and Discussion

In order to obtain a general model for eye behaviour during the social interaction, we considered the average blinking parameters of participants. Considering the average number of blinking per minute or Average Blinking Rate (ABR) computed from the participants' data, we categorized participants to either group: first and second groups. The participants with ABR lower than 15 were defined as first group, and the rest of participant that have the ABR higher than 15 were considered as the second group. The threshold ABR=15 is obtained according to the result of experimental data analysis. The ABR of the first group was 6.9 with standard deviation of 3.4 (ABR1 = 6.9, SD1=3.4), and the ABR of the second group was 23.5 with standard deviation of 7.9 (ABR2 = 23.5, SD2=7.9).

We computed the number of gaze shifts and blinking of all participants over every minute of the experiment (total 7 minutes). Table 1 reports the average of gaze shift and blinking numbers obtained from participants of both groups. Table 2 details the blink parameters corresponding to each group, which were obtained based on blinking numbers of all participants in that group. These parameters are: the average number of *frequent blinks* (Num.Freq.Blk), and the average of participants' minimum, maximum and median duration of all *single blinks*, in millisecond (Min(ms), Max(ms) and Median(ms), respectively).

Table 1. The average blinking and gaze numbers over 7 minutes of the experiment. The Blink 1 and Gaze 1 refer to the participants with lower blinking rate and Blink 2 and Gaze 2 refer to the participants with higher blinking rate. Nob is the abbreviation of number of blink.

Time (min)	1	2	3	4	5	6	7	Total Nob.
Av. Blink 1	8.8	8.7	6.2	6.2	5.7	5.7	6.8	48
Av. Gaze 1	15.7	18	14.3	15.7	18.3	14.3	18.8	115.2
Av. Blink 2	19.8	28.4	23.2	23.2	22.4	20.4	26.8	164.2
Av. Gaze 2	14.4	13.4	12	13.4	15.2	13.2	11.2	92.8

Table 2. The details of blinking behaviours of first and second groups

Average Blink Parameter	Num.Freq.Blk	Min(ms)	Max(ms)	Median(ms)
Group 1	3.5	157.3	520	333.5
Group 2	11	227.8	856	408

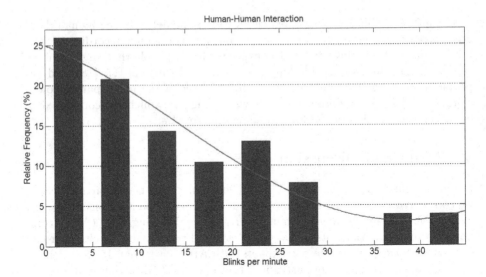

Fig. 5. The distribution of the participants' blink rate during social human-human interaction

The histogram in Fig. 5 represents the distribution of participants' blink rate in the human-human interaction. The obtained fitting curve provides important information about human blinking behaviour during a social interaction. It is the basis for the design of a humanlike blinking model for our humanlike robot. As illustrated, about 25% of sample data of participants blinking over one minute are below 5, while according to the literature people have the blinking rate in this range, when they focus on the object or environment for an extended time. It means that the cognitive task highly influences the blinking rate [19] [20], and thus, to model blinking behaviour, we need to evaluate human behaviour in the same scenario as our human-robot interaction application.

As the second part of data analysis, we analysed the gaze behaviour of participants during the experiment. For example, Fig. 6 shows the gaze behaviour of the participant, when he was looking at the first person (A), the second person (B), and the environment in the video. In this figure, the duration of gaze and the time of gaze shifts are illustrated. The important parameter that should be taken into account is the moment that a gaze shift occurred as well as the participant's gaze direction.

Fig. 7 shows the total number of blinking and gaze shifts performed by each participant during the experiment. As can be seen, in one group, there are more eye blinks than gaze shifts and in the other group it is the opposite. However, in both groups the blinking behaviours of participants involved in a social interaction are correlated to the gaze behaviours. In other words, we observed that for the participants with normal (and lower) blinking rate the higher number of gaze shifts result in an increased blinking rate. This fact is also valid for the participants with higher blinking rate; however, as shown the correlation

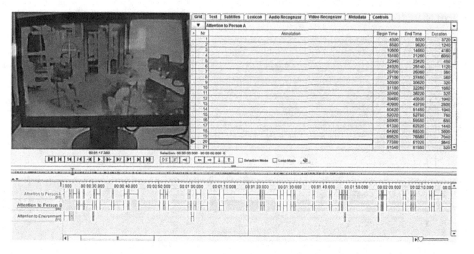

Fig. 6. Through the annotation process, the participants' gaze duration and direction (at the person A, the person B, and the environment) as well as the moment of gaze shifts are obtained. The bottom side of the figure illustrates the participants' gaze occurrence time and their gaze directions (person A/B or the environment). The top side of the figure shows the duration of the gaze.

rate is different for two groups. It should be noted that as can be seen in Fig. 7, there is a qualitative relationship between the number of eye-blinks and the number of gaze shifts however, the proportionality between them has not been yet demonstrated.

As observed from the participants' blinking behaviours, people most likely blink before the moment of shifting the gaze from one point to another. In addition, the psychology literature [9] reports that gaze shifts evokes blinks, which validates our findings. However, the probability of blinking varies with gaze amplitude [10].

Inspiring from human, and based on our finding, we propose a general blinking model for the humanlike robot. In this model, the number of robot's blink and the duration of each blink are adjusted according to the average experimental data. In addition, considering the observed qualitative correlation between gaze shifts and eye blinks, our model should generate extra blinking signals, when the robot should perform a gaze shift larger than a threshold. In order to increase the acceptability of the robot, the model generates a number of frequent blinking in the range of the experimental data reported in Table 2. The frequent blinking signals can be generated based on a time-variant function that is obtained from the average duration of frequent blinking occurrence.

The presented blinking model is proposed as a component of the gaze control system that we have previously proposed [4] [5]. The gaze control system adjusts the amplitude, velocity, and latency of the robot's gaze shifts by generating the proper signals for robot's head-eyes actuators. Furthermore, employing the proposed blinking model, the gaze control system also should be able to adjust

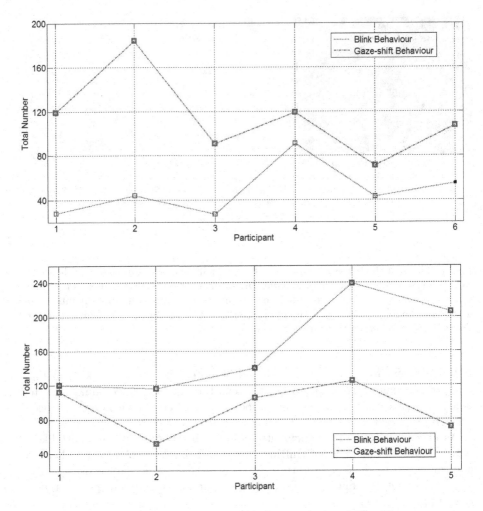

Fig. 7. The participants' total number of blinking and gaze shifts. The upper curve illustrates the behaviours of the first group (participants with normal blinking rate) and the lower curve illustrates the behaviours of the second group (participants with higher blinking rate). As illustrated in the both group, the gaze shifts are qualitatively correlated with the occurrence of the eye blinks.

the blinking behaviour by generating the appropriate control signals for the blinking actuators at the proper time.

4   Conclusion

In this work, a quantitative human gaze tracking study was carried out, to obtain the general blinking behaviour of humans in relation to gaze behaviour in a social human-human interaction. The aim of this study was designing an experimental blinking model as a component of gaze control system for a humanlike robot. In order to obtain human blinking and gaze behaviours, we showed to 11 participants the 7-minute video of a social interaction between two people and we recorded their blinking and gaze behaviours using an eye-tracker device. Analysing the data, we obtained the blinking parameters (blinking rate, minimum, maximum and median of blinking duration, number of frequent blinks, etc.) as well as the relationship between the blinking and gaze behaviours. The participants showed two different blinking behaviours, where some of them performed much higher blinking rate compared to the others. The results of Our study showed that the blinking rate of participants' in both groups in a social human-human interaction were *qualitatively correlated* to their *gaze behaviours*, as the higher number of gaze shift increased the blinking rate. Inspired from human behaviours in a social interaction, we proposed a human-level blinking model as a component of the gaze control system, in order to control the robot's blinking behaviour. Implementing the proposed model, the robot will be able to mimic human gaze and blinking behaviours. The effect of the proposed study in improving the robot ability to mimic human gaze and blinking behaviours (the robot's acceptability), will be considered as the future work.

Acknowledgments. This work was partially funded by the European Commission under the 7th Framework Program projects EASEL, Expressive Agents for Symbiotic Education and Learning, grant agreement No. 611971 - FP7-ICT-2013-10. The authors would like to thank Andreas Haslbeck and the staff of the Institute of Ergonomics at the Technical University of Munich, Germany for providing the eye-tracking system, and Manuel Giuliani for helping us in performing the experiments.

References

1. Breazeal, C.L.: Designing sociable robots. MIT Press (2004)
2. Breazeal, C.: Social interactions in hri: the robot view. IEEE Transactions on Systems, Man, and Cybernetics, Part C: Applications and Reviews 34(2), 181–186 (2004)
3. Lazzeri, N., Mazzei, D., Zaraki, A., De Rossi, D.: Towards a believable social robot. In: Lepora, N.F., Mura, A., Krapp, H.G., Verschure, P.F.M.J., Prescott, T.J. (eds.) Living Machines 2013. LNCS, vol. 8064, pp. 393–395. Springer, Heidelberg (2013)
4. Zaraki, A., Mazzei, D., Giuliani, M., Rossi, D.D.: Designing and evaluating a social gaze-control system for a humanoid robot. IEEE Transactions on Human-Machine Systems PP(99), 1–12 (2014)
5. Zaraki, A., Mazzei, D., Lazzeri, N., Pieroni, M., De Rossi, D.: Preliminary implementation of context-aware attention system for humanoid robots. In: Lepora, N.F., Mura, A., Krapp, H.G., Verschure, P.F.M.J., Prescott, T.J. (eds.) Living Machines 2013. LNCS, vol. 8064, pp. 457–459. Springer, Heidelberg (2013)
6. Ponder, E., Kennedy, W.P.: On the act of blinking. Experimental Physiology 18(2), 89–110 (1927)

7. Bentivoglio, A.R., Bressman, S.B., Cassetta, E., Carretta, D., Tonali, P., Albanese, A.: Analysis of blink rate patterns in normal subjects. Movement Disorders 12(6), 1028–1034 (1997)
8. VanderWerf, F., Brassinga, P., Reits, D., Aramideh, M., de Visser, B.O.: Eyelid movements: behavioral studies of blinking in humans under different stimulus conditions. Journal of Neurophysiology 89(5), 2784–2796 (2003)
9. Evinger, C., Manning, K.A., Pellegrini, J.J., Basso, M.A., Powers, A.S., Sibony, P.A.: Not looking while leaping: the linkage of blinking and saccadic gaze shifts. Experimental Brain Research 100(2), 337–344 (1994)
10. Gandhi, N.J.: Interactions between gaze-evoked blinks and gaze shifts in monkeys. Experimental Brain Research 216(3), 321–339 (2012)
11. Mazzei, D., Lazzeri, N., Hanson, D., De Rossi, D.: Hefes: An hybrid engine for facial expressions synthesis to control human-like androids and avatars. In: 2012 4th IEEE RAS & EMBS International Conference on Biomedical Robotics and Biomechatronics (BioRob), pp. 195–200. IEEE (2012)
12. Mazzei, D., Lazzeri, N., Billeci, L., Igliozzi, R., Mancini, A., Ahluwalia, A., Muratori, F., De Rossi, D.: Development and evaluation of a social robot platform for therapy in autism. In: 2011 Annual International Conference of the IEEE Engineering in Medicine and Biology Society EMBC 2011, pp. 4515–4518. IEEE (2011)
13. Mazzei, D., Billeci, L., Armato, A., Lazzeri, N., Cisternino, A., Pioggia, G., Igliozzi, R., Muratori, F., Ahluwalia, A., De Rossi, D.: The face of autism. In: 2010 IEEE RO-MAN, pp. 791–796. IEEE (2010)
14. Haslbeck: The gaze analytic system dikablis, http://www.ergoneers.com/en/products/dlab-dikablis/overview.html (March 2014)
15. Lausberg, H., Sloetjes, H.: Coding gestural behavior with the neuroges-elan system. Behavior Research Methods 41(3), 841–849 (2009)
16. Sloetjes, H., Wittenburg, P.: Annotation by category: Elan and iso dcr. In: LREC (2008)
17. Stern, J.A., Boyer, D., Schroeder, D.: Blink rate: a possible measure of fatigue. Human Factors: The Journal of the Human Factors and Ergonomics Society 36(2), 285–297 (1994)
18. Barbato, G., Ficca, G., Muscettola, G., Fichele, M., Beatrice, M., Rinaldi, F.: Diurnal variation in spontaneous eye-blink rate. Psychiatry Research 93(2), 145–151 (2000)
19. Fogarty, C., Stern, J.A.: Eye movements and blinks: their relationship to higher cognitive processes. International Journal of Psychophysiology 8(1), 35–42 (1989)
20. Holland, M.K., Tarlow, G.: Blinking and thinking. Perceptual and Motor Skills 41(2), 403–406 (1975)

Motor Learning and Body Size within an Insect Brain Computational Model

Paolo Arena[1,2], Luca Patané[1], and Roland Strauss[3]

[1] Dipartimento di Ingegneria Elettrica, Elettronica e Informatica,
University of Catania, Italy
[2] National Institute of Biostructures and Biosystems (INBB),
Viale delle Medaglie d'Oro 305, 00136 Rome, Italy
[3] Institut für Zoologie III (Neurobiologie), University of Mainz, Germany
{parena,lpatane}@dieei.unict.it

1 Extended Abstract

Nowadays modeling insect brains is also an important source of inspiration to develop learning architectures and control algorithms for applications on autonomous walking robots. Within the insect brain two important neuropiles received a lot of attention: the mushroom bodies (MBs) and the central complex (CX). Recent research activities considered the MBs as a unique architecture where different behavioural functions can be found. MBs are well known in bees and flies for their role in performing associative learning and memory in odor conditioning experiments [4]. They are also involved in the processing of multiple sensory modalities including visual tasks [3], different forms of learning in choice behavior [5] and also in motor learning [6]. The CX is mainly considered as a center for the initiation of behaviors; it is responsible for visual navigation, spatial memory and visual feature extraction [7,8]

To unravel the neural structures involved in these learning processes, among different insect species we focused our attention on the *Drosophila melanogaster* where, through genetic tools, it is possible to inhibit the activity in specific neural sites creating mutant flies, with the aim of better understand which neural circuits are responsible for the different behaviours shown.

An important capability needed by animals to interact with the environment is embodiment: how an animal's decisions depend on the shape of its body and which mechanisms are involved, are important questions that we are starting to unravel. For instance, the concept of peripersonal space is an interesting research topics for psychologists, neurobiologists and also for robotic applications. A living being can learn the representation of its own body and take the correct behavioural decision when interacting with the world. Even simple insects like *Drosophila* can learn to adapt their behaviours depending on their body size. This important information is acquired by flies using multisensory integration strategies: they need to connect visual input with tactile experience. To acquire its own body size, the fly need to generate parallax motion through walking. The average step size, which is proportional to the leg length and therefore to the body size, creates an average parallax motion that allows to evaluate the distance

A. Duff et al. (eds.): Living Machines 2014, LNAI 8608, pp. 367–369, 2014.

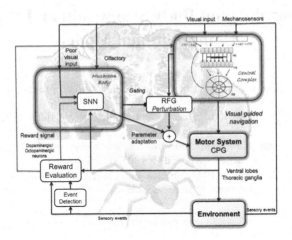

Fig. 1. Block scheme of the key elements of the insect brain involved in the body size and motor learning processes. The Mushroom Bodies and the Central Complex receive inputs from the environment including the internally generated rewarding signals. The motor system where a Central Pattern Generator has been implemented, receives the control commands from the CX for visually guided navigation and the parameter adaptation from MBs to improve the basic behaviors through motor learning. Information related to the body size of the system is acquired using the parallax motion through the CX.

from objects present in the environment in relation with the fly body, creating a form of reachable space around the fly.

Following this strategy, a computational model based on spiking neurons and threshold adaptation learning mechanisms was developed and tested both in a simulated legged robot and in a real roving platform [9]. The results demonstrate the capability of the system to learn how to shape its behaviours depending on its body size. In particular the system was able to evaluate the distance between its position and the relevant objects present in the scene deciding if performing an approaching behaviour. Interesting results were also provided by artificially altering the visual perception of the robot either by emphasizing or compensating the parallax, by moving the visual target. The learning procedure, initially applied to a targeting task, can be extended to other tasks like gap climbing and obstacle overcoming. This process leads the agent to reach a decision on the affordability of a given task. Once this decision has been taken, the agent undertakes a series of attempts which leads it to improve its capabilities capabilities through motor learning. Motor learning is needed to survive in changing environments. It can be defined as the process to acquire precise, coordinated movements needed to fulfil a task. In performing motor learning agents apply operant strategies in which a movement is made and sensory feedback is used to evaluate its accuracy.

On the basis of the known elements acquired through behavioral experiments in a gap climbing scenario [10], we proposed a new architecture for motor learning inspired by specific structures of the insect brain involved in these processes [11]. The proposed model is a nonlinear control architecture based on spiking neurons. According to the actual neurobiological knowledge, the centers involved in such learning mechanisms are the Mushroom Bodies. They are modeled as nonlinear recurrent spiking neural networks with novel characteristics able to memorize time evolution of key parameters of the neural motor controller, so as to learn motor primitives which enable the structure to efficiently acquire new motor skills. Experimental results on a simulated hexapod robot mimicking the structure of the *Drosophila melanogaster* in a dynamic simulation environment were performed. A block scheme illustrating the key elements of the insect brain architecture able to show, among other capabilities [12], both body size and motor learning, is reported in Fig. 1 where the role of MBs and CX is depicted. Preliminary results show interesting cues to be further assessed with improved experimental campaigns on insects and their mutants.

Acknowledgments. The authors acknowledge the support of the European Commission under the project FP7-ICT-2009-6 270182 EMICAB.

References

1. Menzel, M., Giurfa, R.: Dimensions of cognitive capacity in an insect, the honeybee. Behav. Cogn. Neurosci. Rev. 5, 24–40 (2006)
2. Niven, J., Chittka, L.: Are bigger brains better? Current Biology 19, R995R1008 (2009)
3. Liu, G., Seiler, H., Wen, A., Zars, T., Ito, K., Wolf, R., Heisenberg, M., Liu, L.: Distinct memory traces for two visual features in the Drosophila brain. Nature 439, 551–556 (2006)
4. Liu, X., Davis, R.: Insect olfactory memory in time and space. Curr. Opin. Neurobiol. 6, 679–685 (2006)
5. Tang, S., Guo, A.: Choice behavior of drosophila facing contradictory visual cues. Science 294, 1543–1547 (2001)
6. Pick, S., Strauss, R.: Goal-driven behavioral adaptations in gap-climbing drosophila. Curr. Biol. 15, 1473–1478 (2005)
7. Mronz, M., Strauss, R.: Visual motion integration controls attractiveness of objects in walking flies and a mobile robot. In: International Conference on Intelligent Robots and Systems, Nice, France, pp. 3559–3564 (2008)
8. Arena, P., Berg, C., Patané, L., Strauss, R., Termini, P.: An insect brain computational model inspired by drosophila melanogaster: architecture description. In: WCCI 2010 IEEE World Congress on Computational Intelligence, Barcelona, Spain, pp. 831–837 (2010)
9. Arena, P., Mauro, G.D., Krause, T., Patané, L., Strauss, R.: A spiking network for body size learning inspired by the fruit fly. In: Proceedings of International Joint Conference on Neural Networks, Dallas, TX, pp. 1251–1257 (2013)
10. Triphan, T., Poeck, B., Neuser, K., Strauss, R.: Visual targeting of motor actions in climbing Drosophila. Curr. Biol. 20(7), 663–668 (2010)
11. Arena, P., Caccamo, S., Patané, L., Strauss, R.: A computational model for motor learning in insects. In: Proceedings of International Joint Conference on Neural Networks, Dallas, TX, pp. 1349–1356 (2013)
12. Arena, P., Patané, L.: Spatial Temporal Patterns for Action-Oriented Perception in Roving Robots II: an insect brain computational model. Cognitive Systems Monographs, vol. 21. Springer (2014)

Effects of Gaze Synchronization in Human-Robot Interaction

Stavroula Bampatzia[1], Vasiliki Vouloutsi[1], Klaudia Grechuta[1],
Stéphane Lallée[1], and Paul F.M.J. Verschure[1,2]

[1] Laboratory of Synthetic, Perceptive, Emotive and Cognitive Systems – SPECS,
Universitat Pompeu Fabra, Roc Boronat 138, 08018, Barcelona, Spain
[2] ICREA Institució Catalana de Recercai Estudis Avançats, Passeig Lluís Companys 23,
E-08010 Barcelona, Spain

Abstract. In recent years, humanoid robots have been designed to resemble humans and interact with them like human beings interact with each other in a social setting. Joint attention plays an essential role in human-human social interaction. In this study, based on these findings, we want to implement gaze synchronization to a new game scenario, as a part of the communication process between humans and robots. We describe a method, which will be used in order to study the effects of gaze synchronization as a communication channel on human performance and on trustworthiness during a cooperative cognitive task between humans and a robot.

Keywords: gaze synchronization, trustworthiness, cooperative memory task, humanoid robot, experimental methods.

1 Introduction

Building social and anthropomorphic robots requires endowing them with human like skills and capabilities [1],[2]. For this reason numerous researchers who follow the "biologically inspired" approach in robot design, support that non-verbal communication cues such as eye contact and shared gazed should be implemented [1], [3], [4], [5]. These non-verbal communication cues provide necessary information to the joint attention mechanism which is a crucial part of social interaction between human beings [6],[7]. For Bruner [8], joint attention is the ability of an individual to make inferences from observable behaviors of others by attending to objects or events that others attend to. Several studies claim that these social cues form part of joint attention, facilitate the human cognition [9] and, more specifically, allow for mutual comprehension, which can be assisted when a listener is able to exploit the speaker's visual attention [10]. Additionally, it is observed that when robots behave in ways, which support human recognition of joint attention, it leads to more complex behavioral outcomes on the human side [11]. For instance, Mutlu et al. found that subjects were able to better recall stories when the robot looked at them more while telling the story [12]. As suggested above, our aim is to investigate the role of gaze synchronization in human performance during a collaborative game with a robot. Thus, people find individuals who make

A. Duff et al. (eds.): Living Machines 2014, LNAI 8608, pp. 370–373, 2014.

direct eye contact more trustworthy and more attractive than individuals who do not make eye contact [13]. However, some others support that prolonged direct gaze can be seen as threatening [14]. For this reason, our goal is to examine if gaze synchronization between robot and human can lead to an increase trustworthiness towards the robot from the side of human.

2   Materials and Methods

2.1   Experimental Setup

In this study we hypothesize that human performance during a cooperative memory task is more efficient due to the existence of gaze synchronization between the robot and the human. Our second hypothesis is that this gaze synchronization will lead to a higher level of trustworthiness from the human to the robot. To investigate if our hypotheses are valid, we propose an experimental setup with a humanoid robot, iCub [15] and a participant playing at the tabletop tangible interface "Reactable" [16]. Our scenario is a variation of the "Simon Says" game, which is an imitation memory game. Here, the game consists of, on one side, four, differently colored virtual cards on the Reactable. The goal of the game is to remember and repeat the commands of the Reactable (leader). The leader makes commands by highlighting a sequence of cards, then the position of the virtual cards is changed on the Reactable so the Simon effect [17] can be avoided, and finally the follower has to reproduce the sequence at his/her/its side. The follower can be the iCub or a human participant successively until one of them fails to repeat the sequence. As the game progresses, the number of cards in the sequence increases. During the whole process, the human participant has to recite the alphabet, which according to the cognitive load theory [18], will make it harder to memorize the sequence of cards.

2.2   Experimental Procedure

Prior to the final experiments, a series of pilot experiments will be performed. In the pilot experiments, 10 pairs of participants will play the game in a cooperative way on the Reactable. In these tests the participants will have the role of the follower and they will play against the Reactable. Our intention is to measure the Average Time (AT) of shared gaze from human to human, from human to an object and from object to object. The number of successful trials and the Response Time (RT) of the players will be analyzed as additional measurements. We will apply the ATs of shared gaze to the robot's transitional eye fixations.

In the final within-subjects design experiments, we will test twenty five subjects and ask them to play the game as followers with the iCub. In this scenario the iCub will be used to test two different conditions. In the first condition the iCub will help the participant to repeat the sequence by indicating it using its gaze when it is the participant's turn to play. In this condition, the robot will indicate the correct sequence seventy percent (70%) of the times and an incorrect sequence thirty percent (30%) of the times. In the second condition, the robot will not help the participant to identify the sequence, and its gaze will be fixed in a particular point. In this case robot will only act as an opponent of the human participant. Each participant will play at least five games of each condition during one session, in a random order. After the tenth game, the robot will ask the participant whether he wants to continue playing. We will measure the human performance by counting the number of successful trials, the RTs and the number of times the player is willing to keep playing more turns of the game. In order to measure how much the participant trusts the robot, we will count the times the player followed the suggestion of the robot. Additionally, we will measure the ATs of the shared gaze from human to robot and from object to object (video coding). In order to analyze the level of arousal of the participants we will measure the physiological data including the Respiration Rate (RR), the Skin Conductance Level (SCL) and the Heart Rate Variability (HRV). Finally, the participants will be asked to complete a questionnaire about demographic data and their emotions regarding the experienced Human Robot Interaction(HRI). The described research protocol is still being developed. We expect to have the first experimental results in the near future.

References

1. Fong, T., Nourbakhsh, I., Dautenhahn, K.: A survey of socially interactive robots. Robotics and Autonomous Systems 42(3), 143–166 (2003)
2. Breazeal, C., Brooks, A., Gray, J., Hoffman, G., Kidd, C., Lee, H., Lieberman, J., Lockerd, A., Mulanda, D.: Humanoid robots as cooperative partners for people. Int. Jnl. Humanoid Robots (2004a)
3. Breazeal, C., Kidd, C.D., Thomaz, A.L., Hoffman, G., Berlin, M.: Effects of nonverbal communication on efficiency and robustness in human-robot teamwork. In: IEEE/RSJ International Conference on Intelligent Robots and Systems, pp. 708–713 (2005)
4. Argyle, M.: Bodily communication. International Universities Press Inc., New York (1975)
5. Fiske, J.: Introduction to communication studies. Routledge, London (1982)
6. Scassellati, B.M.: Foundations for a Theory of Mind for a Humanoid Robot (Doctoral dissertation, Massachusetts Institute of Technology) (2001)
7. Knoblich, G., Sebanz, N., Böckler, A.: Observing shared attention modulates gaze following. Journal Title Cognition, 292–298 (2011)
8. Bruner, J.: Foreword from joint attention to the meeting of minds: An introduction. In: Moore, C., Dunham, P.J. (eds.) Joint Attention: Its Origins and Role in Development, pp. 1–14. Lawrence Erlbaum Associates (1995)
9. Carpenter, M., Nagell, K., Tomasello, M., Butterworth, G., Moore, C.: Social cognition, joint attention, and communicative competence from 9 to 15 months of age. Monographs of the Society for Research in Child Development (1998)
10. Staudte, M., Crocker, M.W.: Investigating joint attention mechanisms through spoken human–robot interaction. Cognition 120(2), 268–291 (2011)
11. Yu, C., Scheutz, M., Schermerhorn, P.: Investigating multimodal real-time patterns of joint attention in an hri word learning task. In: Proceedings of the 5th ACM/IEEE International Conference on Human-Robot Interaction, pp. 309–316. IEEE Press (2010)
12. Mutlu, B., Forlizzi, J., Hodgins, J.: A storytelling robot: Modeling and evaluation of human-like gaze behavior. In: 2006 6th IEEE-RAS International Conference on Humanoid Robots, pp. 518–523. IEEE (2006)

13. Mason, M.F., Tatkow, E.P., Macrae, C.N.: The look of love gaze shifts and person perception. Psychological Science 16(3), 236–239 (2005)
14. Argyle, M., Cook, M.: Gaze and mutual gaze. Cambridge University Press, Cambridge (1976)
15. Metta, G., Sandini, G., Vernon, D., Natale, L., Nori, F.: The icub humanoid robot: an open platform for research in embodied cognition. In: Proceedings of the 8th Workshop on Performance Metrics for Intelligent Systems, pp. 50–56. ACM (2008)
16. Geiger, G., Alber, N., Jorda, S., Alonso, M.: The reactable: A collaborative musical instrument for playing and understanding music. Her&Mus. Heritage & Museography (4), 36–43 (2010)
17. Borgmann, K.W., Risko, E.F., Stolz, J.A., Besner, D.: Simon says: Reliability and the role of working memory and attentional control in the Simon task. Psychonomic Bulletin & Review 14(2), 313–319 (2007)
18. Sweller, J.: Cognitive load theory, learning difficulty, and instructional design. Learning and Instruction 4(4), 295–312 (1994)

Individual Differences and Biohybrid Societies

Emily C. Collins and Tony J. Prescott

Department of Psychology, The University of Sheffield, UK
{e.c.collins,t.j.prescott}@sheffield.ac.uk

Abstract. Contemporary robot design is influenced both by task domain (e.g., industrial manipulation versus social interaction) as well as by classification differences in humans (e.g., therapy patients versus museum visitors). As the breadth of robot use increases, we ask how will people respond to the ever increasing number of intelligent artefacts in their environment. Using the Paro robot as our case study we propose an analysis of individual differences in HRI to highlight the consequences individual characteristics have on robot performance. We discuss to what extent human-human interactions are a useful model of HRI.

Keywords: individual differences, biohybrid, HRI, attachment, design.

1 Background

As a broad, multidisciplinary field it can be difficult to concisely define Human-Robot Interaction (HRI), but it maybe useful to consider HRI as being to robotics, as ergonomics is to design. Contemporary robot design is influenced both by the task domain (e.g., industrial manipulation versus social interaction) as well as by discrete classifications of the humans with which the robot will interact. For instance, robots designed for autism therapy behave quite differently from robots designed to interact with office workers or groups of people in crowded spaces such as airports [1]. The diversity of robot design reflects the considerable increase over the past few decades in the breadth of robot use.

Given the ongoing integration of intelligent machines into human life, we ask: how will people respond to these new *biohybrid societies*? Perhaps robot designers and engineers should be tackling how to make robots do the intricate things we wish of them before addressing the question of how we finesse a robot's social behaviour, but the speed with which robotics is advancing indicates that the time for thinking about the consequences of technology capable of eliciting complex, lasting emotions from humans is now [2]. This study investigates to what extent theories of human-human interaction are useful to an understanding of HRI as a contributing factor in the development of tools for robot design.

2 Paro and Individual Differences

Paro is a therapeutic device modelled on a baby harp seal (Fig. 1). Its primary use is in dementia-care as a robotic replacement for pets otherwise used in animal-therapy. Already it has become apparent that one size does not fit all even for this

A. Duff et al. (eds.): Living Machines 2014, LNAI 8608, pp. 374–376, 2014.

Fig. 1. Paro's *cute* design is intended to elicit positive responses from humans

early example. Whilst some patients take to Paro, and benefit from its use, some do not [3]. Despite patients' shared condition - dementia - the characteristics of the individual remain highly influential over the robot's performance.

In psychology, the study of these intra-class differences is referred to as *individual differences*. Whilst psychology is ostensibly the study of individuals, modern psychology more generally studies groups defined by shared biology or cognition. The statistical controls upon which empirical work relies are often defined in terms of a comparison between and within these groups, with individual differences treated as deviations from the main dimensions of study.

In order to design robots that will be effective for all people, rather than just some, we propose that within HRI whilst studying groups we also seek to understand dimensions of behaviour upon which individuals differ. This risks leading to a very complex design process - how are we even to begin to understand how humans will vary in their individual responses to robots in particular situations?

One way is to consider human-human interactions as a model of HRI. Robots with some autonomy and a physical presence that are capable of social interaction can be expected to elicit significant emotional responses. Considering the importance of emotions to the development of relationships one possible way to tackle the question of individual human reactions to robots is to consider to what extent human-human (or human-animal) interactions are a useful model of human-robot interactions [4]. To explore this we have chosen to analyse interaction with Paro using the human-human bonding theory of attachment.

3   Attachment Theory and Experiment

Attachments are thought to be driven by an evolutionarily programmed behavioural system which keeps infants close to their caregivers and safe from harm. A list of specific features are required for a relationship to be considered an attachment (in the psychological sense) [5], and attachment styles themselves are well defined in the literature as individual personality traits, that strongly influence emotional bonds and reactions to social partners [6].

The aim of the experiment is to measure whether individuals with attachment avoidance type personalities tend towards deactivating their emotional response towards Paro, in contrast to those with attachment anxiety personalities from whom we expect a hyperactivated emotional response.

Before arrival participants complete an online questionnaire measuring attachment style. Upon arrival a baseline measure of emotional receptiveness is taken. Participants are then led to a room and asked to explore the interactive features of Paro before being let into the room alone. During the interaction participants are covertly video recorded. After five minutes have elapsed the experimenter returns and a measure of emotional receptiveness is taken again. Finally the experimenter conducts a short interview with the participant about their interaction with Paro, and the expectations they had before the interaction.

The questionnaire and interview data is used to cross reference the recorded behavioural data. The video is coded for robot directed displays of affection and interactivity: frequency/duration of touches, type of touch, e.g., stroking, cuddling, etc. As a robot designed to elicit caregiving behaviour in its users, Paro should tap into the caregiving structures of our participants. Thus participants measuring high in attachment avoidance should display less of these affectionate behaviours than those high in attachment anxiety. If Paro's design is effective our hypothesis may well be confirmed. If our hypothesis is not confirmed it is possible that this is due to Paro's design being less efficient at eliciting emotion than it could be. Alternatively it may indicate that a human-human interaction model cannot be directly applied to a human-robot interaction scenario.

However, as our focus here is on variation amongst individuals, and personal attachment style, we expect to obtain some data which will feed into our ideas about the potential for robot customisation. Future robots may, for example, measure an individual's personal attachment style, and then adapt its behaviour appropriately, allowing the achievement of personalised human-robot interactions without a requirement for personalised configuration, an idea which in turn may lend itself to greater advancements in our developing biohybrid societies.

References

1. Goodrich, M.A., Schultz, A.C.: Human-robot interaction: a survey. Foundations and trends in human-computer interaction 1(3), 203–275 (2007)
2. Watanabe, M., Ogawa, K., Ishiguro, H.: Field study: can androids be a social entity in the real world? In: Proceedings of the 2014 ACM/IEEE International Conference on Human-Robot Interaction, pp. 316–317. ACM (2014)
3. Chang, S., Sung, H.: The effectiveness of paro robot therapy on mood of older adults: a systematic review. International Journal of Evidence-Based Healthcare 11(3), 216 (2013)
4. Collins, E.C., Millings, A., Prescott, T.J.: Attachment to Assistive Technology: A New Conceptualisation. In: Proceedings of the 12th European AAATE Conference (Association for the Advancement of Assistive Technology in Europe) (2013)
5. Hazan, C., Zeifman, D.: Pair bonds as attachments: Evaluating the evidence. Guilford Press (1999)
6. Vrtička, P., Andersson, F., Grandjean, D., Sander, D., Vuilleumier, P.: Individual attachment style modulates human amygdala and striatum activation during social appraisal. PLoS One 3(8), e2868 (2008)

Programming Living Machines:
The Case Study of *Escherichia Coli*

Jole Costanza[1], Luca Zammataro[1], and Giuseppe Nicosia[2]

[1] Italian Institute of Technology, Milan
[2] Department of Mathematics and Computer Science, University of Catania
{jole.costanza,luca.zammataro}@iit.it, nicosia@dmi.unict.it

In 1952, Turing outlined computational processes in the morphogenesis [8], thus thinking of the biological evolution of an organism as a consequence of the computation that it can perform. Following Turing's idea on morphogenesis, many biological processes have been recently analysed from a computational standpoint. In 1995, Bray [2] argued that *a single protein is a computational or information carrying element*, being able to convert input signals into an output signal. Evolution had already been associated with computation many years before, by von Neumann and Burks [9], who constructed a self-replicating cellular automaton with the aim of developing synthetic models of a living organism. Starting from this concept, in this work we propose a relation between computation and metabolism.

1   *E. coli* as a Molecular Machine

We associate the structure of the *Escherichia coli* bacterium with the von Neumann architecture, showing that the components of a bacterium can be mapped to a processing unit, a control unit, a memory storing the "program" of the bacterium, and an input-output section. In this way, the bacterium becomes a molecular machine with computation capability. Furthermore, the set of all its chemical reactions represents a processing unit and the entire metabolic network works as a Turing Machine (TM) [1]. The bacterium takes as input chemicals (substrates) necessary for his growth and duplication, and through its biochemical network (coded by the genes of its genome), produces metabolites as output. The genome sequence is thought of as an executable code specified by a set of commands in a sort of ad-hoc low-level programming language. Each combination of genes is coded as a string of bits $y \in \{0,1\}^L$, each of which represents a gene set. So, we can design the genetical code and reprogramm the molecular machine. In this work, turning off a gene set means turning off the chemical reactions associated with it. Each bit in y is a gene set that distinguishes between single and multi-functional enzymes, isoenzymes, enzyme complexes, enzyme subunits. The string y acts as a program stored in the memory unit. The memory unit contains the string y. The control unit is a function g_Φ that defines a partition of the string, and is uniquely determined by the pathway-based clustering of the chemical reaction network. The function g_Φ interprets the binary string y and knocks out gene sets, thus turning syntax into semantics [1].

A. Duff et al. (eds.): Living Machines 2014, LNAI 8608, pp. 377–379, 2014.

2   Optimal Design of *E. coli*

An optimal molecular machine designed for a particular task can be obtained using GDMO algorithm [3]. Indeed, Pareto fronts obtained from GDMO represent optimal organisms that are output of a multi-objective optimisation carried out in the metabolic network. Each point provided by GDMO is a Pareto optimal molecular machine whose computation is aimed at maximising the concentration of two or more metabolites (outputs) simultaneously. A Pareto front is the set of points in a given objective space such that there does not exist any other point that dominates them in all the objectives. It is obtained as a result of a multi-objective optimisation technique [4] needed when a system (a given phenotype) cannot be optimal at all the tasks it performs, and particularly when tasks are in contrast with each other [5]. Formally, given r objective functions $f_1, ..., f_r$ to maximise/minimise, the problem of optimising in a multi-objective fashion can be rephrased as the problem of finding a vector x^* that satisfies all the constraints and optimises the vector function $f(x) = (f_1(x), f_2(x), ..., f_r(x))^\mathsf{T}$, where x is the variable (vector) to be optimised in the search space. The many-objective Pareto optimality is a useful and powerful tool to understand the phenotype of organisms in different environmental conditions and genetic strategies.

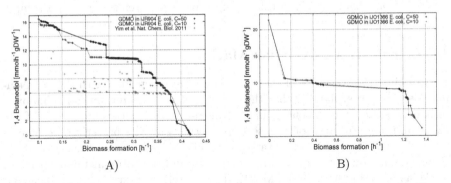

Fig. 1. Pareto fronts obtained with GDMO to maximise BDO and biomass. In A) we used the *i*JR904 model [7] and compared our results (in blue and black) with solutions proposed by Yim et al. [10] (in red). In B) we used the most recent iJO1366 *E. coli* model [6]. In the key, the parameter C represents the maximum allowable knockout.

By using Pareto analysis, we analyzed two genome-scale *E. coli* networks modeled by using Flux Balance Analysis: that of Reed et al. [7] and that of Orth et al. [6]. The first network links 913 biochemical reactions, 625 metabolites and 904 genes, while the second network links 1136 reactions, 1136 metabolites and 1366 genes. We additionally programmed the *E. coli* bacterium machine in order to allow the synthetic production and overproduction of 1,4-butanediol (BDO) and compared our results with those obtained by Yim et al. [10]. BDO is an inorganic compound used industrially as a solvent and in the manufacture of some types of plastics, elastic fibers and polyurethanes. Currently, BDO is manufactured entirely from petroleum-based feedstock. By including the novel synthetic DNA y string, we designed the living machine in order to overproduce BDO by knocking out. By means GDMO, we obtained the results shown in Fig. 1. In A, in blue and black Pareto points obtained maximizing BDO and biomass in the smaller network by setting the maximal knockout number to 10 and 50 respectively. In red, proposed solutions of Yim et al. [10]. We can observe that, by considering the same experimental protocol, the Pareto fronts overcome points in red, providing better results. By analyzing all Pareto points, we found that by turning off 6 genes, BDO production increases from 0 (wild type production) to 15.17 mmol/h/gDW. In Fig. 1-B we repeated the same experiments by considering the most recent network [6]. From a production of 1.42 in wild type, we found a solution that reaches 10.87 mmol/h/gDW by turning off 6 genes.

Running a program in a molecular machine can represent an effective intervention in a cell, driving it towards the modification of its behaviour. More specifically, a code can instruct the cell to make decisions by taking into account external variables, the current cell state or user-imposed goals. Although modelling the whole life of an organism, as a TM would certainly be computationally unfeasible, our approach is aimed at explaining the single operation executed by a bacterium in light of a computational instruction. This approach can be readily used to evaluate the computational effort for a specific task, or the computational capability of the whole organism under investigation.

References

1. Angione, C., et al.: Computing with metabolic machines. In: Voronkov, A. (ed.) Turing-100. EPiC Series, vol. 10, pp. 1–15 (2012)
2. Bray, D., et al.: Protein molecules as computational elements in living cells. Nature 376(6538), 307–312 (1995)
3. Costanza, J., et al.: Robust design of microbial strains. Bioinformatics 28(23), 3097–3104 (2012)
4. Cutello, V., Narzisi, G., Nicosia, G.: A class of pareto archived evolution strategy algorithms using immune inspired operators for ab-initio protein structure prediction. In: Rothlauf, F., et al. (eds.) EvoWorkshops 2005. LNCS, vol. 3449, pp. 54–63. Springer, Heidelberg (2005)
5. Cutello, V., et al.: A multi-objective evolutionary approach to the protein structure prediction problem. Journal of the Royal Society Interface 3(6), 139–151 (2006)
6. Orth, J.D., et al.: A comprehensive genome-scale reconstruction of Escherichia coli metabolism-2011. Molecular Systems Biology 77(Article number 535), 1–9 (2011)
7. Reed, J., et al.: An expanded genome-scale model of escherichia coli k-12 (ijr904 gsm/gpr). Genome Biology 4(9), R54 (2003)
8. Turing, A.M.: The chemical basis of morphogenesis. Bulletin of mathematical biology 52(1), 153–197 (1990)
9. Von Neumann, J., Burks, A.W., et al.: Theory of self-reproducing automata (1966)
10. Yim, H.R., et al.: Metabolic engineering of escherichia coli for direct production of 1,4-butanediol. Nature Chemical Biology 7(7), 445–452 (2011)

Force Contribution of Single Leg Joints in a Walking Hexapod

Chris J. Dallmann and Josef Schmitz

Department of Biological Cybernetics and Cognitive Interaction Technology–Center
of Excellence (CITEC), Bielefeld University,
Universitätsstraße 25, 33615 Bielefeld, Germany
{cdallmann,josef.schmitz}@uni-bielefeld.de

Abstract. We study leg joint torques of a large insect (*Carausius morosus*) to infer the functions of individual joints in closed kinematic chains during unrestrained walking. Leg joints were found to differentially contribute to multiple locomotor functions of a leg, such as body weight support and propulsion. We conclude that quantifying joint torques in freely behaving hexapods may provide a powerful tool in unraveling the feedback control strategies underlying motor flexibility.

Keywords: Motor control, ground reaction force, full body kinematics, joint torque, stick insect.

1 Introduction

The movement of multi-jointed limbs in natural environments requires a considerable control effort for biological and technical walking systems alike because movements of joints in closed kinematic chains have to be highly coordinated. Stick insects have provided important bio-inspiration for the control of hexapod walking [1], yet surprisingly little is known about how their joint controllers deal with motor redundancy and manage to adapt to different walking contexts such as locomotion on unstable ground or climbing [2,3]. As a first step to further elucidate the control strategies underlying this motor flexibility, we characterized the functions of individual leg joints of the stick insect *Carausius morosus* during unrestrained walking by analyzing single joint torques.

2 Methods

Adult female stick insects walked freely on a rigid horizontal path (40 × 500 mm) with integrated force transducers (Fig. 1A). If an animal stepped onto one of the transducers (foothold size: 5 × 5 mm), whole body kinematics were combined with single-leg three-dimensional ground reaction forces to estimate the net joint torques of the respective leg during stance. Kinematics were calculated based on automatically tracked body and leg positions. Tracking was performed at 200 Hz with a marker-based Vicon MX10 motion capture system consisting of eight T10 infrared high-speed cameras with a spatial resolution of ∼0.1 mm (Vicon, Oxford,

A. Duff et al. (eds.): Living Machines 2014, LNAI 8608, pp. 380–382, 2014.

UK). Single leg forces were recorded at 1000 Hz with a resolution of $\sim$0.05 mN using custom-made strain gauge force transducers.

To estimate the net torques at single joints, the leg was treated as a serial three-link manipulator with three degrees of freedom (DoF), one for each of the three main leg joints (thorax-coxa (TC), coxa-trochanter (CT), and femur-tibia (FT) joint). The endpoint of the tibia was considered as the end effector. Owing to the complexity of the actual TC-joint and the short length of the coxa ($\sim$1.5 mm), we relied on two simplifications with regard to the direction of the TC-axis and the position of the CT-axis. The former was considered to be slanted outward by $\theta = 30°$ (cf. [4]), the latter was considered to coincide with the TC-joint (Fig. 1B). These simplifications were estimated to have only a small effect on the magnitudes of the TC- and FT-torque ($<$15%) and essentially no effect on the shape of the respective time courses. The net torques at the three joints ($\tau = [\tau_{TC}, \tau_{CT}, \tau_{FT}]$) were derived analytically for each time step through the principle of virtual work using the manipulator Jacobian J and the three-dimensional force vector acting on the end effector ($F = [F_x, F_y, F_z]$) [5]

$$\tau = J^T F \tag{1}$$

Due to the slanted orientation of the TC-axis, positive torques at the TC-joint protract and supinate the leg plane. Within the leg plane, positive torques at the CT- and FT-joint elevate and extend the leg, respectively (Fig. 1B).

3   Results and Discussion

Our results demonstrate that analyzing complete time courses of joint torques in concert with joint angles and single leg forces yields important insights into the functions of individual joints during unrestrained locomotion.

The middle leg of *C. morosus*, for example, produces forward directed, decelerating forces for the first 70% of the stance phase when walking on level ground ($F_x > 0$, Fig. 1C), yet the TC-torque points in the direction of retraction/pronation ($\tau_{TC} < 0$, Fig. 1E). This seeming contradiction can be resolved by considering that the leg is leaned backward during this time (Sup > 0, Fig. 1D). Hence, a negative torque at the CT-joint ($\tau_{CT} < 0$) can not only support the body weight ($F_z < 0$), but also contribute to deceleration of the body. In turn, when the leg is leaned forward (Sup < 0) toward the end of the stance phase, this torque considerably contributes to acceleration ($F_x < 0$). An almost rectangular leg configuration throughout most of the stance phase (Lev $< 30°$, Ext around 90°) may allow the leg structure itself to transmit the respective forces without requiring strong stabilizing torques at the FT-joint.

Similar torque profiles were found for the hind legs. Here, the CT-torque also functions to support the body weight (the body's center of mass is located between the hind leg coxae), but contributes considerably more to acceleration due to different working ranges of the legs.

The contributions of front leg joints to body weight support and propulsion are weaker and more variable. Front legs may thus mainly serve a tactile exploration function in this nocturnal animal (cf. [3]).

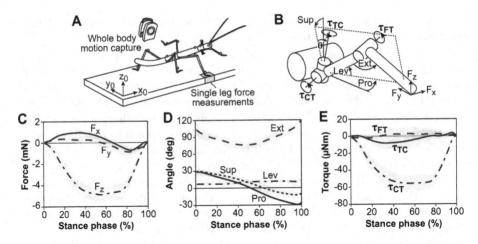

Fig. 1. (A) Marker-based motion capture was combined with single leg force measurements to calculate net joint torques in freely walking stick insects. (B) Net torques were estimated based on a manipulator with three DoFs, one for each of the main leg joints (TC, CT, and FT). Due to the slanted TC-axis ($\theta = 30°$), a positive TC-torque (τ_{TC}) points in the direction of protraction (Pro) and supination (Sup). Within the leg plane (dashed line), positive CT- and FT-torques (τ_{CT} and τ_{FT}) point in the direction of levation (Lev) and extension (Ext), respectively. (C–E) Average time courses of forces, angles, and torques ($\pm$ standard deviation) produced by the right middle leg during the stance phase (n = 201 steps, N = 7 animals).

The task-dependent generation of different joint torques in different legs seems to be under strict feedback control involving dedicated force sensors for each joint. Bio-inspired walking machines might benefit from the transfer of the biological control concepts found in stick insects.

Acknowledgements. This study was supported by EU-FP7, ICT-6-2.1 270182 project "EMICAB".

References

1. Schilling, M., Hoinville, T., Schmitz, J., Cruse, H.: Walknet, a bio-inspired controller for hexapod walking. Biol. Cybern. 107, 397–419 (2013)
2. Bartling, C., Schmitz, J.: Reaction to disturbances of a walking leg during stance. J. Exp. Biol. 203, 1211–1233 (2000)
3. Cruse, H.: The function of the legs in the free walking stick insect, *Carausius morosus*. J. Comp. Physiol. A 112, 235–262 (1976)
4. Cruse, H., Bartling, C.: Movement of joint angles in the legs of a walking insect, *Carausius morosus*. J. Insect Physiol. 41, 761–771 (1995)
5. Spong, M., Hutchinson, S., Vidyasagar, M.: Robot modeling and control. Wiley, London (2006)

High Resolution Tactile Sensors
for Curved Robotic Fingertips

Alin Drimus, Vince Jankovics, Matija Gorsic, and Stefan Mátéfi-Tempfli

Mads Clausen Institute for Product Innovation,
University of Southern Denmark, 6400 Sønderborg, Denmark

Introduction. Tactile sensing is a key element for various animals that interact with the environment and surrounding objects. Touch provides information about contact forces, torques and pressure distribution and by the means of exploration it provides object properties such as geometry, stiffness and texture[5]. For humans, extracting high level information from touch provides a better understanding of the objects manipulated while for insects it is essential for locomotion[3]. While robot designers have been using vision systems to provide the robot with information about its surroundings, this is not always trivial to obtain, dealing with limited accuracy, occlusions and calibration problems. In terms of sensors for static stimuli, such as pressure, there are a range of technologies that can be used to manufacture transducers with various results[5]. A simple approach is to use fingertips with a 6-DOF force-torque sensor for estimating contact conditions[1], but this only allows a single point of contact and is costly. In terms of fingertip and foot tip prototypes, tactile sensors are used for multi modal sensing, similar to biology, for pressure and dynamic stimuli. In this respect Hosoda et al. [4] propose an anthropomorphic fingertip which has randomly distributed straingauges and PVDF (polyvinylidene fluoride) transducers. In [7] a biomimetic tactile array is proposed that shows a low hysteresis and good sensitivity for skin like deformations.

Sensor Development Using Piezoresistive Rubbers. Our approach for development of sensitive foot tip considers the use of piezoresistive rubbers [2], which show a smooth dependence between resistance and applied pressure, with main advantages being flexibility and low cost. They are 0.5 mm thick, flexible and can withstand pressure up to approx 860 psi (60 kg/cm^2) for millions of actuations. To cope with multiple sensing elements, multiplexing algorithms that address matrix structures of n columns and m rows are employed. Iterating through all the combinations of columns and rows can acquire the set of values for such a sensor array at a given moment, or a tactile image. Among various electrode types, our previous work[2] showed best results by working with custom developed flex pcb designs with high conductivity finish and conductive epoxies. Also, there should not be any permanent electrical contact between the electrodes and the piezoresistive rubber patch, as this reduces the sensitivity for the low-forces range. The 3D model for the foot tip base depicted in Figure 1a serves as a solid base for the other layers that cover it. It has two types of grooves on its surface; the deeper is for guiding isolated wires out of the assembly, while the shallow serve as conductive channels.

A. Duff et al. (eds.): Living Machines 2014, LNAI 8608, pp. 383–385, 2014.

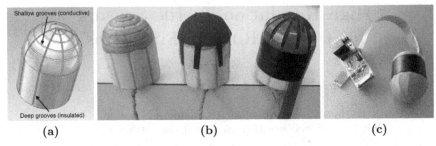

Fig. 1. Sensor fabrication process: a) depiction of the foot tip base, b) resulting prototype after each fabrication step: conductive epoxy layer with the ring electrodes, piezoresistive rubber applied and top flex pcb layer on top and c) finished prototype

An isolating epoxy is then used to fix the wires to the desired position, cover the deep grooves after the wires are in place and smooth out the surface resulted from the 3d printing process. After the fingertip is polished, the result is a spherical tip with five thin conductive rings spaced 1-3 mm apart. Then we apply a thin layer of highly conductive epoxy - approx 0.5 mm thickness (8331 Silver Conductive Epoxy Adhesive manufactured by MGChemicals) that yields a 0.017 Ωcm and it cures in a few hours. After full curing and a second polishing process, it needs to be separated into five sections with a 0.5 mm spacing between by using a lathe turning machine with a fine tip. The end result is a foot tip with 5 smooth electrically conductive rings as shown in Figure 1b.

For the second layer, the piezoresistive rubber is cut in a flower like shape so that it can be applied over the convex surface of the fingertip to ensure a uniform covering. A custom made flex pcb layer with electro-less nickel gold finish and twelve fingers of rectangular shape is applied over the rubber, with the electrodes facing inwards, as it can be seen in Figure 1b. The fingers of this flex pcb are made to match the petals of the flower like rubber layer. The overlapping between the top electrodes from the flex pcb and the bottom conductive rings is perpendicular and the resulting array can be described by a spherical coordinate system. The bottom rings define the polar angles with a 15° angular resolution whereas the top electrodes define the azimuthal angles with an angular resolution of 30° on the surface. A final 0.1 mm layer of nitrile rubber is applied at the end for mechanical protection and isolation, acting as a high pass filter.

The data acquisition electronics, see Figure 1c, uses of the n x m elements as variable resistors in a voltage divider. A matrix of $5 rings$ x $12 columns$ gives the tactile image. The tactile array is scanned at 100 frames per second, with 8 bit data encoding 256 levels of pressure for every taxel.

Experimental Results. The robot setup that was used to apply increasing force patterns under different angles consists of a robotic arm (Universal Robots UR5) equipped with an elastic end effector and a force measurement plate. Figure 2 depicts the results of the experiments for the application of various forces (up to 30 N) and different tilt angles. The sensor shows excellent results with respect to the cells triggered and the force distribution. To compensate for imperfect manufacturing a calibration is required before using it. The images depicted

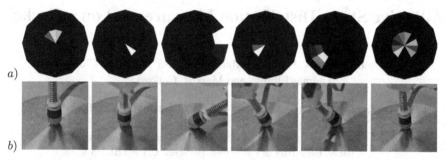

Fig. 2. Example tactile images for different applications of force with respect to the ground plate a) *tactile images from the data acquisition system, where white depicts a high pressure and black shows no contact* and b) position of the robot manipulator for the same events

in Figure 2 also show that there is a very good relationship between the tilt angles and the triggered cells. Bio-inspired hexapod robots like Hector[6] could use such information for understanding the ground contact conditions[3] and devising biologically inspired walking models.

Conclusion. In this paper we have presented a novel approach for developing flexible tactile array sensors that can be used to add the sense of touch to end effectors for biology inspired robots.

Acknowledgments. This work is supported by the EU project EMICAB (FP7-ICT, No. 270182).

References

1. Cutkosky, M.R., Howe, R.D., Provancher, W.R.: Force and Tactile Sensors. In: Springer Handbook of Robotics, pp. 455–476. Springer, Heidelberg (2008)
2. Drimus, A., Kootstra, G., Bilberg, A., Kragic, D.: Design of a flexible tactile sensor for classification of rigid and deformable objects. Robotics and Autonomous Systems, 3–15 (2014)
3. Frigon, A., Rossignol, S.: Experiments and models of sensorimotor interactions during locomotion. Biological Cybernetics 95(6), 607–627 (2006)
4. Hosoda, K., Tada, Y., Asada, M.: Anthropomorphic robotic soft fingertip with randomly distributed receptors. Robotics and Autonomous Systems 54(2), 104 (2006)
5. Lee, M.H., Nicholls, H.R.: Review article tactile sensing for mechatronics–a state of the art survey. Mechatronics 9(1), 1 (1999)
6. Schneider, A., Paskarbeit, J., Schaeffersmann, M., Schmitz, J.: Hector, a new hexapod robot platform with increased mobility - control approach, design and communication. In: Rückert, U., Joaquin, S., Felix, W. (eds.) Advances in Autonomous Mini Robots, pp. 249–264. Springer, Heidelberg (2012)
7. Wettels, N., Santos, V.J., Johansson, R.S., Loeb, G.E.: Biomimetic tactile sensor array. Advanced Robotics 22(8), 829–849 (2008)

Adhesive Stress Distribution Measurement on a Gecko

Eric V. Eason[1], Elliot W. Hawkes[2], Marc Windheim[3], David L. Christensen[2],
Thomas Libby[4], and Mark R. Cutkosky[2]

[1] Dept. of Applied Physics, Stanford University, Stanford, CA 94305, USA
[2] Dept. of Mechanical Engineering, Stanford University, Stanford, CA 94305, USA
[3] Technische Universität München, 85748 Garching, Germany
[4] CiBER, University of California, Berkeley, CA 94720, USA
easone@stanford.edu

Abstract. Gecko adhesion has inspired climbing robots and synthetic adhesive grippers. Distributing loads between patches of adhesive is important for maximum performance in gecko-inspired devices, but it is unknown how the gecko distributes loads over its toes. We report *in vivo* measurements of stress distributions on gecko toes. The results are significantly non-uniform.

Gecko-inspired adhesives are seeing use in applications ranging from climbing robots [1,2] to orbital debris capture [3]. As adhesive systems grow larger and more sophisticated, with multiple independent patches of adhesive, it becomes clear that they must be designed holistically: the mechanisms that apply and distribute loads are as important as the adhesive itself. For example, the gecko-inspired gripper shown in Fig. 1 uses an array of tiles of adhesive, each loaded with a tendon to prevent moments over the tiles, and with load-sharing mechanisms to achieve uniform forces over the entire array [3], [1]. Load-sharing increases the overall performance by preventing individual tiles from becoming overloaded.

We desire to compare these load-sharing mechanisms with the equivalent parts of the gecko adhesive system. Geckos make use of multiple patches of adhesive, but it is not known to what extent they achieve load-sharing between patches. The toes of the gecko are divided into flaps of skin called lamellae (or scansors), which are in turn covered in arrays of microscopic adhesive setae. The lamellae are loaded independently by a

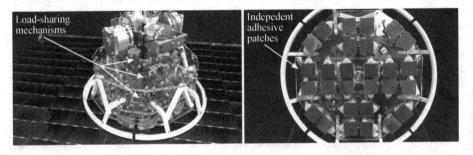

Fig. 1. Prototype of gecko-inspired adhesive gripper for retrieving space debris [3]. Reprinted by permission of the American Institute of Aeronautics and Astronautics, Inc.

A. Duff et al. (eds.): Living Machines 2014, LNAI 8608, pp. 386–388, 2014.

branched system of tendons. In this natural adhesion system, as in synthetic designs, the distribution of adhesive stresses within and between lamellae determines the overall system performance. It is possible that stretching of the lamellar skin or sliding of lamellae relative to each other could contribute to the redistribution of loads.

There is some evidence that the gecko adhesive system may not be ideal. In the tokay gecko (*Gekko gecko*), measurements at different scales of adhesive area show that there is over a 50-fold decrease in adhesive shear stress as the area increases from a single seta to a whole animal [4]. It has been suggested that this decrease in performance as area increases is explained by a reduced contact area or unequal load-sharing between or within lamellae [4,5].

To investigate this question further, we conducted the first known *in vivo* measurements of stress distribution and contact area on tokay gecko toes, using a custom optical tactile sensor optimized for spatial and temporal resolution (100 μm, 60 Hz) rather than stress accuracy. The sensor converts compressive or tensile normal stress into a light signal that is captured by a video camera. It uses frustrated total internal reflection (FTIR) within an acrylic waveguide covered by an 80 μm thick microtextured PDMS film (Fig. 2), molded from an etched Si wafer with pyramidal pits. Light enters the waveguide and is internally reflected until it scatters at points where the waveguide is in contact with the film [6]. Higher pressures cause the scattering to increase as the microtexture is compressed. A vacuum between the film and waveguide adds a compressive stress offset and allows the sensor to measure tensile (adhesive) stresses.

The sensor was oriented vertically, and an adult tokay gecko of body mass 101.2 g was allowed to attach one of its hind feet to the sensor and hang, head downwards. The gecko was then pulled off the sensor manually. This procedure was repeated multiple times and the pull angle was varied to obtain different loading configurations on the foot. Selected test results are shown in Fig. 3.

For measurements of contact area, the microtextured PDMS film was removed and the gecko was allowed to attach to the bare acrylic waveguide (Fig. 3a). It is apparent that a significant fraction of the setal area does not make contact. The most proximal lamellae are totally off the surface, but even for lamellae that touch the surface there are non-contacting regions that appear wrinkled or folded.

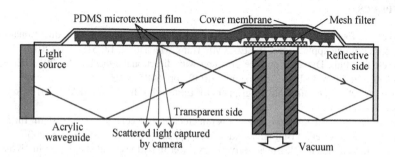

Fig. 2. FTIR-based optical tactile sensor for normal stress distribution measurements. A PDMS film with pyramidal bumps (29 μm height, 54 μm width, 100 μm spacing) produces a scattered light image where the light intensity is proportional to the normal stress.

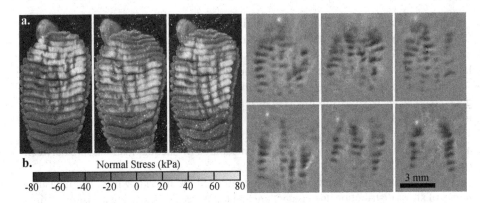

Fig. 3. Third toe of right hind limb of a tokay gecko under varying loading conditions. (a) The toe is placed on an acrylic waveguide illuminated by total internal reflection. The real area of contact is indicated by the white regions. (b) The toe is placed on the custom tactile sensor. Light yellow regions indicate compressive normal stress, and dark blue regions indicate tensile normal stress.

There are also variations in normal stress across the toe (Fig. 3b). The highest tensile stresses are limited to a small fraction of the total area of the adhesive. The locations of the force concentrations vary between tests and depend on the angle of the external load. The data also suggest that some lamellae were under compressive stress.

From these results, the tokay gecko does not achieve full contact of its adhesive to a flat surface, and within the portion of the adhesive that does make contact, there is a nonuniform adhesive stress distribution. This result is consistent with observations of decreasing adhesive performance with increasing area, and demonstrates that the adhesive system of the tokay gecko is non-ideal at the toe scale. The tokay can nevertheless support many times its body weight, so full contact area and equal load-sharing are less critical for geckos than for current synthetic systems.

Acknowledgments. This work was supported by the NSF Graduate Research Fellowship Program, the Stanford Graduate Fellowship Program, and the Hertz Foundation.

References

1. Hawkes, E., Eason, E., Asbeck, A., Cutkosky, M.: The gecko's toe: scaling directional adhesives for climbing applications. IEEE/ASME Trans. Mechatron. 18, 518–526 (2013)
2. Hawkes, E., et al.: Dynamic surface grasping with directional adhesion. In: 2013 IEEE/RSJ Int. Conf. Intelligent Robots and Systems (IROS), pp. 5487–5493 (2013)
3. Parness, A., et al.: On-off adhesive grippers for Earth-orbit. In: AIAA SPACE 2013 Conf. and Exposition (2013)
4. Autumn, K.: Properties, principles, and parameters of the gecko adhesive system. In: Smith, A.M., Callow, J.A. (eds.) Biological Adhesives, pp. 225–256. Springer, Berlin (2006)
5. Bullock, J., Federle, W.: Beetle adhesive hairs differ in stiffness and stickiness: in vivo adhesion measurements on individual setae. Naturwissenschaften 98, 381–387 (2011)
6. Begej, S.: Planar and finger-shaped optical tactile sensors for robotic applications. IEEE J. Robot. Autom. 4, 472–484 (1988)

Design of an Articulation Mechanism for an Infant-like Vocal Robot "Lingua"

Nobutsuna Endo, Tomohiro Kojima, Yuki Sasamoto, Hisashi Ishihara,
Takato Horii, and Minoru Asada

Dep. Adaptive Machine Systems, Graduate School of Engineering, Osaka University,
F1-401, 2-1, Yamadaoka, Suita city, Osaka, Japan, 565-0871
endo@ams.eng.osaka-u.ac.jp

1 Introduction

Spoken language is one of the important means for humans to communicate with others. In developmental psychology, it is suggested that an infant develops it through verbal interaction with caregivers by observation experiments [1]. However, what kind of underlying mechanism works for that and how caregiver's behavior affects on this process has not been fully investigated yet since it is very difficult to control the infant vocalization. On the other hand, there are several constructive approaches to understand the mechanisms by using infant robots with abilities equivalent to those of human infants, as a controllable platform [2].

Sasamoto *et al.* suggest a vocal robot as a platform for constructive investigation of the developmental process of vocalization [3]. Unlike speech conversion and articulation simulators or speech synthesis, robotic platforms have advantages in terms of realtimeness, consonant vocalization by means of flow-acoustics, and interaction with humans. They actually built an infant-like vocal robot that mimics the anatomical shape of the articulator of human infant, and showed that its vocal cords and vocal tract could vocalize vowels in the same range of formant frequencies as that of human infant. However, the driving mechanism could not vocalize the same range because it did not comprise enough degrees of freedom (4-DOFs for tongue, 1-DOF for jaw, 1-DOF for soft palate).

On the other hand, vocal robots which have many degrees of freedom for articulation and can vocalize as well as human adults have been developed [4,5]. Particularly, Fukui *et al.* [4] developed the vocalization robot WT-7RII which could vocalize not only vowels but also consonants by controlling many degrees-of-freedom (7-DOFs for tongue, 1-DOF for jaw, 1-DOF for soft palate, 5-DOFs for lip). However, this robot focused on reproduction of adults' utterance instead of infants'. Between adults and infants, the size of the articulator is different, which is closely related to the difference in their vocalization. The shapes of vocal cord [6] and vocal tract [7] change with the growth. It is necessary to consider the changes in order to understand infant's vocal development [8]. Therefore, in order to reproduce infants' vocalization by means of many degrees of freedom (like WT-7RII), the problem of miniaturization has to be solved.

In this study, aiming at reproducing the infant vocalization, we miniaturized the articulation mechanism of WT-7RII, and developed a new infant-like

A. Duff et al. (eds.): Living Machines 2014, LNAI 8608, pp. 389–391, 2014.

vocal robot named "Lingua". This paper describes the design of its articulation mechanism.

2 Design of the Articulation Mechanism and Preliminary Evaluation

Fig. 1 shows a Lingua's overview, DOF configuration, and structural properties of the vocal cords and tract. This robot consists of a lung, vocal cords, and a vocal tract. The lung and the driving mechanism of the vocal cords are those of WT-7RII. We used the same vocal cords made from soft material as for the infant-like vocal robot by Sasamoto *et al.* [3]. The shape of Lingua's vocal tract is based on the anatomical data of 6-month-old infants [6,7].

The tongue mechanism consists of 7-DOFs that combine rotational and linear movement (Fig. 2). We downsized the linkage mechanism which connects them. The linkage of WT-7RII's tongue consisted of plural shafts by parallel and slider cranks, but we minimized the parallel crank by adopting a coaxial mechanism for it. The movable range of each linkage tip was determined based on simulation results of infants' 3 vowels utterances (/a/, /i/, /u/) by a VLAM (Variable Linear Articulatory Model) articulation simulation [9]. The layout and dimensions of the linkage were determined based on the range calculated by inverse kinematics. The surface of the tongue was molded in silicone rubber. We calculated the elastic coefficient of the tongue based on measuring its stretching when exposed to external load. Then, we designed the mechanical parts such that the minimum yield safety ratio could be 5 by using FEM. Hence, the linkages have enough strength against reciprocating articulatory movements.

We measured formant frequencies of Lingua's vowel vocalization (graph in Fig. 2). This means that Lingua has ability to vocalize the vowels as well as human infant.

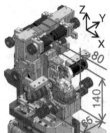

Part	DOF
Upper jaw	1
Lower jaw	8
Vocal cords	2
Lungs	1
Total	**12**

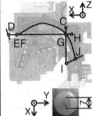

Parameter	Length [mm]
VTL (D to J)	93
VT-V (I-C)	39
PCL (I-G)	27
NPhL (G-C)	12
VT-H (D to H)	65
LTh (D-F)	10
ACL (F to G)	45
OphW (G-H)	10
VT-O (F-H)	55

Fig. 1. Lingua's overview, DOF configuration, and structural properties of the vocal cords and tract

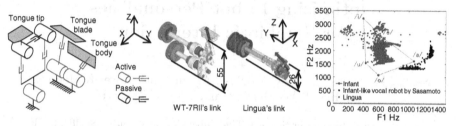

Fig. 2. DOF configuration of Lingua's tongue and comparison of the link mechanisms between WT-7RII and Lingua

3   Conclusion and Future Works

In this paper, we described the design of the articulation mechanism of the infant-like vocal robot "Lingua". Preliminary evaluation shows Lingua's ability to vocalize the vowels as well as human infant. In the future, we will develop the lip mechanism and examine the vocalization performance of the overall mechanism for vowels and consonants. We also aim to reproduce crying and babbling. Moreover, we will conduct interaction experiments between the robot and a caregiver in order to investigate how infants' vocalization develops.

Acknowledgments. This work was partially supported by the MEXT Grant-in-Aid for Specially Promoted Research (24000012), and the MEXT project "Creating Hybrid Organs of the future" at Osaka University. We appreciate the provision of WT-7RII by Atsuo Takanishi Laboratory and Masaaki Honda Laboratory at Waseda University.

References

1. Bates, E., et al.: Individual differences and their implications for theories of language development. The handbook of child language, 96–151 (1995)
2. Asada, M., et al.: Cognitive developmental robotics: A survey. IEEE Transactions on Autonomous Mental Development 1(1), 12–34 (2009)
3. Sasamoto, Y., et al.: Towards understanding the origin of infant directed speech: A vocal robot with infant-like articulation. In: Proceedings of the IEEE Third Joint International Conference on Development and Learning and Epigenetic Robotics (2013)
4. Fukui, K., et al.: Speech Robot Mimicking Human Articulatory Motion. In: Proceedings of INTERSPEECH 2010, pp. 1021–1024 (2010)
5. Sawada, H., et al.: A robotic voice simulator and the interactive training for hearing-impaired people. Journal of biomedicine & biotechnology 2008, 768232 (2008)
6. Eckel, H.E., et al.: Morphology of the human larynx during the first five years of life studied on whole organ serial sections. The Annals of Otology, Rhinology and Laryngology 108(3), 232–238 (1999)
7. Vorperian, et.al.: Anatomic development of the oral and pharyngeal portions of the vocal tract: An imaging study. The Journal of the Acoustical Society of America 125, 1666 (2009)
8. Mugitani, R., Hiroya, S.: Development of vocal tract and acoustic features in children. Acoustical Science and Technology 33(4), 215–220 (2012)
9. Boe, L.J., et al.: Anatomy and control of the developing human vocal tract: A response to Lieberman. Journal of Phonetics 41(5), 379–392 (2013)

Optimising Robot Personalities
for Symbiotic Interaction

Samuel Fernando[1], Emily C. Collins[2], Armin Duff[3], Roger K Moore[1],
Paul F.M.J. Verschure[3,4], and Tony J. Prescott[2]

[1] Department of Computer Science, The University of Sheffield, Sheffield, UK
[2] Department of Psychology, The University of Sheffield, Sheffield, UK
{s.fernando,e.c.collins,r.k.moore,t.j.prescott}@sheffield.ac.uk
[3] SPECS, Department of Technology, Universitat Pompeu Fabra, Barcelona, Spain
[4] Catalan Institute of Advanced Studies (ICREA), Barcelona, Spain
{armin.duff,paul.verschure}@upf.edu

Abstract. The Expressive Agents for Symbiotic Education and Learning (EASEL) project will explore human-robot symbiotic interaction
(HRSI) with the aim of developing an understanding of *symbiosis* over
long term tutoring interactions. The EASEL system will be built upon
an established and neurobiologically grounded architecture - *Distributed
Adaptive Control (DAC)*. Here we present the design of an initial experiment in which our facially expressive humanoid robot will interact with
children at a public exhibition. We discuss the range of measurements
we will employ to explore the effects our robot's expressive ability has
on interaction with children during HRSI, with the aim of contributing
optimal robot personality parameters to the final EASEL model.

Keywords: HRI, symbiosis, neurobiological, robot tutors, facial
expressions.

1 Introduction and Core System

A key challenge in HRSI is to develop robots that can learn from interactions
over time. Successful social engagement requires a robot to develop an engag-
ing personality that is responsive and adaptive to its human user. Within the
EASEL project we aim to explore such robot capabilities by basing our robot
system upon the neurobiologically grounded *DAC* architecture [1], which will
be used as the basis of the Synthetic Tutor Assistant (STA) model. The EASEL
tutor assistant will also have an interaction manager which is responsible for
the multimodal interaction with the student and determines how the STA de-
livers content and feedback to the user and how user questions are processed.
A self-regulation system based on key identified variables will manage low-level
interactions. This regulation system will be based on biologically grounded models such as the motivation and emotion control provided by the reactive layer of
DAC, which has already been tested in large-scale interactive spaces [4] and with
robots. Further stages will develop computational methods to extract knowledge
from long sequences of behavioural data.

A. Duff et al. (eds.): Living Machines 2014, LNAI 8608, pp. 392–395, 2014.
© Springer International Publishing Switzerland 2014

Employing the EASEL tutor assistant we aim to explore and develop a theoretical understanding of symbiosis in HRSI. Symbiosis is defined as the capacity of the robot and its human user to mutually influence each other's behaviour over different time-scales (for instance within encounters and across encounters). Symbiosis requires that the robot has social salience [2]. For instance it should be responsive to the behaviour and emotional state of its human user, and adapt its own behaviour to this in ways that have predictable effects on the person.

Within the EASEL project HRSI will be explored in the context of educational child-robot interactions in school classrooms and at natural history museum family events. We aim to determine what aspects of HRSI are most relevant to providing effective instruction and what level of human-robot affect is appropriate for communicating learning content in an ethical manner. We will frame the evaluative methodologies employed to explore this in a conceptualisation of attachment to humanoid robots taken from a taxonomy in development, which proposes that human-robot bonds can be analysed in terms of their similarities to different types of existing bond with other humans, animals and objects [3].

2 Initial Interaction in the Field

In an initial EASEL experiment we will use the Hanson Robokind Zeno R50 humanoid robot (Fig. 1) in a child-robot interaction activity at a public museum. A distinctive feature of Zeno is the platform's realistic face which is capable of displaying a wide range of facial expressions.

We will explore how Zeno's facial expressions affect a child's perceptions and interactions during a game of 'Simon says' (Fig. 2). During the game Zeno will give spoken commands to one child at a time to perform simple actions, such as 'Wave your arms', which the child is instructed to only perform if Zeno precedes the command with the phrase 'Simon says'. Zeno will then provide verbal feedback confirming whether the child's action was correct or not.

Fig. 1. The Hanson Robokind *Zeno* R50 robot with some facial expressions illustrated

Fig. 2. The museum setup - a child plays Simon Says with the Zeno robot

We will test two different configurations. Zeno will either maintain a static, neutral facial expression throughout game-play, or will alter its facial expressions to convey appropriate emotions alongside verbal feedback: producing a happy smiling expression with positive feedback and a sad/sympathetic expression with negative feedback. Interactions between Zeno and the children will be video recorded, whilst logs of spatial distance data will be gathered using a Kinect sensor, and questionnaires will be used to ask parents/guardians and their child their opinions about the interaction. We predict that children playing with Zeno during the emotionally expressive configuration will enjoy their experience more than children playing with Zeno during the neutral facial expression configuration. We predict that this will manifest in the following ways:

– Interpersonal distances will be smaller (measured with Kinect sensor data)
– The child will participate in game-play for a longer period of time
– Questionnaires will indicate a markedly greater level of game-play enjoyment
– During game-play the child will appear more animated (smile/laugh more)
– During game-play the child will be more likely to mimic *Zeno's* expressions

Zeno's configuration will be informed by recent work investigating the alignment of visual, vocal and behavioural affordances in social robots [7]. This work suggests that the voice, facial expressions and behaviours of the robot should be appropriate for its size and morphology and also consistent with each other - an alignment that is crucial to avoid the 'uncanny valley' effect [8]. Given Zeno's capacity to allow us to explore this alignment of robot facial expression and the response of its human user, our work will contrast with other recent HRSI work which has made use of the Aldebaran NAO robot, a technically advanced but facially non-expressive platform (e.g., [9,10]). Initial EASEL experiments such as that presented here will extend this current HRSI research by measuring exactly what effects emotionally appropriate facial expressions have on human-users during child-robot interactions, allowing us to better understand how to optimise the expressed personalities of robot platforms in symbiotic interaction.

References

1. Verschure, P.F.: Distributed Adaptive Control: a Theory of the Mind, Brain, Body Nexus. Biologically Inspired Cognitive Architectures 1, 55–72 (2012)
2. Inderbitzin, M.P., Betella, A., Lanata, A., Scilingo, E.P., Bernardet, U., Verschure, P.F.: The Social Perceptual Salience Effect. Journal of Experimental Psychology: Human Perception and Performance 41 (2012)
3. Collins, E.C., Millings, A., Prescott, T.J.: Attachment to Assistive Technology: A New Conceptualisation. In: Proceedings of the 12th European AAATE Conference (Association for the Advancement of Assistive Technology in Europe) (2013)
4. Wasserman, K., Manzolli, J., Eng, K., Verschure, P.: Live soundscape composition based on synthetic emotions: Using music to communicate between an interactive exhibition and its visitors. IEEE MultiMedia (2003)
5. Sanchez-Fibla, M., Bernardet, U., Wasserman, E., Pelc, T., Mintz, M., Jackson, J.C., Lansink, C., Pennartz, C., Verschure, P.F.: Allostatic control for robot behavior regulation: a comparative rodent-robot study. Advances in Complex Systems 13(03), 377–403 (2010)
6. Vouloutsi, V., Lallée, S., Verschure, P.F.M.J.: Modulating behaviors using allostatic control. In: Lepora, N.F., Mura, A., Krapp, H.G., Verschure, P.F.M.J., Prescott, T.J. (eds.) Living Machines 2013. LNCS, vol. 8064, pp. 287–298. Springer, Heidelberg (2013)
7. Moore, R.K., Maier, V.: Visual, vocal and behavioural affordances: some effects of consistency. In: 5th International Conference on Cognitive Systems - CogSys 2012 (2012)
8. Moore, R.K.: A Bayesian explanation of the 'Uncanny Valley' effect and related psychological phenomena. Scientific reports 2 (2012)
9. Beck, A., Cañamero, L., Damiano, L., Sommavilla, G., Tesser, F., Cosi, P.: Children interpretation of emotional body language displayed by a robot. Social Robotics, 62–70 (2011)
10. Tielman, M., Neerincx, M., Meyer, J.J., Looije, R.: Adaptive emotional expression in robot-child interaction. In: Proceedings of the 2014 ACM/IEEE International Conference on Human-robot Interaction, HRI 2014 (2014)
11. Costa, S., Soares, F., Santos, C.: Facial expressions and gestures to convey emotions with a humanoid robot. Social Robotics, 542–551 (2013)

A Bio-inspired Wing Driver for the Study of Insect-Scale Flight Aerodynamics

Nick Gravish, Stacey Combes, and Robert J. Wood

SEAS & OEB, Harvard University,
60 Oxford St., Cambridge, MA, 02138
scombes@oeb.harvard.edu,
{gravish,rjwood}@seas.harvard.edu

Abstract. Insect flight studies have advanced our understanding of flight biomechanics and inspire micro-aerial vehicle (MAV) technologies. A challenge of centimeter or millimeter scale flight is that small forces are produced from relatively complex wing motions. We describe the design and fabrication of a millimeter-sized wing flapping mechanism to simultaneously control pitch and stroke of insect and MAV wings. Using micro-fabrication techniques we construct this wing driver and observe that wing motion matches the natural degrees of freedom of insect wings. We actuate wing stroke-position and pitch in open-loop at frequencies relevant to Dipteran and Hymenopteran flight (100-200Hz) and describe the advancements and limitations of this system.

Keywords: Biomimetics, Insect flight, Micro-robotics, Aerodynamics.

1 Introduction

Combined robotic and biological experiments have advanced our knowledge of the world of insect flight, and serve as a source of inspiration for the future design of micro-aerial vehicles (MAVs). Furthermore, the study of robotic and biological flapping wing flight has highlighted a need to understand unsteady aerodynamics [1]. However, despite the myriad laboratory experiments on flapping wing aerodynamics, major open questions of flapping wing flight remain unanswered.

Insect wings are heterogeneous structures with non-linear elasticity and complex structural design [2]. The functional consequences of features such as wing flexibility, shape, or material properties are largely open questions in flapping wing aerodynamics. While recent studies have highlighted the importance of wing flexibility in a dynamically scaled experiment [3], at-scale experimental approaches are necessary since inertial forces cannot be accurately captured by dynamically scaled models. Inertia is a key factor in aeroelastic deformations of wings during flapping and thus at-scale laboratory experiments are needed to explore form and function of insect wings.

Advances in micro-robotic manufacturing, such as the smart-composite manufacturing process [4], have enabled the development of insect-scale actuated

A. Duff et al. (eds.): Living Machines 2014, LNAI 8608, pp. 396–398, 2014.
© Springer International Publishing Switzerland 2014

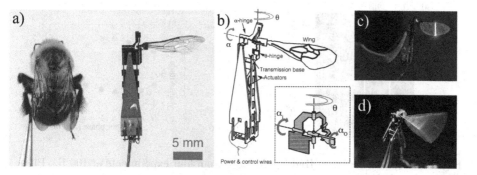

Fig. 1. Insect inspired study of flapping wing aerodynamics. a) A bumblebee (*Bombus impatiens*) (left) and wing driver (right). A bumblebee wing is attached to the driver. b) Schematic of the wing driver. Inset shows spherical 5-bar mechanism. c) Laser sheet illumination. d) Wing motion during flapping at 100Hz.

mechanisms. We present the design of a micro-robotic wing flapping mechanism capable of actuating the millimeter sized wings of MAVs and the bumblebee (*Bombus impatiens*) for aerodynamics study. This wing driver is inspired by the motion of real insect wings during flight. We construct and test the wing driver and discuss the future outcomes of these experiments.

2   Driver Design

Our design goal is a wing driver that generates bumblebee scale (*Bombus impatiens*) wing kinematics (Fig. 1a). During hover, bumblebees flap their wings at approximately 170 Hz with a mean stroke amplitude of 140°. During a typical wing stroke the position (θ), pitch (α), and vertical excursion are all modulated. Previous flapping mechanisms to study bumblebee scale aerodynamics utilized a passive hinge to modulate wing pitch. Here we present a mechanism capable of actuating a bumblebee sized wing in θ and α over ranges of $\theta \in [-70°, 70°]$ and $\alpha \in [-45°, 45°]$. For manufacturing considerations we require this mechanism to be minimally complex, and be lightweight for time resolved force measurements.

We laser cut and fold planar sheets of carbon fiber and polymer flexures into their designed shapes to construct this mechanism [4]. The driver is actuated by two piezo-electric bimorphs (Fig. 1b). A spherical five bar linkage provides two degree of freedom control (Fig. 1b and inset). A feature of this linkage is the minimal coupling between θ and the α output [5].

3   Experiments and Outlook

We tested wing motion under open-loop control in both θ and α. A laser sheet illuminated the mid-plane of the wing normal to the span-wise direction in the neutral position (Fig. 1c). We tracked the motion of this laser line in high-speed

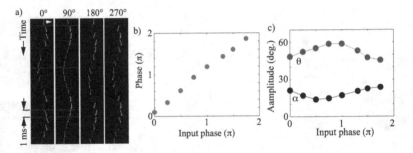

Fig. 2. Wing kinematics. a) High-speed video from four experiments at 100 Hz. Phase offset denoted above. Frames are 1 ms apart. b) Input and measured phase difference in wing stroke and pitch. c) Peak-to-peak amplitude of wing stroke and amplitude of angle of attack.

video. Wings were oscillated sinusoidally at a frequencies of 100 Hz (Fig. 1d) with varied phase-lag between α and θ (Fig. 2a). The mechanism was capable of wing frequencies of up to 200 Hz.

We fit sin curves to the measured α and θ. The output phase lag between α and θ increased with the commanded phase lag approximately linearly, with a small positive offset (Fig. 2b). However, the output amplitudes of α and θ varied with phase lag (Fig. 2c) likely due to the rotational inertia of the wing during stroke reversal. These results highlight the need for closed-loop control to compensate for the non-uniform output kinematics associated with the variation of wing control signals.

The development of a two degree of freedom wing driver capable of matching the wing kinematics of millimeter sized insects will give valuable insight into the functional consequences of insect wing form. Through closed-loop control on θ and α we will systematically vary wing and kinematic parameters in our at-scale experiments in the near future.

References

1. Sanjay, P.: Sane. The aerodynamics of insect flight. Journal of Experimental Biology 206(23), 4191–4208 (2003)
2. Combes, S.A., Daniel, T.L.: Flexural stiffness in insect wings i. scaling and the influence of wing venation. Journal of experimental biology 206(17), 2979–2987 (2003)
3. Zhao, L., Huang, Q., Deng, X., Sane, S.P.: Aerodynamic effects of flexibility in flapping wings. Journal of The Royal Society Interface 7(44), 485–497 (2010)
4. Whitney, J.P., Sreetharan, P.S., Ma, K.Y., Wood, R.J.: Pop-up book mems. Journal of Micromechanics and Microengineering 21(11), 115021 (2011)
5. Teoh, Z.E., Wood, R.J.: A flapping-wing microrobot with a differential angle-of-attack mechanism. In: 2013 IEEE International Conference on Robotics and Automation (ICRA), pp. 1381–1388. IEEE (2013)

Characterizing the Substrate Contact of Carpal Vibrissae of Rats during Locomotion

Thomas Helbig[1], Danja Voges[1], Sandra Niederschuh[2],
Manuela Schmidt[2], and Hartmut Witte[1]

[1] Technische Universität Ilmenau, Group of Biomechatronics,
Institute of Micro- and Nano-Technologies IMN MacroNano®, Ilmenau, Germany
{thomas.helbig,danja.voges,hartmut.witte}@tu-ilmenau.de
[2] Friedrich-Schiller-Universität Jena, Institut für Spezielle Zoologie und Evolutionsbiologie
mit Phyletischem Museum, Jena, Germany
{sandra.niederschuh,schmidt.manuela}@uni-jena.de

Abstract. Excitation of sensors triggered by carpal vibrissae has an influence on the kinematics of legs during locomotion of rats. Via motion studies, anatomic and mechanical characterization of vibrissae – especially in the contact period with the substrate – we try to gain a better understanding of the adaptability of those special sensory organs. This knowledge might lead to new approaches for passive sensor systems in robot locomotion or other tactile tasks.

Keywords: carpal vibrissae, locomotion, pedipulator, touch, sensor.

1 Introduction

Vibrissae are tactile hairs found at multiple locations on the body surfaces of mammals [1], [2]. This contribution deals with carpal vibrissae (cV) at the forelimbs of rats. Although known for over 120 years – cV first were discovered in lemurs [3] – the knowledge about structure and functions of carpal vibrissae is very limited in comparison to that about mystacial vibrissae (mV) – commonly known as whiskers. Carpal vibrissae gain increasing interest, since there is evidence on influences on the kinematics of the segmental chains of legs. One hypothesis is that cV signals serve to adjust the stiffness of legs and to prepare them for contact with different substrates or irregularities in the substrate – a function interesting for robotic locomotion.

2 Characterization of Carpal Vibrissae

2.1 Differences between Carpal and Mystacial Vibrissae

Mystacial vibrissae are well described in different mammalian species – regarding geometric and mechanical parameters [4], [5], modeling [6], neurobiology [7], up to biomimetic transfer of principles into technical sensor systems for object recognition and discrimination [8], [9]. At a first glance, cV seem to be a down-scaled form of

A. Duff et al. (eds.): Living Machines 2014, LNAI 8608, pp. 399–401, 2014.

mV with a similar morphology (Fig 1. left). However due to the lack of muscles able to move the bearings of cV, no (obvious) whisking behavior has been observed so far. In comparison to mV, cV can be a role model for a passive (not actively moved) sensor system (in contrast to e.g. [9]). Furthermore slight differences in the innervation of non-mystacial vibrissae are proven [10].

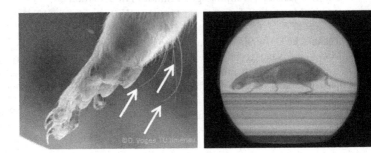

Fig. 1. left: Triple of carpal vibrissae at a forelimb of *Rattus norvegicus*. Regular (arrows) and renewable **right:** sketch of a cineradiography of the walking movement of *Rattus norvegicus*

2.2   Characterization of the Influence of Carpal Vibrissae on Leg Movement

Motion studies with a biplanar high-speed cineradiography system (Neurostar®, Siemens) explored the sensory functions of cV for identification of irregularities in the substrate (Fig. 1. right) in response to locomotion as well as to the spatial spread of mV. Quantifications of head and hip height and of spatial-temporal walking parameters were performed by cineradiographic analyses, while horizontal and vertical motions of mV were recorded by high-speed cameras (Results yet unpublished).

Due to the small size of cV it is not possible in this setup to visualize their mechanical bearing under locomotion suitable for further image processing and quantification of their behavior, nor by cineradiography neither by high frequency videography. In particular questions regarding at which gait phases cV are in contact with the substrate and how they are altered have to be answered by different approaches.

3   Observation of the Substrate Contact Using a Pedipulator

To visualize the behavior of cV during a step cycle, we propose the use of a "pedipulator" – a gearing mechanism which guides a dissected leg of a rat cadaver artificially. While following the observed kinematics from cineradiography this enables us to imitate the natural movement of the forelimb of the rat on an extended time scale. Furthermore the pedipulator allows to focus our gaze on the exact spot of contact between cV and substrate, without being subject to restrictions arising with the depth of focus and uncertain contact points during natural movement of the living rat. First experiments (Fig. 2.) observed with a high frequency microscope (Keyence VW-9000) indicate a contact phase of the carpal vibrissae with the substrate longer than expected. Additionally, cV show a planar contact with the substrate in contrast to most simulations made so far regarding mV (e.g. [6], [11]). Associated with the planar contact, a high influence of friction between vibrissa and substrate on the forces and torques routed to the mechanical receptors in the sensor complexes of the vibrissa has to be expected. By analyzing the deformation of cV during the whole contact phase, regarding different substrates with different friction coefficients, we hope to gain a better understanding of those specific sensory organs, which can be biomimetically transferred for the application in technical developments.

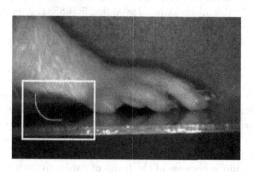

Fig. 2. Contact between a carpal vibrissa and the substrate during provoked movement of the left forelimb of a rat cadaver. Manually enhanced curve: carpal vibrissa, rectangle: contact area.

References

1. Sokolov, V.E., Kulikov, V.E.: The structure and function of the vibrissal apparatus in some rodents. Mammalian Species 51(1), 125–138 (1987)
2. Klauer, G.: Vibrissen – Analyse eines Sinnesorgans. Habilitationsschrift, Universität Essen, Fachbereich Bio- und Geowissenschaften, Landschaftsarchitektur (1999)
3. Sutton, J.B.: On the arm-glands of the lemurs. Proc. of the Zool. Soc. of London 55(2), 369–372 (1887)
4. Quist, B.W., Faruqi, R.A., Hartmann, M.J.: Variation in Young's modulus along the length of a rat vibrissa. Journal of Biomechanics 44(16), 2775–2781 (2011)
5. Voges, D., Carl, K., Klauer, G.J., Uhlig, R., Schilling, C., Behn, C., Witte, H.: Structural characterization of the whisker system of the rat. IEEE Sensors Journal 12(2), 332–339 (2012)
6. Behn, C., Schmitz, T., Witte, H., Zimmermann, K.: Animal vibrissae: modeling and adaptive control of bio-inspired sensors. In: Proc. IWANN, Teneriffa, pp. 159–170 (2013)
7. Guic-Robles, E., Jenkins, W.M., Bravo, H.: Vibrissal roughness discrimination is barrel-cortex-dependent. Behavioural Brain Research 48(2), 145–152 (1992)
8. Prescott, T.J., Pearson, M., Mitchinson, B., Sullivan, J.C., Pipe, A.: Whisking with robots from rat vibrissae to biomimetic technology for active touch. IEEE Robot. Automat. Mag. 16(3), 42–50 (2009)
9. Scholz, G.R., Rahn, C.D.: Profile sensing with an actuated whisker. IEEE Trans. Robot and Automation 20(1), 124–127 (2004)
10. Fundin, B.T., Arvidsson, J., Rice, F.L.: Innervation of nonmystacial vibrissae in the adult rat. Journal of Comparative Neurology 357, 501–512 (1995)
11. Hires, S.A., Pammer, L., Svoboda, K., Golomb, D.: Tapered whiskers are required for active tactile sensation. eLife 2(0) 01350 (2013)

Self-organization of a Joint
of Cardiomyocyte-Driven Robot

Naoki Inoue[1], Masahiro Shimizu[2], and Koh Hosoda[2]

[1] Graduate School of Information Science and Technology, Osaka University, 2-1,
Yamada-oka, Suita, 565-0871, Japan
[2] Graduation School Science, Osaka University,
1-3, Machikaneyama, Toyonaka, 560-8531, Japan
inoue.naoki@ist.osaka-u.ac.jp,
{shimizu,hosoda}@arl.sys.es.osaka-u.ac.jp

Abstract. In this presentation, we intend to spontaneously realize a
joint by utilizing self-organization of cardiomyocytes. The function of
the joint is provided by mechanical structure and cell aggregation. The
robot was built by culturing neonatal rat cardiomyocytes on a thin collagen sheet whose shape is like a butterfly. The robot could move around
the butterfly's hinge because of the beats of cardiomyocytes, and the aggregation of the cells is self-organized through motion-based mechanical
stimulation. After 1 week cultivation of cardiomyocyte-driven robot, cell
aggregation emerged around the hinge, and motion of the joint became
efficient.

Keywords: self-organization, mechanical stimulation, cell aggregation,
cardiomyocyte-driven robot, joint.

1 Introduction

Realization of micro robots by using biological material is recently gathering
increased attention. E.g. Nawroth et al.[1] developed a tissue-engineered jellyfish robot by using rat cardiomyocyte and recreated the jellyfish-like movement. Akiyama et al.[2] developed an actuator driven by insect muscle tissues which are robust over a range of culture conditions. To realize important function like a joint, we have to manually design mechanical structure
and actuation mechanism. In order to resolve this problem, we spontaneously
fabricate a joint mechanism of a micro cardiomyocyte-driven robot through
self-organization by adopting mechanical stimulation. Living organisms change
their structure by self-organization, and they are highly adaptive to changes of
environment[3]. Self-organization occurs in a cellular level, as prominent functions such as self-reproduction, self-repair and self-assembly. Engler et al. [4]
found that MSCs (mesenchymal stem cells) differentiate into bones in response
to the stiffness of the scaffolding: mechanical strength of the substrate to which
they adhere. Hayakawa et al. [5] reported that cells change their orientation perpendicular to mechanical stress by sensing outer forces through actin filaments.

A. Duff et al. (eds.): Living Machines 2014, LNAI 8608, pp. 402–404, 2014.

Shimuzu et al. [6] developed a bio-actuator with oriented muscle fibers induced from mechanical stimulation. Based on these knowledge, this study shows that through self-organization by adopting mechanical stimulation, the cell structure emerges, and as a result, a joint function is realized.

2   Method

We built a joint of a cardiomyocyte-driven robot utilizing self-organization. The function of a joint is provided by mechanical structure and actuation (cell aggregation). We tested two types of shape of bio-robot's mechanical structure, i.e., a butterfly-like and a square-like (a control condition which is not included in this article) sheet. A buttfly-like robot has a hinge in the middle. These robots were built by culturing neonatal rat cardiomyocytes (CMC02, Cosmobio Co.,Ltd, Japan) on a thin collagen sheet (ID-002, Atree Inc., Japan). They can move because of beats of cardiomyocytes.

3   Result

After 1 week cultivation of a butterfly-like cardiomyocyte-driven robots, we found the cells aggregation around a joint structure, i.e., the center of the burtterfly-like shape (see Fig.1E). And we successfully observed the change of bending motion that leads to cells aggregation (see Fig.2). The robot obtained bending motion in such way that the hinge (Fig.1E) is used as a joint. This demonstrates we have achieved a robotic system that has growing ability through self-organization by motion-based mechanical stimulation. In the future work, Utilizing biological devices in real-world situations remains challenging. For example, rat cardiomyocyte are only capable of surviving in cell culture medium, hence it is necessary to maintain good culture environment by changing the cell

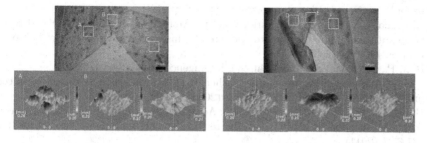

Fig. 1. Cell aggregation comparison between 1st day (left side) and 7th day (right side) after starting culture. Upper figures show photo under microscope. Here, the black scale bar indicates $200[\mu\ m]$. Lower graphs show how the cells aggregate inside each ROI. As in the result of 7 days culturing, a biased aggregate of cells found around the center of a butterfly-like shape.

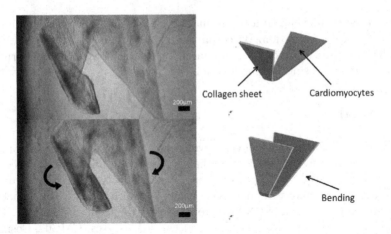

Fig. 2. Bending movement of cardiomyocyte-driven robot. Left side photos indicate snapshots of bending motion, see from upper to lower ones. Right side images show the schematic pictures of the resultant robot's shapes. The cardiomyocytes are cultivated on the red-colored side of sheet.

culture medium regularly. In the future, not only muscle cells as well as neurons will be used as controller in bio-robots. Co-culture system of muscle cell actuators and neurons can be verified.

Acknowledgments. This work was partially supported by Grant-in-Aid for Scientific Research on 24680023, 23220004, 25540117 of Japan.

References

[1] Nawroth, J.C., Lee, H., Feinberg, A.W., Ripplinger, C.M., McCain, M.L., Grosberg, A., Dabiri, J.O., Perker, K.K.: A tissue-engineered jellyfish with biomimetic propulsion. Nature Biotechnorogy 30, 792–797 (2012)
[2] Akiyama, Y., Iwabuchi, K., Furukawaa, Y., Morishima, K.: Long-term and room temperature operable bioactuator powered by insect dorsal vessel tissue. Lab on a Chip 9, 140–144 (2009)
[3] Rolf, P., Christian, S.: Understanding intelligence. MIT Press (2001)
[4] Engler, A.J., Sen, S., Sweeney, H., Discher, D.E.: Matrix elasticity directs stem cell lineage specification. cell 126, 677–689 (2006)
[5] Hayakawa, K., Tatsumi, H., Sokabe, M.: Actin filaments function as a tension sensor by tension-dependent binding of cofilin to the filament. Journal of Cell Biology 195, 721–727 (2011)
[6] Shimizu, M., Yawata, S., Miyamoto, K., Miyasaka, K., Asano, T., Yoshinobu, T., Yawo, H., Ogura, T., Ishiguro, A.: Toward biorobotic systems with muscle cell actuators. In: The Proc. of AMAM 2011, pp. 87–88 (2011)

Development of an Insect Size Micro Jumping Robot

Je-Sung Koh and Kyu-jin Cho[*]

Biorobotics Lab,
Dept. Mechanical and Aerospace Eng., Seoul National University, Seoul, South Korea
{kjs15,kjcho}@snu.ac.kr

Abstract. An insect size micro jumping mechanism is developed and jumps 40cm. The prototype is fabricated with the composite structures cut by precision UV laser. The robot mechanism is bio-mimetic system that is inspired by the small jumping insect, Flea. A single sheet shape memory alloy coil actuator is used for propulsion and energy storage. The compliant mechanism in the body allows to reduce the number of actuators for triggering. The robot mechanism has 36mg weight, 2 cm length and 2mm height except of wire legs.

Keywords: Bio-mimetic robot, SMA actuator, Jumping, Flea.

1 Introduction

Jumping is extreme locomotion for a few living creatures and even fine robots. In this research, the flea inspired catapult mechanism is employed to minimize and simplify the energy storage and the triggering structure [1]. The flea has unique catapult mechanism in its anatomy of jumping legs [2]. The flea inspired catapult mechanism generates the force that shows unique profile which is different from spring based jumping legs as shown in fig.1 [3]. The initial force is zero and the force gradually increases while the initial force is maximum in spring legs as shown in fig. 1 (a). We can expect that this profile enhance the momentum transfer of the robot with a constant maximum force.

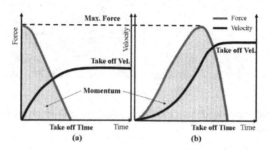

Fig. 1. Schematic diagram of the force profile with (a) spring leg and (b) flea inspired catapult mechanism

---

[*] Corresponding author.

A. Duff et al. (eds.): Living Machines 2014, LNAI 8608, pp. 405–407, 2014.

2   Design of Micro Jumping Mechanism

The ultra-light and small jumping robot is designed and fabricated based on the novel composites based robotic structures and the shape memory alloy (SMA) artificial muscle actuators [1]. The robot is bio-mimetic structures that is inspired by the small jumping insect, Flea. The flea inspired catapult mechanism is the torque reversal mechanism that the stored energy is exploded when the force direction of the muscle shifts to the opposite direction respect to the joint as shown in fig. 2 (a).

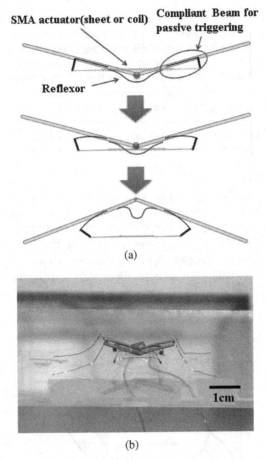

(a)

(b)

Fig. 2. (a) The triggering procedure of the flea inspired catapult mechanism (b) Prototype of the micro jumping robot

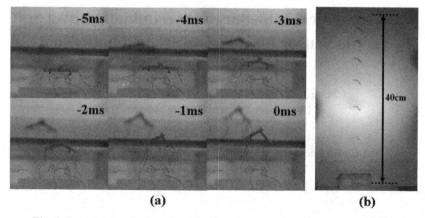

(a)   (b)

Fig. 3. Sequential stop pictures of the jumping motion of the robot prototype

The robot prototype is fabricated by laminating the fiber composite plastic sheets and Polyimide film which are precisely cut by the precision UV laser [4]. Figure 2 (b) shows the prototype of the micro jumping robot. The legs is made of Ni-Cr wires in 200 μm diameter. The prototype has 36mg weight with legs, 2 cm axial length, and 2mm height except the legs.

3   Experimental Result

The robot prototype can jump on the ground with heating by the single heat wire. Figure 3 is sequential stop pictures of jumping motion taken by high speed camera in 5000 frame per second. The stored energy can be controlled by designing the compliant trigger and the sheet SMA coil actuator. The jumping height is improved into 40cm high by increasing stored energy compared to the previous prototype presented in Koh *et al.* [1].

Acknowledgement. This research was supported by the Priority Research Centers Program (2013-055323) through the National Research Foundation of Korea (NRF) funded by the Ministry of Education, Science and Technology (MEST), Republic of Korea.

References

1. Koh, J., Jung, S., Wood, R.J., Cho, K.: A jumping robotic insect based on a torque reversal catapult mechanism. IEEE/RSJ IROS 2013, 3796–3801 (2013)
2. Rothschild, M., Schlein, J.: The jumping mechanism of Xenopsylla cheopis. I. Exoskeletal structures and musculature. Philosophical Transactions of the Royal Society of London. Series B: Biological Sciences 271(914), 457–490 (1975)
3. Noh, M., Kim, S., An, S., Koh, J., Cho, K.: Flea-Inspired Catapult Mechanism for Miniature Jumping Robots. IEEE Transactions on (TRO) Robotics 28(5), 1007–1018 (2012)
4. Wood, R.J., Avadhanula, S., Sahai, R., Steltz, E., Fearing, R.S.: Microrobot design using fiber reinforced composite. Journal of Mechanical Design 130, 52304 (2008)

Soil Mechanical Impedance Discrimination by a Soft Tactile Sensor for a Bioinspired Robotic Root

Chiara Lucarotti[1,2], Massimo Totaro[1], Lucie Viry[1],
Lucia Beccai[1,*], and Barbara Mazzolai[1]

[1] Center for Micro-BioRobotics, Istituto Italiano di Tecnologia, Pontedera, Italy
[2] The BioRobotics Institute, Scuola Superiore Sant'Anna, Pontedera, Italy
lucia beccai@iit.it

Abstract. During the penetration into the soil, plant roots experience mechanical impedance changes and come into contact with obstacles which they avoid and circumnavigate during their growth. In this work, we present an experimental analysis of a sensorized artificial tip able to detect obstacles and discriminate between different mechanical impedances in artificial and real soils. The conical shaped tip is equipped with a soft capacitive tactile sensor consisting of different elastomeric and conductive layers. Experimental results show that the sensor is robust yet sensitive enough to mechanical impedance changes in the experimented soils.

Keywords: Soft tactile sensor, sensorized tip, plant root, soil impedance.

When a plant root comes into contact with obstacles to its growth, it adopts efficient strategies to circumnavigate the barriers and to direct its growth towards low impedance pathways [1]. Therefore, the apex must be able to experience changes in mechanical impedance related to constraints provided by the soil or soil compaction. In a robotic implementation, the artificial root must be equipped with a sensing system able to detect barriers to growth and to discriminate between different mechanical impedances of the soil [2]. To this aim, we developed a soft capacitive tactile sensor, built from a combination of elastomeric and conductive layers. The sensor was integrated in a conical artificial tip made of acrylic resin material by rapid prototyping (Fig. 1D), which shape is conical, as recently suggested for a root apex inspired artefact [3]. The sensor (Fig. 1A, 1C) consists of two parallel circular electrodes (5mm diameter, 70μm thickness), made of soft and unstretchable copper/tin coated woven fabrics, and separated by a spin coated silicone elastomeric dielectric film (300μm thickness). Similar materials were proven appropriate for highly sensitive soft tactile sensing [4]. However, in this work an additional layer made of a compliant and robust silicone material (Sugru©, FormFormForm Ltd, London, UK) (1.5mm thick) was integrated on top of the capacitor in order to increase the robustness of the final device, allowing its correct operation in the soil during fifty repeated trials (Fig. 1D). When a force is applied to the top electrode, the dielectric layer d_0 decreases, resulting

---

[*] Corresponding author.

A. Duff et al. (eds.): Living Machines 2014, LNAI 8608, pp. 408–410, 2014.

in a capacitance variation ΔC respect to its nominal value $C_0 = k_0 A_0 / d_0$, as explained in the following equation:

$$\Delta C = C - C_0 = \frac{k_0 A_0}{d_0 - \Delta d} - \frac{k_0 A_0}{d_0} = k_0 A_0 \frac{\Delta d}{d_0 (d_0 - \Delta d)}$$

where k_0, A_0, are the permittivity and the sensing area, respectively. The readout electronics consists of a 24 bit capacitance-to-digital converter, and a 32 bit PIC microcontroller connected to a PC via USB. A differential configuration is implemented in the electronics to minimize the effect of parasitic capacitances, and, also, connection between sensor and electronics is achieved by shielded cables. At first, the tip was characterized in air by indentation tests. The top end of an aluminium shaft (20mm diameter, 500mm length) was fixed to an electromechanical equipment (Instron 4464, Instron Corporation, Canton, MA, USA) integrating a load cell able to detect forces up to ±1kN. The sensorized tip was installed at the bottom free end of the shaft and indentations were performed on a rigid substrate. The sensor response, in terms of change in capacitance vs. force (up to 80N), is shown in Fig. 1B. In a second phase, penetration tests in both real and artificial soils were performed. Granular media was used as artificial soil, i.e. PolyOxyMethylene (POM) beads with average grain diameter of 4mm. Moreover, a circular Delrin obstacle (30mm diameter) was placed into the soil at a 45mm depth. For experiments in real soil, two different soils, i.e. sand (1.72kg/dm^3 density) and loam (1.4kg/dm^3 density), were used and compacted manually to evaluate the response of the sensorized tip to different mechanical impedances during the penetration. Experimental results (Fig. 1E) show an abrupt variation both for the force measured by the reference load cell, and for the output capacitance sensor signal when the tip comes into contact with the obstacle in granular soil. On the other hand, experimental results in real soil (Fig. 1F) show that the sensorized tip is capable of discriminating between the two different soils while the Instron load cell is not able to experience any change in the soil composition. This means that the tactile sensor (detecting the local pressure on the tip) is sensitive to an impedance variation of about 0.3kg/dm^3 due to the presence of two different soils, while the load cell (measuring the total friction of both tip and shaft with the soil) is not able to detect such changes. Considering that the force is the integral of the pressure on the whole surface, even if the pressure presents a local discontinuity, the force measured by the load cell is continuous. To have an abrupt variation in the force measured by the load cell, a sharp mechanical impedance variation (able to stop the penetration) is required, as in the previous case. However, in real soil (without obstacles) the density variation is not abrupt, due to the partial mixing at the loam-sand interface. Additionally, the penetration, imposed by the Instron, continues at the same velocity. In both experiments, the developed tactile sensor shows proper robustness during soil penetration. These results are promising and demonstrate the validity of the soft sensing approach for implementing thigmotropism in a robotic root. Future work will concentrate on the investigation of a new sensor layout for spatial detection of different mechanical stimulations of the tip (i.e., lateral obstacles, contact angles).

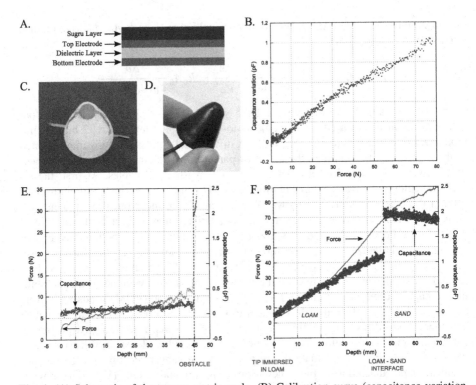

Fig. 1. (A) Schematic of the sensor, not in scale. (B) Calibration curve (capacitance variation vs. force) of the sensorized tip outside the soil. (C) Integration of the sensor in the artificial tip (height of 15 mm, base diameter of 20mm). (D) Artificial tip after the integration of the Sugru© layer. (E) Tests in POM soil at 20mm/min rate with obstacle at 45mm depth. Response of the sensor (black) and force measured by the load cell (grey) vs. penetration depth. (F) Tests in real soil at 20mm/min rate. The sensor response (black) shows its capability to experience mechanical impedance change, in contrast to the load cell (grey).

Acknowledgments. This work was supported by the FET programme within the 7th FP for Research of the European Commission, under the PLANTOID project FETO-pen n. 293431.

References

1. Monshausen, G.B., Gilroy, S.: Feeling Green: Mechanosensing in Plants. Trends in Cell Biology 19(5), 228–235 (2009)
2. Sadeghi, A., Tonazzini, A., Popova, L., Mazzolai, B.: A Novel Growing Device Inspired by Plant Root Soil Penetration Behaviour. PLoS ONE 9(2), 1–10 (2014)
3. Tonazzini, A., Popova, L., Mattioli, F., Mazzolai, B.: Analysis and Characterization of a Robotic Probe Inspired by the Plant Root Apex. In: Proceeding of the 4th IEEE/RAS-EMBS International Conference on Biomedical Robotics and Biomechatronics, pp. 1134–1139 (2012)
4. Viry, L., Levi, A., Totaro, M., Mondini, A., Mattoli, V., Mazzolai, B., Beccai, L.: Flexible Three-Axial Force Sensor for Soft and Highly Sensitive Artificial Touch. Advanced Materials 26(17), 2659–2664 (2014)

Fetusoid35: A Robot Research Platform for Neural Development of Both Fetuses and Preterm Infants and for Developmental Care

Hiroki Mori, Daii Akutsu, and Minoru Asada

Osaka University, Graduate School of Engineering,
Adaptive Machine Systems,
Yamada-oka 2-1, Suita City, Osaka Prefecture, Japan
{hiroki,daii.akutsu,asada}@ams.eng.osaka-u.ac.jp
http://www.er.ams.eng.osaka-u.ac.jp

Abstract. We have been developing a robot called Fetusoid35 that resembles a human fetus or preterm infant. We suppose that the robot could contribute to developmental science by shedding a new insight on the understanding the developmental process of fetuses and preterm infants. Based on the mechanism, we would expect to improve a developmental care of preterm infants. This extended abstract briefly introduces the design policy of Fetusoid35 with its specifications and the current status.

Keywords: Fetus, Preterm infant, Robot platform, Developmental care.

1 Introduction

To understand the development of human fetuses is not only a fundamental issue in developmental science but also in developmental care for preterm infants who have less experience in the uterus. This issue has become more and more serious since it recently turned out that preterm infants might have higher risk of being developmental disorders than the term infants. Developmental care including nesting and swaddling care, that try to imitate a maternal womb, and massage therapy are used in almost cases in Neonatal Intensive Care Unit (NICU), whereas the scientific evidence for the effectiveness of these methods has not been fully found [3]. If we understand a mechanism of fetal development, we can improve the care based on the mechanism.

To investigate a mechanism of fetal development, Mori and Kuniyoshi 2010 [2] constructed a whole body musculoskeletal system of a human fetus as a computer simulation with rigid body dynamics. Using the simulation, we reproduce fetal behavioral development in the first half of the pregnancy with the fetal body model and the nervous system model consisting of a brain stem, a spinal cord and mechanoreceptors from a perspective of self-organization. However, the simulation cannot realize realistic interaction in developmental care.

In this paper, we propose a robot called Fetusoid35 resembling a human fetus and preterm infant having a 35 gestational weeks body, air actuators and somatic

A. Duff et al. (eds.): Living Machines 2014, LNAI 8608, pp. 411–413, 2014.
© Springer International Publishing Switzerland 2014

sensors including whole body tactile sensors. The real robot in conjunction with the simulator can contribute to understanda mechanism of fetal development, and we improve developmental care by observing the sensory information within the physical interaction with physiotherapist or special equipment for a nesting care.

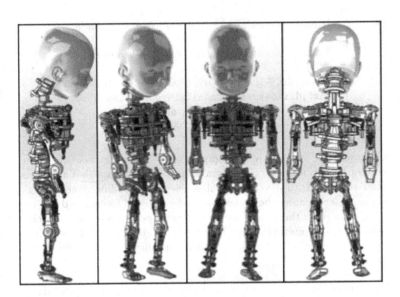

Fig. 1. The design of the Fetusoid35

2   The Specifications and the Appearance of the Robot

The appearance of the robot is shown in Fig. 1. The size of the robot is equivalent to a 35 gestational weeks fetus or preterm infant [1]. The age is appropriate for a study of fetuses and preterm infants because 35 weeks preterm infants mostly survive and have a risk to have a developmental disorder.

The robot has 30 DoFs; The spine including the neck and back bone has 4 Ball joints (3DoFs × 4), each shoulder has 3 DoFs (3DoFs × 2), each hip joint has 3 DoFs (3DoFs × 2), each knee, elbow and ankle has 1 DoF, respectively (1 × 4). All joints of the robot are driven by McKibben type air actuators [4]. Electric bulbs are embedded in the chest and an air tube, a signal cable and a power cable are bundled analogous to an umbilical cord.

Now we are planing to implement tactile sensors on a whole body skin. Tactile sensation in the uterus might affect development significantly because it responds directly to the fetal movement in the womb.

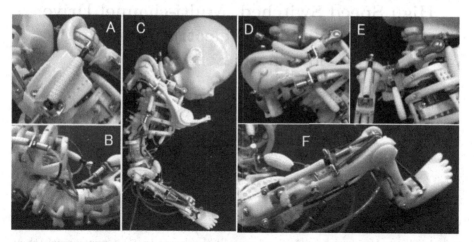

Fig. 2. Fetusoid35. A: Right shoulder joint from a back view. B: Spine from a lateral view. C: Whole body from a lateral view. D: Right shoulder joint (Adduction). E: Right shoulder joint (Abduction). F: Right leg.

3   Conclusion

In this paper, we proposed a robot resembling a fetus and a preterm infant: Fetusoid35. We think that the robot can be used as a moving sensor of the fetal experience in the uterus and of preterm infants in a developmental care, and also a platform for examination of nervous system models.

References

1. Archie, J.G., Collins, J.S., Lebel, R.R.: Quantitative standards for fetal and neonatal autopsy. American Journal of Clinical Pathology 126, 256–265 (2006)
2. Mori, H., Kuniyoshi, Y.: A human fetus development simulation: Self-organization of behaviors through tactile sensation. In: IEEE 9th International Conference on Development and Learning, pp. 82–97 (2010)
3. Sizun, J., Westrup, B., Committee, t.E.N.C.: Early developmental care for preterm neonates: a call for more research. Archives of Disease in Childhood Fetal & Neonatal Edition 89, F384–F388 (2004)
4. Tondu, B., Boitier, V., Lopez, P.: Naturally compliant robot-arms actuated by mckibben artificial muscles. In: 1994 IEEE International Conference on Systems, Man, and Cybernetics, pp. 2635–2640 (1994)

High Speed Switched, Multi-channel Drive for High Voltage Dielectric Actuation of a Biomimetic Sensory Array

Martin J. Pearson and Tareq Assaf

Bristol Robotics Laboratory, University of Bristol
and University of the West of England, Bristol, UK
martin.pearson@brl.ac.uk
http://www.brl.ac.uk/bnr

Introduction: Electro-Active Polymers (EAP) have been described as artificial muscles due to their composition and muscle-like dynamics [1]. Consequently they have attracted a lot of attention from the biomimetic robotics research community and heralded as a potential alternative to conventional electromagnetic, pneumatic or hydraulic actuation technologies [2]. However, in practice there are a number of technical barriers to overcome before they gain widespread acceptance as robotic actuators [3]. Here we focus on overcoming one of those limiting factors for a type of EAP referred to as Dielectric Electro-Active Polymers (DEAP).

DEAPs work by exploiting the electrostatic force across two parallel conductive plates to deform a polymer based dielectric placed between them. For this to work the density of the electric field must be high ($\sim100V/\mu m$). The polymer that constitutes the dielectric, therefore, must have a high electrical breakdown strength in addition to the desirable elastic properties that constrain the actuation. We have found that the range of VHB^{TM}acrylic elastomer tapes from 3M^{TM}provide the best compromise between high strains generated and longevity of material by pre-strectching by 400% and limiting peak applied voltage to $\sim$3-4KV. For this there are a number of compact high voltage power supplies available that can be incorporated into realisable robotic systems [4][5]. In this study we propose a high-speed switching circuit and micro-controller interface to reduce the response time and number of such supplies required to drive multiple DEAP actuators. This is to enable a study of a cerebellar inspired adaptive control approach to the motion of an array of flexible whisker-like sensors and an eyeball-like camera using multiple coupled DEAP based actuators.

Background: Functional roles for the cerebellum have been proposed through extensive anatomical, electro-physiological, behavioural and pathological studies. The broad consensus is that cerebellum is instrumental in the acquisition of actions or skills that require sensori-motor coordination [6]. From this consensus a number of computational models have also been proposed to emulate some aspect of the inferred functionality and to generate comparable behaviour in artificial sensori-motor systems [7], [3]. Of particular interest to this study is that such algorithms can accommodate time varying and non-linear dynamics of the plant through a process of adaptive filtering. EAP based actuation is also

A. Duff et al. (eds.): Living Machines 2014, LNAI 8608, pp. 414–416, 2014.
© Springer International Publishing Switzerland 2014

subject to control issues such as creep, wide tolerances between actuators and non-linear response. Recently this bioinspired approach to adaptive filtering has been shown to be more effective at controlling a single degree of freedom DEAP actuator than a conventional filtered-x LMS (FXLMS) adaptive control scheme[3]. We wish to extend this principle toward the control and sensori-motor integration of a multi-degree of freedom robotic system. This system will incorporate the same principles of "cerebellar-like" learning at three levels of sensori-motor co-ordination; In the non-linear control of the motor plant itself, i.e., DEAP actuation; To negate the sensory response generated by self-motion (re-afferent noise); And, in the calibration of a multi-modal ego-centric sensory map. The sensory apparatus will consist of an array of 20 flexible whisker-like tactile sensors arranged around a centrally mounted eye-like camera. The whiskers and eye will be actuated using custom built DEAP technology [8] to emulate rapid eye movements and the whisking behaviour observed in the facial whiskers of many mammals [9]. This active sensory assembly will be mounted as the end-effector on a standard 5 degree-of-freedom industrial manipulator such that it can freely explore a controlled workspace. To enable this research we require a large number of individually controllable high-voltage supplies with a response time to meet the desired performance from the 2 model motor systems, i.e., $\sim$10Hz for whisking and $\sim$20Hz for gaze stabilisation[10].

Description of Hardware: The PCB developed (hereafter referred to as a *blade*) incorporates 4 groups of 3 solid-state relay ICs connected in series to safely accommodate the fixed 4KV applied across them by a compact high voltage power supply from HVM Technology[4]. The relays chosen provide electrical isolation of up to 5KV from their inputs which are driven in parallel by a Pulse Width Modulated (PWM) signal generated by a Microchip dsPIC33fMC802 microcontroller located on the blade. This microcontroller drives the 4 individual switching circuits, maintaining the voltage across each series chain of relays by setting an appropriate duty cycle to each of the PWM signals. A simple potential divider circuit, that incorporates the 40MΩ bleed resistors required for the DEAPs, is used to reduce the voltages to an appropriate level for sampling by the Analogue to Digital Converter (ADC) module of the microcontroller enabling local closed-loop feedback control. The desired voltages across each DEAP are passed to the blade via a CAN bus, which in turn is bridged to a High Speed USB 2.0 interface with a standard PC. This ensures a real-time update rate of 250Hz for upto 48 separate DEAP actuators driven by an array of 12 blades connected to the CAN bus.

Experimental Results: To compare the response times (rise and fall) of a blade to a stand-alone HVMTech. supply we recorded the full voltage step response of each, loaded with a fixed 300pF capacitor model of a typical DEAP actuator. As is clear from the plots shown in figure 1 the rise and fall times of the blade are significantly shorter than the HVMTech. supply (10ms/375ms rise time and 18ms/315ms fall time for blade and HVMTech. supply respectively). Using the switched system of the blade, therefore, enables the potential for drive

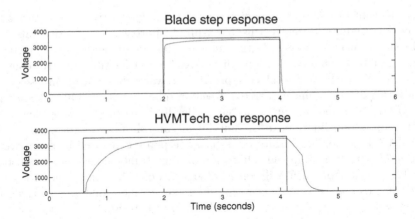

Fig. 1. Step response of an HVMTech. high voltage supply and the solid-state relay based switching circuit on the blade. The desired step input is the black trace in each plot with voltage response measured across the capacitive load in red

signals of upto 55Hz to be reproduced, compared to 2.7Hz using a stand-alone HVMTech. supply.

References

1. Bar-Cohen, Y.: Electoactive Polymer Actuators as Artificial Muscles- realisty, potential, and challenges. SPIE press, Bellingham (2001)
2. Carpi, F., Kornbluh, R., Sommer-Larsen, P., Alici, G.: Electroactive polymer actuators as artificial muscles are they ready for bioinspired applications? Bioinspir.Biomim. 6(4) (2011)
3. Wilson, E.D., Assaf, T., Pearson, M.J., Rossiter, J.M., Anderson, S.R., Porrill, J.: Bioinspired adaptive control for artificial muscles. In: Lepora, N.F., Mura, A., Krapp, H.G., Verschure, P.F.M.J., Prescott, T.J. (eds.) Living Machines 2013. LNCS, vol. 8064, pp. 311–322. Springer, Heidelberg (2013)
4. HVMTechnology: Ultra-miniature high voltage power supply datasheet, revision 4718e (2014), http://www.hvmtech.com/pdf/umhvspecsheet.pdf
5. EMCO: AG series, isolated, proportional DC to high voltage converter datasheet, revision 100715 (2014), http://www.emcohighvoltage.com/pdfs/agseries.pdf
6. Ghez, C., Krakauer, J.: The organisation of movement. In: Principles of neural science, 4th edn., pp. 653–673. McGraw-Hill (2000)
7. Anderson, S.R., Pearson, M.J., Pipe, A., Prescott, T., Dean, P., Porrill, J.: Adaptive cancelation of Self-Generated sensory signals in a whisking robot. IEEE Transactions on Robotics 26(6), 1–12 (2010)
8. Assaf, T., Rossiter, J., Pearson, M.: Contact sensing in a bio-inspired whisker driven by electroactive polymer artificial muscles. In: 2013 IEEE Sensors, pp. 1–4 (November 2013)
9. Gao, P., Bermejo, R., Zeigler, H.: Whisker deafferation and rodent whisking patterns: behavioural evidence for a central pattern generator. The Journal of Neuroscience 21(14), 5374–5380 (2001)
10. Grossman, G.E., Leigh, R.J., Abel, L.A., Lanska, D.J., Thurston, S.E.: Frequency and velocity of rotational head perturbations during locomotion. Experimental Brain Research 70(3), 470–476 (1988) PMID: 3384048

A Combined CPG-Stretch Reflex Study on a Musculoskeletal Pneumatic Quadruped

Andre Rosendo, Xiangxiao Liu, Shogo Nakatsu,
Masahiro Shimizu, and Koh Hosoda

Graduate School of Information Science and Technology, Osaka University
Osaka, Japan

Abstract. Quadruped animals combine the versatility of legged locomotion with extra stability from the additional number of limbs. Cats and dogs can walk in different gait patterns, outperforming current robotic technology while exploiting interactions between brain, spine, muscle and environment, which inner workings are not fully understood. We propose a controller which combines a rhythmic sinusoidal pattern (feed-forward) with a muscular stretch reflex (feedback) on a biomimetic musculoskeletal robot.

Keywords: quadruped, musculoskeletal, air muscle, pneumatic robot, stretch reflex, CPG.

1 Introduction

Understanding and applying animal locomotion knowledge to robots are challenges which faces biologists and roboticists for years. In this vein, in [1] electromyographic signals from walking cats are analyzed, while in [2] joint angles are compared over different locomotion patterns. Although biological experiments enhance our understanding of locomotion, the inner workings of such complex living machines can be better understood when replicated.

Simulations can, to some extent, predict the dynamic behavior of a system interacting with its environment. Seyfarth *et al.* [3] studied leg adjustment strategies in an artificial three-dimensional environment. In [4], Ekeberg *et al.* proved that a musculoskeletal structure is capable of producing a very adaptive behavior, performing well against computer programmed disturbances in a simulation environment.

From a robotic approach, dealing with real world disturbances and noise, a few quadruped robots have been developed in the last few years. From a biomimetic point of view, Puppy [5] embodies a biomimetic dog with a monoarticular muscular system. Owaki *et al.* [6] proposed an uncoupled central pattern generator (CPG), which halts the stance phase when ground reaction force is present on the leg. Similar CPG-based approach is found in [7], using coupled CPG with reflex signal solely as a corrective measure in cases of stumbling upon an obstacle.

In this work, we propose a sinusoidal coupled CPG which considers ground reaction forces during walking on a treadmill to increase muscular activation on

A. Duff et al. (eds.): Living Machines 2014, LNAI 8608, pp. 417–419, 2014.

limbs. The joint coordination from proximal and distal links of fore and hindlimbs are influenced to different extents by the ground feedback, with proximal members immune to such signals.

2   Robot Design

We approach the problem with a deeper biomimetic approach. Differently from usual quadruped robots, we created a lightweight pneumatic quadruped robot with 14 artificial muscles, 3 degrees-of-freedom per hindlimb and 4 degrees-of-freedom per forelimb (shown in Fig. 1). We decided to simulate scapular function on the forelimb to better understand the relationship between this extra link and forepaws placement.

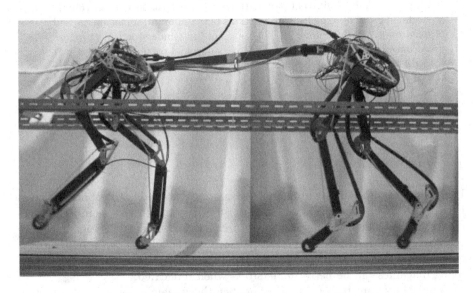

Fig. 1. Pneupard walking on a treadmill

Our controller is based on a sinusoidal CPG function where each joint is controlled by a compliant pneumatic muscle, and feedback contribution from force sensors increase the amount of muscular contraction on distal part of limbs, on an approach suggested by avian research [8]. Due to its pneumatic origin, the robot is named Pneupard [9] and intends to perform adaptive locomotion through compliant muscles.

3   Discussion

This ongoing work aims to prove that, as observed by Daley *et al.* [8] the presence of a proximal-distal gradient may generate adaptive behavior. Proximal links of birds do not have their rhythmic motion affected by unexpected ground changes (constant feed-forward, similarly to a CPG), while the distal limbs adapt according to the situation (stretch reflex response representing muscular feedback).

Here we suggest that distal links receive lower amplitude CPG signals to their muscle activation, having their amplitude incremented by a value proportional to the force sensor output. This way, during an "air walk" the gait would still be visible, and during actual walking it would adapt to the ground reaction force.

So far, many approaches use virtual environments to validate control methods [7], but similar methods cannot deal with noisy physical environments. Additional discussions concerning the effectiveness of artificial muscles should be made, considering the recent growth of biomimetic robots. Although the majority of robots opt for electric motors as main source of traction, we argue that the compliance present on muscles aid on producing self-stable behavior, as studied by [3] and shown in a version of this robot's hindlimb [9].

References

1. Engberg, I., Lundberg, A.: An electromyographic analysis of muscular activity in the hindlimb of the cat during unrestrained locomotion. Acta Physiol. Scand. 75, 614–630 (1969)
2. Goslow Jr., G.E., Reinking, R.M., Stuart, D.G.: The cat step cycle: hind limb joint angles and muscle lengths during unrestrained locomotion, J. J. Morphol. 141, 1–42 (1973)
3. Peuker, F., Maufroy, C., Seyferth, A.: Leg-adjustment strategies for stable running in three dimensions. Bioinspir. Biomim. 7, 36002 (2012)
4. Ekeberg, O., Pearson, K.: Computer simulation of stepping in the hindlegs of the cat: an examination of mechanisms regulating the stance-to-swingtransition. J. Neurophysiol. 94, 4256–4268 (2005)
5. Aschenbeck, K., Kern, N., Bachmann, R., Quinn, R.: Design of a quadruped robot driven by air muscles. In: Proc. Intl. Conf. Biomed. Robot. and Biomechatronics, pp. 875–880 (2006)
6. Owaki, D., Kano, T., Nagasawa, K., Tero, A., Ishiguro, A.: An electromyographic analysis of forelimb muscles during overground stepping in the cat. J. Exp. Biol. 76, 105–122 (1978)
7. Ajallooeian, M., Pouya, S., Sproewitz, A., Ijspeert, A.J.: Central Pattern Generators augmented with virtual model control for quadruped rough terrain locomotion. In: Proc. Intl. Conf. on Robot. and Autom., pp. 3321–3328 (2013)
8. Daley, M.A., Felix, G., Biewener, A.A.: Running stability is enhanced by a proximo-distal gradient in joint neuromechanical control. J. Exp. Biol. 210, 383–394 (2007)
9. Rosendo, A., Nakatsu, S., Narioka, K., Hosoda, K.: Producing alternating gait on uncoupled feline hindlimbs: muscular unloading rule on a biomimetic robot. Advanced Robotics 28, 351–365 (2014)

Swimming Locomotion of *Xenopus Laevis* Robot

Ryo Sakai[1], Masahiro Shimizu[2], Hitoshi Aonuma[3], and Koh Hosoda[2]

[1] Graduate School of Information Science and Technology, Osaka University,
2-1 Yamada-oka, Suita, 565-0871, Japan
[2] Graduate School of Engineering Science, Osaka University,
1-3, Machikaneyama, Toyonaka, 560-8531, Japan
[3] Research Institute for Electronic Science, Hokkaido University,
Kita 8 Nishi 5, Sapporo, 060-0808, Japan
{ryo.sakai,shimizu,hosoda}@arl.sys.es.osaka-u.ac.jp,
aon@es.hokudai.ac.jp

Abstract. An adaptive swimming locomotion of *Xenopus laevis* are mainly generated through hydrodynamic interaction between its musculoskeletal system and water environments. To understand the mechanism of frog locomotion, therefore, it is a promising approach to copy morphology of the frog and let it swim in the water. We developed a swimming robot that had a similar musculoskeletal structure as a frog driven by living muscles. We realized kick motion that generated propulsion for swimming locomotion by exciting the living muscles. The robot is expected to be a powerful tool to understand the swimming locomotion of *Xenopus laevis*.

Keywords: *Xenopus laevis*, bio-mechanic hybrid system, living muscles, musculoskeletal structure.

1 Introduction

Animals realize adaptive locomotion in various environments. In this study, we focus on adaptive swimming locomotion of *Xenopus laevis*[1]. Underwater locomotion is mainly generated through hydrodynamic interaction between the musculoskeletal system and water environments. We have to clarify how the structure of muscles and skeletons plays an role in this interaction. An approach to solve this problem is to build a hybrid musculoskeletal robot that has similar structure as *Xenopus laevis*, and that excite living muscles to drive the body like in vivo. This approach allows us to modify the function in the natural context. Based on the same idea, Herr et al.[2] developed a fish robot driven by frog's muscles in water, and realized swimming locomotion. Richards et al.[3] developed bio-robotic platform of *Xenopus laevis*, and investigated hydrodynamic of swimming locomotion. In these experiments, however, a musculoskletal structure was too simple to investigate muscle function for *Xenopus*'s locomotion. The reaction force from the water emerging from the interaction should be realized.

A. Duff et al. (eds.): Living Machines 2014, LNAI 8608, pp. 420–422, 2014.

Fig. 1. *Xenopus laevis* and the robot

2 Method

We developed a swimming robot that has the similar musculoskeletal structure as a *Xenopus laevis* driven by living muscles. We use the gastrocnemius muscle as the actuator because it is known that this muscle generates propulsion mainly in *Xenopus*'s swimming locomotion. Moreover, it is important to reconstruct the musculoskletal structure. Thus, we built musculoskeletal structure of the hind leg by a 3D printer copying the anatomy of *Xenopus laevis*. Finally, we developed the swimming robot modeled *Xenopus laevis* by driven the gastrocnemius muscles. In order to understand swimming locomotion, this constructive robot is expected to be a powerful tool to investigate the functions of each involved bone and muscle, and also their mutual relationships.

Figure 1 shows the robot that models *Xenopus*'s musculoskeletal structure driven by living gastrocnemius muscle. Based on our past research[1], we use the gastrocnemius muscle as the robot's actuator for generating propulsion for swimming. We constructed the fore and hind legs by a 3D printer, and the body by PET. The body length, width and mass are designed to imitate a real of *Xenopus laevis*. Figure 2 shows the structure the hind leg in the robot. The hind leg is designed to imitate *Xenopus*'s musculoskeletal structure that has 4 degree of freedom, i.e., hip, knee, ankle and claw joints. The both legs have totally 8 degree of freedom, and are actuated by the gastrocnemius muscle dissected out from a *Xenopus laevis*. By measuring lengths of the femur and tibiofibular of a real frog, we determine the parameters of the robot's hind leg. The range of movement of each joint is similar to Richards et al.[4]. We are is designed to imitate three passive muscles of *Xenopus laevis*, these muscles are hip-flection mono-articular muscle, hamstrings and tibialis anterior muscles[5].

To fix the gastrocnemius muscle on the ankle joint, we use the Achilles tendon of a dissected out *Xenopus laevis* together with a gastrocnemius. For fixing it on the knee joint of the robot, we use a part of the tibiofibula. We use a suction electrode to apply electrical stimulation to the ischiadic nerve. The stimulation is generated by a hand-made stimulator, consisting of a microcontroller (Arduino Fio, ATmega328V at 8MHz) and a wireless communication system (XBee, Bandwidth 2.4GHz). A sequence of stimulation consists of several square pulses, and we apply several sequences for the robot to swim.

422   R. Sakai et al.

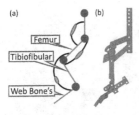

Fig. 2. (a) is musculoskeletal structure model of the hind leg. Blue lines are bones as femur, tibiofibular and web bone's. Red points are hinge joints as hip, knee, ankle and claw joints. Three muscles are passive muscles as hip-flection mono-articular muscle, hamstrings and tibialis anterior muscles. (b) is CAD data of the hind leg in the robot.

Fig. 3. Swimming locomotion by the robot. These are snapshots of swimming locomotion in every 367 msec. The stimulation voltage was 1 V, width of the square pulse was 1 msec, and the pulse period was 21 msec. A sequence of stimulation consisted of 30 pulses and interval between sequences was 1000 msec.

3   Result

Figure 3 shows one trial of swimming locomotion by the robot. The maximum speed of swimming was 55.1 mms^{-1} when we applied 1 V stimulation in which width of the square pulse was 1 msec and the pulse periods were 21 msec. A sequence of stimulation consisted of 7 pulses and interval between sequences was 200 msec. We could successfully make the robot swim in the water, and found that the behavior looks the same as a real frog. We should further investigate behavior of the robot by changing the excitation patterns, so that we can understand the functions of the musculoskeletal structure.

Acknowledgments. This work was partially supported by Grant-in-Aid for Scientific Research on 24680023, 23220004, 23300113, 25540117 of Japan.

References

[1] Richards, C.T., Biewener, A.A.: Modulation of *in vivo* muscle power output during swimming in the African clawed frog (*Xenopus laevis*). The Journal of Experimental Biology 210, 3147–3159 (2007)
[2] Herr, H., Dennis, R.G.: A swimming robot actuated by living muscle tissue. Journal of NeuroEngineering and Rehabilitation 1, 1–9 (2004)
[3] Richards, C.T., Clemente, C.J.: A bio-robotic platform for integrating internal and external mechanics during muscle-powered swimming. Bioinspiration and Biomimetics 7(1) (2012)
[4] Richards, C.T.: Kinematics and hydrodynamics analysis of swimming anurans reveals striking interspecific differences in the mechanism for producing thrust. The Journal of Experimental Biology 213, 621–634 (2010)
[5] Duellman, W., Trueb, L.: Biology of Amphibians. Johns Hopkins University Press (1994)

Empathy in Humanoid Robots

Marina Sardà Gou[1], Vasiliki Vouloutsi[1], Klaudia Grechuta[1],
Stéphane Lallée[1], and Paul F.M.J. Verschure[1,2]

[1] Laboratory of Synthetic, Perceptive, Emotive and Cognitive Systems – SPECS,
Universitat Pompeu Fabra, Roc Boronat 138, 08018, Barcelona, Spain
[2] ICREA Institució Catalana de Recercai Estudis Avançats, Passeig Lluís Companys 23,
E-08010 Barcelona, Spain

Abstract. Humanoid robots should be able to interact with humans in a familiar way since they are going to play a significant role in the future. Thus, it is necessary that Human-Robot Interaction (HRI) is designed in such a way that allows humans to communicate with robots effortlessly and naturally. Emotions play an important role in this interaction since humans feel more predisposed to interact with robots if they are able to create an affective bond with them. In this study, we want to know whether humans are able to empathize with a humanoid robot. Therefore, in the present research, we are going to recreate a Milgram experiment in which we expect participants to empathize with the robot while playing a matching game. Like in Milgram's experiment, they will have to give fake electrical shocks to the robot thinking that they are punishing it. In that way, an empathic state, which we expect to see in our results, may be induced.

Keywords: Milgram experiment, empathy, humanoid robot, Human-Robot Interaction (HRI).

In the future, humanoid robots will undoubtedly play an important role in our society since they can perform a variety of practical activities in peoples' everyday life. The robots which are meant to interact with people should be therefore designed to provide a smooth and comfortable Human-Robot Interaction (HRI). Studying and incorporating emotions is a crucial element of HRI research which aims at achieving that goal since robots, which are able to display emotions through facial expression and/or gestures are perceived as more familiar and warm. In that way, humans feel more likely to interact with them. (Fong, 2003) This may also result in the increased level of empathy towards the robot, and consequently in an affective bond from the human to the robot. The purpose of this study is to investigate whether a person can feel empathy towards a humanoid robot, and if so, to which extent.

In order to answer our question, first, it is necessary to define and understand what "empathy" means and how we can measure it. According to Hoffman, empathy is "an affective response more appropriate to another's situation than to one's own." (Hoffman 2001). We can feel empathy in two different ways; one being cognitive and another being affective. Cognitive empathy is the understanding of another's emotions,

A. Duff et al. (eds.): Living Machines 2014, LNAI 8608, pp. 423–426, 2014.

while affective empathy is the possession of that emotion. (D'Ambrosio, 2009) It is "an important contributor to successful social interaction, allowing us to predict and understand others' behaviour and react accordingly." (Engen et al., 2012) However, it is not something that happens every time we perceive emotions from another person. Empathy depends on several factors such as the social context, culture or the characteristics of both the empathizer and the object of empathy (Engen et al., 2012). Therefore, in order to feel empathy, one must feel, in some way, identified with the person having an emotional response.

Humans do not only feel empathy for human beings. Indeed, we can also empathize with animals as well as physical objects once we perceive that they are expressing emotions. For instance, in fiction, there are quite a few films about inanimate objects which show feelings through facial expressions or gestures and, consequently, people empathize with them. Since we are aware that objects have no emotions at all, it is our imagination that plays an important role in the process. Humans can empathize with an inanimate object when, through their imagination, they can perceive that a given object has/displays an emotional response, and this perception causes humans to feel empathy for this particular object. (Misselhorn, 2009) That is to say, with our imagination, we can attribute affective behaviour to some inanimate objects and, as a consequence, we may feel empathy towards them.

In order to test whether humans can feel empathy for a humanoid robot, we will run an adaptation of Milgram experiment (Milgram, 1963). Milgram did his experiments in the 1960's in which he proved that ordinary people are able to cause a huge amount of pain to another person if there is an authority figure that forces them to do so. However, this does not mean that they do not feel empathy for the victim. In fact, as Milgram claimed, the participants were under extreme stress during the experimental procedure. "An unanticipated effect was the extraordinary tension generated by the procedures. [...] In a large number of cases the degree of tension reached extremes that are rarely seen in socio-psychological laboratory studies." (Milgram 1963) Therefore, the participants did suffer because of the victim's pain. Two of the variations Milgram did of his own experiment were bringing the victim closer and removing physically the authority figure from the laboratory. There, he could see that the obedience of the participants dropped drastically. (Milgram, 1974) This also supports the idea of participants being empathic towards the victim. In that way, Milgram's experiment can be used in order induce this emotional state to participants.

In fact, there are already some studies in which researchers performed Milgram's experiment with robots and virtual humans. Slater et al. found that, using an avatar as the victim, participants tend to respond, physiologically and behaviourally, in a similar way that they do with a human victim. Some of them hesitated to continue the experiment or, sometimes, did not want to proceed. In a different way, using less anthropomorphic robots than a virtual human, participants felt compassion for the robot, but the urges of the authority figure were enough to make them proceed. (Bartneck et al., 2005)

In the present study we will apply a similar method. The robot we are going to use is the humanoid robot, iCub. Our hypothesis is that when a robot uses eye contact and shows emotions, humans feel more compassion towards it. As a consequence, they will administer less amount of shock to the robot. Like in Milgram experiment, participants won't know the actual purpose of the experiment. They will be able to interact with the robot through a "matching game" where the goal is to teach the robot the colours in Spanish. The participant and the iCub will be playing the game at the Reactable.

In the game, first, the participant says one colour to the robot and, after that, the robot touches the correspondent colour in the Reactable. There will be 40 trials and, like in Milgram's experiment, three wrong answers to one correct answer. Every time the matching is incorrect, the human needs to punish the robot by administering simulated electrical shocks. The shocks are not real but we want the participant to think that they are real because, in that way, the subject might feel that he is causing pain to the robot and may empathize with it. The shock generator is a crucial issue that needs to be explained since we are not going to use the one that Milgram used. Instead, in the present scenario there is a regulator button on which the user can choose what amount of pain s/he wants to give to the robot. However, we needed to constrain the participant's choice to prevent from choosing the same shock amount. We will therefore force him to increase the shocks as the experiment proceeds. Since we allow the participant to stop the experiment at any time we need an authority figure. It will not be a human but a set of pre-recorded sentences that will instruct the participant. If the participant does not want to proceed anyway, the experiment will end. Since we want to know how the robot's behaviour can affect participants' empathy, we will have four conditions in which we will change the iCub responses. Two independent variables will be the iCub's eye contact and its emotional responses expressed through speech and facial expressions. In the first condition, which is the control group, the robot will not have any of them; in the second condition, it will use the eye contact; in the third condition, it will show emotional responses, and in the fourth condition it will have the both of them.

In every condition, we will have several dependent variables. We will measure participants' heart rate and since it has been proved that, when humans feel empathy, the heart rate increases (Miu 2012, Silva 2011). We will also record participants' performance and a naïve judge will further analyze the eye contact, facial expressions and the speech of all the subjects. We will store the Reaction Times (RTs) between the robot making a mistake and the participant giving the shock, and measure the amount of shock. If the participant hesitates to continue, the system will measure the number of audio pre-recorded sentences s/he needs to hear in order to proceed. Finally, every participant will answer a questionnaire about empathy in which, among other issues, they are going to be asked if they think that the robot can feel pain. If that is the case, this would help us to understand any empathic response participants may have.

When the robot shows emotional responses and eye contact, and as the amount of shock increases, we expect to see the following participants' reactions: an increasing heart rate, less eye contact with the iCub (Milgram showed that participants refused to look at the victim when giving the shocks), more reaction time, less amount of shock administered, the participants' necessity to hear more audio instructions to continue, and finally, the answers in the questionnaire, which show that the participant feels empathy for the robot.

The goal of the present study is to investigate whether humans are able to empathize with a humanoid robot and, if so, what are the behaviours, which cause empathy in humans. It is also necessary to say that the experimental procedure described above should be tested in a pilot experiment that we are going to perform in the time coming.

References

Bartneck, C., et al.: Robot abuse—a limitation of the media equation. In: Proceedings of the Interact 2005 Workshop on Agent Abuse (2005)

D'Ambrosio, F., Olivier, M., Didon, D., Besche, C.: The basic empathy scale: a French validation of a measure of empathy in youth. Personality and Individual Differences 46, 160–165 (2009)

Engen, H.G., Singer, T.: "Empathy circuits.". Current Opinion in Neurobiology 23, 275–282 (2012)

Fong, T., Nourbakhsh, I., Dautenhahn, K.: "A survey of socially interactive robots". Robotics and Autonomous Systems 42, 143–166 (2003)

Milgram, S.: Behavioral study of obedience. Journal of Abnormal and Social Psychology 67, 371–378 (1963)

Milgram, S.: Obedience to Authority. Harper & Row, New York (1974)

Catrin, M.: Empathy with Inanimate Objects and the Uncanny Valley. Minds and Machines 19, 345–359 (2009)

Miu, A.C., Balteş, F.R.: Empathy Manipulation Impacts Music-Induced Emotions: A Psychophysiological Study on Opera. PloS ONE 7(1) (2012)

Silva, P.O., Gonçalves, Ó.F.: Responding empathically: a question of heart, not a question of skin. Appl Psychophysiol Biofeedback 36(3), 201–207 (2011)

Slater, M., Antley, A., Davison, A., Swapp, D., Guger, C., et al.: A virtual reprise of the Stanley Milgram obedience experiments. PLoS ONE 1(1) (2006)

HECTOR, A Bio-Inspired and Compliant Hexapod Robot

Axel Schneider, Jan Paskarbeit, Malte Schilling, and Josef Schmitz

University of Bielefeld, Center of Excellence Cognitive Interaction Technology,
Bielefeld, Germany
{axel.schneider,jan.paskarbeit,malte.schilling,
josef.schmitz}@uni-bielefeld.de

Abstract. The newly built and currently tested hexapod robot HEC-
TOR is introduced. The robot consists of 18 embedded, custom designed
and compliant joint drives based on an integrated elastomer coupling.

Keywords: hexapod robot, compliant joint, elastic actuation.

1 Introduction

Bio-inspired robotic research focusses on the abstraction and application of abilities as found in biological systems. Especially the ability to adapt their movements in a complex environment allows animals to traverse a terrain more effectively than most mobile robots. In this context, the interaction between the
animal and its environment seems to play an important role in the motion control
scheme. However, to allow this level of reciprocal physical interaction, biological
systems as well as robotic systems have to be compliant. In addition but unlike
technical systems, biological systems are enriched with a vast amount of proprioceptive sensors to perceive compliant interaction and use this information
in the active movement generation process. This extended abstract reports on
the newly developed and built hexapod robot HECTOR (**HE**xapod **C**ognitive
au**T**onomously **O**perating **R**obot) which combines compliant joint drives, a rich
sensorization and decentral control approaches to develop bio-equivalent walking
capabilities.

2 Robot Setup and Compliant Drive Configuration

The general setup of HECTOR is shown in the photographs of Fig. 1. Figure 1(a)
shows a top view of the insectoid robot. The robot consists of three body segments (front = prothorax, middle = mesothorax and hind = metathorax). The
robot's front is pointing downwards. Like in insects, a body segment carries two
legs, each equipped with three compliant joint drives. The three body segments
possess additional intersegmental drives which allow panning and tilting movements between each pair of body segments. Figure 1(b) shows a side view of

A. Duff et al. (eds.): Living Machines 2014, LNAI 8608, pp. 427–429, 2014.
© Springer International Publishing Switzerland 2014

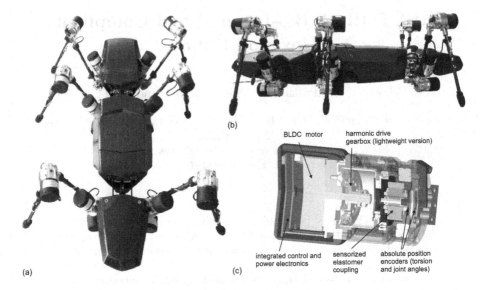

Fig. 1. (a) Top view of the six-legged robot HECTOR with the front part at the bottom. (b) Side view of HECTOR with the front part aligned to the right. (c) One of HECTOR's 18 compliant drives.

HECTOR with the front segment pointing to the right. A section view of a compliant drive is shown in Fig. 1(c). The main features are its small size (diameter 50 mm, length 90 mm), the lightweight construction (weight 0.39 kg), its high power/weight ratio (170 W/kg), an integrated, sensorized elastomer coupling [2] and an integrated onboard electronics which handles BLDC-motor commutation, impedance control and communication via a flexible, small scale serial-bus system which was also developed for HECTOR. Since the robot mainly operates in sprawled postures, walking requires a high torque with as little weight of the drives as possible.

3   HECTOR's Sensor and Actuator Framework

The concept of walking in HECTOR is based on the decentral WALKNET approach [1]. This decentral control structure is also reflected in the setup of the sensor and actuator framework. This framework is shown as a block diagram in Fig. 2. The body segments from front to back are shown from left to right. Within each body segment, a bus master board is used to distribute communication packets for all clients in each leg. The bus master board has two channels, one for each leg. Each bus master board is connected to the PC/104 system in the mesothorax via USB. The illustration of the left front leg in Fig. 2 was replaced by an itemised depiction of the three rotatory joint drives [cmp. Fig. 1(c)] to show those sensors which are included in the drives and whose values can be retrieved via the bus. These sensors measure the joint and motor angle, motor currents and the torsion of the elastomer coupling (mechanical load of the

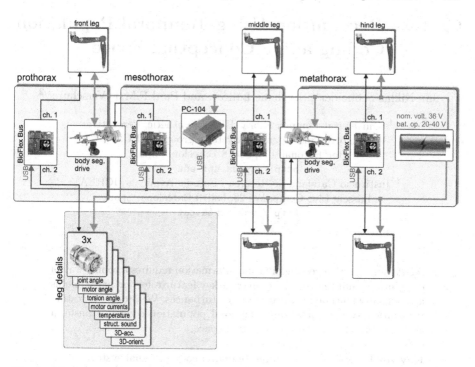

Fig. 2. Block diagram of HECTOR. Three body segments - prothorax. mesothorax and metathorax - are shown from left to right. A body segment possesses a pair of legs which are built up of three compliant drives each.

drive), temperature, structural sound, 3D-acceleration and 3D-orientation. A future goal is the integration of local load information in the control of walking.

Acknowledgments. This work has been supported by the DFG Center of Excellence "Cognitive Interaction Technology" (EXC277) and by an EU-FP7 grant (ICT-2009.2.1, No. 270182).

References

1. Cruse, H., Kindermann, T., Schumm, M., Dean, J., Schmitz, J.: Walknet - a biologically inspired network to control six-legged walking. Neural Networks 11(7-8), 1435–1447 (1998)
2. Paskarbeit, J., Annunziata, S., Basa, D., Schneider, A.: A self-contained, elastic joint drive for robotics applications based on a sensorized elastomer coupling - design and identification. Sensors and Actuators A: Physical 199, 56–66 (2013)

Gesture Recognition Using Temporal Population Coding and a Conceptual Space

Jan Niklas Schneider[1], Stéphane Lallée[1], and Paul F.M.J. Verschure[1,2]

[1] Universitat Pompeu Fabra (UPF), Synthetic, Perceptive,
Emotive and Cognitive Systems group (SPECS)
Roc Boronat 138, 08018 Barcelona, Spain
http://specs.upf.edu
[2] Institució Catalana de Recerca i Estudis Avançats (ICREA)
Passeig Llus Companys 23, 08010 Barcelona, Spain
http://www.icrea.cat

Abstract. The processing of visual information requires a robust way of encoding stimuli that is able to extract key features for further stimulus assessment while staying invariant to disturbances. We propose a gesture recognition system based on a temporal population code to transform gestures into elements in a conceptual space.

Keywords: gesture recognition, temporal coding, visual system.

1 Introduction

The mammalian vision processing capabilities have been studied extensively in the past. However, still little is known about the underlying way of encoding visual stimuli and the mechanisms employed by the visual cortex to relate those to abstract meaningful concepts [1]. The processing of raw visual information requires a way of encoding that is able to extract decisive features while being robust to disturbances such as shifts in position or viewing angle. These requirements can be sufficiently fulfilled by a temporal population code [2] which encodes stimuli with the collective spike response of a neuronal population over time.

This is achieved by structuring a set of neurons in fixed-sized local neighbourhoods and connecting all members of such neighbourhoods with excitatory synapses which reflect the spatial distance of coupled neurons by a transmission delay.

Previously temporal population coding has already been successfully used in tasks such as digit differentiation and face recognition [3]. Mammalian visual systems encounter a constant stream of visual information that also poses the requirement of processing stimuli changes over time such as movements.

Here we propose a movement encoding and recognition system for human arm gestures based on a temporal population code to translate gestures into abstract, comparable elements in a conceptual space.

A. Duff et al. (eds.): Living Machines 2014, LNAI 8608, pp. 430–432, 2014.

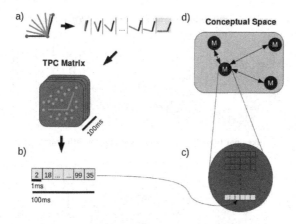

Fig. 1. Overview of the gesture recognition and storage system. **a)** Frames of a gesture are recorded if they surpass a movement detection stage. Projection onto the neuronal population is conducted frame-by-frame. **b)** Every projected frame induces activation into the neuronal population. The emerging temporal spiking pattern is saved as a vector of total spike counts per millisecond in a window of 100 ms. **c)** Once all frames of the gesture are processed they are stored as a collection of vectors holding spike counts and saved in the conceptual space of all movements. **d)** Movements can be compared and clustered in conceptual space by an appropriate metric.

2   Implementation

The population of neurons responsible for the encoding is composed of a three-dimensional grid of 10x10x10 Izhikevich neurons [4]. Each neuron is connected to all other neurons within a spherical neighbourhood with radius 2.
Synapses operate with a transmission delay proportional to the Euclidian distance between post- and presynaptic neuron.

Figure 1 illustrates the gesture storing and identification process. Skeleton data captured by a Microsoft Kinect® system is used to project the upper and lower right arm frame-by-frame onto the grid of neurons. Each frame of the input, respectively being a posture of the human arm, leads to a temporal pattern of spike counts (Figure 2) after which the neuron matrix is resetted and projection of the next frame begins. A movement, itself being composed of a sequence of frames, is therefore stored as a vector of vectors of spike counts. Movements can be compared by calculating the minimal average Euclidian distance between their respective postures by sliding the shorter movement vector alongside the longer one. Movements can then be clustered by distances. Therefore clusters represent gesture categories in a conceptual sense. Results (Table 1) show a high separability of encoded gestures belonging to different categories and thus indicate that the encoding scheme is adequate for this task.

432   J.N. Schneider, S. Lallée, and P.F.M.J. Verschure

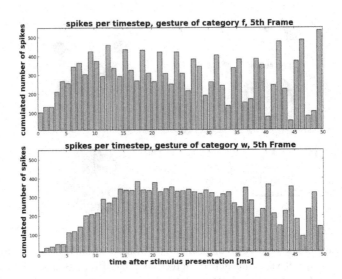

Fig. 2. Spiking response examples for the fifth frame each for two movements belonging to different gesture categories. f: forward arm movement, w: waving.

Table 1. Results for clustering 49 gestures taken from four gesture categories. b: arm bending, f: forward arm movement, p: drawing the letter 'p', w: waving.

Cluster	Content	Average Silhouette Value [5]
1	w,w,w,w,w,w,w,w,w,w	0.728
2	b,b,b,b,b,b,b,b,b,b,b,b,b,b	0.702
3	f,f,f,f,f,f,f,f,f,f,f	0.391
4	p,p,p,p,p,p,p,p,p,p,p,p,p	0.295

References

1. Gärdenfors, P.: Conceptual spaces: The geometry of thought. MIT Press (2004)
2. Wyss, R., König, P., Verschure, P.F.: Invariant representations of visual patterns in a temporal population code. Proceedings of the National Academy of Sciences 100(1), 324–329 (2003)
3. Luvizotto, A., Rennó-Costa, C., Pattacini, U., Verschure, P.F.M.J.: The encoding of complex visual stimuli by a canonical model of the primary visual cortex: Temporal population code for face recognition on the icub robot. In: 2011 IEEE International Conference on Robotics and Biomimetics (ROBIO), pp. 313–318 (2011)
4. Izhikevich, E.M., et al.: Simple model of spiking neurons. IEEE Transactions on neural networks 14(6), 1569–1572 (2003)
5. Rousseeuw, P.J.: Silhouettes: a graphical aid to the interpretation and validation of cluster analysis. Journal of computational and applied mathematics 20, 53–65 (1987)

Roving Robots Gain from an Orientation Algorithm of Fruit Flies and Predict a Fly Decision-Making Algorithm

Roland Strauss[1], Stefanie Flethe[1], José Antonio Villacorta[2,3],
Valeri A. Makarov[3], Manuel G. Velarde[2], Luca Patané[4], and Paolo Arena[4]

[1] Johannes Gutenberg-Universität, Institut für Zoologie III – Neurobiologie, Mainz, Germany
{rstrauss,flethes}@uni-mainz.de
[2] Universidad Complutense de Madrid, Instituto Pluridisciplinar, Spain
[3] Universidad Complutense de Madrid, F. CC. Matemáticas, Applied Mathematics, Spain
vmakarov@mat.ucm.es, {mgvelarde,joseavillacorta}@pluri.ucm.es
[4] DIEEI - University of Catania, Catania, Italy
{lpatane,parena}@dieei.unict.it

Abstract. Simple organisms like bacteria are directly influenced by momentary changes in concentration or strength of sensory signals. In noisy sensory gradients frequent zigzagging reduces the performance of the cell or organism. *Drosophila melanogaster* flies significantly deviate from a direct response to sensory input when orienting in gradients. A dynamical model has been derived which reproduces fly behaviour. Here we report on an emergent property of the model. Implemented in a robot, the algorithm is sustaining decisions between visual targets. The behaviour was consequently found in wild-type flies, which stay with a once-chosen visual target for considerable longer times than mutant flies with a specific brain defect. This allowed the localisation of the integrator. Flies were tested in a virtual-reality arena with two alternatingly visible target objects under different visibility regimes. The finding exemplifies how basic research and technical application can mutually benefit from close collaboration.

Keywords: insect orientation, working memory, biomimetic robots.

1 Extended Abstract

Biological Background. Bacteria and other single- or multi-cellular organisms direct their movements according to chemicals in their environment, up the gradient e.g. for finding food, or down-gradient for avoiding toxic substances. Momentary changes in strength of the sensory signal directly influence their direction of heading. Chemotaxis in noisy gradients causes frequent and costly zigzagging. *Drosophila* flies seek humidity dependent on their state of satiation: sated flies prefer dry places; thirsty flies wet ones. In linear humidity gradients sated wild-type flies run reliably to the dry side. But when a steep increase in humidity occurs (saw-tooth profile), the probability to continue in the previous down-gradient direction through the 'humidity barrier' is dependent on the length of the path during which environmental humidity had improved. The history seems to govern the probability to overcome short stretches of

A. Duff et al. (eds.): Living Machines 2014, LNAI 8608, pp. 433–435, 2014.
© Springer International Publishing Switzerland 2014

aversive humidity [1]. Subsequently the potential influence of the path length towards preferred conditions was evaluated in a non-linear temperature gradient [2]. Again, a dependency was found on the length of the path towards the preferred condition. The longer this path was, the longer the flies tolerated to walk up a gradient towards no-preferred conditions. Finally, a dependency on the history was found also in visual orientation. In an otherwise featureless, bright environment, wild-type flies readily approach a dark vertical object. If the target disappears during approach, the flies can continue their journey in the former direction [3]. The length of the continuation in the former direction was found to be dependent on the length of the path during which the object was visible. The history determines the durability of the memory effect.

Model and Robot Implementation. The model network consists of neurons responsible for sensory-behaviour integration, short-term integration, and generation of behaviours. The memory effect is built on top of chemotaxis as a complementary strategy that may correct chemotactic decisions. It is modelled as a two-stage process. Short-term integration (STI or low-pass filtering) of the sensory information smoothes zigzagging in noisy gradients and improves performance [4], but the gain depends strongly on the stochastic nature of the sensory signal. In time-evolving or space-dependent environments STI offers little if any performance gain. In order to account for fly-behaviour results and to provide the robot with adaptability to changing environments, a prediction-based reinforcement following the classical delta-rule learning [5] is included into the network. One unit performs sensory-behavioural integration and compares the basic chemotactic behaviour with the experience-induced behaviour on the preceding time step. When both behaviours coincide, the sensory-behaviour integration changes the neuronal state. This process reinforces the optimal behaviour (but no motor action) by an implicit feedback from the environment to the robot's behaviour, which enables self-adaptation of the robot to environmental situations.

The robotic platform is a 30x30 cm^2 differential drive wheeled machine [6]. The setup was kept deliberately simple to study memory-based in comparison to chemotaxis-based orientation. A rectangular arena of 3x2 m^2 had been devised with two identical LCD screens on the small sides facing each other. The roving robot carried an actuated camera (visual field of 300°). A visual landmark appeared, for each step of the experiment, on either one of the screens. The robot started each trial at the middle of the long side. When a landmark is perceived, the robot can either move forward in the same direction as its previous choice, or turn and move towards the new target. When the landmark leaves the visual field of the camera, the robot turns randomly with a uniform angular distribution between -180° and 180° to explore the environment. Once the sensory information is generated, it is processed by the decision-making layer. If chemotaxis is chosen for control, the sensory information is transformed directly into a motor action. If the memory algorithm is activated, the robot behaviour is determined by STI and sensory-motor integration as described above.

The comparison between chemotaxis and memory-based control was performed under varying levels of noise by changing the probability of appearance of the landmark on either one of the two screens. The main advantage of memory-based orientation becomes evident in the presence of high noise. Zigzagging behaviour, which is typical for chemotaxis, is greatly reduced. The orientation strategies are compared in terms of covered area, number of steps spent in each location while trying to reach the landmark, and number of steps needed to reach the landmark - all at different noise levels. The performance burst introduced by the memory is maximal at high levels of noise whereas it is comparable with chemotaxis when the noise level is low. Benefits at high noise are: the robot takes decisions faster than in chemotaxis; the dwelling times in the central region and at local minima are reduced; the robot reaches the target faster than in chemotaxis.

Back to Fly Behaviour. Would an algorithm found in gradient orientation govern also object choice of flies? We reproduced the robot's two-object choice situation in a virtual-reality arena for flies. Dark vertical stripes were alternatingly visible at opposite sides of the cylindrical LED screen surrounding the walking platform (Ø 85mm). The stripe switched sides, e.g. every 1s, but wild-type flies stay with their first-chosen side, when the stripe was initially shown there just a little longer (e.g. for 2s before alteration starts). Consistently in different time regimes tested, normal flies stay with the slightly more salient side and suppress zigzagging. The memory effect is missing in mutant *ellipsoid body open* flies with a dysfunctional ellipsoid body in the central complex of the fly brain [7]. Zigzagging is observed, consistent with the expected outcome of a direct, unfiltered influence of acute visual information on motor output. Thus, the outcome of the robot experiments triggered identification and localisation of a fly decision making mechanism.

Acknowledgements. Supported by EU FP7, Project EMICAB, grant no. 270182.

References

1. Zaepf, B., Regenauer, C., Strauss, R.: The fruit flies' basic strategies of humidity-orientation and a new challenge for mushroom bodies. Neuroforum XVII, Suppl. T25-15C (2011)
2. Berg, C., Villacorta, J.A., Makarov, V., Velarde, M.G., Arena, P., Patané, L., Termini, P.S., Strauss, R.: An advanced orientation strategy in fruit flies and its consequences in visual targeting and temperature orientation. Neuroforum XVII, Suppl. T25-12B (2011)
3. Strauss, R., Pichler, J.: Persistence of orientation toward a temporarily invisible landmark in *Drosophila melanogaster*. J. Comp. Physiol. A 182, 411–423 (1998)
4. Castellanos, N.P., Lombardo, D., Makarov, V.A., Velarde, M.G., Arena, P.: Quimiotaxis, infotaxis y memotaxis: estrategias de exploración y supervivencia. Rev. Esp. Fisica 22, 42–46 (2008) ISSN 0213-862X
5. Makarov, V.A., Song, Y., Velarde, M.G., Hubner, D., Cruse, H.: Elements for a general memory structure: Properties of recurrent neural networks used to form situation models. Biol. Cybern. 98, 371–395 (2008)
6. Arena, P., De Fiore, S., Fortuna, L., Patané, L.: Perception-action map learning in controlled multiscroll systems applied to robot navigation. Chaos 18, 1–16 (2008)
7. Thran, J., Poeck, B., Strauss, R.: Serum Response Factor (SRF) mediated gene regulation in a *Drosophila* visual working memory. Curr. Biol. 23, 1756–1763 (2013)

The Si elegans Project – The Challenges and Prospects of Emulating Caenorhabditis elegans[*]

Axel Blau[1], Frank Callaly[4], Seamus Cawley[4], Aedan Coffey[4], Alessandro De Mauro[3], Gorka Epelde[3], Lorenzo Ferrara[1], Finn Krewer[4], Carlo Liberale[1], Pedro Machado[2], Gregory Maclair[3], Thomas-Martin McGinnity[2], Fearghal Morgan[4], Andoni Mujika[3], Alessandro Petrushin[1], Gautier Robin[3], and John Wade[2]

[1] Dept. of Neuroscience and Brain Technologies (NBT) and Nanostructures Unit (NAST), Fondazione Istituto Italiano di Tecnologia (IIT), 16163 Genoa, Italy, iit.it
[2] Intelligent Systems Research Centre (ISRC), University of Ulster, Derry BT487JL, Northern Ireland
isrc.ulster.ac.uk
and School of Science and Technology, Nottingham Trent, University, UK
ntu.ac.uk/sat/
[3] Dept. of eHealth & Biomedical Applications (eHBA) and 3D animation and Interactive Virtual environments, Vicomtech-IK4, San Sebastián, Spain
vicomtech.org
[4] Bio-Inspired Electronics and Reconfigurable Computing Group (BIRC), National University of Ireland, Galway, Ireland
birc.nuigalway.ie
si-elegans@outlook.com

Abstract. *Caenorhabditis elegans* features one of the simplest nervous systems in nature, yet its biological information processing still evades our complete understanding. The position of its 302 neurons and almost its entire connectome has been mapped. However, there is only sparse knowledge on how its nervous system codes for its rich behavioral repertoire. The EU-funded *Si elegans* project aims at reverse-engineering *C. elegans'* nervous system function by its emulation. 302 in parallel interconnected field-programmable gate array (FPGA) neurons will interact through their sensory and motor neurons with a biophysically accurate soft-body representation of the nematode in a virtual behavioral arena. Each FPGA will feature its own reprogrammable neural response model that researchers world-wide will be able to modify to test their neuroscientific hypotheses. In a closed-feedback loop, any sensory experience of the virtual nematode in its virtual environment will be processed by sensory and subsequently interconnected neurons to result in motor commands at neuromuscular junctions at the hardware-software interface to actuate virtual muscles of the virtual nematode. Postural changes in the virtual world will lead to a new sensory experience and thus close the loop. In this contribution we present the overall concepts with special focus on the virtual embodiment of the nematode. For further information and recent news please visit http://www.si-elegans.eu.

---

[*] The members of the *Si elegans* consortium* are listed in alphabetical order; contributing authors are underlined.

A. Duff et al. (eds.): Living Machines 2014, LNAI 8608, pp. 436–438, 2014.
© Springer International Publishing Switzerland 2014

Keywords: Biomimicry, Brain-Inspired Computation, Nervous System Emulation, Soft Body Simulation, Virtual World Embodiment, Neurocomputational Response Models on FPGAs.

1   Introduction

The *C. elegans* hermaphrodite, a soil-dwelling worm, is comprised of exactly 959 cells, including 95 body wall muscle cells and 302 neurons. The morphology, arrangement and connectivity of each cell including neurons have been completely described and are found to be almost invariant across different individuals. There are approximately 7000 chemical synaptic connections, 2000 of which occur at neuromuscular junctions, and approximately 600 gap junctions [2]. Despite its simplicity, the nervous system of *C. elegans* does not only sustain vital body function, but generates a rich variety of behavioral patterns in response to internal and external stimuli. The *Si elegans* project aims at providing a comprehensive artificial *C. elegans* emulation system from which the principles of neural information processing underlying its behavior can be derived.

2   Si elegans Concepts

The *Si elegans* project will provide the 302 neurons of the *C. elegans* nervous system as highly reconfigurable FPGA-based hardware modules. In contrast to popular serial communication protocols (*e.g.*, address event representation (AER), network-on-chip (NoC)), individual FPGA neurons will be linked by free-space electro-optical or acoustic interconnection concepts. They will not only replicate the known connectome of *C. elegans* to result in the complete and correctly wired neural circuitry of the nematode, but allow for a genuinely parallel axo-synaptic information flow between neurons. A single FPGA will be utilised per *C. elegans* neuron, thus allowing for highly detailed and biologically plausible neuron and synaptic models to be implemented. Intuitive drag-and-drop configuration from primitives (*e.g.*, synapses, integration algorithms) as well as import modules for existing models from common simulation engines (*e.g.*, NEURON, BRIAN, NeuroML) will be provided. This hardware nervous system will be embodied by a virtual representation of the nematode (including correct biophysics, sensors and actuators) being situated in a virtual environment with programmable stimuli for real-time behavioural studies. Web-accessible services will include the software for defining neural response models, for designing virtual behavioural experiments and for transcoding models into HDL code. They will connect to a dedicated computer running the virtual arena (VA) and interfacing to the *Si elegans* hardware framework for its programming and for the real-time, closed-loop run-time streaming of sensory input stimuli and motor output commands between the virtual worm and the respective sensory and motor neurons of its 'hardwired' nervous system. The system will be designed for scalability to allow for the emulation of larger nervous systems.

3   The Virtual Arena

The VA is devoted to the realistic virtualization of the nematode including its muscles, its environment, all relevant chemical and mechanosensory stimuli, and the resulting behaviors. The simulation is based on the real biophysics of the *in vivo* model. Its interaction with the simulated environment will map experience (stimuli) onto a sensory input matrix of the emulation hardware. The virtual arena is interactive. All parameters can be defined by researchers through the same web portal where the simulation output will be rendered as well. This will permit to define experiments and validate neurocomputational and behavioral hypotheses. The environment description includes different terrains or fluids which influence the locomotion of the nematode. The complex biomechanical and fluid dynamic-based simulation will be based on the high-level middleware SOFA [1] in a workstation directly connected with the FPGA architecture (Fig. 1).

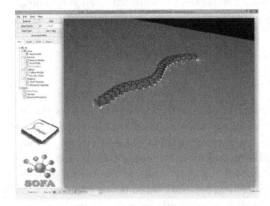

Fig. 1. First prototype of the virtual arena in SOFA. The visual output of the simulation is rendered in the *Si elegans* (Fig.1) web browser using WebGL. This allows for a fast rendering (animation) on the research client of the virtual arena simulation. The SOFA engine, which uses CUDA, allows the realistic interaction between deformable objects and has the ability to simulate fluid dynamics. The modeling strategy for the 95 muscles is based on mass spring dampers.

The rendering of the simulation in SOFA runs at 65 frames per second (FPS) while the rendering of the WebGL animation will be slower. In a future version, we will implement more complex locomotion models and more environmental features (*i.e.* terrains features, obstacles, chemical gradients, etc.).

Acknowledgements. This project 601215 is funded by the 7th Framework Programme (FP7) of the European Union under FET Proactive, call ICT-2011.9.11: Neuro-Bio-Inspired Systems (NBIS).

References

1. Allard, J., Cotin, S., Faure, F., et al.: SOFA - An open source framework for medical simulation. Stud. Health Technol. Inform. 125, 13–18 (2007)
2. White, J.G., Southgate, E., Thomson, J.N., et al.: The Structure of the Nervous System of the Nematode Caenorhabditis elegans. Philosophical Transactions of the Royal Society of London. B, Biological Sciences 314, 1–340 (1986)

A Self-organising Animat Body Map

Hiroki Urashima and Stuart P. Wilson

University of Sheffield, Sheffield, UK
s.p.wilson@sheffield.ac.uk

Abstract. Self-organising maps can recreate many of the essential features of the known functional organisation of primary cortical areas in the mammalian brain. According to such models, cortical maps represent the spatial-temporal structure of sensory and/or motor input patterns registered during the early development of an animal, and this structure is determined by interactions between the neural control architecture, the body morphology, and the environmental context in which the animal develops. We present a minimal model of pseudo-physical interactions between an animat body and its environment, which includes each of these elements, and show how cortical map self-organisation is affected by manipulations to each element in turn. Initial simulation results suggest that maps robustly self-organise to reveal a homuncular organisation, where nearby body parts tend to be represented by adjacent neurons.

Keywords: self-organisation, cortical maps, brain-body interaction, animat model.

Input driven self-organisation represents the dominant theory for how cortical maps emerge in the mammalian neocortex. According to this theory, three main ingredients are required for topological maps to emerge; Hebbian learning between neurons that are active together in time, local interactions between neurons that are located together in space, and spatial-temporal contingency in the typical patterns of sensory input experienced during development [1]. These contingencies depend on the morphology of the sensory receptor surfaces on the body, the physical environment in which the organism moves, and the control strategy used by the organism to move the body with respect to its environment. To explore how each factor affects cortical map development we have been developing a novel computational model and simulation platform. The model comprises i) a pseudo-physical simulation for moving a simple animat body in its environment, used to generate patterns of thermo-tactile input for driving simulated cortical development, ii) a simple control architecture for controlling movements of the body to maximise (thermotatile) reward, and iii) a self-organising map model with a set of Hebbian-modifiable weighted connections to thermo-tactile sensors on the animat body, and a set of Hebbian-modifiable weighted connections to units that direct the animat body towards thermo-tactile reward.

To generate input patterns to the self-orgnanising map model we use the influential shape-matching algorithm [2], which is used by animators to efficiently simulate realistic-looking deformations of a body surface under gravitational and other environmental forces. We implemented the cluster-based

A. Duff et al. (eds.): Living Machines 2014, LNAI 8608, pp. 439–441, 2014.
© Springer International Publishing Switzerland 2014

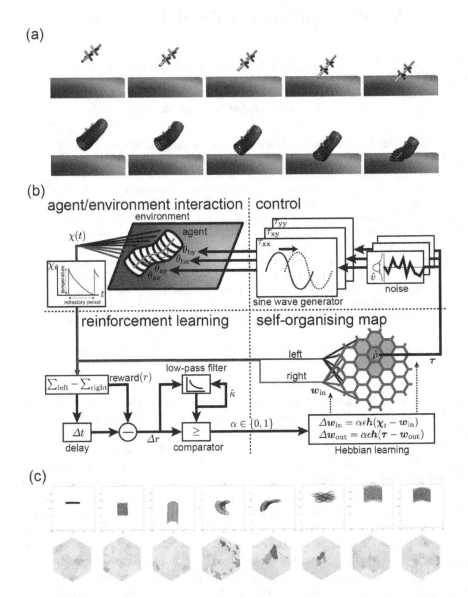

Fig. 1. (a) An animat model generates patterns of tactile stimulation. (b) A model for investigating the factors affecting tactile map self-organisation. (c) A smooth progression of colour across the cortical sheet in self-organised (hexagonal) map plots indicates the emergence of a continous map representation of the taxels along the central axis of the animat body. For these early proof-of-concept simulations, no reward function was used to mediate unsupervised learning, and in each simulation the animat was moved using a fixed value of the movement control parameter. Each value of the control parameter τ_{xx} gives rise to a different animat trajectory (shown as a line drawing for $\tau_{xx} \in \{0\frac{\pi}{4}, 1\frac{\pi}{4}, \ldots, 8\frac{\pi}{4}\}$) and a different continous map organisation.

quadratic-deformation variant of the shape-matching algorithm reported by [2], which allows maximal deformation of the body surface under the influences of gravity and within the contraints imposed by contact between the animat skin surface and planar surfaces that constitute its physical environment. The basic version of our animat body is a simple point-cloud, arranged so as to form an open-ended cylindrical 'skin' surface from a contiguous array of spherical thermo-tactile elements (taxels), which enclose three rigid bones that form a rudimentary spinal cord. When the bones are rotated, the shape-matching algorithm is used to reposition the taxels such that in the absence of environmental contraints the skin would (by gradient descent) tend towards the original undeformed point-cloud configuration. The result is an apparently realistic pseudo-physical simu-lation of skin deformation that we can use to generate patterns of thermotactile stimulation to drive cortical map self-organisation. In the basic simulation, the animat moves in an environment comprising a solid floor with heat distributed across it following a 2D Guassian pattern. As the animat simulation runs, we detect when each taxel makes contact with the floor, and upon contact trigger an exponentially decaying spike of activity, with a spike amplitude that is scaled by the temperature of the floor at the point of contact.

To simulate input-driven cortical map self-organisation, and thus to derive pre-dictions about how the form of cortical maps is affected by the movement of the animat relative to its environment, we adapt a model originally outlined by [3]. This model is itelf an extension to Kohonen's self-organising map model (see [4] for an overview) to include an additional set of Hebbian-modifiable weighted 'out-put' connections from a cortical sheet to a small set of control units, and a mech-anism for reward-based learning. In order to actively deform the animat body in response to envrionmental reward, we articulate the animat spinal cord in 3D by using offset sine-waves to rotate the peripheral bones relative to the central bone, on two (initially aligned) orthogonal axes. The aim is for the cortical network to learn a set of weights that lead to rotations of the bones, and thereby deforma-tions of the skin, that increase the overall temperature of the animat. Hence in an environment where temperature increases gradually in the direction of a heat source, the animat should learn a set of weights that allow it to climb the tem-perature gradient and move towards a rewarding heat source. In the next steps of this modelling effort we will use the full scheme outlined in Figure 1 to close the sensor/motor loop, by allowing reward levels to affect map self-organisation, and allowing this organisation to control the animat behaviour. Implementation of the full model should allow us to generate testable predictions about the relationship between cortical map representations of sensory versus motor spaces.

References

1. Miikkulainen, R., Bednar, J.A., Choe, Y., Sirosh, J.: Computational maps in the visual cortex. Springer, Berlin (2005)
2. Muller, M., Heidelberger, B., Teschner, M., Gross, M.: Meshless deformations based on shape matching. ACM Transactions on Graphics (TOG) 24(3) (2005)
3. Ritter, H., Martinez, T., Schulten, K.: Neural computation on self-organising maps: An introduction. Addison-Wesley, New York (1992)
4. Kohonen, T., Honkela, T.: Kohonen network. Scholarpedia 21(5), 1568 (2007)

A Novel Bio-inspired Tactile Tumour Detection Concept for Capsule Endoscopy

Benjamin Winstone[1], Chris Melhuish[1], Sanja Dogramadzi[1],
Tony Pipe[1], and Mark Callaway[2]

[1] Bristol Robotics Laboratory, University of the West of England
Benjamin.Winstone@brl.ac.uk
[2] Department of Radiology, Bristol Royal Infirmary
Mark.Callaway@UHBristol.nhs.uk

1 Introduction

Examination of the gastrointestinal(GI) tract has traditionally been performed using endoscopy tools that allow a surgeon to see the inside of the lining of the digestive tract. Endoscopes are rigid or flexible tubes that use fibre-optics or cameras to visualise tissues in natural orifices. This can be an uncomfortable and very invasive procedure for the patient.

Modern advances in optical diagnostics have developed wireless capsule endoscopy(WCE) such as the PillCam®from Given Imaging, which is a passive imaging system in the form of a large pill swallowed by the patient. WCE is used to inspect the GI tract by use of an internal imaging camera. A capsule endoscopy allows a more comfortable examination for the patient that can reach further in to the GI tract than traditional endoscopy, however with up to 8 fps only and direction being determined by the peristaltic motion of the gut observation of tumours is not guaranteed.

Here we propose using the TACTIP, a tactile sensor to be placed on the outer surface of a WCE so that it can trace along the wall of the intestines during travel through the GI tract. Using established signal processing algorithms the TACTIP can identify raised bumps which deform the surface of the sensor. Previous work by Roke et al.[1], has shown that the TACTIP is capable of tumour detection in the context of remote palpation of artificial flesh. What differs here is that the TACTIP sensor will be dragged along the target surface rather than palpated.

In previous work, [2] we have shown that the TACTIP is capable of texture analysis whilst being dragged along a textured surface. It is the application of tumour detection and lateral movement along a target surface that we propose to enrich the detection capabilities of WCE.

2 Previous Work

TACTIP is a biologically-inspired sensing device, based upon the deformation of the epidermal layers of the human skin. Deformation from device-object interaction is measured optically by tracking the movement of internal papillae pins

A. Duff et al. (eds.): Living Machines 2014, LNAI 8608, pp. 442–445, 2014.

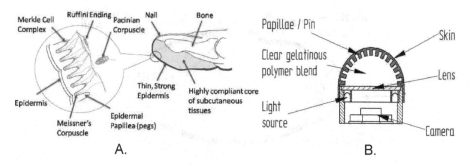

Fig. 1. A. Biological inspiration for TACTIP, B. TACTIP tactile sensing device

on the inside of the device skin. These papillae pins are representative of the intermediate epidermal ridges of the skin, whose static and dynamic displacement are normally detected through the skin's mechanoreceptors, see Fig.1.

PillCam® capsule endoscopy is a commercial product used within the medical industry as a minimally invasive endoscopy tool. The pill is swallowed by the patient and an internal camera records the journey through the GI tract. Spada et al. report a case study [3] identifying the advantages of the PillCam over traditional endoscopy. The work concludes that the PillCam can be a feasible and safe diagnostic tool, and may represent an alternative in cases of difficult or incomplete colonoscopies, however Triantafyllou et al. [4] evaluate whether WCE can complete colon examination after failure of conventional colonoscopy. Their findings conclude that in their series of patients with incomplete colonoscopy, WCE did not always satisfactorily examine the colon due to insufficient exposure to the entire region at risk.

3   Methods and Results

Fig.2 demonstrates the concept of placing a TACTIP sensor around the surface of a capsule endoscope. As the capsule passes naturally through the GI tract, the walls of the capsule push against the intestinal wall which will stimulate the TACTIP sensor. As far as we know little or no work has been published exploring such a tactile sensor on a capsule endoscope with the intention of tumour detection.

An initial proof of concept experiment has been performed whereby a TACTIP sensor has been laterally dragged along the surface of an artificial tissue sample. Embedded within the tissue sample is an artificial tumour. The density of the tumour is much greater than the tissue and so it leaves an impression on the TACTIP surface which is easily identifiable with image processing as presented by Assaf et al. [5]. Fig.3 demonstrates the internal TACTIP view with the output from the image processing algorithm overlaid over the raw camera image. The three images show progression of the TACTIP across the tissue sample with the identified tumour moving right to left.

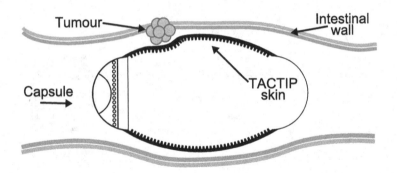

Fig. 2. Tactile sensing capsule endoscope

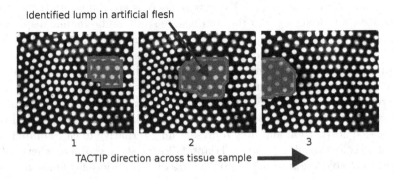

Fig. 3. TACTIP demonstrating lump detection. The TACTIP is laterally pushed along an artificial tissue sample from left to right and the embedded tumour is shown moving right to left.

4 Summary

Our initial experiment has shown that the TACTIP sensor has potential as a tactile sensor for capsule endoscopy with the ability to detect tumour like objects embedded in soft tissue wall. With some changes to the TACTIP design the device could be reshaped to fit on to the exterior of a capsule. This would offer either an alternative to the standard image camera, or an additional sensing technology to enrich overall data acquisition.

References

[1] Roke, C., Spiers, A., Pipe, T., Melhuish, C.: The effects of laterotactile information on lump localization through a teletaction system. In: 2013 World Haptics Conference (WHC), pp. 365–370 (April 2013)

[2] Winstone, B., Griffiths, G., Pipe, T., Melhuish, C., Rossiter, J.: TACTIP - tactile fingertip device, texture analysis through optical tracking of skin features. In: Lepora, N.F., Mura, A., Krapp, H.G., Verschure, P.F.M.J., Prescott, T.J. (eds.) Living Machines 2013. LNCS, vol. 8064, pp. 323–334. Springer, Heidelberg (2013)

[3] Spada, C., Riccioni, M.E., Petruzziello, L., Marchese, M., Urgesi, R., Costamagna, G.: The new PillCam Colon capsule: difficult colonoscopy? No longer a problem? Gastrointestinal endoscopy 68(4), 807–808 (2008)
[4] Triantafyllou, K., Tsibouris, P., Kalantzis, C., Papaxoinis, K., Kalli, T., Kalantzis, N., Ladas, S.D.: PillCam Colon capsule endoscopy does not always complement incomplete colonoscopy. Gastrointestinal endoscopy 69(3, Pt. 1), 572–576 (2009)
[5] Assaf, T., Chorley, C., Rossiter, J., Pipe, T., Stefanini, C., Melhuish, C.: Realtime Processing of a Biologically Inspired Tactile Sensor for Edge Following and Shape Recognition. In: TAROS (2010)

Electro-communicating Dummy Fish Initiate Group Behavior in the Weakly Electric Fish *Mormyrus rume*

Martin Worm[1], Tim Landgraf[2], Hai Nguyen[2], and Gerhard von der Emde[1]

[1] University of Bonn, Institute of Neuroethology and Sensory Ecology,
Endenicher Allee 11-13, 53115 Bonn, Germany
`{mworm,vonderemde}@uni-bonn.de`
[2] Freie Universität Berlin, FB Mathematik u. Informatik, Arnimallee 7, 14195 Berlin, Germany
`tim.landgraf@fu-berlin.de`

Abstract. The mechanisms that underlie collective behavior in groups of fish have been the subject of numerous quantitative and theoretical studies. We use a robotic platform to investigate social interactions in weakly electric fish by exploiting their unique electro-sensory modality for animal-robot communication. Our results demonstrate that weakly electric fish interact with a mobile dummy fish based on species-specific electrical playback signals and are therefore proposed to be a promising model organism for establishing a mixed-society of real and artificial fish.

Keywords: collective behavior, playback, robotic platform, video-tracking.

1 Introduction

Mormyrid weakly electric fish possess the ability to probe their environment by means of self-generated electric pulses that are produced by an electric organ located in their tail. These electric organ discharges (EOD) are generated at different rates and in patterns that depend on behavioral context. The resulting inter-discharge interval (IDI) is therefore highly variable and can convey information in intra-specific communication [1]. At the same time, mormyrids display complex social interactions that range from aggressive and territorial to collective behavior in small groups. Studies that have investigated the underlying mechanisms of shoaling in fish in general have concluded that this group behavior is mainly mediated by vision and the lateral line system [2], but there is also evidence that there is a strong influence of electric signaling on group formation and coherence in mormyrids [3]. Recent advances have made it possible to introduce mobile dummy fish into groups of real fish to investigate collective behavior under controlled conditions [4]. Here we employ a robotic platform which allows moving a mobile dummy fish along an arbitrary trajectory while it emits electrical playback signals. Our first result show, that weakly electric fish of the genus Mormyrus follow this dummy fish for an extended period of time while responding to electric playbacks with patterns that are typical for intra-specific electro-communication.

A. Duff et al. (eds.): Living Machines 2014, LNAI 8608, pp. 446–448, 2014.
© Springer International Publishing Switzerland 2014

2   Experimental Results

Playback signals were generated by sampling the EOD waveform of a specimen of *Mormyrus rume*. Artificial EODs were then connected to playback sequences using behavior-specific intervals. Using an A/D-converter and a battery driven stimulus-isolation unit, these signals were fed to a pair of carbon electrodes incorporated in a mobile dummy fish (Fig.1.). This dummy fish was made from a commercial fishing bite, which was mounted with a stick onto a magnetic base on the tank floor to allow steering from below the tank.

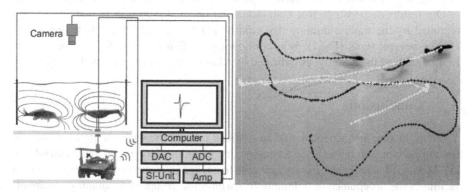

Fig. 1. Left: Schematic drawing of the experimental setup. Right: Topview of the experimental tank showing the trajectories of two individuals of *Mormyrus* (blue) and the mobile dummy fish (yellow).

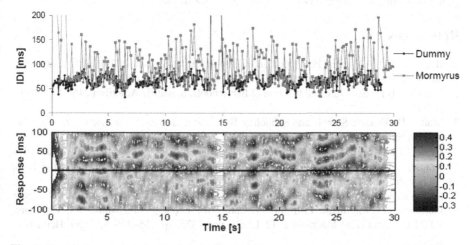

Fig. 2. Electrical response of a single individual of *Mormyrus* to a moving dummy fish playing an electrical IDI-sequence taken from behavioral context. (A): IDI-patterns of playback (black) and electric fish (blue). (B): The cross-correlation diagram reveals synchronization with correlation coefficients color-coded. The upper trace of the diagram depicts the response time of the fish towards the dummy, while the opposite situation is shown in the lower trace.

The robotic platform was used as described in [5]. The wheeled robot was moved manually on an arbitrary trajectory on a plane underneath the experimental tank using wireless communication. The connection between robot and dummy fish was established via a set of strong polarized magnets, which are attached to the robot and the base of the dummy fish within the tank, respectively (Fig.1).

All experiments were filmed with video cameras from above for subsequent video tracking based on background subtraction in Ctrax [6]. Simultaneously, the EODs of the fish and the playback EODs were recorded differentially and digitized via a multi-electrode array. Given the dipole characteristics of the electric signals, this allowed us to assign each EOD to the respective signaler based on its current position in the tank.

Our results demonstrate that both individual fish and groups of two *Mormyrus* are recruited by the mobile dummy that emitted playbacks and follow its trajectory including several turns (Fig.1). While following the fish dummy, fish respond to the EOD signals of the dummy, as indicated by adaptive cross-correlations (Fig.2).

3   Conclusion

The introduction and acceptance of an artificial conspecific within a group of animals is an important step towards the formation of a mixed-society of real and artificial fish [7]. Mormyrid weakly electric fish are able to engage in electrocommunication, which facilitates the acceptance of a dummy fish within the group. If the dummy communicates certain patterns representing specific types of behaviors, fish behavior may as well be modulated in a group. Based on these capabilities as well as the results presented above, we propose mormyrid weakly electric fish to be promising model organisms for the investigation of collective behavior in fish shoals.

References

1. Hopkins, C.D.: Electrical Perception and Communication. In: Squire, L.R. (ed.) Encyclopedia of Neuroscience, vol. 3, pp. 813–831. Academic Press, Oxford (2009)
2. Partridge, B.L.: Structure and function of fish schools. Scientific American 246(6), 112–114 (1982)
3. Moller, P.: Electric Signals and Schooling Behavior in a Weakly Electric Fisch, Marcusenius cyprinoides L (Mormyriformes). Science 193, 697–699 (1976)
4. Faria, J.J., Dyer, J.R., Clément, R.O., Couzin, I.D., Holt, N., Ward, A.J., Waters, D., Krause, J.: A novel method for investigating the collective behaviour of fish: introducing 'Robofish'. Behavioral Ecology and Sociobiology 64(8), 1211–1218 (2010)
5. Landgraf, T., Akkad, R., Nguyen, H., Clément, R.O., Krause, J., Rojas, R.: A Multi-agent Platform for Biomimetic Fish. In: Prescott, T.J., Lepora, N.F., Mura, A., Verschure, P.F.M.J. (eds.) Living Machines 2012. LNCS, vol. 7375, pp. 365–366. Springer, Heidelberg (2012)
6. Branson, K., Robie, A.A., Bender, J., Perona, P., Dickinson, M.H.: High-throughput ethomics in large groups of Drosophila. Nature Methods 6(6), 451–457 (2009)
7. Mondada, F., Martinoli, A., Correll, N., Gribovskiy, A., Halloy, J.I., Siegwart, R., Deneubourgh, J.-L.: A general methodology for the control of mixed natural- artificial societies. In: Kernbach, S. (ed.) Handbook of Collective Robotics, pp. 399–428. Pan Stanford (2013)

A Concept of Exoskeleton Mechanism for Skill Enhancement

Tomoyuki Yamamoto[1,2] and Hiroshi Ishiguro[2]

[1] CiNet, National Institute of Communications Technology (NICT)
1-4 Yamadaoka, Suita, 565-0871 Osaka Japan
[2] Graduate School of Engineering Science, Osaka University
1-3 Machikaneyama, Toyonaka, 560-8531 Osaka Japan
yamamoto@irl.sys.es.osaka-u.ac.jp, ishiguro@sys.es.osaka-u.ac.jp

While common implementation of the exoskeleton mechanism is "powered suit" with onboard power plant, it suffers low power-to-weight ratio and often fails to inefficacy problem. On the other hand, quasi-passive mechanism can exploit human efficiency further, which can be regarded as a "wearable bicycle". In this approach, we are developing an exoskeleton mechanism as a hybrid system for improving users' skills.

Previous works on quasi-passive mechanisms are mainly focus rehabilitation (e.g., [1]) or increase efficiency in locomotion (e.g., [2]). These researches do not explicitly aim enhancement of skills, where diverse possibilities of application exist. Our current target is the lifting movement in industry, in which workers that are able bodied but not highly trained. Bad strategies of lifting, such as stooping or leaning back are commonly seen in novice workers and they are potential cause of injury or chronic disorder, which are making a huge losses in the industry.

Our study on biomechanical analysis of lifting [3] showed that difference between experts (professional workers; caregivers and a fire fighter) and novice persons is seen organization of coordinative structure between body parts. Fig.1 A depicts the difference of motion pattern between novice and experts. There were no instruction is given to the subjects and this differentiation of motion paths. We have found that novices tend to synchronously move the arms and legs upward, resulting leaning back posture. In other words, their movements are 1 DoF (Degree of Freedom)-like, while experts motion paths are 2 DoF-like, where the arms and the torso are independent. See Fig.1B.

Our idea is to improve novices' movements by inducing that of experts.A possible reason for entering novice route is that novice tend to actuate the arms, which is not necessary. This makes posture unstable and back learning posture is induced. To prevent this movement, the exoskeleton constrain the user's movement: fixing of the arms by braking devices is thought to induce the expert routes (see also Fig. 2A). By inducing expert's movement pattern, user can acquire the skill and increase of performance is expected.

The controlling, e.g., activation of brake in the arm, is based on analysis and prediction of the user's movements. The prediction is based on the diagram based on the previous analysis, such as shown in Fig.1 A. Using motion sensors, specific

A. Duff et al. (eds.): Living Machines 2014, LNAI 8608, pp. 449–451, 2014.
© Springer International Publishing Switzerland 2014

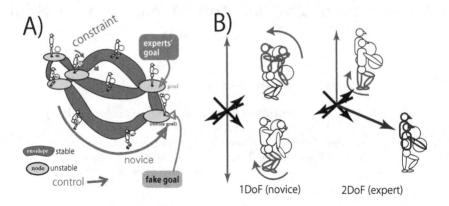

Fig. 1. A:Schematic representation of lifting movements. Novice subject tends to move the arms in synchronous to the leg and trapped into fake goal. B: Difference in coordinative structure between novice (left) and expert (right).

postures and coordination pattern are analyzed for prediction. In our working prototype model, accelerometers, rate gyros and angle sensors are developed and dynamical variables (acceleration, velocity and coordinates) are measured without adopting differentiation or integration. Using a custom FPGA program, synchronous measurement in 1 kHz for 26 nodes is realized.

Using above sensor system, the coordination pattern can be evaluated online by calculation of mutual angular velocities. Correlation analysis and principal component analysis will be used. Our previous studies showed that the coordinative structure is strongly correlated to skill level (e.g., [4]) and this motion sensor system also can be used portable motion capture system. It is well studied that convergence of coordinative structure (i.e., decrease of DoF) as an index of skill acquisition (e.g., [5]).

The control to the user is given in the forms of constraints. For static constraints (i.e., joint to be fixed), posture is constrained by braking mechanism. This can be done by ratchet with clutch or air cylinder with valve, where active parts are minimized. For dynamic constraints (i.e., joint to be moved), vibrators or auditory signals may be used to notify initiating movement.

In addition, quasi-passive mechanism increase human performance further without producing positive work. See Fig. 2. The rigid frame can provide pivoting point of lever-arm mechanism and user's force can be increased. Also, using lever-arm effect more explicitly, offsetting the handling point to make posture suited for producing appropriate force can be possible. In the latter, the braking mechanism of an ankle is used as a pivoting point and body weight is the power source. After the load is lifted, usual lifting begins from offseted position and total effort is thought to decrease. This method comes from the fact that some of experts uses open their legs fore-aft direction and used only one of the leg to lift.

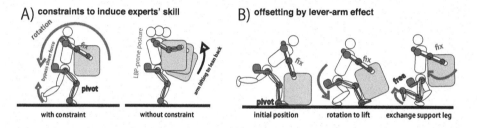

Fig. 2. A: possible use of constraints to enhance skill of lifting. B: offsetting the load by lever-arm effect to change initial position. After exchanging support leg, lifting became easy.

As for implementation, we are currently developing two prototypes. One is sensor-rich exoskeleton for movement analysis and the other is specific for enhancing lifting skill. The former is used for developing algorithm for the latter and pilot study for other skills. The latter integrates essential sensor nodes and quasi-passive actuators. Since one of the merits of quasi-passive design is compactness of the system, our goal is to adopt this mechanism for mobile situation, such as delivery drivers in the logistics industry where user do not have to take off the mechanism when driving a vehicle.

Acknowledgments. This work is partially supported by JSPS Core-to-Core Program, A. Advanced Research Networks.

References

1. Agrawal, S.K., Banala, S.K., Fattah, A., Sangwan, V., Krishnamoorthy, V., Scholz, J.P., Hsu, W.L.: Assesment of Motion of a Swing Leg and Gait Rehabilitation With a Gravity Balancing Exoskeleton. IEEE trans. on Neural Systems and Rehabilitation Engineering 15, 410–420 (2007)
2. Walsh, C., Endo, K., Herr, H.: A quasi-passive leg exoskeleton for load-carrying augmentation. International Journal of Humanoid Rbotics 4, 487–506 (2007)
3. Yamamoto, T., Terada, K., Kuniyoshi, Y.: Lifting Techniques for the Humanoid Robots: Insights from Human Movements. In: Proceedings of 2008 IEEE-RAS International Conference on Humanoid Robotics, pp. 251–258 (2008)
4. Yamamoto, T., Fujinami, T.: Hierarchical organization of the coordinative structure of the skill of clay kneading. Human Movement Science 27, 812–822 (2008)
5. Vereijken, B., van Emmerik, R.E.A., Whiting, H.T.A., Newell, K.M.: Free(z)ing degrees of freedom in skill acquisition. Journal of Motor Behavior 24, 133–142 (1992)

Author Index